Lecture Notes in Computer Science   13947

More information about this series at https://link.springer.com/bookseries/558

Amitabha Bagchi · Rahul Muthu (Eds.)

Algorithms and Discrete Applied Mathematics

9th International Conference, CALDAM 2023
Gandhinagar, India, February 9–11, 2023
Proceedings

Springer

Editors
Amitabha Bagchi (iD)
Department of Computer Science
and Engineering
Indian Institute of Technology Delhi
New Delhi, India

Rahul Muthu
Dhirubhai Ambani Institute of Information
and Communication Technology
Gandhinagar, India

ISSN 0302-9743   ISSN 1611-3349 (electronic)
Lecture Notes in Computer Science
ISBN 978-3-031-25210-5   ISBN 978-3-031-25211-2 (eBook)
https://doi.org/10.1007/978-3-031-25211-2

This Springer imprint is published by the registered company Springer Nature Switzerland AG
The registered company address is: Gewerbestrasse 11, 6330 Cham, Switzerland

Preface

This volume contains the papers presented at CALDAM 2023 (the 9th International Conference on Algorithms and Discrete Applied Mathematics) held during February 9–11, 2023 at DA-IICT, Gandhinagar, Gujarat, India. CALDAM 2023 was organised by the Dhirubhai Ambani Institute of Information & Communication Technology (DA-IICT), Gandhinagar, Gujarat, and the Association for Computer Science and Discrete Mathematics (ACSDM), India. The program committee consisted of 30 highly experienced and active researchers from various countries.

The conference had papers in the areas of algorithms and optimization, computational geometry, game theory, graph coloring, graph connectivity, graph domination, graph matching, and graph partition and graph covering. We received 73 submissions with authors from all over the world. Each paper was extensively reviewed by program committee members and other expert reviewers. The committee decided to accept 34 papers for presentation. The program included two Google invited talks by Professors Mark de Berg (of Eindhoven University of Technology) and Ignasi Sau (Université de Montpellier).

As volume editors, we would like to thank the authors of all submissions for considering CALDAM 2023 for potential presentation of their works. We are very much indebted to the program committee members and the external reviewers for providing serious reviews within a very short period of time. We thank Springer for publishing the proceedings in the Lecture Notes in Computer Science series. Our sincerest thanks to the invited speakers Mark de Berg and Ignasi Sau for accepting our invitation to give a talk. We thank the organizing committee chaired by Sunitha Vadivel Murugan of the Dhirubhai Ambani Institute of Information & Communication Technology, Gandhinagar, for the smooth conduct of CALDAM 2023 and the Dhirubhai Ambani Institute of Information & Communication Technology, Gandhinagar and the Indian Institute of Technology, Gandhinagar for providing the necessary facilities. We are very grateful to the chair of the steering committee, Subir Ghosh, for his active help, support, and guidance throughout. We thank the previous edition's chairs R. Inkulu and Niranjan Balachandran for assistance in various details in the processes. We thank our sponsors Google Inc. for their financial support. We also thank Springer for its support for the best paper presentation awards. We thank the EasyChair conference management system, which were very effective in handling the entire process. We thank Springer for detailed guidelines for the process of preparing the proceedings.

February 2023

Amitabha Bagchi
Rahul Muthu

Organization

Steering Committee

Subir Kumar Ghosh (Chair)	Ramakrishna Mission Vivekananda Educational and Research Institute, India
Gyula O. H. Katona	Alfréd Rényi Institute of Mathematics, Hungarian Academy of Sciences, Hungary
János Pach	École Polytechnique fédérale de Lausanne (EPFL), Lausanne, Switzerland
Nicola Santoro	Carleton University, Canada
Swami Sarvattomananda	Ramakrishna Mission Vivekananda Educational and Research Institute, India
Chee Yap	Courant Institute of Mathematical Sciences, New York University, USA

Program Committee

Yasmeen S. Akhtar	BITS Pilani Goa, India
Amitabha Bagchi (PC Co-chair)	IIT Delhi, India
Niranjan Balachandran	IIT Mumbai, India
Manoj Changat	University of Kerala, India
Keerti Choudhary	IIT Delhi, India
Sandip Das	ISI Kolkata, India
Srikrishnan Divakaran	Ahmedabad University, India
Florent Foucaud	Université Clermont Auvergne, France
Daya Gaur	University of Lethbridge, Canada
Iqra Altaf Gillani	NIT Srinagar, India
Sathish Govindarajan	IISC Bangalore, India
Andrzej Lingas	Lund University, Sweden
Anil Maheshwari	Carleton University, Canada
Bodo Manthey	University of Twente, The Netherlands
Rogers Mathew	IIT Hyderabad, India
Gervais Mendy	Université Cheikh Anta Diop de Dakar, Senegal
Neeldhara Misra	IIT Gandhinagar, India
Rahul Muthu (PC Co-chair)	DA-IICT, India
N. S. Narayanaswamy	IIT Madras, India
Arti Pandey	IIT Ropar, India
Maria Cristina Pinotti	Università degli Studi di Perugia, Italy
Giuseppe Prencipe	University of Pisa, Italy

S. Francis Raj	Pondicherry University, India
Deepak Rajendraprasad	IIT Palakkad, India
Vlady Ravelomanana	Université de Paris, France
Sagnik Sen	IIT Dharwad, India
Rishi Ranjan Singh	IIT Bhilai, India
Uéverton Souza	Universidade Federal Fluminense, Brazil
C. R. Subramanian	IMSc Chennai, India
Anuj Tawari	DA-IICT, India

Organizing Committee

Puneet Bhateja	DA-IICT, India
Bireswar Das	IIT Gandhinagar, India
Swami Dhyanagamyananda	RKMVERI Belur, India
Manish Gupta	DA-IICT, India
Manoj Gupta	IIT Gandhinagar, India
Mahipal Jadeja	MNIT Jaipur, India
Pritee Khanna	IIITDM Jabalpur, India
Anish Mathuria	DA-IICT, India
Supantha Pandit	DA-IICT, India
Manoj Raut	DA-IICT, India
Sunitha Vadivel Murugan (Chair)	DA-IICT, India

Additional Reviewers

Sheila Almeida	Antoine Dailly
N. R. Aravind	Arun Kumar Das
Pradeesha Ashok	Soura Sena Das
T. Asir	Tapas Das
John Augustine	Hiranya Dey
Manu Basavaraju	Amit Kumar Dhar
Deepu Benson	Stoyan Dimitrov
Srimanta Bhattacharya	Mitre Dourado
Sushil Bhunia	Yan Gerard
Márcia Cappelle	Barun Gorain
Dipayan Chakraborty	Dishant Goyal
Prerona Chatterjee	Luciano Gualá
Juhi Chaudhary	Manish Kumar Gupta
Alessio Conte	Anni Hakanen
Federico Coró	Jesper Jansson
Rahul C. S.	Sangram Kishor Jena
Supraja D K	Ce Jin
Clément Dallard	Lijo Jose

Rajesh Kannan
Nilesh Khandekar
Miroslaw Kowaluk
R. Krithika
Christos Levcopoulos
Carlos Vinicius Lima
Carla Negri Lintzmayer
Gokulnath M.
Raghunath Reddy Madireddy
Soumen Maity
Atrayee Majumder
Tapas Kumar Mishra
Kaushik Mondal
Zin Mar Myint
Alfredo Navarra
Francis P.
Sajith P.
Pavan P. D.

Sreejith K. Pallathumadam
Alexandre Pinlou
Mauro Passacantando
Veena Prabhakaran
Vijayaragunathan Ramamoorthi
Vinod Reddy
Aniket Basu Roy
Taruni S.
Lakshay Saggi
Abhishek Sahu
Brahadeesh Sankarnarayanan
Tarkeshwar Singh
Francesco Betti Sorbelli
Kavaskar T.
Rakesh Venkat
S. Venkitesh
Shaily Verma

Abstracts of Invited Talks

Stable Approximation Schemes

Mark de Berg

Department of Computer Science, TU Eindhoven, the Netherlands
M.T.d.Berg@tue.nl

In a dynamic optimization problem, the goal is to maintain a solution to an optimization problem under insertions and deletions. We are interested in trade-offs between the stability of the solution and its approximation ratio. To formalize this, we introduce the concept of *k-stable algorithms*, which are algorithms that apply at most k changes to the solution upon each insertion and deletion. We are particularly interested in *stable approximation schemes*, which are update algorithms that, for any given parameter $\varepsilon > 0$, are $k(\varepsilon)$-stable and maintain a solution with approximation ratio $1 + \varepsilon$, where the stability parameter $k(\varepsilon)$ only depends on ε and not on the size of the current input. In this talk I will discuss stable approximation schemes for two problems: the RANGE-ASSIGNMENT PROBLEM and MAXIMUM INDEPENDENT SET.

The RANGE-ASSIGNMENT PROBLEM is defined as follows. Let P be a set of points in R^d, where each point $p \in P$ has an associated transmission range, denoted $\rho(p)$. The range assignment ρ induces a directed communication graph $\mathcal{G}_\rho(P)$ on P, which contains an edge (p, q) iff $|pq| \leq \rho(p)$. In the broadcast range-assignment problem, the goal is to assign the ranges such that $\mathcal{G}_\rho(P)$ contains an arborescence rooted at a designated root node and the cost $\sum_{p \in P} \rho(p)^2$ of the assignment is minimized. For this problem, the stability of a dynamic algorithm is the number of ranges that are modified upon the insertion or deletion of a point in P. We will show that the RANGE-ASSIGNMENT PROBLEM admits a stable approximation scheme in R^1, but not in R^2.

For MAXIMUM INDEPENDENT SET on a graph $\mathcal{G}$, we consider the dynamic problem where each insertion adds a single vertex plus all its incident edges to the graph $\mathcal{G}$, and a deletion removes a vertex and all its incident edges. The stability of a dynamic algorithm is then defined as the number of vertices that are added to or deleted from the maintained independent set I upon the insertion or deletion of a vertex into $\mathcal{G}$. We show that for graphs that admit sublinear clique-based separators—examples of such graphs are planar graphs and disk graphs—a stable approximation scheme exists, and we show that for certain expander graphs a stable approximation scheme does not exist.

The talk is based on joint work with Arpan Sadhukhan and Frits Spieksma.

Keywords: Approximation algorithms · Online algorithms · Stable approximation schemes

Supported by the Dutch Research Council (NWO) through Gravitation-grant NETWORKS-024.002.003.

Graph Modification Problems with Forbidden Minors

Ignasi Sau

LIRMM, Université de Montpellier, CNRS, Montpellier, France

In a generic graph modification problem, given an input graph G, the goal is to apply some modifications to it, belonging to a prescribed set $\mathcal{M}$ (say, vertex deletion or edge contraction), in order to obtain a graph that belongs to a target graph class $\mathcal{C}$ (say, a planar graph or a 3-regular graph). Different instantiations of $\mathcal{M}$ and $\mathcal{C}$ yield a number of well-studied problems such as VERTEX COVER or FEEDBACK VERTEX SET. A very active line of research studies the parameterized complexity of this family of problems for various choices of the parameter. Of particular relevance is the case where the target graph class $\mathcal{C}$ excludes some graph as a minor. The objective of this talk is to survey recent work in this direction, along with some of the most common techniques used in the literature, including the strong interplay of this family of problems with logic.

Contents

Algorithms and Optimization

Efficient Reductions and Algorithms
for Subset Product

Pranjal Dutta[1] and Mahesh Sreekumar Rajasree[2]([⊠])

[1] Chennai Mathematical Institute, Chennai, India
pranjal@cmi.ac.in
[2] Indian Institute of Technology, Kanpur, Kanpur, India
mahesr@cse.iitk.ac.in

Abstract. Given positive integers $a_1, \ldots, a_n$ and a target integer t, the Subset Product problem asks to determine whether there exists a subset $S \subseteq [n]$ such that $\prod_{i \in S} a_i = t$. It differs from the Subset Sum problem where the multiplication operation is replaced by addition. There is a pseudopolynomial-time dynamic programming algorithm which solves the Subset Product in $O(nt)$ time and $\Omega(t)$ space.

In this paper, we present a simple and elegant randomized algorithm for Subset Product in $\tilde{O}(n + t^{o(1)})$ expected-time. Moreover, we also present a poly(nt) time and $O(\log^2(nt))$ space deterministic algorithm.

In fact, we solve a more general problem called the SimulSubsetSum. This problem was introduced by Kane 2010. Given k instances of Subset Sum, it asks to decide whether there is a 'common' solution to all the instances. Kane gave a logspace algorithm for this problem. We show a polynomial-time reduction from Subset Product to SimulSubsetSum and also give efficient algorithm for the latter. Our algorithms use multivariate FFT, power series and number-theoretic techniques, introduced by Jin and Wu (SOSA 2019) and Kane (2010).

Keywords: simultaneous · power series · subset product · logspace · FFT · pseudo-prime-factor

1 Introduction

The Subset Sum problem (in short, SSUM) is a well-known NP-complete problem [18, p. 226], where given $(a_1, \ldots, a_n, t) \in \mathbb{Z}_{\geq 0}^{n+1}$, the problem is to decide whether there exists $S \subseteq [n]$ such that $\sum_{i \in S} a_i = t$. In the recent years, this problem has gained significant attractions due to applications in provable-secure cryptosystems [10,19] and remarkable algorithmic improvements both in classical and quantum world [3,5,8,9,12–14]. In this paper, we study a well-known variant of the subset sum, called Subset Product.

Problem 1 (Subset Product). Given $(a_1, \ldots, a_n, t) \in \mathbb{Z}_{\geq 1}^{n+1}$, the Subset Product problem asks to decide whether there exists an $S \subseteq [n]$ such that $\prod_{i \in S} a_i = t$.

The full version is available at https://drive.google.com/file/d/1xUX29eVZ_2J1zv062d
H4V8Yc8NhqwV1X/view?usp=share_link.

Subset Product is known to be NP-complete [11, p. 221]. It has a trivial $O(nt)$ time (*pseudo-polynomial* time) dynamic programming algorithm which requires $\Omega(t)$ space [2].

Subset Product has been studied and applied in many different forms. For e.g. 1) constructing a *smooth hash* (VSH) by Contini et al. [6], 2) attack on the Naccache-Stern Knapsack (NSK) public key cryptosystem [7]. Similar problem has also been studied in optimization, in the form of product knapsack problem [21], multiobjective knapsack problem [1].

Next, we define a seemingly unrelated problem. It asks to decide whether there is a 'common' solution to the given many instances of subset sum. This was first introduced by [15, Section 3.3] (but no formal name was given).

*Problem 2 (*SimulSubsetSum*).* Given subset sum instances $(a_{1j}, \ldots, a_{nj}, t_j) \in \mathbb{Z}_{\geq 0}^{n+1}$, for $j \in [k]$, where k is some parameter, the Simultaneous Subset Sum problem (in short, SimulSubsetSum) asks to decide whether there exists an $S \subseteq [n]$ such that $\sum_{i \in S} a_{ij} = t_j, \forall j \in [k]$.

Remarks.

1. When k is fixed parameter (independent of n), we call this $k -$ SimulSubsetSum. There is a trivial $O(n(t_1 + 1) \ldots (t_k + 1))$ time deterministic algorithm for the SimulSubsetSum problem with k subset sum instances (k not necessarily a constant) by extending the dynamic programming algorithm for SSUM.
2. It suffices to work with $t_j \geq 1, \forall j \in [k]$. To argue that, let us assume that $t_j = 0$ for some $j \in [k]$ and $I_j := \{i \in [n] \,|\, a_{ij} = 0\}$. Observe that if SimulSubsetSum has a solution set $S \subseteq [n]$, then $S \subseteq I_j$. Therefore, for every $\ell \in [k]$, instead of looking at $(a_{1\ell}, \ldots, a_{n\ell}, t_\ell)$, it suffices to work with $\{a_{i,\ell} \,|\, i \in I_j\}$ with the target t_ℓ. Thus, we can trivially ignore the jth SSUM instance.

Hardness Depends on k. Linear algebraically, Problem 2 is asking to solve a system of k-linear equations, in n-variables with $0/1$ constraints on the variables. If we assume that the set of vectors $\{(a_{1j}, \ldots, a_{nj}) \,|\, \forall j \in [k]\}$ is linearly independent; then we can perform Gaussian elimination to find a relation between the free variables (exactly $n - k$) and dependent/leading variables. Then, by enumerating over all possible 2^{n-k} values of the free variables and finding the corresponding values for leading variables, we can check whether there is a $0/1$ solution, hence solving it in $\text{poly}(n, k) \cdot 2^{n-k}$ time. This implies that when $k \geq n - O(\log(n))$, SimulSubsetSum (with assuming linear independence) has a polynomial time solution. Whereas, we showed (see [Theorem 6,full version]) that given a subset sum instance, we can convert this into a SimulSubsetSum instance in polynomial time even with $k = O(\log(n))$.

1.1   Our Contributions

Theorem 1 (Time-efficient algorithm for Subset Product). *There exists a randomized algorithm that solves* Subset Product *in* $\tilde{O}(n + t^{o(1)})$ *expected-time.*

Remarks.

1. The result in the first part of the above theorem is reminiscent of the $\tilde{O}(n+t)$ time randomized algorithms for the subset sum problem [4,14], although the time complexity in our case is the expected time, and ours is better.
2. The expected time is because to factor an integer t takes expected $\exp(O(\sqrt{\log(t)\log\log(t)}))$ time [17]. If one wants to remove expected time analysis (and do the worst case analysis), the same problem can be solved in $\tilde{O}(n^2 + t^{o(1)})$ randomized-time. For details, see the end of Subsect. 3.1.
3. While it is true that Bellman's algorithm gives $O(nt)$ time algorithm, the state-space of this algorithm can be improved to (expected) $nt^{o(1)}$-time for Subset Product, using a similar dynamic algorithm with a careful analysis. For details, see [Appendix D,full version].

Theorem 2 (Space-efficient algorithm for Subset Product). *Subset Product can be solved deterministically in* $O(\log^2(nt))$ *space and* poly(nt)-*time.*

Remark. We *cannot* directly invoke the theorem in [15, Section 3.3] to conclude Theorem 2 since the reduction from Subset Product to SimulSubsetSum requires $O(n\log(nt))$ space. Essentially, we use the same identity lemma as [15] and carefully use the space; for details see [Appendix A,full version].

Using a pseudo-prime-factorization decomposition, we show that given a target t in Subset Product, it suffices to solve SimulSubsetSum with at most $\log t$ many instances, where each of the targets is also 'small', at most $O(\log\log t)$ bits.

Theorem 3 (Reducing Subset Product to SimulSubsetSum). *There is a deterministic polynomial time reduction from* Subset Product *to* SimulSubsetSum.

Remark. The reduction uses $\tilde{O}(n\log t)$ space as opposed to the following chain of reductions: Subset Product $\leq_P$ SSUM $\leq_P$ SimulSubsetSum. The first reduction is a *natural* reduction, from an input $(a_1,\ldots,a_n,t)$, which takes log both sides and adjusts (multiply) a 'large' M (it could be $O(n\log t)$ bit [16,21]) with $\log a_i$, to reduce this to a SSUM instance with $b_i := \lfloor M\log a_i\rfloor$. Therefore, the total space required could be as large as $\tilde{O}(n^2\log t)$. The second reduction follows from [Theorem 6,full version]. Thus, ours is more space efficient. Motivated thus, we give an efficient randomized algorithm for SimulSubsetSum.

We also show that SimulSubsetSum, even with 2 instances, is as hard as SSUM. Though the proof is very standard, we sketch this for the completeness. For details, see [Appendix B,full version].

1.2 Prior Works and Limitation of the Obvious Attempts

There have been a very few attempts to classically solve Subset Product or its variants. It is known to be NP-complete and the reduction follows from the Exact Cover by 3-Sets (X3C) problem [11, p. 221]. Though the knapsack and its approximation versions have been studied [16,21], we do not know many classical algorithms and attempts to solve this, unlike the recent attention for the subset sum problem [4,5,13,14]. In this paper, we start investigating similar questions in the Subset Product regime.

Why the Obvious Methods Fail. Since subset sum can be solved in randomized $\tilde{O}(n + t)$ time [14], as mentioned before, one obvious way to solve Subset Product would be to work with $b_i := \lfloor M \log a_i \rfloor$ and a $\mathcal{R}$, a range of target values t' which could be as large as $M \log t$ such that Subset Product is YES iff subset sum instance with b_i and $t' \in \mathcal{R}$ is YES. But M could be as large as $O(n \cdot (\prod_i a_i)^{1/2})$. Therefore, although there is a randomized near-linear time algorithm for subset sum, when one reduces the instance of Subset Product to a subset sum instance, the target becomes very large, failing to give an $\tilde{O}(n+t)$ algorithm.

Similarly, Theorem 3 along with the reduction in [Theorem 7,full version], which reduces the Subset Product to SSUM, actually blows up the target, and fails to give near-linear time algorithm.

We also mention that it is not clear if non-algebraic techniques, as used in [4], could be extended for SimulSubsetSum or not. Moreover, the general techniques, used for subset sum [4,13,14] seem to fail to 'directly' give algorithms for Subset Product. This is exactly why, in this work, the efficient algorithms have been indirect, *via* solving SimulSubsetSum instances.

2   Preliminaries

Notations. $\mathbb{N}, \mathbb{Z}$ and $\mathbb{Q}$ denotes the set of all natural numbers, integers and rational numbers respectively. Let a, b be two m-bit integers. Then, $a//b$ denotes a/b^e where e is the largest non-negative integer such that $b^e \mid a$. Observe that $a//b$ is not divisible by b and the time to compute $a//b$ is $O(m \log(m) \cdot \log(e))$. Also, $\tilde{O}(N)$ denotes $N \cdot \text{poly}(\log N)$.

For any positive integer $n > 0$, $[n]$ denotes the set $\{1, 2, \ldots, n\}$ while $[a, b]$ denotes the set of integers i s.t. $a \leq i \leq b$. $\mathbb{F}[x_1, \ldots, x_k]$ denotes the ring of k-variate polynomials over field $\mathbb{F}$ and $\mathbb{F}[[x_1, \ldots, x_k]]$ is the ring of power series in k-variables over $\mathbb{F}$. We will use the short-hand notation x to denote the collection of variables $(x_1, \ldots, x_k)$ for some k. For any non-negative integer vector $e \in \mathbb{Z}^k$, x^e denotes $\prod_{i=1}^{k} x_i^{e_i}$. Using these notations, we can write any polynomial $f(x) \in \mathbb{Z}[x]$ as $f(x) = \sum_{e \in S} f_e \cdot x^e$ for some suitable set S. We denote $\text{coef}_{x^e}(f)$, as the coefficient of x^e in the polynomial $f(x)$ and $\deg_{x_i}(f)$ as the highest degree of x_i in $f(x)$.

Lemma 1 (Kane's Identity [15]). *Let $f(x) = \sum_{i=0}^{d} c_i x^i$ be a polynomial of degree at most d with coefficients c_i being integers. Let $\mathbb{F}_q$ be the finite field of order $q = p^k > d + 2$. For $0 \leq t \leq d$, define*

$$r_t = \sum_{x \in \mathbb{F}_q^*} x^{q-1-t} f(x) = -c_t \in \mathbb{F}_q$$

Then, $r_t = 0 \iff c_t$ is divisible by p.

Theorem 4 ([20]). *For $n \geq 25$, there is always a prime in $[n, 6/5 \cdot n]$.*

The following is a naive bound, but it is sufficient for our purpose.

Lemma 2. *For integers $a \geq b \geq 1$, we have $(a/b)^b \leq 2^{2\sqrt{ab}}$.*

Proof. Let $x = \sqrt{a/b}$. We need to show that $x^{2b} \leq 2^{2bx}$, which is trivially true since $x \leq 2^x$, for $x \geq 1$.

3 Time-Efficient Algorithm for **Subset Product**

In this section, we give a randomized $\tilde{O}(n + t^{o(1)})$ expected time algorithm for Subset Product. Essentially, we factor all the entries in the instance in $\tilde{O}(n+t^{o(1)})$ expected time. Once we have the exponents, it suffices to solve the corresponding SimulSubsetSum instance. Now, we can use the efficient randomized algorithm for SimulSubsetSum (Theorem 5) to finally solve Subset Product. So, first we give an efficient algorithm for SimulSubsetSum.

Theorem 5 (Algorithm for SimulSubsetSum). *There is a randomized $\tilde{O}(kn + \prod_{i \in [k]}(2t_i + 1))$-time algorithm that solves SimulSubsetSum, with target instances $t_1, \ldots, t_k$.*

Proof. Let us assume that the input to the SimulSubsetSum problem is k SSUM instance of the form $(a_{1j}, \ldots, a_{nj}, t_j)$, for $j \in [k]$. Define a k-variate polynomial $f(\boldsymbol{x})$, where $\boldsymbol{x} = (x_1, \ldots, x_k)$, as follows:

$$ f(\boldsymbol{x}) = \prod_{i=1}^{n} \left(1 + \prod_{j=1}^{k} x_j^{a_{ij}} \right). $$

Here is an immediate but important claim. We denote the monomial $\boldsymbol{m} := \prod_{i=1}^{k} x_i^{t_i}$ and $\mathrm{coef}_{\boldsymbol{m}}(f)$ as the coefficient of $\boldsymbol{m}$ in the polynomial $f(\boldsymbol{x})$.

Claim 1. There is a solution to the SimulSubsetSum instance, i.e., $\exists S \subseteq [n]$ such that $\sum_{i \in S} a_{ij} = t_j, \forall j \in [k]$ iff $\mathrm{coef}_{\boldsymbol{m}}(f(\boldsymbol{x})) \neq 0$.

Therefore, it is enough to compute the coefficient of $f(\boldsymbol{x})$. The rest of the proof focuses on computing $f(\boldsymbol{x})$ efficiently, to find $\mathrm{coef}_{\boldsymbol{m}}(f)$.

Let p be prime such that $p \in [N + 1, (n + N)^3]$, where $N := \prod_{i=1}^{k}(2t_i + 1)$. Define an ideal $\mathcal{I}$, over $\mathbb{Z}[\boldsymbol{x}]$ as follows: $\mathcal{I} := \langle x_1^{t_1+1}, \ldots, x_k^{t_k+1}, p \rangle$. Since, we are interested in $\mathrm{coef}_{\boldsymbol{m}}(f)$, it suffices to compute $f(\boldsymbol{x})$ mod $\langle x_1^{t_1+1}, \ldots, x_k^{t_k+1} \rangle$, and we do it over a field $\mathbb{F}_p$ (which introduces error); for details, see the proof in the end (Randomness and error probability paragraph).

Using [Lemma 6,full version], we can compute all the coefficients of $\ln(f(\boldsymbol{x}))$ mod $\mathcal{I}$ in time $\tilde{O}(kn + \prod_{i=1}^{k} t_i)$. It is easy to see that the following equalities hold.

$$ f(\boldsymbol{x}) \mod \mathcal{I} \equiv \exp(\ln(f(\boldsymbol{x}))) \mod \mathcal{I} \equiv \exp(\ln(f(\boldsymbol{x})) \mod \mathcal{I}) \mod \mathcal{I}. $$

Since, we have already computed $\ln(f(\boldsymbol{x}))$ mod $\mathcal{I}$, the above equation implies that it is enough to compute the exponential which can be done using [Lemma 5,full version]. This also takes time $\tilde{O}(kn + \prod_{i=1}^{k}(2t_i + 1))$.

Randomness and Error Probability. Note that there are $\Omega(n+N)^2$ primes in the interval $[N+1, (n+N)^3]$. Moreover, since $\mathrm{coef}_m(f) \leq 2^n$, at most n prime factors can divide $\mathrm{coef}_m(f(\boldsymbol{x}))$. Therefore, we can pick a prime p randomly from this interval in $\mathsf{poly}(\log(n+N))$ time and the probability of p dividing the coefficient is $O(n+N)^{-1}$. In other words, the probability that the algorithm fails is bounded by $O((n+N)^{-1})$. This concludes the proof.   □

We now compare the above result with some obvious attempts to solve SimulSubsetSum, before moving into solving Subset Product.

A Detailed Comparison with Time Complexity of [15]. Kane [15, Section 3.3] showed that the SimulSubsetSum problem can be solved deterministically in $C^{O(k)}$ time and $O(k \log C)$ space, where $C := \sum_{i,j} a_{ij} + \sum_j t_j + 1$, which could be as large as $(n+1) \cdot (\sum_{j \in [k]} t_j) + 1$, since a_{ij} can be as large as t_j. As argued in [13, Corollary 3.4 and Remark 3.5], the constant in the exponent, inside the order notation, can be as large as 3 (in fact directly using [15] gives a larger constant; but modified algorithm as used in [13] gives 3). Use AM-GM inequality to get

$$\left((n+1) \cdot \left(\sum_j t_j \right) + 1 \right)^{3k} > \left(\frac{2}{k} \cdot \sum_j t_j + 1 \right)^{3k} \overset{\text{AM-GM}}{\geq} \prod_{j=1}^k (2t_j + 1)^3 .$$

Assuming $N = \prod_{j=1}^k (2t_j + 1)$, our algorithm is near-linear in N while Kane's algorithm [15] takes $O(N^3)$ time; thus ours is almost a cubic improvement.

Comparison with the Trivial Algorithm. It is easy to see that a trivial $O(n \cdot (t_1+1)(t_2+1)\ldots(t_k+1))$ time *deterministic* algorithm for SimulSubsetSum exists. Since, $t_i \geq 1$, we have

$$\frac{n}{2} \cdot \prod_{i \in [k]} (1 + t_i) \geq \frac{n}{2} \cdot 2^k \geq kn, \quad \text{and} \quad \frac{n}{2} \cdot \prod (1 + t_i) \geq \frac{n}{2^{k+1}} \cdot \prod (2t_i + 1).$$

Here, we used $2(1 + x) > (2x + 1)$, for any $x \geq 1$. Therefore, $n \cdot \prod_{i \in [k]} (1 + t_i) \geq kn + n/2^{k+1} \cdot \prod (2t_i + 1)$. Thus, when $k = o(\log n)$, our complexity is better.

3.1   Proof of Theorem 1

Once we have designed the algorithm for SimulSubsetSum, we design a time-efficient algorithm for Theorem 1.

Proof. Let $(a_1, \ldots, a_n, t) \in \mathbb{Z}_{\geq 0}^{n+1}$ be the input for Subset Product problem. Without loss of generality, we can assume that all the a_i divides t because if some a_i does not divide t, it will never be a part of any solution and we can discard it. Let us first consider the prime factorization of t and a_j, for all $j \in [n]$. We will discuss about its time complexity in the next paragraph. Let

$$t = \prod_{j=1}^k p_j^{t_j}, \quad a_i = \prod_{j=1}^k p_j^{e_{ij}}, \forall i \in [n],$$

where p_j are distinct primes and t_j are positive integers and $e_{ij} \in \mathbb{Z}_{\geq 0}$. Since, $p_i \geq 2$, trivially, $\sum_{i=1}^{k} t_i \leq \log(t)$, and $\sum_{i=1}^{k} e_{ij} \leq \log(t), j \in [n]$. Also, the number of distinct prime factors of t is at most $O(\log(t)/\log\log(t))$; therefore, $k = O(\log(t)/\log\log(t))$.

Time Complexity of Factoring. To find all the primes that divide t, we will use the factoring algorithm given by Lenstra and Pomerance [17] which takes expected $t^{o(1)}$[1] time to completely factor t into prime factors p_j (including the exponents t_j). Using the primes p_j and the fact that $0 \leq e_{ij} \leq \log(t)$, computing e_{ij} takes $\log^2(t)\log\log(t)$ time, by performing binary search to find the largest x such that $p_j^x \mid a_i$. So, the time to compute all exponents $e_{i,j}, \forall i \in [n], j \in [k]$ is $O(nk\log^2(t)\log\log(t))$. Since, $k \leq O(\log t/\log\log(t))$, the total time complexity is $\tilde{O}(n + t^{o(1)})$.

Setting Up. SimulSubsetSum Now suppose that $S \subseteq [n]$ is a solution to the Subset Product problem, i.e., $\prod_{i \in S} a_i = t$. This implies that

$$\sum_{i \in S} e_{ij} = t_j, \quad \forall j \in [k] .$$

In other words, we have a SimulSubsetSum instance where the jth SSUM instance is $(e_{1j}, e_{2j}, \ldots, e_{nj}, t_j)$, for $j \in [k]$. The converse is also trivially true. We now show that there exists an $\tilde{O}(kn + \prod_{i \in [k]}(2t_i + 1))$ time algorithm to solve SimulSubsetSum.

Randomized Algorithm for Subset Product Using Theorem 5, we can decide the SimulSubsetSum problem with targets $t_1, \ldots, t_k$ in $\tilde{O}(kn + \prod_{i \in [k]}(2t_i + 1))$ time (randomized) while working over $\mathbb{F}_p$ for some suitable p (we point out towards the end). Since $k \leq O(\log(t)/\log\log(t))$, we need to bound the term $\prod_{i \in [k]}(2t_i + 1)$. Note that,

$$\prod_{i \in [k]} (2t_i + 1) = \sum_{S \subseteq [k]} 2^{|S|} \cdot \left(\prod_{i \in S} t_i \right)$$

$$\leq 2^{2k} \cdot \left(\prod_{i \in [k]} t_i \right) .$$

---

[1] Expected time complexity is $\exp(O(\sqrt{\log t \log\log t}))$, which is smaller than $t^{O(1/\sqrt{\log\log t})} = t^{o(1)}$, which will be the time taken in the next step. Moreover, we are interested in *randomized* algorithms, hence expected run-time is.

We now focus on bounding the term $\prod_{i \in [k]} t_i$. By AM-GM,

$$\prod_{i \in [k]} t_i \leq \left(\frac{\sum_{i \in [k]} t_i}{k} \right)^k \leq \left(\frac{\log(t)}{k} \right)^k$$

$$\leq 2^{O\left(\sqrt{k \log(t)}\right)} \qquad [Lemma\ 2]$$

$$\leq 2^{O\left(\sqrt{\log(t)^2 / \log\log(t)}\right)}$$

$$\leq t^{O(1/\sqrt{\log\log(t)})} = t^{o(1)}$$

Note that the prime p in the Theorem 5 was $p \in [N+1, (n+N)^3]$, where $N := \prod_{i=1}^{k}(2t_i + 1) - 1$. As shown above, we can bound $N = t^{o(1)}$. Thus, $p \leq O((n + t^{o(1)})^3)$, as desired. Therefore, the total time complexity is $\tilde{O}(n \log(t) / \log\log(t) + t^{o(1)}) = \tilde{O}(n + t^{o(1)})$. This finishes the proof. $\qquad \square$

Removing the Expected-Time. If one wants to understand the worst-case analysis, we can use the polynomial time reduction from Subset Product to SimulSubsetSum in Sect. .4. Of course, we will not get prime factorization; but the pseudo-prime factors will also be good enough to set up the SimulSubsetSum with similar parameters as above, and the SimulSubsetSum instance can be similarly solved in $\tilde{O}(n + t^{o(1)})$ time. Since the reduction takes $n^2 \text{poly}(\log t)$ time, the total time complexity becomes $\tilde{O}(n^2 + t^{o(1)})$.

4   An Efficient Reduction from Subset Product to SimulSubsetSum

In this section, we will present a deterministic polynomial time reduction from Subset Product to SimulSubsetSum. In Sect. 3, we have given a pseudo-polynomial time reduction from Subset Product to SimulSubsetSum by performing prime-factorization of the input $(a_1, \ldots, a_n, t)$. The polynomial time reduction also requires to factorize the input, but the factors are not necessarily prime. To be precise, we define pseudo-prime-factorization which can be achieved in polynomial time.

Definition 1 (Pseudo-prime-factorization). *A set of integers $\mathcal{P} \subset \mathbb{N}$ is said to be pseudo-prime-factor set of $(a_1, \ldots, a_n) \in \mathbb{N}^n$ if*

1. *the elements of $\mathcal{P}$ are pair-wise coprime, i.e., $\forall p_1, p_2 \in \mathcal{P}, gcd(p_1, p_2) = 1$,*
2. *there are only non-trivial factors of a_i's in $\mathcal{P}$, i.e., $\forall p \in \mathcal{P}, \exists i \in [n]$ such that $p \mid a_i$,*
3. *every a_i's can be uniquely expressed as product of powers of elements of $\mathcal{P}$, i.e., $\forall i \in [n], a_i = \prod_{p \in \mathcal{P}} p^{e_p}, \forall i \in [n]$ where $e_p \geq 0$.*

For a given $(a_1, \ldots, a_n)$, $\mathcal{P}$ may not be unique. A trivial example of a pseudo-prime-factor set of $\mathcal{P}$ for $(a_1, \ldots, a_n)$ is the set of all distinct prime factors of $\prod_{i=1}^{n} a_i$. The following is an important claim which will be used to give a *polynomial* time reduction from Subset Product to SimulSubsetSum.

Claim 2. For any pseudo-prime-factor set $\mathcal{P}$ of $(a_1, \ldots, a_n)$, we have $|\mathcal{P}| \leq k$ where k is the number of distinct prime factors of $\prod_{i=1}^{n} a_i$.

Proof. The proof uses a simple pigeonhole principle argument. Let $g_1, \ldots, g_k$ be the distinct prime factors of $\prod_{i=1}^{n} a_i$. From the definition of $\mathcal{P}$, we know that $g_1, \ldots, g_k$ are the only distinct prime factors of $\prod_{p \in \mathcal{P}} p$. Therefore, if there are more than k numbers in $\mathcal{P}$, then there must exist $p_1, p_2 \in \mathcal{P}$ such that $\gcd(p_1, p_2) \neq 1$ which violates pair-wise coprime property of $\mathcal{P}$.   □

Constructing $\mathcal{P}$ suffices. We now show that having a pseudo-prime-factor set $\mathcal{P}$ for $(a_1, \ldots, a_n, t)$ helps us to reduce a Subset Product instance $(a_1, \ldots, a_n, t)$ to SimulSubsetSum with number of instances $|\mathcal{P}|$, in polynomial time. Wlog, we can assume that $a_i \mid t$ and $a_i, t \leq 2^m, \forall i \in [n]$ for some m. Trivially, $m \leq \log t$. So, using Claim 2, we have $|\mathcal{P}| \leq (n+1) \cdot m = \mathsf{poly}(n \log t)$.

From Definition 1, we have unique non-negative integers e_{ij} and t_j such that $t = \prod_{j \in |\mathcal{P}|} p_j^{t_j}$ and $a_i = \prod_{j \in |\mathcal{P}|} p_j^{e_{ij}}, \forall i \in [n]$. Since, $a_i \mid t$, we have $e_{ij} \leq t_j \leq m, \forall i \in [n], j \in [|\mathcal{P}|]$ and they can be computed in $\mathsf{poly}(m, n)$ time.

Let us consider the $|\mathcal{P}| - \mathsf{SimulSubsetSum}$ instance where the ith SSUM instance is $(e_{1i}, e_{2i}, \ldots, e_{ni}, t_i)$. Then, due to factorization property (the third property in Definition 1) of $\mathcal{P}$, the Subset Product instance is YES, i.e., $\exists S \in [n]$ such that $\prod_{i \in S} a_i = t$ iff the SimulSubsetSum instance with number of instances $|\mathcal{P}|$, is a YES.

4.1   Polynomial Time Algorithm for Computing Pseudo-Prime-Factors

We will now present a deterministic polynomial time algorithm for computing a pseudo-prime-factor set $\mathcal{P}$ for $(a_1, \ldots, a_n)$. We will use the notation $\mathcal{P}(a_1, \ldots, a_n)$ to denote a pseudo-prime-factor set for $(a_1, \ldots, a_n)$. Also, let $\mathcal{S}(a_1, \ldots, a_n)$ be the set of all pseudo-prime-factor sets; this is a finite set.

The following lemma is a crucial component in algorithm 1. We use $a//b$ to denote a/b^e such that $b^{e+1} \nmid a$.

Lemma 3. *Let $(a_1, \ldots, a_n)$ be n integers. Then,*

1. *If a_1 is coprime with $a_i, \forall i > 1$, then for any $\mathcal{P}(a_2, \ldots, a_n) \in \mathcal{S}(a_2, \ldots, a_n)$, $\mathcal{P}(a_2, \ldots, a_n) \cup \{a_1\} \in \mathcal{S}(a_1, \ldots, a_n)$.*
2. *$\mathcal{P}(g, a_1//g, a_2//g, \ldots, a_n//g) \in \mathcal{S}(a_1, \ldots, a_n)$, for given a_i, $i \in [n]$ and any factor g of some a_i.*

Proof. The first part of the lemma is trivial. For the second part, let g be a non-trivial factor of some a_i and

$$\mathcal{P} := \{p_1, \ldots, p_k\} \in \mathcal{S}(g, a_1//g, a_2//g, \ldots, a_n//g),$$

be any pseudo-prime-factor set. Then, p_i's are pair-wise coprime and since each p_i divides either g or $a_i//g$ for some $i \in [n]$, it also divides some a_i because g is a factor of some a_i. Also, we have *unique* non-negative integers e_{ip}, e_{gp} s.t.

$$a_i//g = \prod_{p \in \mathcal{P}} p^{e_{ip}}, \forall i \in [n] \text{ and } g = \prod_{p \in \mathcal{P}} p^{e_{gp}}.$$

Combining these equation, we get $a_i = a_i//g * g^{f_{ig}} = \prod_{p \in \mathcal{P}} p^{e_{ip}+e_{gp}*f_{ig}}$. Here f_{ig} is the maximum power of g that divides a_i. Therefore, $\{p_1, \ldots, p_k\}$ is also a pseudo-prime-factor set for $(a_1, \ldots, a_n)$. □

Pre-processing. Using Lemma 3, Algorithm 1 performs a divide-and-conquer approach to find $\mathcal{P}(a_1, \ldots, a_n)$. Observe that we can always remove duplicate elements and 1's from the input since it *does not* change the pseudo-prime-factors. Also, we can assume without loss of generality that $a_i//a_1 =: a_i, \forall i > 1$ because of the second part in Lemma 3, with $g = a_1$, since it gives us $\mathcal{P}(a_1, a_2//g, \ldots, a_n//g)$ and we know it suffices to work with these inputs.

If a_1 is coprime to the rest of the a_i's, then the algorithm will recursively call itself on $(a_2, \ldots, a_n)$ and combine $\mathcal{P}(a_2, \ldots, a_n)$ with $\{a_1\}$. Else, there exists an $i > 1$ such that $gcd(a_1, a_i) \neq 1$. So, the algorithm finds a factor g of a_1 using Euclid's GCD algorithm and computes $\mathcal{P}(g, a_1//g, \ldots, a_n//g)$. At every step we remove duplicates and 1's. Hence, the correctness of Algorithm 1 is immediate assuming it terminates.

To show the termination and time complexity of Algorithm 1, we will use the *'potential function'* $\mathbb{P}(I) := \prod_{a \in I} a$, where I is the input and show that at each recursive call, the value of the potential function is halved. Initially, the value of the potential function is $\prod_{i=1}^n a_i$. We also remark that since the algorithm removes duplicates and 1's; the potential function can *never* increase by the removal step and so it never matters in showing the decreasing nature of $\mathbb{P}$.

1. a_1 is coprime to the rest of the a_i's: In this case, the recursive call has input $(a_2, \ldots, a_n)$. Since, $a_1 \geq 2$, the value of potential function is

$$\mathbb{P}(a_2, \ldots, a_n) = \prod_{i=2}^n a_i < (\prod_{i=1}^n a_i)/2 = \mathbb{P}(a_1, \ldots, a_n)/2.$$

2. a_1 shares a common factor with some a_i. Let $g = gcd(a_1, a_i) \neq 1$. Since, we have assumed $a_i//a_1 = a_i$, this implies that a_i is not a multiple of a_1. This implies that $2 \leq g \leq a_1/2$. Therefore, the new value of potential function is

$$\mathbb{P}(g, a_1//g, \ldots, a_n//g)$$
$$= g \prod_{j=1}^n a_j//g$$
$$\leq (a_1//g) \times ((a_i//g) \times g) \times \prod_{j \in [n] \setminus \{1,i\}} a_j$$
$$\leq \frac{a_1}{g} \cdot \prod_{j=2}^n a_j$$
$$\leq (\prod_{j=1}^n a_j)/2 = \mathbb{P}(a_1, \ldots, a_n)/2.$$

We used the fact that since, $2 \leq g \mid a_i$, therefore, $g \times (a_i//g) \leq a_i$.

Time Complexity. In both the cases, the value of the potential function is halved. So, the depth of the recursion tree (in-fact, it is just a line) is at most $\log(\prod_{i=1}^n a_i) \leq m \cdot n$. Also, in each recursive call, the input size is increased at most by one but the integers are still bounded by 2^m. This implies that input size, for any recurrence call, can be at most $(m+1) \cdot n$. Since there is no branching, the total number of operations is $(m+1) \cdot n \times m \cdot n = O((mn)^2)$. Therefore, the total time complexity is $n^2 \cdot poly(m)$.

Algorithm 1: Algorithm for Pseudo-prime-factor set

Input: $(a_1, a_2 \ldots, a_n) \in \mathbb{N}^n$ where each a_i is an m-bit integer such that
$a_i//a_1 = a_i > 1, \forall i > 1$
Output: Pseudo-prime-factor set $\mathcal{P}$ for $(a_1, a_2, \ldots, a_n)$

1 **if** $n == 0$ **then**
2 | **return** $\emptyset$;
3 **end**
4 **if** $\exists i > 1$ *such that* $gcd(a_1, a_i) \neq 1$ **then**
5 | $g = gcd(a_1, a_i)$;
6 | $I = \{g\}$;
7 | **for** $i \in [n]$ **do**
8 | | $a_i' = a_i//g$;
9 | | **if** $a_i' \notin I$ *and* $a_i' \neq 1$ **then**
10 | | | $I = I \cup \{a_i'\}$
11 | | **end**
12 | **end**
13 | **return** $\mathcal{P}(I)$;
14 **end**
15 **else**
16 | **return** $\mathcal{P}(a_2, \ldots, a_n) \cup \{a_1\}$;
17 **end**

---

5 Conclusion

In this paper, we give efficient algorithms for Problem 1-2 which are variants of SSUM problem. We also present an efficient reduction from Subset Product to SimulSubsetSum. Here are some immediate questions to investigate.

1. Can we improve the complexity of Theorem 5 to $\tilde{O}(n + \sum_{i=1}^k t_i)$?
2. What can we say about the hardness of SimulSubsetSum with k subset sum instances where $k = \omega(\log(n))$?

References

1. Bazgan, C., Hugot, H., Vanderpooten, D.: Solving efficiently the 0–1 multi-objective knapsack problem. Comput. Oper. Res. **36**(1), 260–279 (2009)

2. Bellman, R.E.: Dynamic programming (1957)
3. Bernstein, D.J., Jeffery, S., Lange, T., Meurer, A.: Quantum algorithms for the subset-sum problem. In: Gaborit, P. (ed.) PQCrypto 2013. LNCS, vol. 7932, pp. 16–33. Springer, Heidelberg (2013). https://doi.org/10.1007/978-3-642-38616-9_2
4. Bringmann, K.: A near-linear pseudopolynomial time algorithm for subset sum. In: Proceedings of the Twenty-Eighth Annual ACM-SIAM Symposium on Discrete Algorithms, pp. 1073–1084. SIAM (2017)
5. Bringmann, K., Wellnitz, P.: On near-linear-time algorithms for dense subset sum. In: Proceedings of the 2021 ACM-SIAM Symposium on Discrete Algorithms (SODA), pp. 1777–1796. SIAM (2021)
6. Contini, S., Lenstra, A.K., Steinfeld, R.: VSH, an efficient and provable collision-resistant hash function. In: Vaudenay, S. (ed.) EUROCRYPT 2006. LNCS, vol. 4004, pp. 165–182. Springer, Heidelberg (2006). https://doi.org/10.1007/11761679_11
7. Draziotis, K.A., Martidis, V., Tiganourias, S.: Product subset problem: applications to number theory and cryptography. arXiv preprint arXiv:2002.07095 (2020)
8. Dutta, P., Rajasree, M.S.: Algebraic algorithms for variants of subset sum. In: Balachandran, N., Inkulu, R. (eds.) CALDAM 2022. LNCS, vol. 13179, pp. 237–251. Springer, Cham (2022). https://doi.org/10.1007/978-3-030-95018-7_19
9. Esser, A., May, A.: Low weight discrete logarithm and subset sum in 20. 65n with polynomial memory. memory 1, 2 (2020)
10. Faust, S., Masny, D., Venturi, D.: Chosen-ciphertext security from subset sum. In: Cheng, C.-M., Chung, K.-M., Persiano, G., Yang, B.-Y. (eds.) PKC 2016. LNCS, vol. 9614, pp. 35–46. Springer, Heidelberg (2016). https://doi.org/10.1007/978-3-662-49384-7_2
11. Garey, M.R., Johnson, D.S.: Computers and Intractability, vol. 174. Freeman, San Francisco (1979)
12. Helm, A., May, A.: Subset sum quantumly in 1.17ⁿ n. In: 13th Conference on the Theory of Quantum Computation, Communication and Cryptography (TQC 2018). pp. 1–15. Schloss Dagstuhl-Leibniz-Zentrum fuer Informatik (2018)
13. Jin, C., Vyas, N., Williams, R.: Fast low-space algorithms for subset sum. In: Proceedings of the 2021 ACM-SIAM Symposium on Discrete Algorithms (SODA), pp. 1757–1776. SIAM (2021)
14. Jin, C., Wu, H.: A simple near-linear pseudopolynomial time randomized algorithm for subset sum. arXiv preprint arXiv:1807.11597 (2018)
15. Kane, D.M.: Unary subset-sum is in logspace. arXiv preprint arXiv:1012.1336 (2010)
16. Kovalyov, M.Y., Pesch, E.: A generic approach to proving np-hardness of partition type problems. Discret. Appl. Math. 158(17), 1908–1912 (2010)
17. Lenstra, H.W., Pomerance, C.: A rigorous time bound for factoring integers. J. Am. Math. Soc. 5(3), 483–516 (1992)
18. Lewis, H.R.: Computers and Intractability. A Guide to the Theory of NP-Completeness, Freeman, San Francisco (1983)
19. Lyubashevsky, V., Palacio, A., Segev, G.: Public-key cryptographic primitives provably as secure as subset sum. In: Micciancio, D. (ed.) TCC 2010. LNCS, vol. 5978, pp. 382–400. Springer, Heidelberg (2010). https://doi.org/10.1007/978-3-642-11799-2_23
20. Nagura, J.: On the interval containing at least one prime number. Proc. Jpn. Acad. 28(4), 177–181 (1952)
21. Pferschy, U., Schauer, J., Thielen, C.: Approximating the product knapsack problem. Optim. Lett. 15, 2529–2540 (2021)

Optimal Length Cutting Plane Refutations of Integer Programs

K. Subramani[✉] and P. Wojciechowski

LDCSEE, West Virginia University, Morgantown, WV, USA
{k.subramani,pwojciec}@mail.wvu.edu

Abstract. In this paper, we discuss the computational complexities of determining optimal length refutations of infeasible integer programs (IPs). We focus on three different types of refutations, namely read-once refutations, tree-like refutations, and dag-like refutations. For each refutation type, we are interested in finding the length of the shortest possible refutation of that type. For our purposes, the length of a refutation is equal to the number of inferences in that refutation. The refutations in this paper are also defined by the types of inferences that can be used to derive new constraints. We are interested in refutations with two inference rules. The first rule corresponds to the summation of two constraints and is called the ADD rule. The second rule is the DIV rule which divides a constraint by a positive integer. For integer programs, we study the complexity of approximating the length of the shortest refutation of each type (read-once, tree-like, and dag-like). In this paper, we show that the problem of finding the shortest read-once refutation is **NPO PB-complete**. Additionally, we show that the problem of finding the shortest tree-like refutation is **NPO-hard** for IPs. We also show that the problem of finding the shortest dag-like refutation is **NPO-hard** for IPs. Finally, we show that the problems of finding the shortest tree-like and dag-like refutations are in **FPSPACE**.

1 Introduction

This paper examines the problems of finding optimal length refutations of infeasible integer programs (IPs). We study three different types of refutations, each of which is characterized by how the reuse of constraints is permitted.

The first type of refutation examined in this paper is read-once refutation. In a read-once refutation, for the most part, constraints cannot be reused. However, if a constraint can be rederived without reusing constraints from the original system, then it can be used as many times as it can be derived.

The second type of refutation examined is tree-like refutation. In a tree-like refutation, constraints from the original system can be reused, and derived

This research was supported in part by the Air-Force Office of Scientific Research through Grant FA9550-19-1-0177 and in part by the Air-Force Research Laboratory, Rome through Contract FA8750-17-S-7007.

© The Author(s), under exclusive license to Springer Nature Switzerland AG 2023
A. Bagchi and R. Muthu (Eds.): CALDAM 2023, LNCS 13947, pp. 15–27, 2023.
https://doi.org/10.1007/978-3-031-25211-2_2

constraints cannot. As with read-once refutations, constraints can be rederived. However, for tree-like refutations, these rederivations *can* reuse constraints from the original system.

The third type of refutation examined is dag-like refutation. In a dag-like refutation, both constraints in the original system and derived constraints can be reused. This means that constraints do not need to be rederived.

For each refutation type, we are interested in finding the length of the shortest possible refutation of that type. For our purposes, the length of a refutation is equal to the number of inferences in that refutation. For both read-once refutations and tree-like refutations, the rederivation of a constraint increases the length of the refutation. This does not matter in the case of dag-like refutations, since rederivation is unnecessary. Each refutation system is associated with a set of inference rules that can be used to produce refutations of a given type of constraint system. For integer programs, we examine refutations that allow for two types of inferences.

The first rule corresponds to the summation of two constraints and is called the ADD rule. The second rule is the DIV rule which divides a constraint by a positive integer. For IPs, we study the complexity of approximating the length of the shortest refutation of each type (read-once, tree-like, and dag-like). In this paper, we show that the problem of finding the shortest read-once refutation is **NPO PB-complete**. Additionally, we show that the problem of finding the shortest tree-like refutation and the problem of finding the shortest dag-like refutation are both **NPO-hard**. Finally, we show that the problems of finding the shortest tree-like and dag-like refutations are in **FPSPACE**.

2   Statement of Problems

In this section, we introduce the concepts examined in this paper and define the problems under consideration.

Definition 1. *A* **polyhedral constraint system** *is a conjunction of constraints in which each constraint in* $\mathbf{C}$ *is an inequality of the form* $\mathbf{a}_j \cdot \mathbf{x} \leq b_j$ *where* $\mathbf{a}_j \in \mathbb{Q}^n$ *and* $b_j \in \mathbb{Q}$.

Note that $\mathbf{C}$ can be represented in matrix form as $\mathbf{A} \cdot \mathbf{x} \leq \mathbf{b}$. In the constraint $\mathbf{a}_j \cdot \mathbf{x} \leq b_j$, b_j is referred to as the defining constant.

Definition 2. *An* **integer** *polyhedral constraint system is a polyhedral constraint system in which for each variable* x_i, *the corresponding domain* $D_i = \mathbb{Z}$.

Such a constraint system is known as an integer program (IP).

Example 1. System (1) is an integer program.

$$3 \cdot x_1 + 5 \cdot x_2 - 4 \cdot x_3 \leq -2 \quad -2 \cdot x_2 + 7 \cdot x_3 \leq 4 \qquad (1)$$
$$x_1 \in \{0,1\} \quad x_2 \in \{-1,0,1\} \quad x_3 \in \{0,1,2\}$$

Refutations are defined by the inference rules that can be used to deduce a contradiction. Refutations of integer programs can use two inference rules.

The first rule corresponds to the summation of two constraints and is defined as follows:

$$\textbf{ADD} : \frac{\sum_{i=1}^{n} a_i \cdot x_i \leq b_1 \quad \sum_{i=1}^{n} a_i' \cdot x_i \leq b_2}{\sum_{i=1}^{n} (a_i + a_i') \cdot x_i \leq b_1 + b_2} \tag{2}$$

We refer to Rule (2) as the ADD rule.

Example 2. Consider the constraints $3 \cdot x_1 + 5 \cdot x_2 - 4 \cdot x_3 \leq -2$ and $-2 \cdot x_2 + 7 \cdot x_3 \leq 4$. Applying the ADD rule to these constraints results in the constraint $3 \cdot x_1 + 3 \cdot x_2 + 3 \cdot x_3 \leq 2$.

It is easy to see that Rule (2) is sound in that any assignment satisfying the hypotheses **must** satisfy the consequent.

Refutations of integer programs also use an additional rule. This is referred to as the DIV rule and is defined as follows:

$$\textbf{DIV} : \frac{\sum_{i=1}^{n} a_i \cdot x_i \leq b \quad k \in \mathbb{Z}^+ : \frac{a_i}{k} \in \mathbb{Z}}{\sum_{i=1}^{n} \frac{a_i}{k} \cdot x_i \leq \lfloor \frac{b}{k} \rfloor} \tag{3}$$

Example 3. Consider the constraint $3 \cdot x_1 + 3 \cdot x_2 + 3 \cdot x_3 \leq 2$. Applying the DIV rule to this constraint with $k = 3$ results in the constraint $x_1 + x_2 + x_3 \leq 0$.

Rule (3) corresponds to dividing a constraint by a common divisor of the left-hand coefficients and then rounding the right-hand side. Since each $\frac{a_i}{k}$ is an integer. This inference preserves integer solutions but does necessarily preserve linear solutions. A constraint derived using the DIV rule is also known as a Chvátal-Gomory cut [10].

Definition 3. *An* **integer refutation** *is a sequence of applications of the ADD and DIV rules that results in a contradiction of the form $0 \leq b, b < 0$.*

In this paper, we study several types of refutations. These are read-once refutations, tree-like refutations, and dag-like refutations. Our focus is on determining the optimal number of inferences in a refutation. Note that both the ADD rule and the DIV rule contribute to the length of the refutation.

Definition 4. *A* **read-once** *refutation is a refutation in which each constraint C can be used in only one inference. This applies to constraints present in the original formula and those derived as a result of previous inferences.*

Note that in a read-once refutation, a constraint can be reused if it can be rederived. However, it must be rederived from a different set of input constraints.

Definition 5. *A* **tree-like** *refutation is a refutation in which each derived constraint can be used at most once.*

Note that in tree-like refutations, the input constraints can be used multiple times. Thus any derived constraint can be derived multiple times as long as it is rederived each time it is used. This rederivation can reuse derived constraints. However, those constraints also need to be rederived. Tree-like refutation is a **complete** refutation system [5].

Definition 6. *A* **dag-like** *refutation is a refutation in which each constraint can be used multiple times.*

It follows that dag-like refutations procedures are **complete** as well.

We now define the notion of length of a refutation.

Definition 7. *The* **length** *of a refutation R of a constraint system is the number of inferences (both ADD and DIV) made in R.*

For each type of refutation, there are two associated problems. These are the decision problem, asking if a system has the desired type of refutation, and the optimization problem, asking for the length of the shortest refutation of the desired type. Note that every infeasible constraint system has a tree-like refutation and a dag-like refutation. Thus, the decision problems for these two refutation types are trivial. Furthermore, we have shown that the problem of determining if an IP has a read-once refutation is **NP-hard** even when the constraints are UTVPI (Unit Two Variable Per Inequality) constraints [26,27].

In this paper, we examine the following optimization problems:

1. The **Integer Programming Optimal Length Read-once Refutation (IP-OLRR)** problem: Given an infeasible IP **I**, what is the length of the shortest read-once refutation of **I**?
2. The **Integer Programming Optimal Length Tree-like Refutation (IP-OLTR)** problem: Given an infeasible IP **I**, what is the length of the shortest tree-like refutation of **I**?
3. The **Integer Programming Optimal Length Dag-like Refutation (IP-OLDR)** problem: Given an infeasible IP **I**, what is the length of the shortest dag-like refutation of **I**?

Note that the problem of determining if an IP has a refutation is only interesting if the IP is infeasible. Thus, the problems studied in this paper are promise problems [17]. That is, the problems are only defined on a subset of possible inputs. Observe that the IP-OLRR, IP-OLTR, and IP-OLDR problems are only defined on infeasible IPs. Additionally, the problem of determining if an integer program is infeasible is **coNP-complete**. The reductions used in this paper are guaranteed to generate infeasible IPs. Thus, the complexity results we obtain only apply to the set of infeasible IPs. Note that a feasible IP trivially lacks any refutation. Since the set of infeasible IPs is a subset of all IPs, our results can easily be generalized to all IPs. Thus, we can consider the non-promise versions of each problem.

3   Motivation and Related Work

In this section, we motivate our work and describe existing work for related problems.

Constraint systems are heavily used in the field of software verification [11,12]. Corresponding to a piece of software, a constraint system can be derived and then combined with constraints corresponding to the negation of the specifications. If the resultant system of constraints is infeasible, then the software is consistent with its specifications. Although this approach is intuitive and straightforward, it may become impractical because of the large number of constraints that are generated. A constraint-based approach to program verification has also been attempted for rule-based programming [6]. Rule-based programming has gained interest in the software industry over the past years, because of the growing use of Business Rules Management Systems. Hence, a demand for the verification of rule programs has emerged. Also, in [19] it is shown how the constraint-based approach can be used to model a wide spectrum of program analysis using disjunctions and conjunctions of linear inequalities. Linear programs have also been used as a finer grained abstraction for sequential programs offering an effective model checking procedure [2].

For integer programs, we are interested in cutting plane based refutations. Cutting planes are often used to refute integer programs constructed from CNF formulas [14]. When applied to such systems, cutting plane based refutations can be exponentially more compact than resolution based refutations [7,15,20,25]. Several restricted versions of cutting planes have been examined [28]. These restrictions included limiting addition to cases where a variable is canceled, and replacing the division rule with a saturation rule. It was shown that these restricted versions of cutting planes can be simulated by resolution when the coefficients are small [28]. Every infeasible integer program with m constraints over n variables has a cutting plane refutation of length $O(n^{3 \cdot n})$ that can be computed using polynomial workspace [13]. Workspace is defined as the amount of space used to store the intermediate constraints. Once an intermediate constraint is no longer necessary, it is removed from the workspace [13]. Recently, Cheung et al. discussed the verification of integer programming results using cutting plane refutations, but from an empirical perspective [9]. To the best of our knowledge, our paper is the first of its kind to focus on approximation complexity for the problem of determining optimal length refutations.

Closely related to the problem of finding the length of the shortest refutation of a system S under a proof system P, is the problem of automatizability [1]. A proof system P is automatizable if there exists a deterministic algorithm that, when given an infeasible system S, generates a refutation of S in time polynomial in the length of shortest refutation of S [3]. It was shown that resolution as a proof system is not automatizable, unless $\mathbf{P} = \mathbf{NP}$ [4]. In the case of integer programming, cutting planes are not automatizable, unless $\mathbf{P} = \mathbf{NP}$ [18]. In [18] it is also shown that it is **NP-hard** to approximate the minimum length of a dag-like cutting plane proof length to within $2^{n^{\epsilon}}$. In this paper, we show that this problem is **NPO-hard** for a different proof system.

4   Optimal Length Read-Once Refutations

In this section, we show that the problem of finding the shortest read-once refutation of an IP is **NPO PB-complete**.

Theorem 1. *The IP-OLRR problem is* **NPO PB-complete**.

Proof. A read-once refutation R of an integer program $\mathbf{I}$ is polynomially sized in terms of the size of $\mathbf{I}$. Additionally, the length of a read-once refutation can be computed in polynomial time. Finally, since each constraint in $\mathbf{I}$ is used at most once, the length of the read-once refutation is linear in terms of the size of $\mathbf{I}$. Thus, the IP-OLRR problem is in **NPO PB**. Now we need to show **NPO PB-hardness**.

This is accomplished by a reduction from the Minimum 0-1 Programming problem. This problem is formulated as follows:

Given an integer program $\mathbf{A} \cdot \mathbf{x} \geq \mathbf{b}$, $\mathbf{x} \in \{0,1\}^n$, find the minimum value of $\mathbf{c} \cdot \mathbf{x}$ for some integer valued vector $\mathbf{c} \geq \mathbf{0}$.

In the general case, this problem is known to be **NPO-complete** [24]. However, for this reduction, we are only interested in the case where $\mathbf{c} = \mathbf{1}$. This specific form of Minimum 0-1 Programming is known to be **NPO PB-complete** [22].

Consider the following instance of the Minimum 0-1 programming problem: $\min \sum_{i=1}^{n} x_i$ $\mathbf{A} \cdot \mathbf{x} = \mathbf{b}$ $\mathbf{x} \in \{0,1\}^n$. Even in this form, the Minimum 0-1 programming problem is **NPO PB-complete** [22].

Corresponding to this system, we can construct the following linear program $\mathbf{L}$:

$\mathbf{y} \cdot \mathbf{A} \leq \mathbf{0}$ $-\mathbf{y} \cdot \mathbf{b} \leq -1$.

From $\mathbf{L}$, we can construct the IP $\mathbf{I}$ as follows:

1. For each variable y_i in $\mathbf{L}$, add the variable y_i to $\mathbf{I}$. Additionally, add the new variable x_0 to $\mathbf{I}$. 2. Add all of the constraints in $\mathbf{L}$ to $\mathbf{I}$. 3. Let p, be the first prime larger than $\max_{i=1...n}(\sum_{i=1}^{m} |a_{ij}|)$ where a_{ij} is an element of the matrix $\mathbf{A}$. 4. Add the term $p \cdot x_0$ to the constraint in $\mathbf{I}$ with defining constant -1. By construction, there is exactly one such constraint. We will refer to this new constraint as I_1. 5. Add the constraint $-x_0 \leq 0$ to $\mathbf{I}$.

We will now show that $\mathbf{I}$ has a read-once integer refutation of length $(k+2)$ if and only if $\mathbf{L}$ has a read-once linear refutation of length k.

First, assume that $\mathbf{L}$ has a read-once linear refutation of length k. By construction, any read-once refutation of $\mathbf{L}$ corresponds to a read-once derivation of the constraint $p \cdot x_0 \leq -1$. We can then divide this constraint by p and sum the result with the constraint $-x_0 \leq 0$ to obtain a contradiction. Since the original refutation had length k, the new refutation has length $(k+2)$.

Now assume that $\mathbf{I}$ has a read-once integer refutation of length $(k+2)$. As mentioned above, we can assume without loss of generality that the constraint I_1 is the only constraint in $\mathbf{I}$ with negative defining constant. Thus, it must be used in any refutation of $\mathbf{I}$.

The only other constraint in $\mathbf{I}$ with the variable x_0 is $-x_0 \leq 0$. Thus, this constraint must be used to cancel the variable x_0. Since the refutation is read-once, we must first apply the DIV rule to the constraint derived from I_1. By construction, the DIV rule must be applied to this constraint with coefficient p. However, by construction, p is larger than any coefficient in any constraint derived from I_1. Thus, the DIV rule cannot be applied until all other variables are eliminated from I_1. Thus, we must derive the constraint $p \cdot x_0 \leq -1$. This derivation takes k steps and corresponds to a read-once linear refutation of $\mathbf{I}$. This means that $\mathbf{I}$ has a read-once linear refutation of length k. Thus, the IP-OLRR problem is **NPO PB-complete**.   □

Since the IP-OLRR problem is **NPO PB-complete**, there exists an $\epsilon > 0$ such that the IP-OLRR cannot be approximated to within a factor of $O(n^\epsilon)$, unless $\mathbf{P} = \mathbf{NP}$ [21]. Thus, the IP-OLRR cannot be approximated to within a polylogarithmic factor, unless $\mathbf{P} = \mathbf{NP}$.

5   Optimal Length Tree-Like and Dag-Like Refutations

In this section, we show that the IP-OLTR and IP-OLDR are **NPO-hard**. Note that for these problems, we are not guaranteed polynomial length refutations. Thus we do not have **NPO-completeness**.

Theorem 2. *The IP-OLTR problem is* **NPO-hard.**

Proof. This will be accomplished by a reduction from the Traveling Salesman Path Problem. This problem is **NPO-complete** [24].

Let $\mathbf{G}$ be a complete undirected graph with n vertices. From $\mathbf{G}$ we create an IP $\mathbf{I}$ as follows: 1. For each vertex v_i in $\mathbf{G}$, create the variable x_i. 2. Create the constraint $x_1 + 2 \cdot x_2 + 2 \cdot x_3 + \ldots + 2 \cdot x_{n-1} + 2 \cdot x_n + p \cdot x_0 \leq -1$. Where p is the fist prime such that $p > 2 \cdot n \cdot \sum_{i=1}^{n-1} \sum_{j=i+1}^{n} w(e_{i,j})$. Additionally, create the constraint $-x_0 \leq 0$. 3. For each edge $e_{i,j}$ in $\mathbf{G}$, create the variables $y_{i,j}$ and $z_{i,j,l}$ for $l = 1, \ldots, n-1$. Additionally, create the constraint $-y_{i,j} \leq 0$. 4. For each edge $e_{i,j}$ such that $i, j \in \{2, \ldots, n\}$, and each $l = 2, \ldots, n-2$, create the constraint $-x_i - x_j + 2 \cdot n \cdot w(e_{i,j}) \cdot y_{i,j} + 2 \cdot z_{i,j,l} \leq 0$. 5. For each edge $e_{i,n}$, create the constraint $-x_i - x_n + 2 \cdot n \cdot w(e_{i,n}) \cdot y_{i,n} + z_{i,n,n-1} \leq 0$. 6. For each edge $e_{1,j}$, create the constraint $-x_1 - x_j + 2 \cdot n \cdot w(e_{1,j}) \cdot y_{1,j} + z_{1,j,1} \leq 0$. 7. For each pair of edges $e_{i,j}$ and $e_{j,k}$ that share an endpoint, and each $l = 1, \ldots, n-2$, create the constraint $-z_{i,j,l} - z_{j,k,l+1} \leq 0$. This construction forms the function f for our PTAS reduction.

First, assume that $\mathbf{G}$ has a Traveling Salesman Path P of length W from x_1 to x_n. Let P traverse the vertices in the order $v_{P(1)}$ through $v_{P(n)}$. We can construct a tree-like refutation R of length $2 \cdot n \cdot (W + 1)$ for $\mathbf{I}$ as follows: 1. Start with the constraint $x_1 + 2 \cdot x_2 + 3 \cdot x_2 + \ldots + 2 \cdot x_{n-1} + 2 \cdot x_n + p \cdot x_0 \leq -1$. 2. Add the constraint $-x_1 - x_{P(2)} + 2 \cdot n \cdot w(e_{1,P(2)}) \cdot y_{1,P(2)} + z_{1,P(2),1} \leq 0$ to R. 3. For $i = 2 \ldots n-2$, add the constraint $-x_{P(i)} - x_{P(i+1)} + 2 \cdot n \cdot w(e_{P(i),P(i+1)}) \cdot y_{P(i),P(i+1)} + 2 \cdot z_{P(i),P(i+1),i} \leq 0$ to R. 4. Add the constraint $-x_{P(n-1)} - x_n +$

$2 \cdot n \cdot w(e_{P(n-1),n}) \cdot y_{P(n-1),n} + z_{P(n-1),n,n-1} \leq 0$ to R. 5. For $i = 2 \ldots n - 2$, add $2 \cdot n \cdot w(e_{P(i),P(i+1)})$ copies of the constraint $-y_{P(i),P(i+1)} \leq 0$ to R. 6. For $i = 1 \ldots n - 2$, the constraint $-z_{P(i),P(i+1),i} - z_{P(i+1),P(i+2),i+1} \leq 0$ to R. Observe that summing the constraints in R results in the constraint $p \cdot x_0 \leq -1$. Applying the DIV rule with $d = p$ to this constraint results in the constraint $x_0 \leq -1$. Adding the constraint $x_0 \leq 0$ results in the contradiction $0 \leq -1$. Note that, R contains a total of $2 \cdot n \cdot (W + 1)$ inferences. Thus R is a tree-like refutation of length $2 \cdot n \cdot W$ for $\mathbf{I}$.

Now assume that $\mathbf{I}$ has a tree-like refutation R of length $2 \cdot n \cdot (W + 1)$. We can construct a set of edges P as follows: For each edge $e_{i,j}$ if R contains the constraint $-y_{i,j} \leq 0$, add $e_{i,j}$ to P. This forms the function g for our PTAS reduction. Observe the following:

1. The constraint $x_1 + 2 \cdot x_2 + 3 \cdot x_2 + \ldots + 2 \cdot x_{n-1} + 2 \cdot x_n + p \cdot x_0 \leq -1$, is the only constraint in the system with a negative defining constant. Thus, it must be part of R. We will refer to this constraint as C. By construction of $\mathbf{I}$, $p > 2 \cdot n \cdot (W + 1)$. Thus, if the DIV rule is not applied to C, then the constraint $x_0 \leq 0$ will need to be used at least $p > 2 \cdot n \cdot (W + 1)$ times. In this case, the length of R is more than $2 \cdot n \cdot (W + 1)$. Thus, the DIV rule must be applied to C. Due to the value chosen for p, this can only happen after everything else is canceled from C.

2. To cancel x_1 from C, R must include a constraint of the form $-x_1 - x_j + 2 \cdot n \cdot w(e_{1,j}) \cdot y_{1,j} + z_{1,j,1} \leq 0$. Let $P(2) = j$. Note that this constraint also cancels a copy of $x_{P(2)}$ from C.

3. To cancel the other copy of $x_{P(2)}$ from C, R must include a constraint of the form $-x_{P(2)} - x_j + 2 \cdot n \cdot w(e_{P(2),j}) \cdot y_{P(2),j} + 2 \cdot z_{P(2),j,1} \leq 0$. Let $P(3) = j$. Note that this constraint also cancels a copy of $x_{P(3)}$ from C.

4. We can continue this process until $P(h) = n$ for some $h \leq n$. Due to the structure of C, the vertices $v_1, v_{P(2)}, v_{P(3)}, \ldots, v_{P(h)}$ are all distinct.

5. Consider the constraint $-x_{P(h-1)} - x_{P(h)} + 2 \cdot n \cdot w(e_{P(h-1),P(h)}) \cdot y_{P(h-1),P(h)} + z_{P(h-1),P(h),1} \leq 0$ in R. Since $P(h) = n$, by construction of $\mathbf{I}$, this constraint must be $-x_{P(h-1)} - x_{P(h)} + 2 \cdot n \cdot w(e_{P(h-1),P(h)}) \cdot y_{P(h-1),P(h)} + z_{P(h-1),P(h),n-1} \leq 0$. Note that this constraint introduces the variable $z_{P(h-1),P(h),n-1}$ to R

6. Consider the constraint $-x_{P(h-2)} - x_{P(h-1)} + 2 \cdot n \cdot w(e_{P(h-2),P(h-1)}) \cdot y_{P(h-2),P(h-1)} + 2 \cdot z_{P(h-2),P(h-1),1} \leq 0$ in R. Recall that R contains the variable $z_{P(h-1),P(h),n-1}$. To cancel this variable, R must contain a constraint of the form $-z_{j,P(h-1),n-2} - z_{P(h-1),P(h),n-1} \leq 0$. By construction, $z_{P(h-2),P(h-1),1} = z_{j,P(h-1),n-2}$. Thus, $l = n - 2$.

7. Continuing this process, we see that $1 = n - (h - 1)$. Thus, $h = n$. As shown previously, the vertices $v_1, v_{P(2)}, v_{P(3)}, \ldots, v_{P(n)}$ are all distinct. Thus P is a Traveling Salesman Path in $\mathbf{G}$. For each edge $e_{P(i),P(i+1)}$ in P, R contains $2 \cdot nw(e_{P(i),P(i+1)})$ copies of the constraint $-y_{P(i),P(i+1)} \leq 0$. From the observations above, R contains an additional $2 \cdot n$ constraints. Thus, R contains a total of $2 \cdot n \cdot (W' + 1)$ constraints where W' is the total length of P. Since R has length $2 \cdot n \cdot (W + 1)$, P has length W.

All that remains is to establish that a **PTAS** reduction exists from the Minimum 0-1 Programming problem to the IP-OLTR problem. This will be done by establishing the existence of the functions f, g, and α.

1. The function f: We provided a method for constructing an integer program **I** from a graph **G**. This forms the function f required for the **PTAS** reduction.
2. The function g: We provided a method to take a tree-like refutation of **I** and construct a Traveling Salesman Path in **G**. This forms the function g required for the **PTAS** reduction.
3. The function α: Let W^* be the shortest Traveling Salesman Path in **G**. **I** has a tree-like refutation of length $2 \cdot n \cdot (W^* + 1)$. Additionally, if **I** had a shorter tree-like refutation, then **G** would have a shorter path. Thus, the IP-OLTR of **I** has length $2 \cdot n \cdot (W^* + 1)$. Let $\alpha(\epsilon) = \frac{\epsilon - 1}{2}$.
   Let R be a tree-like refutation of **I** of length $2 \cdot n \cdot (W + 1)$. The function g produces a Traveling Salesman Path of length W. If $\frac{2 \cdot n \cdot (W+1)}{2 \cdot n \cdot (W^*+1)} \leq 1 + \alpha(\epsilon) = \frac{\epsilon + 1}{2}$, then

$$\frac{W}{W^*} \leq \frac{2 \cdot W}{2 \cdot W^*} \leq \frac{2 \cdot (W + 1)}{W^* + 1} \leq \frac{2 \cdot (\epsilon + 1)}{2} = 1 + \epsilon.$$

Thus, the IP-OLTR problem for linear programs is **NPO-hard**. ☐

Since the IP-OLTR problem is **NPO-hard**, there exists an $\epsilon > 0$ such that the IP-OLTR cannot be approximated to within a factor of $O(2^{n^\epsilon})$, unless **P** = **NP** [21]. Thus, the IP-OLTR cannot be approximated to within a polynomial factor, unless **P** = **NP**.

Theorem 3. *The IP-OLDR problem is* **NPO-hard**.

Proof. This will be accomplished by a reduction from the Minimum Integer Programming problem.

Consider the following instance of the Minimum 0-1 programming problem:

$$\min \sum_{i=1}^{n} (2 \cdot \log c_i + 1) \cdot x_i \quad \mathbf{A} \cdot \mathbf{x} = \mathbf{b} \quad \mathbf{x} \in \{0, 1\}^n.$$

Assume without loss of generality that $\mathbf{c} \geq 1$.

While in general Minimum 0-1 programming is **NPO-complete**, the values of the coefficients in the optimization function are polynomial in the size of the input. Thus, the final value of the objective function is polynomial in the size of the input. Consequently, this problem is **NPO PB-complete** [22,24].

Let **D** be the $n \times n$ matrix such that $d_{i,i} = c_i - 1$ and $d_{i,j} = 0$ for $i \neq j$. Corresponding to the Minimum 0-1 programming instance, we can construct the following linear program **L**:

$$\mathbf{y} \cdot \mathbf{A} + \mathbf{z} \cdot \mathbf{D} \leq 0 \quad -\mathbf{z} \leq 0 \quad -\mathbf{y} \cdot \mathbf{b} \leq -1$$

From **L**, we can construct the IP **I** as follows:

1. For each variable y_i in $\mathbf{L}$, add the variable y_i to $\mathbf{I}$. Additionally, add the variable x_0 to $\mathbf{I}$. 2. Add all of the constraints in $\mathbf{L}$ to $\mathbf{I}$. 3. Let p, be the first prime larger than $\max_{i=1\ldots n}(\sum_{i=1}^{m}|a_{ij}|)$. 4. Add the term $p \cdot x_0$ to the constraint L_1 in $\mathbf{L}$ with defining constant -1. By construction, there is exactly one such constraint. We will refer to this new constraint as I_1. 5. Add the constraint $-x_0 \leq 0$ to $\mathbf{I}$.

We will now show that $\mathbf{I}$ has a dag-like integer refutation of length $(k+2)$ if and only if $\mathbf{L}$ has a dag-like linear refutation of length k.

First, assume that $\mathbf{L}$ has a dag-like linear refutation of length k. By construction, any dag-like refutation of $\mathbf{L}$ corresponds to a dag-like derivation of the constraint $c_L \cdot p \cdot x_0 \leq -c_L$ where c_L is the number of times constraint L_1 was used in the dag-like refutation. We can then divide this constraint by $c_L \cdot p$ and sum the result with the constraint $-x_0 \leq 0$ to obtain a contradiction. Since the original refutation had length k, the new refutation has length $(k+2)$.

Now assume that $\mathbf{I}$ has a dag-like integer refutation of length $(k+2)$. As mentioned previously, we can assume without loss of generality that the constraint I_1 is the only constraint in $\mathbf{I}$ with negative defining constant. Thus, it must be used in any refutation of $\mathbf{I}$.

The only other constraint in $\mathbf{I}$ with the variable x_0 is $-x_0 \leq 0$. Thus, this constraint must be used to cancel the variable x_0. We want to avoid using the constraint $-x_0 \leq 0$ p times. Thus, we must first apply the DIV rule to the constraint derived from I_1. By construction, the DIV rule must be applied to this constraint with a coefficient divisible by p. However, by construction, p is larger than any coefficient in any constraint derived from I_1 by a dag-like derivation of length k. Thus, the DIV rule cannot be applied until all other variables are eliminated from I_1. Thus, we must derive the constraint $c_L \cdot p \cdot x_0 \leq -c_L$. This derivation takes k steps and corresponds to a dag-like linear refutation of $\mathbf{I}$. This means that $\mathbf{I}$ has a dag-like linear refutation of length k. Thus, the IP-OLDR problem is **NPO-hard**. □

Since the IP-OLDR problem is **NPO-hard**, there exists an $\epsilon > 0$ such that the IP-OLDR cannot be approximated to within a factor of $O(2^{n^\epsilon})$, unless $\mathbf{P} = \mathbf{NP}$ [21]. Thus, the IP-OLDR cannot be approximated to within a polynomial factor, unless $\mathbf{P} = \mathbf{NP}$.

Note that in general, tree-like and dag-like cutting plane based refutations can be exponentially long [7]. Thus, these problems do not belong to the class **NPO**. However, we can show that both the IP-OLDR problem and the IP-OLTR problem belong to the class **FPSPACE**.

Theorem 4. *The IP-OLTR problem is in* **FPSPACE**.

Proof. Let $\mathbf{I}$ be an infeasible integer program with m constraints over n variables. We will show that a tree-like integer refutation of $\mathbf{I}$ can be constructed by a non-deterministic Turing Machine using working space polynomial in the size of $\mathbf{I}$. We know that any integer program has a tree-like integer refutation of length at most $O(n^{3 \cdot n})$ [13]. Additionally, this refutation only needs to store polynomially many constraints at a time. Thus, each inference in a tree-like integer refutation can be identified using a number of bits polynomial in the size of the input integer program.

For each inference in a possible refutation R, we can non-deterministically guess the following:

1. The constraints used by the inference – Note that we can assume without loss of generality that the coefficients in these constraints have an absolute value of at most $(n \cdot C_1 + C_2)$ where C_1 is the largest coefficient of any constraint in $\mathbf{I}$ and C_2 is the largest defining constant [16]. Thus, each constraint can be represented using space polynomial in the size of $\mathbf{I}$.
2. The constraint produced by the inference.
3. The source of each constraint used by the inference – This is either the inference used to derive the constraint or the original integer program $\mathbf{I}$.
4. The inference that will use the derived constraint – Note that since the refutation is tree-like, there is at most one such inference.

Thus, each inference can be generated in polynomial space. Once each inference is generated, the space can then be reused to generate the next inference. Thus, the entire refutation can be generated using at most polynomial space.

The correctness of R can similarly be verified in polynomial space as follows:

1. Non-deterministically guess an inference in the refutation.
2. Verify that the derived constraint is correct for the given input constraints.
3. Verify that the input constraints come from the specified sources.
4. Verify that each source is either $\mathbf{I}$ or a previous inference.
5. For each constraint derived by a previous inference, verify that inference lists the current inference as using its derived constraint. Note that this ensures that derived constraints are not repeated.

This can be easily done in space polynomial in the size of $\mathbf{I}$. Once every inference in the refutation is verified, we know that the constraint derived by the last inference is derivable from the constraints in $\mathbf{I}$. If the last inference of R derives a contradiction, then R is a tree-like integer refutation of $\mathbf{I}$. By performing both the construction and verification procedures for each possible refutation length, the first tree-like integer refutation generated in this way is IP-OLTR of $\mathbf{I}$.   □

Note that the refutations generated in the proof of Theorem 4 are tree-like because we ensure that each derived constraint is used by at most one future inference. If we remove this restriction, then the procedure instead generates dag-like integer refutations. This gives us the following corollary.

Corollary 1. *The IP-OLDR problem is in* **FPSPACE**.

6   Conclusion

In this paper, we studied the problems of finding optimal length refutations of infeasible integer programs (IPs). We looked at three different types of refutations, namely read-once refutations, tree-like refutations, and dag-like refutations.

Constraint systems are heavily used in the field of software verification [8, 12]. Refutations of these constraint systems provide evidence that the system is infeasible. Thus, refutations, especially short refutations, are also very useful in this field. As a result, the contributions in this paper will provide insights useful in software verification.

Specifically, we showed that the IP-OLRR problem is **NPO PB-complete** while the IP-OLTR and IP-OLDR problems are **NPO-hard** and in **FPSPACE**.

This paper only examined general forms of integer programs. However, restricting the form of the program can change the complexity of the problems examined. For example, in systems of difference constraints, the OLRR, OLTR, and OLDR problems for integer feasibility can be solved in polynomial time. Thus, future work can examine the complexity of these problems for other restricted IPs.

References

1. Alekhnovich, M., Razborov, A.: Resolution is not automatizable unless W[P] is tractable. In: Proceedings of the 42nd IEEE Symposium on Foundations of Computer Science, FOCS 2001, pp. 210, USA. IEEE Computer Society (2001)
2. Armando, A., Castellini, C., Mantovani, J.: Software model checking using linear constraints. In: Davies, J., Schulte, W., Barnett, M. (eds.) ICFEM 2004. LNCS, vol. 3308, pp. 209–223. Springer, Heidelberg (2004). https://doi.org/10.1007/978-3-540-30482-1_22
3. Atserias, A., Luisa Bonet, M.: On the automatizability of resolution and related propositional proof systems. Inf. Comput. **189**(2), 182–201 (2004)
4. Atserias, A., Müller, M.: Automating resolution is NP-hard. CoRR, abs/1904.02991 (2019)
5. Beame, P., Pitassi, T.: Simplified and improved resolution lower bounds. In: 37th Annual Symposium on Foundations of Computer Science, pages 274–282, Burlington, Vermont, 14–16 October 1996. IEEE (1996)
6. Berste, B., Leconte, M.: Using constraints to verify properties of rule programs. In: Proceedings of the 2010 International Conference on Software Testing, Verification, and Validation Workshops, pp. 349–354 (2008)
7. Bonet, M.L., Pitassi, T., Raz, R.: Lower bounds for cutting planes proofs with small coefficients. J. Symb. Log., **62**(3), 708–728 (1997)
8. Ceberio, M., Acosta, C., Servin, C.: A constraint-based approach to verification of programs with floating-point numbers. In: Proceedings of the 2008 International Conference on Software Engineering Research and Practice, pp. 225–230 (2008)
9. Cheung, K.K.H., Gleixner, A.M., Steffy. D.E.: Verifying integer programming results. In: Integer Programming and Combinatorial Optimization - 19th International Conference, IPCO 2017, vol. 10328 of Lecture Notes in Computer Science, pp. 148–160 (2017)

10. Chvatal, V.: Edmonds polytopes and a hierarchy of combinatorial problems. Discret. Math. **4**(10–11), 886–904 (1973)
11. Collavizza, H., Reuher, N.: Exploration of the capabilities of constraint programming for software verification. In: Proceedings of the 2006 International Conference on Tools and Algorithms for the Construction and Analysis of Systems (2006)
12. Collavizza, H., Rueher, M., Van Hentenryck, P.: CPBPV: a constraint-programming framework for bounded program verification. In: Stuckey, P.J. (ed.) CP 2008. LNCS, vol. 5202, pp. 327–341. Springer, Heidelberg (2008). https://doi.org/10.1007/978-3-540-85958-1_22
13. Cook, W.: Cutting-plane proofs in polynomial space. Math. Program. **47**(1), 11–18 (1990)
14. Cook, W., Coullard, C.R., Turan, Gy. On the complexity of cutting-plane proofs. Discrete Appl. Math. **18**, 25–38 (1987)
15. Fleming, N., Pankratov, D., Pitassi, T., Robere, R.: Random $\Theta(\log n)$-CNFs are hard for cutting planes. In: 58th IEEE Annual Symposium on Foundations of Computer Science, FOCS 2017, pp. 109–120 (2017)
16. Galesi, N., Pudlák, P., Thapen, N.: The space complexity of cutting planes refutations. In: Proceedings of the 30th Conference on Computational Complexity, CCC 2015, pp. 433–447 (2005)
17. Goldreich, O.: On promise problems (a survey in memory of Shimon even [1935-2004]). Electron. Colloquium Comput. Complex **61**(018) (2005)
18. Sajin Koroth, M., Mertz, I., Pitassi, T.: Automating cutting planes is NP-hard. In: Proceedings of the 52nd Annual ACM SIGACT Symposium on Theory of Computing, STOC 2020, pp. 68–77 (2020)
19. Gulwani, S., Srivastava, S., Venkatesan, R.: Program analysis as constraint solving. In: Proceedings of the 2008 ACM SIGPLAN Conference on Programming language design and implementation, New York, NY. ACM (2008)
20. Hrubes, P., Pudlák, P.: Random formulas, monotone circuits, and interpolation. In: 58th IEEE Annual Symposium on Foundations of Computer Science, FOCS 2017, pp. 121–131 (2017)
21. Kann, V.: On the approximability of np-complete optimization problems. Ph.D. thesis, Royal Institute of Technology Stockholm (1992)
22. Kann, V.: Polynomially bounded minimization problems that are hard to approximate. Nordic J. of Compu. **1**(3), 317–331 (1994)
23. Krentel, M.W.: The complexity of optimization problems. J. Comput. Syst. Sci. **36**(3), 490–509 (1988)
24. Orponen, P., Mannila, H.: On approximation preserving reductions: complete problems and robust measures. Technical report, Department of Computer Science, University of Helsinki (1987)
25. Pudlák, P.: Lower bounds for resolution and cutting plane proofs and monotone computations. J. Symbol Logic **62**(3), 981–998 (1997)
26. Subramani, K., Wojciechowski, P.: Integer feasibility and refutations in UTVPI constraints using bit-scaling. Algorithmica (Accepted In Press 2022)
27. Subramani, K., Wojciechowski, P.J.: A bit-scaling algorithm for integer feasibility in UTVPI constraints. In: Combinatorial Algorithms - 27th International Workshop, IWOCA 2016, vol. 9843, pp. 321–333 (2016)
28. Vinyals, M., Elffers, J., Giráldez-Cru, J., Gocht, S., Nordström, J.: In between resolution and cutting planes: a study of proof systems for pseudo-Boolean SAT solving. In: Beyersdorff, O., Wintersteiger, C.M. (eds.) SAT 2018. LNCS, vol. 10929, pp. 292–310. Springer, Cham (2018). https://doi.org/10.1007/978-3-319-94144-8_18

Fault-Tolerant Dispersion of Mobile Robots

Prabhat Kumar Chand$^{(\boxtimes)}$ (ID), Manish Kumar (ID), Anisur Rahaman Molla (ID),
and Sumathi Sivasubramaniam (ID)

Indian Statistical Institute, Kolkata, India
pchand744@gmail.com, manishsky27@gmail.com, anisurpm@gmail.com,
sumathivel89@gmail.com

Abstract. We consider the mobile robot dispersion problem in the presence of
faulty robots (crash-fault). Mobile robot dispersion consists of $k \leq n$ robots in
an n-node anonymous graph. The goal is to ensure that regardless of the initial
placement of the robots over the nodes, the final configuration consists of having at most one robot at each node. In a crash-fault setting, up to $f \leq k$ robots
may fail by crashing arbitrarily and subsequently lose all the information stored
at the robots, rendering them unable to communicate. In this paper, we solve the
dispersion problem in a crash-fault setting by considering two different initial
configurations: i) the rooted configuration, and ii) the arbitrary configuration.
In the rooted case, all robots are placed together at a single node at the start. The
arbitrary configuration is a general configuration (a.k.a. arbitrary configuration
in the literature) where the robots are placed in some $l < k$ clusters arbitrarily
across the graph. For the first case, we develop an algorithm solving dispersion
in the presence of faulty robots in $O(k^2)$ rounds, which improves over the previous $O(f \cdot \min(m, k\Delta))$-round result by [23]. For the arbitrary configuration, we
present an algorithm solving dispersion in $O((f + l) \cdot \min(m, k\Delta, k^2))$ rounds,
when the number of edges m and the maximum degree Δ of the graph is known
to the robots.

Keywords: Distributed algorithm · Mobile robot · Dispersion · Fault-tolerant
algorithm · Crash-fault · Round complexity · Memory complexity

1 Introduction

The dispersion of autonomous mobile robots to spread them out evenly in a region is a
problem of significant interest in distributed robotics, e.g. [9, 10]. Initially, this problem
was formulated by Augustine and Moses Jr. [2] in the context of graphs. They defined
the problem as follows: Given any arbitrary initial configuration of $k \leq n$ robots positioned on the nodes of an n-node anonymous graph, the robots reposition autonomously
to reach a configuration where each robot is positioned on a distinct node of the graph.
Mobile robot dispersion has various real-world and practical applications, such as the
relocation of self-driving electric cars (robots) to recharge stations (nodes). Assuming

The work of A.R. Molla and S. Sivasubramaniam were supported, in part, by ISI DCSW/TAC
Project (file no. G5446 and G5719). A full version of the paper can be found at [4].

that the cars have smart devices to communicate with each other, the process to find a free or empty charging station, coordination including exploration (to visit each node of the graph in minimum possible time), scattering (spread out in an equidistant manner in symmetric graphs like rings), load balancing (nodes send or receives loads, and distributes them evenly among the nodes), covering, and self-deployment can all be explored as mobile robot dispersion problems [11, 13–15].

The problem has been extensively studied in different graphs with varying assumptions since its conceptualization [11–16, 19–22]. In this paper, we continue the study about the trade-off of memory requirement and time to solve the dispersion problem. Recently, Pattanayak et al. [23] explored the problem of dispersion in a set-up where some of these mobile robots are prone to crash faults. Whenever a robot crashes, it loses all its information immediately, as if the robot has vanished from the network. This makes the problem more challenging and also makes it more realistic in terms of real-world scenarios, where faulty robots can crash at any moment. In this paper, we have continued to study the efficacy of the problem in the same faulty environment. We studied the dispersion problem with both the rooted and arbitrary configuration of the robots in a faulty setup. Both algorithms maintain the optimal level of memory requirement for each robot.

The following table (Table 1) lists up the major notations used throughout the paper.

Table 1. List of major notations

Symbols	Meaning
G	The arbitrary graph acting as the underlying network for the robots
n	The number of nodes (vertices) in G
m	The number of edges in G
Δ	The highest degree of the nodes in G
k	Number of robots
f	Number of faulty robots among the k robots
l	Number of initial clusters of robots in the *arbitrary* configuration
r_i	A robot with ID i
R_c	*root* node in the *rooted* configuration

1.1 Our Results

We consider a team of $k \leq n$ mobile robots placed on an arbitrary, undirected simple graph, consisting of n anonymous, memory-less nodes and m edges. The ports at each node are labelled. The robots have unique IDs and a restricted amount of memory (measured in the number of *bits*). These robots have some computing capability and can communicate with the other robots, only when they are at the same node. We consider two different starting scenarios, based on the initial configuration of the robots. When the robots start from a single node, we call the configuration as *rooted*, otherwise, we call it an *arbitrary* configuration. We further assume that $f \leq k$ faulty robots in the network are prone to crash at any point of time. Our first algorithm for the rooted configuration crucially uses depth first search (DFS) traversal and improves the round complexity from $O(f \cdot \min(m, k\Delta))$ [23] rounds to $O(k^2)$. The second algorithm for the arbitrary configuration is an entirely new result whose complexity depends upon the factors: the number of faulty robots (f), number of robot clusters (l), total number of edges in the graph (m), number of robots (k) and the highest degree of the graph (Δ). In this case, the round complexity is $O((f + l) \cdot \min(m, k\Delta, k^2))$. The results are summarized in the following two theorems:

Theorem 1 (Crash Fault with Rooted Initial Configuration). *Consider any rooted initial configuration of $k \leq n$ mobile robots, out of which $f \leq k$ may crash, positioned on a single node of an arbitrary, anonymous n-node graph G having m edges, in synchronous setting* DISPERSION *can be solved deterministically in $O(k^2)$ rounds with $O(\log(k + \Delta))$ bits memory at each robot, where Δ is the highest degree of the graph.*

Theorem 1 improves over the previously known algorithm (in the worst case, improvement is from cubic to quadratic) that takes $O(f \cdot \min(m, k\Delta))$ rounds for f faulty robots [23]. The theorem also matches the optimal memory bound $(\Omega(\log(\max(k, \Delta)))$ [13]) with $O(\log(k + \Delta))$ bit memory and can handle any number of crashes.

Theorem 2 (Crash Fault with arbitrary Initial Configuration). *Consider any arbitrary initial configuration of $k \leq n$ mobile robots, out of which $f \leq k$ may crash and positioned on $l \leq k/2$ nodes of an arbitrary and anonymous n-node graph G having m edges, in synchronous setting* DISPERSION *can be solved deterministically in $O((f + l) \cdot \min(m, k\Delta, k^2))$ time with $O(\log(k + \Delta))$ bits memory at each robot.*

Theorem 2 solves the dispersion for arbitrary configuration with optimal memory per robot. The time complexity matches the one conjectured by Pattanayak et al. [23]. When f, l and Δ are constants, the time complexity matches the lower bound of $\Omega(k)$. Moreover, the algorithm can handle any number of faulty robots. The results are summarized in the Table 2.

2   Related Work

The problem of dispersion was first introduced in [2] by Moses Jr. et al., where they solved the problem for different types of graphs. They had given a lower bound of $\Omega(\log n)$ on the memory of each robot (later, made more specific with $\Omega(\log(\max(k, \Delta)))$ in [13]) and of $\Omega(D)$ on the time complexity, for any deterministic algorithm on arbitrary graphs. They also proposed two algorithms on arbitrary graphs, one requiring $O(\log n)$ memory and running for $O(mn)$ time while the other needing a $O(n \log n)$ memory and having a time complexity of $O(m)$.

Kshemkalyani and Ali [11] provided several algorithms for both synchronous and asynchronous models. In the synchronous model, they solved the dispersion problem in $O(\min(m, k\Delta))$ rounds with $O(k \log \Delta)$ memory. For the asynchronous cases, they proposed several algorithms, one particularly requiring $O(\Delta^D)$ rounds and $O(D \log \Delta)$ memory, while another requiring $O(\max(\log k, \log \Delta))$ memory and having a time complexity of $O((m - n)k)$. In a later work, Kshemkalyani et al., in [13] improved the time complexity to $O(\min(m, k\Delta) \log k)$ keeping the memory requirement to $O(\log n)$, while requiring that the robots know the parameters m, n, k, Δ beforehand. In a subsequent work, [24] kept the time and memory complexity of [13] intact while dropping the requirement of the robots having to know m, k, Δ beforehand. Recently, Kshemkalyani and Sharma [16] improved the time complexity to $O(\min(m, k\Delta))$. Works of [6,22] used randomization, which helped to reduce the memory requirement for each robot.

In [12], Kshemkalyani et al., studied the problem in the *Global Communication Model*, in which the robots can communicate with each other irrespective of their positions in the graph[1]. The authors obtained a time complexity of $O(k\Delta)$ rounds when $O(\log(k + \Delta))$ bits of memory were allowed at each robot. Whereas, when robots were allowed $O(\Delta + \log k))$ bits, the number of rounds reduced to $O(\min(m, k\Delta))$. Both were for arbitrary initial configuration of robots. They also used BFS traversal techniques for investigating the dispersion problem. The BFS traversal technique yielded a time of $O((D + k)\Delta(D + \Delta))$ rounds with $O(\log D + \Delta \log k)$ bits of memory at each robot, using global communication, for arbitrary starting configuration of robots. Here D denotes the diameter of the graph. The problem was also studied on *dynamic* graphs in [1,15,17]. *Graph Exploration*, which is a related problem, has also been intensively studied in literature [3,5,7,8]

The dispersion problem has also been recently studied for configurations with faulty robots. In [18], Molla et al., considered the problem for anonymous rings, tolerating weak Byzantine faults (robots that behave arbitrarily but cannot change their IDs). They gave three algorithms (i) the first one being memory optimized, requiring $O(\log n)$ bits of memory, $O(n^2)$ rounds and tolerating up-to $n - 1$ faults.(ii) the second one is time optimized with $O(n)$ rounds, but require $O(n \log n)$ bits of memory, tolerating up-to $n - 1$ faults. (iii) the third one runs in $O(n)$ time and $O(\log n)$ memory but cannot tolerate more than $\lceil \frac{n-4}{17} \rceil$ faulty robots. In [20], the authors proposed several algorithms for dispersion with some of them tolerating strong Byzantine robots (robots that behave arbitrarily and can tweak their IDs as well). Their algorithms are mainly based on the idea of gathering the robots at a root vertex, using them to construct an isomorphic map of G and finally dispersing them over G according to a specific protocol. However, their algorithms take exponential rounds for strong Byzantine robots starting from an arbitrary configuration. For the rooted configuration, their algorithm takes $O(n^3)$ rounds, but tolerates no more than $\lceil n/4 - 1 \rceil$ strong Byzantine robots. Dispersion under *crash faults* has been dealt with in [23]. In [23], Pattanayak et al., have considered the problem for a team of robots starting at a rooted configuration, with some robots being crash prone. Their algorithm handles an arbitrary number of crashes, with each robot requiring $O(\log(k + \Delta))$ bits of memory. The algorithm completes in $O(f \cdot \min(m, k\Delta))$ rounds. In our paper, we improve this time complexity while keeping the memory requirement to optimal and also extend the problem for the robots starting in arbitrary configuration. A comparison between our results and the most aligned works is shown in Table 2.

---

[1] In the *Local Communication Model* robots can communicate with each other only when they are at the same node.

Table 2. Results on Dispersion of $k \leq n$ robots with $f \leq k$ faulty robots on n-node arbitrary anonymous graphs having m edges such that Δ is the highest degree of the graph in the local communication model. Each uses an optimal memory of $O(\log(k + \Delta))$ bits on each robot.

Algorithm	Initial Config.	Crash handling	Time
Kshemkalyani et al. [16]*	Arbitrary	No	$O(\min(m, k\Delta))$
Pattanayak et al. [23]	Rooted	Yes	$O(f \cdot \min(m, k\Delta))$
Algorithm in Sect. 4	Rooted	Yes	$O(k^2)$
Algorithm in Sect. 5	Arbitrary	Yes	$O((f + l) \cdot \min(m, k\Delta, k^2))$

*The best known result as of now for fault-free dispersion.

3   Model

We now elaborate our model in detail.

Graph: The underline graph G is connected, undirected, unweighted and anonymous with n vertices and m edges. The vertices of G (also called nodes) do not have any distinguishing identifiers or labels. The nodes do not possess any memory and hence cannot store any information. The degree of a node $i \in V$ is denoted by δ_i and the maximum degree of G by Δ. Edges incident on i are locally labelled using a port number in the range $[1, \delta_i]$. A single edge connecting two nodes receives two independent port numbers at either end. The edges of the graph serve as *routes* through which the robots can commute. Any number of *robots* can travel through an edge at any given time.

Robots: We have a collection of $k \leq n$ robots $\mathbb{R} = \{r_1, r_2, ..., r_k\}$ residing on the nodes of the graph. Each robot has a unique ID and has some memory to store information. The robots cannot stay on an edge, but one or more robots can be present at a node at any point of time. A group of such robots at a node is called $co - located$ robots. Each robot knows the port number through which it has entered and exited a node.

Crash Faults: The robots are not fault-proof and a faulty robot can *crash* at any time during the execution of the algorithm. Such *crashes* are not recoverable and once a robot *crashes* it immediately loses all the information stored in itself, as if it was not present at all. Further, a crashed robot is not visible or sensible to other robots. We assume there are f faulty robots such that $f \leq k$.

Communication Model: Our paper considers a local communication model where only the co-located robots can communicate among themselves.

Time Cycle: Each robot r_i, on activation, performs a $Communicate - Compute - Move$ (CCM) cycle as follows.

- Communicate: r_i reads its own memory along with the memory of other robots co-located at a node v_i.

- Compute: Based on the gathered information and subsequent computations, r_i decides on several parameters. This includes, deciding whether to settle at v_i or otherwise determine an appropriate exit port, choosing the information to pass/store at the settled robot and the information to carry along-with, if, exiting v_i.
- Move: r_i moves to the neighbouring node using the computed exit port.

We consider a synchronous system, where every robot is synchronized to a common clock and becomes active at each time cycle or round.

Time and Memory Complexity: We evaluate the time in terms of the number of discrete rounds or cycles before achieving DISPERSION. Memory is the number of bits of storage required by each robot to successfully execute DISPERSION. Our goal is to solve DISPERSION using optimal time and memory.

Let us now formally state the problem of fault-tolerant dispersion below.

Definition 1 (Fault-Tolerant Dispersion). *Given $k \leq n$ robots, up to f of which are faulty (which may fail by crashing), initially placed arbitrarily on a graph of n nodes, the non-faulty robots, i.e., the robots which are not yet crashed must re-position themselves autonomously to reach a configuration where each node has at most one (non-faulty) robot on it and subsequently terminate.*

4 Crash-Fault Dispersion for Rooted Configuration

In this section, we present a deterministic algorithm that disperses the robots with single-source (rooted configuration) in adaptive crash fault. Our goal is to minimize the round complexity as well as keep the memory of the robots low. The pseudocode and a pictorial description of the algorithm can be found in the full version of the paper [4].

4.1 Algorithm

In the absence of faulty nodes, one can run the DFS (depth first search) algorithm to solve the robot dispersion problem in $O(min(m, k\Delta))$ rounds. But in the presence of crash faults, due to crashes, it becomes challenging to explore the graph. Classic dispersion algorithms rely on the robots themselves to keep track of the paths during exploration. The presence of a crashed robot in this instance may lead to an endless cycle. Therefore, our goal is to ensure the dispersion of mobile robots despite the presence of faulty robots.

In the rooted configuration, to manage the presence of faults, we avoid exploring the graph together with all the robots. That is, the graph is explored sequentially such that each robot r_i ($1 \leq i \leq k$) does not begin exploring the graph, until the previous robot r_{i-1} is guaranteed to have settled. During exploration, whenever a robot r_i finds an empty node it settles down at that spot. Let us call this algorithm as ROOTED-CRASH-FAULT-DISPERSION. Below, we explain the algorithm in detail.

Functionality: For simplicity, let us assume that the robot's ID lies in the range of $[1, k]$. Otherwise, the robots can map their IDs from the actual range to the range $[1, k]$, since the IDs are distinct. We denote the rooted configuration by R_c. We slightly abuse notation and use R_c to indicate both the root and the initial gathering of robots. Robots at R_c traverse the graph via the DFS (Depth First Search) approach, where the decision of which edge to traverse first is based on the port numbers. The process proceeds in increasing order of IDs, starting with the robot with the minimum ID at R_c. R_c then sends each robot to explore the graph via DFS.

Let the robot with the current minimum ID be r_i. Then r_i begins to explore the graph via DFS (starting with the minimum port number at R_c). Once it leaves R_c, it has $3i$ rounds within which it can either i) settle at the first empty node it finds or ii) return to R_c if it does not find an empty node to settle within $2i$ rounds. If r_i reports to R_c within $3i$ rounds, then R_c ensures that it does not release the robot with the next lowest ID, say r_{i+1}. This can be guaranteed as r_i needs to traverse at most $(i-1)$ edges to explore the sub-graph traversed by r_{i-1}. r_i requires at most i rounds to return to the base R_c since the next traversed edge might lead to the already visited node which is not empty. As r_i requires i rounds to report at the R_c, therefore, r_i explores the graph for only $2i$ rounds. Notice that a robot will not traverse the distance of more than $(i+1)$, before that, there will be an empty edge at a distance (distance from the root) of $(i+1)$ and the robot will settle down there. If r_i did not find the empty node within $2i$ rounds then it starts to traverse towards R_c. In this way, r_i reports to R_c within $3i$ rounds so that R_c does not send another robot to explore the graph. R_c re-sends r_i to explore the graph. In this way, any r_i traverses the graph until it finds an empty node. Note that in our process, we ensure that there are no two robots that are exploring the graph at the same time.

To maintain the protocol, each r_i maintains the following fields. Its ID (r_i), a parent pointer $(r_i.parent)$ that represents the edge it traversed, a current direction pointer $(r_i.cdr)$ which indicates the direction it is required to follow. And finally, a backward traversal value $(r_i.B)$ which is initially 0, and is set to 1 once the backward traversal is complete. Here, our procedure performs the traditional DFS protocol but one-by-one, that is, the robots do not explore the graph simultaneously. A detailed account of the DFS traversal is provided in the full version [4].

Note that we have not addressed the case where a robot finds an empty node when returning to R_c (because the previously settled robot has crashed). In such an instance, the newly settled robot has a $r_i.parent$ and $r_i.cdr$ that point in the inappropriate direction. We address this condition below.

Decision: If r_i encounters an unexpected child, r_u i.e., a child whose parent and current pointer direction are set in the inappropriate direction w.r.t the perspective of r_i, it considers (correctly) that r_u replaced a robot that has previously crashed. In such a situation, r_i changes the parent of r_u appropriately, i.e., minimum available port number other than $r_u.parent$.

Lemma 1. *In the non-faulty setup, round complexity is $O(k^2)$.*

Proof. In a non-faulty setup, each robot behaves robustly and there are no crashes. Therefore, after the backtracking flag is set on a node, an edge is not traversed again during the DFS traversal. In traversing a graph from R_c, two kinds of situations may arise, either a robot r_i reaches an empty node after $O(i)$ edge traversals, or it traverses

$O(i^2)$ edges. In the first case, there is an empty node at a distance of $O(i)$. Therefore, r_i settles at the empty node after $O(i)$ rounds. If such kind of situation arises repeatedly, then the algorithm takes $O(1) + O(2) + \cdots + O(k) = O(k^2)$ rounds. In the second case, there might be a situation such that r_i traverses $O(i^2)$ edges to find the empty node and only encounters previously settled nodes (at most $i(i-1)/2$ edges). More preciously, $i/2$ new edges are traversed in $3i$ rounds. Notice that a robot will traverse only earlier traversed nodes at the distance $(i + 1)$, if not, then there will be an empty edge at a distance (distance from the root) of $(i + 1)$ and the robot will settle down there. Therefore, r_i covers $O(i^2)$ edges in $O(i^2)$ rounds and future robots (i.e., robots having ID r_j; $\forall j > i$) will not traverse these edges again. Hence, we can conclude that the non-faulty setup takes $O(k^2)$ rounds in the given model.   □

Lemma 2. *In the faulty setting, a crashed robot may bring about an extra cost of $O(k)$ rounds in comparison to the non-faulty setting.*

Proof. In the faulty setup, a robot might crash at any time and the respective node becomes empty, say node v_i. As a consequence, the information held by that robot (at the node v_i) is also lost. Accordingly, the next robot that discovers v_i, say r_i, settles down at v_i. A robot possesses the information of current direction, parent node and backtracking status apart from its own ID. For that reason, the current direction pointer is pointing towards the edge based on its least labelled edge. But there might be the case (in the worst case) that the last crashed node has traversed up to $(i - 2)$ edges which should be traversed again by the r_{i+1}. This takes extra $O(i)$ rounds. Also, in the worst case, this value can be $O(k)$ since the number of robots is k. Hence, the lemma.   □

Lemma 3. *There is (at most) one robot moving (neither settled at its respective node, nor at rooted configuration R_c) at any instance.*

Proof. Proof by contradiction, let us suppose there exist two robots in moving condition, say r_i and r_{i+1}. Also, assume r_i started before, r_{i+1}. Now, as r_i has not settled, r_i reports to R_c every $3i$ rounds. But if r_i reports every $3i$ rounds then R_c does not release the next robot which is contradictory to our assumption.   □

Lemma 4. *A loop or cycle may be formed by the current direction pointer (cdr pointer). The algorithm* ROOTED-CRASH-FAULT-DISPERSION *successfully avoids any loop during dispersion.*

Proof. During the execution of the algorithm, a loop or cycle may be formed if a robot r_i crashes at a node n_i then the current direction pointer (cdr pointer) is set by the upcoming robot r_{i+1} with the lowest port. That lowest port might have been traversed earlier. Therefore, a loop is formed. From Lemma 3, we know that only one robot is moving at any instance, say r_{i+1}. Therefore, r_{i+2} (the next robot) starts after r_{i+1} settles. If r_{i+2} encounters any robot with an unexpected *cdr pointer* then r_{i+1} changes the *cdr pointer* appropriately. Thus, loops are avoided in the network.   □

Lemma 5. *The algorithm* ROOTED-CRASH-FAULT-DISPERSION *takes at most* $7k^2$ *rounds and* $O(\log(k + \Delta))$ *bits memory.*

Proof. In case of round complexity, a non-faulty set-up from Lemma 1, the total number of rounds are $3(1 + 2 + \cdots + k) < 3k^2$ (in the best case where r_i finds the empty node within $3i$ rounds). Additionally, a robot can traverse at most $i/2$ new edges in $3i$ rounds (in a particular phase) without settling down on an empty node (in the worst case). Therefore, round complexity for $k(k-1)/2$ edges in the non-faulty setup is $< 3k^2$. Moreover, from Lemma 2, we know that the extra cost incurred for f robot's crashing is at most fk. Hence, overall round complexity is at most $3k^2 + 3k^2 + k^2 = 7k^2$.

In case of memory complexity, each robot stores its ID which takes $O(\log k)$ bit space. Along with that parent pointer and current direction pointer takes $O(\log \Delta)$ bit memory each. While the backward pointer takes a single bit. Therefore, the memory complexity is $O(\log(k + \Delta))$. □

From the above discussion, we conclude the following result.

Theorem 1. *Consider any rooted initial configuration of $k \leq n$ mobile robots, out of which $f \leq k$ may crash, positioned on a single node of an arbitrary, anonymous n-node graph G having m edges, in synchronous setting DISPERSION can be solved deterministically in $O(k^2)$ time with $O(\log(k + \Delta))$ bits memory at each robot, where Δ is the highest degree of the graph.*

5   Crash-Fault Dispersion for Arbitrary Configurations

In this configuration setting, the robots are distributed across the graph in clusters such that there are $C = \{C_1, \ldots, C_l\}$ groups of robots at l different nodes at the start such that $\sum_i C_i = k$. The goal of the dispersion is to ensure that the robots are dispersed among the graph vertices such that each node has at most one robot. In this setting, we assume that the robots are aware of k, f, Δ, l and m.

Procedure: Our protocol runs in phases, in which each phase consists of $\min(m, k\Delta, k^2)$ rounds. At the start of each phase, each cluster begins a *counter* that counts down from $\min(m, k\Delta, k^2)$. Each cluster C_i then begins exploring the network simultaneously via the traditional DFS algorithm (in the trivial case of a singleton cluster consisting of only one robot, it considers itself dispersed). Unlike the rooted configuration, individual robots do not explore and return, the entire cluster moves together. Whenever a cluster encounters a new (empty) node in the network, the robot with the current highest ID in the cluster settles, and sets its pointers appropriately. At the end of each round, the counter is decreased by 1. When the counter becomes zero, it signals the end of the phase, and all flags are reset. That is all pointers become null, including the pointers of already settled robots. After that, each cluster starts exploring the network with its current node as a point of origin. This continues until all robots in the cluster settle. The pseudocode and a pictorial description of the algorithm can be found in the full version of the paper [4].

Detailed Procedure. There are two main parts to the protocol, i) exploration, ii) encounter. Exploration deals with the general procedure involved in exploring the graph, while encounter deals with the details involved when robots from different clusters meet.

Let's begin with all the information stored at a robot. Each robot r in a cluster C_i consists of the following pointers cid, $parent$, cdr, $priority$, and B (backtrack). The pointer cid denotes the ID of the cluster it belonged to when a robot settles. cid of a cluster C_i is determined at the start of the phase, and it is the ID of the robot with the highest ID. When a robot decides to settle at a node, the $parent$ pointer keeps track of the port through which it entered the node. Similarly, the cdr pointer is used to keep track of the port through which a cluster leaves the node in which it is settled. The $priority$ pointer of a settled robot keeps track of its priority in various clusters, originally this is simply the cid of the cluster it was part of, that is the priority of a cluster is simply its cid. However, a robot's priority may change if a higher priority cluster discovers it and updates its priority pointer. In our work, priority is decided by the cluster's ID, that is, between two clusters, the cluster with the higher cid has higher precedence. And of course, the backtrack pointer B keeps track of the backtrack status of its DFS. In addition to all of these, each robot also has a field called $counter$, which is set to $\min(m, k\Delta, k^2)$ at the beginning of each phase. Note that since all robots set the counter at the beginning of the phase simultaneously, the counter has the same value across all robots.

Exploration. As mentioned before, as long as a cluster is non-empty, at the beginning of each phase, each cluster begins exploring the graph via the traditional DFS until the cluster is empty or it encounters a robot from a higher priority cluster (more on this in the encounter section). In each phase, each robot in any cluster C_i sets its cid and $priority$ to the highest ID in the cluster, and its counter to $\min(m, k\Delta, k^2)$. We consider the node in which C_i is at the start of the phase, to be its root. C_i then follows the traditional DFS format for exploration. It leaves the node via the smallest unexplored port. If the node is empty, the robot with the current minimum ID, say r, sets its $parent$ and cdr pointers and settles at the node.

The update function for the cdr pointer is exactly the same as the one in the rooted case, i.e., it follows the traditional DFS procedure, except that all the robots in the cluster move through the exit port. All robots in C_i decrease their counter by one and C_i leaves through the port in $r.cdr$. If a cluster ever finds itself returning to a node with a robot r from its own cluster and it has exhausted all of the ports in which r has settled, then it sets r's backtrack flag. Once a phase has finished, if the cluster is non-empty it resets all flags and counter and begins DFS once again. During exploration, if the cluster C_i reaches a node u whose degree is k, then they use BFS to explore the neighbourhood of u and settle the robots of C_i in at most $O(k)$ rounds. However, here we have not explored what happens if a robot from a cluster C_i meets a robot from C_j. That brings us to the next important part of the protocol, the encounters.

Encounter. This section contains the explanation of the encounter part of the protocol. When a robot (or cluster) meets, that is $encounters$ a robot from a different cluster, the next step in exploration is decided based on priority. Simply put, the robot with higher priority always takes precedence as follows. There are two distinct scenarios, i) a cluster finds an already settled node or ii) multiple clusters meet on the same node. In the first case, if a cluster with a higher priority (say C_i) finds a robot r_p from a lower priority cluster (say C_j) on a node, it sets r_p's priority to its own priority, resets r_p's parent and cid to its own, and finally sets r_p's cdr (to the minimum unexplored port the higher priority cluster has not explored so far) and continues its DFS. If on the other hand, a lower priority cluster finds a robot r_p with a higher priority, it stops its exploration and just continues decreasing its counter at every round till the end of the phase, and begins the exploration in the next phase. Note, if a cluster finds a settled robot whose flags have been reset (i.e., set to $null$), then it's the same scenario as that of finding a robot from a lower priority cluster. The settled robot takes the priority and ID of the newly arrived cluster.

In the second scenario, if two (or more) clusters meet, the clusters merge and take on the priority of the cluster with the highest priority among them. They stop and count-down and begin exploration as a merged cluster in the next phase.

Note that the number of clusters is non-increasing between two consecutive phases. At any phase, a cluster may either (i) disperse over the nodes completely, or (ii) survive to explore in the next phase, or (iii) merge with a higher priority cluster. Thus, the number of clusters either remains the same or decreases at the end of every phase. Now we show that after $(l + f)$ phases, dispersion is achieved.

Lemma 6. *The effects of a robot crash, that is time delay caused by the presence of a crash are limited to the phase it occurs in. After that, it ceases to have effect.*

Proof. Since at the end of every phase, all robots reset their flags, including the parent and cdr pointers, previously explored paths are equivalent to new unexplored paths in the current phase, as their pointers are set by the currently exploring clusters. Hence, previous phases do not have any impact on the DFS running in the current phase.   □

Lemma 7. *Let C_i be the cluster with the highest priority in Phase j. C_i is guaranteed dispersion by the end of j if j is fault-free.*

Proof. From Lemma 6 we know that crashes in previous rounds do not have an effect on exploration in the current phase. And, in the absence of faults during the phase itself, we see that C_i exploration is equivalent to a rooted single cluster exploration of the network. Thus it is able to complete its dispersion using DFS without any delays or interference from other clusters, which takes less than $O(\min(m, k\Delta, k^2))$ rounds to complete.   □

Lemma 8. *Each cluster $C_i \in C$ is guaranteed to have at least a single fault-free phase in which it has the highest priority.*

Proof. Quite trivially, since there are $(l + f)$ phases, each cluster is guaranteed at least one phase in which no faults occur, and in which they are the highest priority.   □

Lemma 9. *At the end of $(l + f)$ phases, all clusters are guaranteed to have dispersed.*

Proof. This follows directly from Lemmas 7 and 8. Each cluster is guaranteed to have at least one fault-free phase in which it has the highest priority. From 6 we know, in that phase, there is guaranteed dispersion. Hence, in $(l + f)$ phases, we are guaranteed to have complete dispersion of all clusters.   □

Thus, we have the following theorem.

Theorem 2. *In the synchronous setting, the crash-tolerant algorithm for the arbitrary configuration (algorithm* ARBITRARY-CRASH-FAULT-DISPERSION*) ensures dispersion of mobile robots in an arbitrary graph from an arbitrary initial configuration in* $O((f + l) \cdot min(m, k\Delta, k^2))$ *rounds with each robot requiring* $O(\log(k + \Delta))$ *bits of memory.*

Remark 1. If only the number of robots (k) is known and all other factors are unknown to the network then the algorithm for arbitrary configuration takes $O(k^3)$ rounds.

6 Conclusion and Future Work

In this paper, we studied Dispersion for distinguishable mobile robots on anonymous port-labelled arbitrary graphs under crash faults. We presented a deterministic algorithm that solves robot dispersion in two different settings, i) a rooted configuration of robots and ii) an arbitrary configuration of robots. We achieved the $O(k^2)$ round complexity in rooted configuration while $O((f + l)min(m, k\Delta, k^2))$ round complexity in arbitrary setting. In both cases, we used $O(\log(k+\Delta))$ bits of memory. Some open questions that are raised by our work: i) What is the non-trivial lower bound for the round complexity in both the setting by keeping the memory to $O(\log(k+\Delta))$. ii) If it is possible to give a similar round complexity for the case of arbitrary configuration as we achieved in rooted configuration. iii) If it is possible to get the same bound in the arbitrary configuration without the knowledge of f, l, Δ and m. iv) Finally, whether similar bounds hold in the presence of Byzantine failures.

References

1. Agarwalla, A., Augustine, J., Moses Jr., W.K., Madhav K.S., Sridhar, A.K.: Deterministic dispersion of mobile robots in dynamic rings. In: ICDCN (2018)
2. Augustine, J., Moses Jr., W.K.: Dispersion of mobile robots: a study of memory-time trade-offs. In: ICDCN (2018)
3. Bampas, E., Gasieniec, L., Hanusse, N., Ilcinkas, D., Klasing, R., Kosowski, A.: Euler Tour Lock-in Problem in the Rotor-Router Model. In: DISC (2009)
4. Chand, P.K., Kumar, M., Molla, A.R., Sivasubramaniam, S.: Fault-tolerant dispersion of mobile robots (2022). https://arxiv.org/abs/2211.12451
5. Cohen, R., Fraigniaud, P., Ilcinkas, D., Korman, A., Peleg, D.: Label-guided graph exploration by a finite automaton. ACM Trans, Algorithms 4, 42 (2008)
6. Das, A., Bose, K., Sau, B.: Memory optimal dispersion by anonymous mobile robots. In: CALDAM (2021)
7. Dereniowski, D., Disser, Y., Kosowski, A., Pajak, D., Uznanski, P.: Fast collaborative graph exploration. In: ICALP (2013)
8. Fraigniaud, P., Ilcinkas, D., Peer, G., Pelc, A., Peleg, D.: Graph Exploration by a Finite Automaton. Theorr. Comput. Sci. 345, 331–344 (2005)
9. Hsiang, T., Arkin, E.M., Bender, M.A., Fekete, S.P., Mitchell, J.S.B.: Algorithms for rapidly dispersing robot swarms in unknown environments. In: WAFR (2002)

10. Hsiang, T., Arkin, E.M., Bender, M.A., Fekete, S.P., Mitchell, J.S.B.: Online dispersion algorithms for swarms of robots. In: Fortune, S. (ed.) SCG 2003
11. Kshemkalyani, A.D., Ali, F.: Efficient dispersion of mobile robots on graphs. In: ICDCN (2019)
12. Kshemkalyani, A.D., Molla, A.R., Sharma, G.: Dispersion of mobile robots on grids. In: WALCOM (2020)
13. Kshemkalyani, A.D., Molla, A.R., Sharma, G.: Fast dispersion of mobile robots on arbitrary graphs. In: ALGOSENSORS (2019)
14. Kshemkalyani, A.D., Molla, A.R., Sharma, G.: Dispersion of mobile robots using global communication. J. Parallel Distributed Comput. **121**, 100–117 (2022)
15. Kshemkalyani, A.D., Rahaman Molla, A., Sharma, G.: Efficient dispersion of mobile robots on dynamic graphs. In: ICDCS (2020)
16. Kshemkalyani, A.D., Sharma, G.: Near-optimal dispersion on arbitrary anonymous graphs. In: OPODIS (2021)
17. Luna, G.A.D., Flocchini, P., Pagli, L., Prencipe, G., Santoro, N., Viglietta, G.: Gathering in dynamic rings, pp. 79–98. Theor. Comput, Sci. **811**, 79–98 (2020)
18. Mandal, S., Molla, A.R., Jr., W.K.M.: Live exploration with mobile robots in a dynamic ring, revisited. In: ALGOSENSORS (2020)
19. Molla, A.R., Jr., W.K.M.: Dispersion of mobile robots. In: ICDCN (2022)
20. Molla, A.R., Mondal, K., Moses Jr., W.K.: Byzantine dispersion on graphs. In: IPDPS (2021)
21. Molla, A.R., Mondal, K., Moses Jr., W.K.: Efficient dispersion on an anonymous ring in the presence of weak byzantine robots. In: ALGOSENSORS (2020)
22. Molla, A.R., Moses Jr., W.K.: Dispersion of mobile robots: The power of randomness. In: TAMC (2019)
23. Pattanayak, D., Sharma, G., Mandal, P.S.: Dispersion of mobile robots tolerating faults. In: ICDCN (2021)
24. Shintaku, T., Sudo, Y., Kakugawa, H., Masuzawa, T.: Efficient dispersion of mobile agents without global knowledge. In: SSS (2020)

Resource Management
in Device-to-Device Communications

Leila Karimi$^{(\boxtimes)}$, Vijay Adoni, and Daya R. Gaur

Mathematics and Computer Science, University of Lethbridge, Lethbridge T1K 3M4,
AB, Canada
{l.karimi,vijay.adoni}@uleth.ca, gaur@cs.uleth.ca

Abstract. We study the spectrum sharing in device-to-device (D2D)
communications in an underlay cellular network. Our model maximizes the
total sum rate such that i) at most one sub-channel is used for each D2D
pair, and ii) the total interference is at most the required maximum. Our
model can also minimize the interference subject to a guaranteed total sum
rate. We give a branch-n-cut algorithm. We provide an iterative rounding
algorithm that achieves at least a quarter of the optimal sum rate and no
more than the required maximum of the total interference when the objective is to maximize the sum rate. Our experiments establish the effective-
ness of the branch-n-cut approach for resource management.

Keywords: Device-to-Device communication · Iterative rounding
algorithm · Branch and cut

1 Introduction

Device-to-device (D2D) communication occurs when neighbouring cellular
devices communicate directly, typically over a shared spectrum. D2D communication improves the transmission rate and frequency reuse and reduces the hop
count. But can lead to interference with a cellular user and even more substantial interference with another D2D pair in a neighbouring cell. Therefore new
approaches are needed for radio resource allocation [7]. The new methods should
increase the system sum rate without too much deterioration in the signal, and
the total interference should be limited.

In this paper, we study the knapsack-based model for resource allocation
first proposed in [12]. Given a set of cellular users and a set of D2D pairs (from
the cellular users), the model allows for the radio resource of a cellular user to
be used by at most one D2D pair (1–1 resource). If used, this radio allocation
will generate some quantifiable interference given the system model and provide
some sum rate. The model seeks to find those allocations which simultaneously
meet the sum rate and the interference requirements from the systems. For a
cellular user and a D2D receiver pair, the model specifies the signal interference
noise ratio (SINR) at a cellular use when the base station transmits and the
SINR at the D2D receiver. The SINR at the cellular user and D2D receiver is
considered the minimum level of detail needed to examine interference [15].

A. Bagchi and R. Muthu (Eds.): CALDAM 2023, LNCS 13947, pp. 41–55, 2023.
https://doi.org/10.1007/978-3-031-25211-2_4

We can also model the resource allocation problem as a minimum knapsack problem with side constraints. If a target sum rate is required, then the objective is to minimize the total interference. The knapsack constraint models the requirement that the target sum rate is met, and the side constraints model the 1–1 resource requirement; this model was first studied in [14].

OUR CONTRIBUTIONS: We give the first branch and cut method to maximize the sum rate while limiting the total interference for D2D communications. Extensive experiments are performed on hard instances arising from knapsack [23]. We give an approximation algorithm based on iterative rounding in Theorem 1.

We describe the system model briefly in Sect. 2. Section 3 presents the work that is directly related. The branch-n-cut algorithm is described in Sect. 4. An iterative rounding algorithm for the maximization of the sum rate objective and the proof of the quality of the approximation is in Sect. 5. The results of the empirical study of the branch and cut algorithm are in Sect. 6. Finally, we conclude with a discussion and some open problems in Sect. 7.

2   System Model

The system model below is from [8,12]. Let D be the set of D2D pairs and C the group of all cellular users. G_{bc} is the channel gain between the base station b and cellular user $c \in C$. G_{dc} is the channel gain between a transmitter for a D2D pair $d \in D$ and the cellular user $c \in C$. G_{tr} denotes the channel gain between D2D transmitter t and D2D receiver r (for D2D pair). G_{bd} is the channel gain between the base station b and d's receiver. The transmission power of cellular user c and D2D pair d are P^c and P^d, respectively. If $x_{(c,d)} \in \{0,1\}$ indicates whether a D2D pair (d) accepts resources from a cellular user (c), and $s_{(c,d)}$ represents the sum rate, then the total sum rate of the cellular users which share resources with some D2D pair is given by,

$$S = \sum_{c \in C} \sum_{d \in D} x_{(c,d)} s_{(c,d)}$$

The sum rate for a pair (c,d) is given by Shannon's formula.

$$s_{(c,d)} = B \, \log_2 \left(1 + \gamma_{(c,d)}\right) + B \, \log_2 \left(1 + \gamma_{(d,b)}\right)$$

where B is the bandwidth of the channel in Hz, $\gamma_{(c,d)}$ is the SINR at cellular user with signal transmitting from the base station and $\gamma_{(d,b)}$ is the SINR at the D2D receiver end. $\gamma_{(c,d)}$ and $\gamma_{(d,b)}$ are given by,

$$\gamma_{(c,d)} = \frac{P^c G_{bc}}{T + P^d G_{dc}} \quad \text{and} \quad \gamma_{(d,b)} = \frac{P^d G_{tr}}{T + P^c G_{bd}}$$

T above is the thermal noise at the receiver end, also known as the energy of Additive White Gaussian Noise (AWGN). The interference for pair (c,d) is given by,

$$I_{(c,d)} = P^d G_{dc} + P^c G_{bd}.$$

In this model, each cellular user and D2D pair can share resources with at most one D2D pair (and one cellular user). Therefore, $\sum_{c \in C} x_{(c,d)} \leq 1, \forall d \in D$. Similarly, $\sum_{d \in D} x_{(c,d)} \leq 1, \forall c \in C$. Let I be the total interference allowed then $\sum_{c \in C} \sum_{d \in D} x_{(c,d)} I_{(c,d)} \leq I$. The model can now be written as

$$\max \sum_{c \in C} \sum_{d \in D} x_{(c,d)} s_{(c,d)} \tag{1}$$

$$\sum_{c \in C} \sum_{d \in D} x_{(c,d)} I_{(c,d)} \leq I \tag{2}$$

$$\sum_{c \in C} x_{(c,d)} \leq 1 \qquad \forall d \in D \tag{3}$$

$$\sum_{d \in D} x_{(c,d)} \leq 1 \qquad \forall c \in C \tag{4}$$

$$x_{(c,d)} \in \{0,1\} \qquad \forall c \in C, \forall d \in D \tag{5}$$

The objective function (1) maximizes the total sum rate. Constraint (2) ensures that the interference is less than the target interference. Finally constraints (3) and (4) indicates that each D2D and the user cellular can share resources with at most one user cellular and D2D, respectively. We can think of cellular users and D2D pairs as two sides of a complete bipartite graph ($V = C \cup D, E$) and (c, d) as an edge in this graph. Let $\mathcal{M}$ be the set of matchings in the bipartite graph. Then, by a simple change of variable $(c, d) = e$, we can write the above model as:

$$\max \sum_{e \in E} x_e s_e \tag{6}$$

$$\sum_{e \in E} x_e I_e \leq I \tag{7}$$

$$x \in \mathcal{M} \tag{8}$$

where constraint (8) states that the set of edges in $\{e : x_e = 1\}$ form a matching and constraint (7) is the knapsack constraint. Therefore, we say that model is a knapsack with a matching side constraint. We write constraint (8) as

$$\sum_{e \in \delta(v)} x_e \leq 1 \qquad \forall v \in V$$

at times where $\delta(v)$ is the set of edges incident on v (either cellular user or a D2D pair).

3   Related Work

With the emergence of 4G and 5G networks, resource allocation for D2D communication underlaying cellular networks has been studied extensively. The focus has been on the minimization of interference, maximization of sum rate, fair distribution of resources, and restricted allocation of resources. We describe some of the results that are immediately relevant to this study.

One of the early papers in the area is due to Doppler et al. [4], which proposed a mechanism for session setup and management for D2D communications in LTE-A networks. They showed that D2D communication increases the total throughput. Janis et al. [16] gave the scheme to monitor interference between cellular users and D2D pairs and use it to minimize interference. Their simulations demonstrate substantial gains in the sum rate.

Islam et al. [13] proposed a minimum knapsack-based resource allocation model and algorithm for D2D communication (MIKIRA) underlaying cellular networks. This algorithm which takes $O(n^2 log(n))$ time, is a knapsack-based approach to maintain a target sum rate while minimizing interference. This approach is unfair as the algorithm ceases to assign resources when the target sum rate is met. Islam et al. [14] addressed this fairness issue with a two-phase auction-based resource allocation algorithm (TAFIRA). The algorithm starts with a solution with minimum interference and tries to obtain a better sum rate via an auction. A subsequent study by Hassan et al. [8] showed that MIKIRA doesn't provide a feasible solution in most cases and that TAFIRA has an unbounded integrality gap. To overcome these limitations, the authors present a two-phase algorithm. A maximum weight matching is found in the first phase using the Hungarian method, and a local search algorithm is used to find another matching with reduced interference and at least the target sum rate in phase two.

Hussain et al. [11] extended the resource-sharing model where many D2D pairs may share the radio resource of many cellular users. A graph-based approach for resource allocation was proposed in Zhang et al. [29]. Zhang et al. [30] formulated the interference relationships among different D2D and cellular communication links into a novel interference graph and proposed an algorithm (InGRA) that can compute near-optimal solutions with low computational complexity in practice.

Xu et al. [28] developed a second price auction for allocating spectrum resources where the D2D pairs bid in sequence. Their method shows improved sum rate and fairness in simulations. A reverse iterative combinatorial auction in which spectrum resources are auctioned off as goods were proposed by Xu et al. [27]. They prove that the auction is cheat-proof and converges. Simulations show that the method yields a reasonable sum rate.

An admissibility-based approach was developed by Feng et al. [6]. The scheme has three phases. Admissible D2D pairs are identified based on the distance from the base station in the first phase. This ensures that the sum-rate requirement is not violated. In the second stage, power is allocated to each D2D pair and cellular user. The third stage identifies the resource allocation by solving a weighted matching problem.

A two-stage semi-distributed framework for managing D2D communications was given by Lee et al. [19]. Energy efficient resource allocation was studied by Hoang et al. [9] where the objective is to maximize the weighted energy efficiency while guaranteeing a minimum sum rate. The authors gave a dual decomposition algorithm.

A semi-centralized control method for selecting cellular relays is due to Li and Cai [22]. The interference is modelled using stochastic geometry, and two traffic models are analyzed. Lee and Lee [20] is a theoretical study that determines the analytic forms for outage probability for a cellular user and the achievable rate for a D2D pair. Using this analytical form, they determine the optimal spectrum and power allocation to maximize the average rate. Li et al. [21] studies the resource allocation, which has minimum energy consumption that obeys both service qualities and transmission rate requirements. They use the statistical information about the channels, give a non-convex mixed integer non-linear programming model, and propose heuristic methods.

Hoang et al. [10] consider the channel allocation for each D2D pair to maximize the sum rate while guaranteeing the desired sum rate for each D2D pair. Their model leads to a three-dimensional matching problem which is solved using a branch-n-bound algorithm and iterative rounding. The work in [10] is methodologically the closest to our work. However, there are key differences. First, the underlying graph problem that is solved is different. We use a knapsack model instead of a three-dimensional matching in [10]. Therefore, a direct experimental comparison of our results with them is not fair. Secondly, we augment the branch-n-bound method with a branch-n-cut algorithm. Thirdly, iterative rounding is used as a tool in both papers. However, the proof of the approximation ratio of the iterative rounding method in [10] uses the local ratio technique, whereas our guarantee relies on the structure of the extreme point solution to the LP and uses the Rank Lemma.

The constant/uniform interference case was first studied by Saha et al. [24]. They proposed a two-phase polynomial-time algorithm when interferences are uniform when the objective is to minimize the interferences while satisfying a target sum rate. Phase one is similar to the first phase in [8], and phase two improves on phase one by iteratively finding unique triples. For recent in-depth reviews of the device to device communications in underlay cellular networks see [3, 25].

4   Branch-n-Cut

Branch and bound is a tree search method to solve integer programs. If we add linear inequalities known as cuts at every search node, then the method is called branch-n-cut [2]. All feasible solutions satisfy the cuts and reduce the search space. We use linear inequalities known as cover-cuts [17] that are generated from Eq. (7) and the current LP solution as explained next.

4.1   Cover Cuts

A *cover* S is a set of pairs (c, d) where $c \in C, d \in D$ such that the total interference on the pairs in S is larger than the allowed interference, i.e.,

$$\sum_{(c,d) \in S} I_{(c,d)} > I \tag{9}$$

S is a *minimal cover* if no proper subset of S is also a cover. Any feasible solution can contain only at most $|S| - 1$ elements from a minimal cover S. Therefore, for any feasible solution x,

$$\sum_{(c,d) \in S} x_{(c,d)} \leq |S| - 1 \tag{10}$$

Equivalently, for a minimal cover S and a feasible solution x^*

$$\sum_{(c,d) \in S} (1 - x^*_{(c,d)}) \geq 1 \tag{11}$$

Therefore, if there is an S such that the following equations hold,

$$\sum_{(c,d) \in S} (1 - x^*_{(c,d)}) < 1 \tag{12}$$

$$\sum_{(c,d) \in S} I_{(c,d)} > I \tag{13}$$

Then inequality (10) can be added to the node in the search tree during branch and bound. The existence of a minimal cover S that satisfies inequality (12) is determined by solving the following knapsack problem.

$$\max \sum_{c \in C} \sum_{d \in D} y_{(c,d)} I_{(c,d)} \tag{14}$$

$$\sum_{c \in C} \sum_{d \in D} y_{(c,d)} (1 - x^*_{(c,d)}) < 1 \tag{15}$$

Where $y_{c,d} \in \{0, 1\}$ is the decision variable in the knapsack problem. We use JuMP v 0.21 [5] library in Julia [1] to implement branch and cut. The cuts are discovered by solving an integer program. Gurobi 9.1.2 is used as the solver for JuMP. The experimental results are reported in Sect. 6.

5   Approximation Algorithm

Saha et al. [24] studied the problem of interference minimization when the sum rates were arbitrary, and the interference was uniform and gave a polynomial time algorithm. In this section, we give an iterative rounding algorithm that, in each round, will solve an LP and set the value of one of the variables. At most, the number of rounds is the number of edges, and the work done in each round is polynomial. The algorithm will work with extreme point solutions of the LP. We will show that a variable with a value in a prescribed set exists in each round. We will prove a bound on the approximation ratio.

The input is a bipartite graph $B = (V, E)$ and an interference limit I. $V(E)$ is the set of vertices (edges). The two sides of the partition are the D2D pairs on one side and the cellular users on the other. Each edge, e, has an interference I_e and a sum rate s_e. The set of edges incident on $v \in V$ is denoted $\delta(v)$. The goal is to find a maximum weight matching such that the total sum of the interference on the edges in the matching is at most I. We have the following integer program (IP).

$$IP = \max \sum_{e \,\in\, E} s_e x_e \tag{16}$$

$$\sum_{e \in \delta(v)} x_e \leq 1 \qquad \forall v \in V \tag{17}$$

$$\sum_{e \,\in\, E} I_e x_e \leq I \tag{18}$$

$$x_e \in \{0, 1\} \qquad \forall e \in E. \tag{19}$$

The linear programming relaxation (LP) is:

$$LP = \max \sum_{e \,\in\, E} s_e x_e \tag{20}$$

$$\sum_{e \in \delta(v)} x_e \leq 1 \qquad \forall v \in V \tag{21}$$

$$\sum_{e \,\in\, E} I_e x_e \leq I \tag{22}$$

$$x_e \geq 0 \qquad \forall e \in E. \tag{23}$$

5.1 Algorithm

We give a procedure (Algorithm 1) that will construct in polynomial time (depends on the time needed to solve a linear program) an integer solution such that

– the interference is no more than interference capacity I,
– the sum rate is at least a quarter the maximum sum rate in an optimal solution to the LP.

Algorithm 1 relies on an optimal solution to the LP relaxation and is iterative. Given an optimal solution, x^* to the LP, one of the two things happens: in each step, a reduced problem (smaller in size) is obtained, which is solved iteratively.

Algorithm 1

1: $F = \{\}$.
2: Solve the LP relaxation for graph B. If the optimal LP solution value is 0, then GOTO END.
3: If there is e such that $x_e = 0$ then remove edge e from B. GOTO 2.
4: If there is e such that $x_e = 1$ then add e to F and remove edge e and edges incident on e from B and $I = I - I_e$. GOTO 2.
5: If $0 < x_e < 1$ for all the edges, decompose the graph into two matchings and select the matching with a larger total sum rate.
6: Use the Greedy Algorithm 2 on the selected matching, add edges determined by the greedy algorithm to F. GOTO 7.
7: END

---

1. An edge e exists such that $x_e^* = 0$. Edge e is removed from the graph B. This gives a reduced problem B'.
2. If an edge e exists such that $x_e^* = 1$ then e is added to the solution. The edge e and any edges incident on e are removed from B to obtain a reduced problem B' with interference bound $I' = I - I_e$.
3. For all edges $0 < x_e^* < 1$. In this case, the graph induced by edges in $\{e \mid x_e^* > 0\}$ is a "near" cycle. We decompose this graph into two matchings, select the matching with the larger total sum rate, and return the edges selected from Algorithm 2, which is a greedy algorithm for the knapsack problem, as the solution.

Note that when we remove an edge, we delete its endpoints and any edge incident on the endpoints. We prove the following.

Theorem 1. *Algorithm 1 computes a solution with the sum rate at least a quarter of the sum rate of the optimal LP solution and the total interference less than equal to the target interference.*

Proof of Theorem 1 relies on the following Lemmas 2, 3, and 4. The proof of Lemma 2 relies on the structure of the extreme point solutions captured by the following Lemma.

Lemma 1 (Rank Lemma). *Let $P = \{x \mid Ax \geq b, x \geq 0\}$ and let x be an extreme point solution to P such that each component $x_i > 0$. If C is any maximal set of linearly independent tight constraints $(A[i, :]x = c[i])$ then $|C| = |X|$, where $|X|$ is the number of variables.*

The intuition behind the rank lemma can be summarized as follows: because x is an extreme point solution, the columns of A are linearly independent. Also, the column rank of A equals its row rank. Therefore any maximal set of linearly independent tight constraints equals the number of variables. See [18, Chapter 2] for proof of the Rank Lemma. First, we note that in any optimal solution (extreme point or not) to the LP, Constraint (7) is satisfied at equality in any optimal solution. As a direct application of the rank lemma, we get Lemma 2.

Lemma 2. *Let x^* be a basic feasible solution to the LP such that $0 < x_e^*$ for all $e \in E$ and $W = V_1 \cup V_2$ be the set of vertices such that $\sum_{e \in \delta(v)} x_e^* = 1$ and $\sum_{e \in E} I_e x_e^* \leq I$, then $|W| + 1 \geq |E|$.*

Note that if constraint (7) is in the maximal set of linearly independent tight constraints, then $|W| + 1 = |E|$. In the other case $|W| = |E|$. We now prove that if $0 < x_e^* < 1$ for every edge, the bipartite graph is near a cycle. One example of an optimal solution to LP that is a "near cycle" is shown in Fig. 1. Next, lemma is used to characterize the structure of the extreme point solution when all variables have fractional values.

Lemma 3 (Structure). *Let x^* be an optimal solution to the LP defined by the graph $B' = (V', E')$ such that $0 < x_e^* < 1$ for all $e \in E'$. Then, the number of vertices of degree ≥ 3 is at most 2.*

Proof. Given an optimal solution x^*, call a constraint in the LP tight if it is satisfied with equality.

If W is the set of vertices at which the matching constraint is tight $(\sum_{e \in \delta(v)} x_e = 1)$ then by Lemma 2, $|W| + 1 \geq |E|$. Let the number of vertices in W with a degree of at least 3 be k. We know

$$2W| + 2 \geq 2|E| \geq \sum_{v \in W} \delta(v) \geq 3k + (|W| - k)2 = (2W| + k). \tag{24}$$

The first inequality in Eq. (24) follows from Lemma 2. The second inequality follows as $W \subseteq V$. The third inequality is due to the assumption that $\delta(v) \geq 3$ for k vertices $\in W$. We infer, $k \leq 2$, i.e. there are $|W| - 2$ vertices with degree ≤ 2.

As $x_e < 1$ for all $e \in E$, every vertex in W has a degree at least 2. Therefore there are $|W| - 2$ vertices with degree 2. There are at most two vertices of degree more than 2. Suppose there are only degree three vertices (either 0, 1 or 2). Then the graph is either a cycle, a cycle plus one edge, or a cycle plus two edges. The set of edges E is nearly a cycle, and We call such a graph a "near cycle." We assume that the sum rates are low enough compared to the optimal LP solution value, so we ignore these extra edges in the analysis in the next part. There cannot be any vertex of degree four in W else Eq. (24) is not satisfied.   □

We focus only on the even cycle in the rest of the section. An example of the structure and then how this structure is used in Algorithm 1 is addressed next. The edges in the cycles can be partitioned into two sets M_1, M_2, each of which is a matching such that for any $e \in M_1$, the value in the optimal solution to LP, $x_e^* < 1/2$.

5.2   Structure

As a consequence of Lemmas 2 and 3, we have seen that the edges in any optimal solution x^* in which every variable takes on a fractional value form a near cycle.

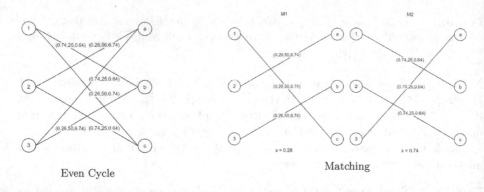

Fig. 1. Decomposition of "near" cycle

In each iteration, we ensure that the interference at each edge in the current solution at most the remaining interference capacity I_r. This is achieved by removing edges e for which $I_e > I_r$ after step 4 of Algorithm 1.

Figure 1 shows an example of such a cycle in an optimal solution of the LP. The values associated with every edge are of the form (x_e, S_e, I_e). M_1, M_2 are two matchings that this cycle is decomposed to. The edges in M_1 take value x, and the edges in M_2 take value $1 - x$ as shown in Fig. 1. now prove Theorem 1.

5.3   Proof of Theorem 1

Proof (Proof of Theorem 1). Let B be the graph at the start of Algorithm 1 and B' the reduced graph obtained after steps 3/4/5/6. Let x^* be the optimal solution to LP for graph B. If x' is the optimal solution to the LP for B'. We use induction on the number of iterations.

BASE CASE: If there is only a single iteration, then three possibilities arise. The subproblem B' is empty in each case.

- An edge $x_e^* = 0$ is removed. The value of the objective function does not change. The LHS in the sum-rate constraint does not change.
- An edge $x_e^* = 1$ is added to the solution, $I = I - I_e$.
- For all the edges $0 < x_e^* < 1$. By Lemma 3 the graph is a near cycle. We decompose the cycle into two matchings. Select the matching with the larger total sum rate and use Algorithm (2). This gives the total sum rate at least a quarter of the optimal LP solution, which is proved in Eq. (33) next, and the total interference is less than equal to the target interference.

Let S^* be the optimal value to LP problem with the variable values x_e^*. Let M_1 be the matching that $x_e^* < \frac{1}{2}$ for all $e \in M_1$ and, M_2 be the matching that $x_e^* \geq \frac{1}{2}$ for all $e \in M_2$. If S^* is the optimal solution of LP-relaxation (16). By definition, we have:

$$S^* = \sum_{i=1}^{2} S^*(M_i)$$

We select the matching with the larger total sum rate between $S^*(M_1)$ and $S^*(M_2)$. If the matching M_i, $i \in \{1,2\}$ has the highest total sum-rate, then $S^*(M_i) \geq \frac{S^*}{2}$.

We can select a subset of edges from M_i in the solution. This is modelled as the following knapsack problem:

$$KS - IP = \max \sum_{e \in M_1} s_e x_e \tag{25}$$

$$\sum_{e \in M_1} I_e x_e \leq I_r \tag{26}$$

$$x_e \in \{0,1\} \qquad \forall e \in M_i \tag{27}$$

The LP relaxation of the KS is:

$$KS - LP = \max \sum_{e \in M_1} s_e x_e \tag{28}$$

$$\sum_{e \in M_1} I_e x_e \leq I_r \tag{29}$$

$$x_e \geq 0 \qquad \forall e \in M_i \tag{30}$$

Let S_o^* be the optimal value of $KS - LP$, therefore $S_o^* \geq S^*(M_i)$ as the optimal solution for matching M_i is a feasible solution for LP of knapsack. Therefore,

$$S_o^* \geq S^*(M_i) \geq \frac{S^*}{2} \tag{31}$$

We use Algorithm 2 which computes an integer solution S_{OI} to maximum knapsack that is the least half in value of the optimal solution to the LP relaxation (KS-LP). By Lemma 4,

$$S_{OI} \geq \frac{S_o^*}{2} \tag{32}$$

From Eqs. (31) and (32), we get:

$$S_{OI} \geq \frac{S_o^*}{2} \geq \frac{S^*(M_i)}{2} \geq \frac{S^*}{4}. \tag{33}$$

INDUCTIVE STEP: The restriction of x^* to B' is denoted x^r. By induction hypothesis B' has a solution x' which is integral and satisfies i) $\sum_{e \in E'} s_e x'_e \geq 1/4 \sum_{e \in E} s_e x_e^r$ and ii) $\sum_{e \in E} x_e I_e \leq I'$ where I' is the interference in B'. We use $I(x)$ to mean $\sum_{e \in E} I_e x_e$ and $s(x)$ to mean $\sum_{e \in E} x_e$. We examine the two cases.

- If some $e : x_e = 0$ was removed in B then $I' = I$ and $x^r = x^* - \{x_e^*\}$. Note that by induction hypothesis $I(x') \leq I' = I$, $s(x') \geq s(x^*)/4$ and x' is a solution to B.
- If some $e : x_e = 1$ and edges incident on e was removed in B then $I' = I - I_e$. The sum-rate for the solution to B is $s(x') + s_e \geq 1/4s(x^*)$. The knapsack constraint is $I(x') + I_e \leq I' + I_e \leq I$.

□

5.4 Density Ordered Greedy for KS

We provide a density-ordered greedy algorithm and the proof that it is a two approximation which is from [26].

Algorithm 2. Density ordered Greedy

$G \leftarrow \emptyset$
while $I(G \cup argmax_{i \notin G}\{\frac{s_i}{I_i}\}) \leq I_c$ **do**
    $G \leftarrow G \cup argmax_{i \notin G}\{\frac{s_i}{I_i}\}$
end while
$a \leftarrow argmax_{i \notin G}\{\frac{s_i}{I_i}\}$
$G \leftarrow argmax\{S(G), s_a\}$
return G

Lemma 4. *Algorithm 2 computes a solution with the sum rate at least half the sum rate of the optimal LP solution [26].*

6 Experiments and Results

We study the efficacy of the cover cuts described in Sect. 4 on computationally hard instances. Cuts are added at each node in the branch and bound search tree. The algorithm uses cover cuts and default cuts that are part of the Gurobi solver, notably mixed-integer rounding cuts, generalized upper bounds cover cuts, Chvatal-Gomory strong cuts, and minimal cover cuts.

6.1 Instance Generation

Profit-ceiling instances from [23] are used to model the knapsack constraint. The instances are known to be computationally hard for Knapsack. The interferences are randomly generated between 1 and 50. The sum-rate is generated by the formula, $s_e = 3\lceil c_e/3 \rceil$ for every D2D and CU pair e.

6.2 Methodology

We use the JuMP library [5] in Julia [1] to program branch and cut algorithms. We use Gurobi as the backend, which can be replaced with any other open-source or commercial solver. To evaluate the benefits of adding cover-cuts (from Sect. 4). We initialize the Gurobi solver with the pre-solve reductions turned off. The number of threads is one, no internal cuts are used (MIR, GUC, strong CG), heuristics, and the generation of equivalent models is also turned off at the start.

Some of the models we solve are large (10,000 or more variables). The basic branch and bound can take hours on these instances, so we use CEDAR, the Compute Canada cluster, with a limit of 30 h on each solve and a maximum memory requirement of 12GB. There are no time outs with this limit and all the instances reported are solved optimally (see Table 1).

6.3   Results

Table 1. Branch and cut result (Hard Instances)

Number of variables	Gurobi configuration	Cover cuts	Nodes explored	Cutting planes	
2500	1 thread	325	371	User	25
	Optimized	1966	3262	User	316
				MIR	1
				GUB	1
3600	1 thread	160	209	User	4
	optimized	2362	4239	User	111
				CC	1
				MIR	4
				GUB	2
4900	1 thread	327	1570	User	6
	Optimized	3773	6940	User	290
				MIR	1
6400	1 thread	4763	5816	User	573
	Optimized	4241	7992	User	233
				MIR	2
				StrongCG	1
8100	1 thread	5927	6712	User	668
	Optimized	4104	5596	User	32
				MIR	1
10000	1 thread	507	712	User	11
	Optimized	5617	11492	User	397
				GUB	1

The branch and cut algorithm gives us an exact optimal solution. We examine the total number of cuts added during the branch and bound process. The number of cuts added for hard cases is in Table 1. Not all the cover cuts that the solver discovers are added to the nodes due to design implementation issues in JuMP and Gurobi. The fourth column, labelled cover cuts, lists the total number of cover cuts discovered during the entire search process, whereas the last column, called Cutting Planes, lists the number of cuts the solver used.

7   Conclusion and Future Work

This paper studied the D2D resource allocation model, which maximizes the sum rate while capping the interference. We gave a branch and cut algorithm for computing optimal solutions. Our technique works for the case when the objective is to minimize the interference subject to a guaranteed sum rate for D2D communications (not shown here). We performed a detailed empirical evaluation of the branch and cut algorithm on instances that are computationally hard. Although it takes time to solve hard instances using the branch and cut, we can have the optimal solution for the model. We give an iterative rounding algorithm with constant factor approximation bound. We prove that the iterative rounding method provides a near-optimal feasible solution that is four approximate.

References

1. Bezanson, J., Edelman, A., Karpinski, S., Shah, V.B.: Julia: a fresh approach to numerical computing. SIAM Rev. **59**(1), 65–98 (2017)
2. Caccetta, L., Hill, S.P.: Branch and cut methods for network optimization. Math. Comput. Model. **33**, 517–532 (2001)
3. Chakraborty, C., Rodrigues, J.J.: A comprehensive review on device-to-device communication paradigm: trends, challenges and applications. Wirel. Pers. Commun. **114**(1), 185–207 (2020)
4. Doppler, K., Rinne, M., Wijting, C., Ribeiro, C.B., Hugl, K.: Device-to-device communication as an underlay to LTE-advanced networks. IEEE Commun. Mag. **47**(12), 42–49 (2009). ISSN 0163–6804
5. Dunning, I., Huchette, J., Lubin, M.: JuMP: a modeling language for mathematical optimization. SIAM Rev. **59**(2), 295–320 (2017)
6. Feng, D., Lu, L., Yuan-Wu, Y., Li, G.Y., Feng, G., Li, S.: Device-to-device communications underlaying cellular networks. IEEE Trans. Commun. **61**(8), 3541–3551 (2013). ISSN 0090–6778
7. Fodor, G., Dahlman, E., Mildh, G., Parkvall, S., Reider, N., Miklós, G., Turányi, Z.: Design aspects of network assisted device-to-device communications. IEEE Commun. Mag. **50**(3), 170–177 (2012)
8. Hassan, Y., Hussain, F., Hossen, S., Choudhury, S., Alam, M.M.: Interference minimization in D2D communication underlaying cellular networks. IEEE Access **5**, 22471–22484 (2017)
9. Hoang, T.D., Le, L.B., Le-Ngoc, T.: Dual decomposition method for energy-efficient resource allocation in D2D communications underlying cellular networks. In: 2015 IEEE Global Communications Conference (GLOBECOM), pp. 1–6. IEEE (2015)
10. Hoang, T.D., Le, L.B., Le-Ngoc, T.: Resource allocation for D2D communication underlaid cellular networks using graph-based approach. IEEE Trans. Wirel. Commun. **15**(10), 7099–7113 (2016)
11. Hussain, F., Hassan, M.Y., Hossen, M.S., Choudhury, S.: System capacity maximization with efficient resource allocation algorithms in d2d communication. IEEE Access **6**, 32409–32424 (2018)
12. Islam, M.: Radio Resource Allocation for Device-to-Device Communications Underlaying Cellular Networks. Ph.D. thesis, Queens University, Canada (2016)
13. Islam, M.T., Taha, A.E.M., Akl, S.: Reducing the complexity of resource allocation for underlaying device-to-device communications. In: 2015 International Wireless Communications and Mobile Computing Conference (IWCMC), pp. 61–66. IEEE (2015)
14. Islam, M.T., Taha, A.E.M., Akl, S., Choudhury, S.: A two-phase auction-based fair resource allocation for underlaying D2D communications. In: 2016 IEEE International Conference on Communications (ICC), pp. 1–6. IEEE (2016)
15. Iyer, A., Rosenberg, C., Karnik, A.: What is the right model for wireless channel interference? IEEE Trans. Wirel. Commun. **8**(5), 2662–2671 (2009)

16. Janis, P., Koivunen, V., Ribeiro, C., Korhonen, J., Doppler, K., Hugl, K.: Interference-aware resource allocation for device-to-device radio underlaying cellular networks. In: VTC Spring 2009 - IEEE 69th Vehicular Technology Conference, pp. 1–5 (2009). ISSN 1550–2252
17. Kaparis, K., Letchford, A.: Cover Inequalities (2011). ISBN 9780470400531. https://doi.org/10.1002/9780470400531.eorms0204
18. Lau, L.C., Ravi, R., Singh, M.: Iterative Methods in Combinatorial Optimization, vol. 46. Cambridge University Press, Cambridge (2011)
19. Lee, D.H., Choi, K.W., Jeon, W.S., Jeong, D.G.: Two-stage semi-distributed resource management for device-to-device communication in cellular networks. IEEE Trans. Wirel. Commun. **13**(4), 1908–1920 (2014)
20. Lee, J., Lee, J.H.: Performance analysis and resource allocation for cooperative D2D communication in cellular networks with multiple D2D pairs. IEEE Commun. Lett. **23**(5), 909–912 (2019)
21. Li, R., Hong, P., Xue, K., Zhang, M., Yang, T.: Energy-efficient resource allocation for high-rate underlay D2D communications with statistical CSI: a one-to-many strategy. IEEE Trans. Veh. Technol. **69**(4), 4006–4018 (2020)
22. Li, Y., Cai, L.: Cooperative device-to-device communication for uplink transmission in cellular system. IEEE Trans. Wirel. Commun. **17**, 3903–3917 (2018)
23. Pisinger, D.: Where are the hard knapsack problems? Comput. Oper. Res. **32**(9), 2271–2284 (2005)
24. Saha, P.R., Choudhury, S., Gaur, D.R.: Interference minimization for device-to-device communications: a combinatorial approach. In: 2019 IEEE Wireless Communications and Networking Conference (WCNC), pp. 1–6. IEEE (2019)
25. Siddiqui, M.U.A., Qamar, F., Ahmed, F., Nguyen, Q.N., Hassan, R.: Interference management in 5G and beyond network: requirements, challenges and future directions. IEEE Access **9**, 68932–68965 (2021)
26. Singer, Y.: Advanced Optimization (2016). http://people.seas.harvard.edu/yaron/AM221-S16/lecture_not. Accessed 23 Nov 2022
27. Xu, C., Song, L., Han, Z., Zhao, Q., Wang, X., Cheng, X., Jiao, B.: Efficiency resource allocation for device-to-device underlay communication systems: a reverse iterative combinatorial auction based approach. IEEE J. Sel. Areas Commun. **31**(9), 348–358 (2013). ISSN 0733–8716
28. Xu, C., Song, L., Han, Z., Zhao, Q., Wang, X., Jiao, B.: Interference-aware resource allocation for device-to-device communications as an underlay using sequential second price auction. In: IEEE International Conference on Communications (ICC), Ottawa, Canada, pp. 445–449, June 2012. ISSN 1938–1883
29. Zhang, R., Cheng, X., Yang, L., Jiao, B.: Interference-aware graph based resource sharing for device-to-device communications underlaying cellular networks. In: IEEE Wireless Communications and Networking Conference (WCNC), Shanghai, China, pp. 140–145, April 2013. ISSN 1525–3511
30. Zhang, R., Cheng, X., Yang, L., Jiao, B.: Interference graph-based resource allocation (InGRA) for D2D communications underlaying cellular networks. IEEE Trans. Veh. Technol. **64**(8), 3844–3850 (2014)

Computational Geometry

Algorithms for k-Dispersion for Points in Convex Position in the Plane

Vishwanath R. Singireddy[1], Manjanna Basappa[1(✉)], and Joseph S.B. Mitchell[2]

[1] BITS Pilani, Hyderabad Campus, Hyderabad 500078, Telangana, India
{p20190420,manjanna}@hyderabad.bits-pilani.ac.in
[2] Stony Brook University, Stony Brook, NY 11794-3600, USA
joseph.mitchell@stonybrook.edu

Abstract. In this paper, we consider the following k-dispersion problem. Given a set S of n points placed in the plane in convex position and an integer k $(0 < k < n)$, the objective is to compute a subset $S' \subset S$ such that $|S'| = k$ and the minimum distance between a pair of points in S' is maximized. Based on the bounded search tree method, we propose an exact fixed-parameter algorithm in $O(2^k n^2 \log^2 n)$ time for this problem, where k is the parameter. The proposed exact algorithm improves on the algorithm of Akagi et al. (2018), which requires time $n^{O(\sqrt{k})}$, whenever $k < c \log^2 n$ for some constant c. We then give an exact polynomial-time $(O(n^4 k^2))$ algorithm, for any $k > 0$, thus answering the open question about the complexity of this restricted dispersion problem. For $k = 3$, there is an $O(n^2)$-time algorithm by Kobayashi et al. (2021).

Keywords: Obnoxious facility location · Max-min dispersion · Fixed parameter tractable · Delaunay triangulation · Dynamic programming

1 Introduction

In many variants of the facility location problems that are studied in the literature [6,7], we are given a set of n points, and, among them, we need to locate k facilities such that some objective function is minimized. In contrast, in the obnoxious facility location problems, we need to maximize an objective function. In the literature, this wider class of facility location problems that aim to maximize some diversity measures are called dispersion problems. In the case of the max-min k-dispersion problem, we need to maximize the minimum distance between the selected k facilities. The applications of k-dispersion problems arise in many areas. Consider a specific application where the k-dispersion problem can be used in which the given points are in convex position, as discussed below. For example, consider a convex island where some oil storage plants are to be

M. Basappa—Partially supported by the Science and Engineering Research Board (SERB), Govt. of India, under Sanction Order No. TAR/2022/000397.
J. Mitchell—Partially supported by the National Science Foundation (CCF-2007275) and the US-Israel Binational Science Foundation (project 2016116).

established on the shore for transport using ships. Moreover, these plants should be kept as far away from each other as possible so that any accident in one plant should not affect the other plant. We can model this problem as the k-dispersion problem, in which the plants are placed on the island's boundary to maximize the distance between any pair of plants. We define the problem formally below. *Discrete k-dispersion on a Convex Polygon (*DKCONP*):* Given a set S of n points in convex position, assume that the points in S are ordered in a clockwise order around the centroid of S, forming a convex polygon $\mathcal{P}$. Then, observe that the k-dispersion problem on the set S can be equally stated as packing k congruent disks of maximum radius, with their centers lying at the vertices of the convex polygon $\mathcal{P}$.

1.1   Literature Survey

The discrete k-dispersion problem for $k \geq 3$ is known to be NP-complete even when the triangle inequality is satisfied [8]. The Euclidean k-dispersion problem is proved NP-hard by Wang and Kuo [16]. Akagi et al. [1] gave an algorithm to solve the k-dispersion problem in the Euclidean plane exactly in $n^{O(\sqrt{k})}$ time. They also gave an $O(n)$-time algorithm to solve the special cases of the problem in which the given points appear in order on a line or on the boundary of a circle. Later, Araki and Nakano [2] improved the running time of [1] to $O(\log n)$ for the line case. Ravi et al. [14] proved that for the max-min k-dispersion problem on an arbitrary weighted graph, we cannot give any constant factor approximation algorithm within polynomial time unless P=NP. If the triangle inequality is satisfied by the edge weights, then we cannot approximate the problem with a better factor than $\frac{1}{2}$ in polynomial time unless P=NP. They also gave a polynomial time $\frac{1}{2}$-approximation algorithm for the problem in graph metric.

Horiyama et al. [10] solved the max-min 3-dispersion problem in $O(n)$ time in both L_1 and L_∞ metrics when the given points are in 2-dimensional plane. They also designed an $O(n^2 \log n)$ time algorithm for the 3-dispersion problem in L_2 metric. The 1-dispersion problem is trivial when the points are in a convex position, and we can solve the 2-dispersion problem in $O(n \log n)$ time by computing the diameter of the convex polygon formed by these points [15]. Recently, Kobayashi et al. [11] gave $O(n^2)$-time algorithm for the 3-dispersion problem on a convex polygon. In the literature, to the best of our knowledge, the k-dispersion problem on a convex polygon for any $k > 3$ has remained open; it is resolved in our work here. When the points are arbitrarily placed in the Euclidean plane, the current best approximation algorithm is still the $\frac{1}{2}$-approximation algorithm proposed by Ravi et al. [14] for the metric case. Hence, also from the point of designing ρ-approximation algorithm for $\rho > \frac{1}{2}$, the problem is open. Other related results in the literature are the following. Baur and Fekete [4] studied the problem of maximizing the rectilinear distance between a selected set of n points within a polygon and they showed that this problem cannot be approximated with the factor $\frac{13}{14}$ unless P = NP. Fekete and Meijer [9] studied the discrete k-dispersion problem with a constraint of maximizing the average rectilinear distance between k facilities in d-dimensional space. They solved the problem in linear time when k is fixed and gave a polynomial-time approximation scheme when k is part of the input.

2   Preliminaries

This section introduces some terminologies and observations useful in discussing our solution for the DKCONP problem.

Let us use $\overline{v_i v_j}$ to denote the line segment connecting the points v_i and v_j. We use $|.|$ (i) to denote the length $|v_i v_j|$ of the line segment $\overline{v_i v_j}$, (ii) to denote the absolute value $|x|$ of a real number $x \in \mathbb{R}$, and also, (iii) to denote the cardinality $|S|$ of any set S. The center of any disk d is denoted by $\mathcal{C}(d)$, and the diameter of any convex polygon $\mathcal{P}$ is denoted by $\mathcal{D}(\mathcal{P})$. Let r_{max} be the (maximum) radius of the disks in an optimal solution to the DKCONP problem. Let $\overline{v_i v_j}$ be a chord of $\mathcal{P}$ corresponding to the pair (v_i, v_j) of vertices of $\mathcal{P}$, where $1 < |i - j| < n - 1$ and $i, j \in \{1, 2, \ldots, n\}$. Let $C = \{\overline{v_i v_j} : 1 < |i - j| < n - 1, i, j \in \{1, 2, \ldots, n\}\} \cup \{\overline{v_1 v_2}, \overline{v_2 v_3}, \ldots, \overline{v_n v_1}\}$ be the set of chords and edges of $\mathcal{P}$, where $\overline{v_i v_{i+1}}$, for $i = 1, 2, \ldots, n - 1$, are the edges of $\mathcal{P}$ and $\overline{v_n v_{n+1}} = \overline{v_n v_1}$. Clearly, $|C| = \frac{n(n-1)}{2}$. Now, let $C' = \{|v_i v_j| \mid i, j = 1, 2, \ldots, n$ and $\overline{v_i v_j} \in C\}$ be the set of all distinct distances between pairs of vertices of $\mathcal{P}$.

Observation 1. $2r_{max} \in C'$ and $|C'| = O(n^2)$.

Due to Observation 1, we can find r_{max} in at most $\lceil 2 \log n \rceil$ stages of the binary search, provided that for any given r we can decide whether $r > r_{max}$ or $r \leq r_{max}$. Based on a bounded search tree, we propose a fixed-parameter algorithm to answer this decision question, where k is the parameter.

3   An Exact Fixed-Parameter Algorithm

For the DKCONP problem, here we aim to develop a fixed-parameter algorithm using the bounded search tree method. To do this, we first consider the following decision problem:

DECISION($\mathcal{P}, k, r$): Given a convex polygon $\mathcal{P}$ with n vertices and a positive integer $k < n$ and a radius r, is it possible to pack k (non-overlapping) congruent disks of radius r, with centers lying at the vertices of $\mathcal{P}$?

Observe that the answer to DECISION($\mathcal{P}, k, r$) is YES if the radius r is less than or equal to the radius r_{max} of the disks in an optimal solution of the DKCONP problem. Now, we shall design an algorithm that solves DECISION($\mathcal{P}, k, r$) in $O(f(k) \cdot n^{O(1)})$ time and returns a set of k disks of radius r packed on the boundary of $\mathcal{P}$ if the answer is YES, and returns NO otherwise, where $f(k)$ is an arbitrary exponential function in k.

3.1 Decision Algorithm

The outline of the algorithm is as follows. First, we align the polygon $\mathcal{P}$ such that its leftmost vertex v_1 is placed at the origin. Then, we place the disk d_1 of radius r centered at v_1. From the vertex v_1 in clockwise direction along the boundary of $\mathcal{P}$, we find the first vertex u at which we can center a r-radius disk d_2 that does not overlap with any previously placed disks. Now, we again have two ways to place the next disk d_3, namely, moving in clockwise direction from the center of d_2 or moving in counter-clockwise direction from the center of d_1. Similarly, from the vertex v_1 in counter-clockwise direction along the boundary of $\mathcal{P}$, we find the first vertex u' at which we could center the disk d_2. In this way, our search for finding all $k - 1$ vertices of $\mathcal{P}$ (as the centers) to pack the disks proceeds like a 2-way search tree. The depth of the search tree is k because we stop after placing k disks and return the disks. At any point along a path of the 2-way search tree, if we can not place a disk, we backtrack to placing a disk in the other direction. Thus, the branching factor of every node is at most 2, resulting in $O(2^k)$ nodes in total. We repeat the above procedure by placing the disk d_1 at each of the n vertices of $\mathcal{P}$. Note that the disks corresponding to the vertices of any path of length $\geq k$ in the 2-way search tree together form a feasible solution for the DKCONP problem.

Now, we shall describe how to pack the next disk d_{j+1} after having packed the disks $d_1, d_2, \ldots, d_j$. In the above 2-way search tree procedure, for the value of $j = 1, 2, \ldots, k - 1$, after packing the disk d_j centered at some vertex of $\mathcal{P}$, the candidate vertices u and u' for the center of the disk d_{j+1} can be computed as follows: let u be the first vertex at a distance of at least $2r$ from the center of the most recently packed disk (d_j or d_{j-1}) clockwise from the center of d_1. Similarly, let u' be the first vertex at a distance at least $2r$ from the center of the most recently packed disk (d_j or d_{j-1}) counter-clockwise from the center of d_1. Note that u and u' are the two candidate vertices for packing the next disk d_{j+1}, which will be centered at one of them. However, it is required to ensure that the distance between the candidate center vertex u or u' and each of the vertices at which the already-packed disks $d_1, d_2, \ldots, d_j$ are centered is at least $2r$. Observe that for a convex polygon $\mathcal{P}$, the distances between a fixed vertex and the remaining vertices of $\mathcal{P}$ form a multi-modal function. Therefore, we can not directly employ binary search to find the center vertices u and u' for packing the next disk d_{j+1}, $j = 1, 2, \ldots, k - 1$. For a vertex v_i of $\mathcal{P}$, there are γ vertices that are the modes or local maxima [3], where $\gamma \leq n/2$. We will exploit this property to identify all the candidate center vertices and to quickly locate the one among them for centering the disk d_{j+1}. Hence, given a candidate radius r, we do some preprocessing before we shall call the decision algorithm.

3.2 The Optimization Scheme

To solve the optimization problem, i.e., to find the maximum value r_{max} of r, we solve DECISION($\mathcal{P}, k, r$) repeatedly while performing a binary search on C'. In each stage of the binary search, the radius r will be the median element of

C' divided by 2. The median will be found using a linear-time median finding algorithm [5]. We then perform the above 2-way search tree-based procedure to find an answer to DECISION$(\mathcal{P}, k, r)$. If the answer is YES, then we update C' by removing all the elements of it that are smaller than $2r$. Otherwise, we update by removing all the elements that are at least $2r$. In either case, the size of the updated C' will be half of the previous C'. The main routine of the algorithm is outlined in Algorithm 1.

Preprocessing: Here we describe how to precompute all the candidate center vertices for the next disk d_{j+1} ($j = 1, 2, \ldots, k-1$) once an element $2r \in C'$ is fixed, where a candidate center vertex is a vertex of $\mathcal{P}$ at which a disk is likely to be centered in the packing computed by Algorithm 1. We also see how to use this precomputed information in every $(j+1)$th step of the decision algorithm given that the disks $\{d_1, d_2, \ldots, d_j\}$ are packed on $\partial\mathcal{P}$ with centers $\mathcal{C}(d_1) = v_{\alpha_1}, \mathcal{C}(d_2) = v_{\alpha_2}, \ldots, \mathcal{C}(d_j) = v_{\alpha_j}$, where $j = 1, 2, \ldots, k-1$, and $\partial\mathcal{P}$ is the boundary of $\mathcal{P}$.

In Algorithm 1, we initially set $\mathcal{X}_1 = C'$, the set of all distances between the points in S. In the ith stage of the binary search (while loop in Algorithm 1), we have that $|\mathcal{X}_i| \leq \frac{|\mathcal{X}_{i-1}|}{2}$, where the set $\mathcal{X}_i$ always contains the element $2r_{max}$ along with possibly some other candidate radii for $i = 2, 3, \ldots, \lceil 2\log n \rceil$. Now, consider a straight line ℓ_s (with any orientation) through any vertex v_s of $\mathcal{P}$, that splits $\mathcal{P}$ into two parts, each with at least one vertex other than v_s. Given a median $2r \in \mathcal{X}_i$, for each vertex v_s of $\mathcal{P}$ the candidate center vertices $u_{\theta_1}, u_{\theta_2}, \ldots, u_{\theta_\gamma}$ lying above any straight line ℓ_s through v_s (and $u_{s'_1}, u_{s'_2}, \ldots, u_{s'_{\gamma'}}$ lying below ℓ_s) are such that for $1 \leq \beta \leq \gamma$ we have that $|v_s u_{s_\beta-1}| < 2r$, $|v_s u_{s_\beta+1}| > 2r$ and $|v_s u_{s_\beta}| \geq 2r$ or $|v_s u_{s_\beta-1}| > 2r$, $|v_s u_{s_\beta+1}| < 2r$ and $|v_s u_{s_\beta}| \geq 2r$, where $\max(\gamma, \gamma') \leq \frac{n}{2}$. These candidate vertices can be pre-computed by doing the distance checks while linearly scanning through $\partial\mathcal{P}$ both in clockwise and counter-clockwise from every vertex v_i. Hence, this pre-computation at the beginning of each stage (line 4) will take $O(n^2)$ time. The overall time across all stages of the binary search for these pre-computations will be $O(n^2 \log n)$. These candidate center vertices are stored in the array A_i for each vertex v_i (see line 4 of Algorithm 1). Let $v_{\alpha_{up}}$ be the vertex in clockwise order from v_{α_1} ($= \mathcal{C}(d_1)$), at which the recently packed disk is centered (see Fig. 1). Let $v_{\alpha_{low}}$ be the vertex in counter-clockwise from v_{α_1}, at which the recently packed disk is centered. Let $u_{i_1}, u_{i_2}, \ldots, u_{i_\gamma}$ be the candidate center vertices in clockwise order from $v_{\alpha_{up}}$ for packing the next disk d_{j+1}. Similarly, the vertices $u_{i'_1}, u_{i'_2}, \ldots, u_{i'_{\gamma'}}$ are the candidate center vertices in counter-clockwise order from $v_{\alpha_{low}}$.

Computation of a Center Vertex for d_{j+1} by the Decision Algorithm: Now consider the $(j+1)$th iteration in the ith stage of the binary search. Let us denote the right most disks in the packing $\{d_1, d_2, \ldots, d_j\}$ on both upper and lower boundaries of $\partial\mathcal{P}$ by $d_{\alpha_{up}}$ and $d_{\alpha_{low}}$ centered respectively at $v_{\alpha_{up}}$ and $v_{\alpha_{low}}$ (see Fig. 1). The candidate center vertices from the center of $d_{\alpha_{up}}$ in clockwise order are $u_{i_1}, u_{i_2}, \ldots, u_{i_\gamma}$ and from the center of $d_{\alpha_{low}}$ in counter-clockwise order are $u_{i'_1}, u_{i'_2}, \ldots, u_{i'_{\gamma'}}$. Merge these two lists into one single list in convex position

---

Algorithm 1: Exact-fixed-parameter

Input: A convex polygon $\mathcal{P}$ with V vertices and an integer k
Output: Radius r_{max} of k disks packed
$\mathcal{X}_1 \leftarrow C', i \leftarrow 1$
while $|\mathcal{X}_i| \geq 2$ **do**
    $r \leftarrow median(\mathcal{X}_i)/2$ // invoke the linear-time median finding algorithm [5]
    Based on the value of $2r$, precompute the candidate center vertices for each
    $v_s \in V$ and store in a global array A_s, $s = 1, 2, \ldots, n$.
    if DECISION$(\mathcal{P}, k, r)$ **then**
        $\mathcal{X}_{i+1} \leftarrow \mathcal{X}_i \setminus \{e \in \mathcal{X}_i | e \leq 2r\}$
    else
        $\mathcal{X}_{i+1} \leftarrow \mathcal{X}_i \setminus \{e \in \mathcal{X}_i | e \geq 2r\}$
    $i \leftarrow i + 1$
$r = \min(\mathcal{X}_i)/2$
return r

---

order (i.e., respecting the initial given order in S) by discarding the candidate centers lying to the left of the line $\ell_{low,up}$ through $v_{\alpha_{low}}$ and $v_{\alpha_{up}}$ (see Fig. 1). This merging will take $\gamma + \gamma' - 1 = O(n)$ time (as we need to check at most n vertices in the order of S). Observe that due to the convexity of $\mathcal{P}$ each of the vertices $\mathcal{C}(d_1), \mathcal{C}(d_2), \ldots, \mathcal{C}(d_j)$ lie either on $\ell_{low,up}$ or to the left of $\ell_{low,up}$. Assume that the vertices (in clockwise order) between u_{i_1} and $u_{i'_{\gamma'-1}}$ all have distances at least $2r$ from both $v_{\alpha_{low}}$ and $v_{\alpha_{up}}$, and that $u_{i'_{\gamma'}}$ appears before u_{i_1} in clockwise order from $v_{\alpha_{up}}$. For each $p = i_1, i_1 + 1, \ldots, i'_{\gamma'-1}$ consider the line $\ell_{p,up}$ through $v_{\alpha_{up}}$ and v_p, and the line $\ell_{p,low}$ through v_p and $v_{\alpha_{low}}$, respectively. Note that the distances from the line $\ell_{p,up}$ to the vertices $\mathcal{C}(d_1), \mathcal{C}(d_2), \ldots, \mathcal{C}(d_j)$ satisfy unimodality because $\mathcal{P}$ is in convex position. Similarly the distances from $\ell_{p,low}$ to $\mathcal{C}(d_1), \mathcal{C}(d_2), \ldots, \mathcal{C}(d_j)$ satisfy unimodality property. Hence, we can use binary search to discard a vertex v_p if it is of distance strictly less than $2r$ from one of $\mathcal{C}(d_1), \mathcal{C}(d_2), \ldots, \mathcal{C}(d_j)$, as follows: Let $\mathcal{C}(d_1), \mathcal{C}(d_2), \ldots, \mathcal{C}(d_j)$ be in convex position order (in clockwise along $\partial\mathcal{P}$), find contiguous subsequences (if exists) of these points that have distance strictly less than $2r$ from the lines $\ell_{p,up}$ and $\ell_{p,low}$, by doing a binary search over the latter ordered list. Then check if there is a point in any of these sequences that is of distance strictly less than $2r$ from v_p, by using binary search again. Discard v_p if so. Also we linearly search the contiguous subsequence $u_{i_1}, u_{i_1+1}, \ldots, u_{i'_{\gamma'-1}}$ to find the first vertex v_p clockwise from $v_{\alpha_{up}}$ such that $\mathcal{C}(d_{j+1}) = v_p$. This process takes $O(n + n \log j)$ amortized time, where $O(n)$ is for merging and $O(n \log j)$ for finding the center for the disk d_{j+1} by doing binary search on $\mathcal{C}(d_1), \mathcal{C}(d_2), \ldots, \mathcal{C}(d_j)$. Similarly, we spend the same time if we are packing d_{j+1} in the counter-clockwise direction from $v_{\alpha_{low}}$. Then we have the following claim.

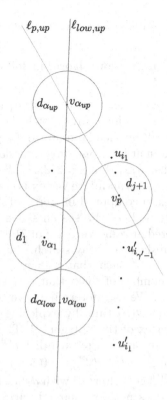

Fig. 1. Preprocessing and computation of a center vertex for d_{j+1}

Claim. If DECISION$(\mathcal{P}, k, r)$ =YES then there exists a vertex v_{α_1} of $\mathcal{P}$ with $v_{\alpha_1} = \mathcal{C}(d_1)$ that results in the following: there is a root-leaf path of a 2-way search tree with the root corresponding to v_{α_1} such that at every node along the path we are able to pack the next disk d_{j+1} centered at one of the candidate vertices.

Proof. Suppose the disk d_{j+1} in the optimal packing is not centered at one of the candidate vertices in the $(j+1)$th iteration. Also, consider that previous j disks in the optimal packing are centered at the candidate vertices (let us call them optimal centers). Then it must be the case that the optimal center vertex at which d_{j+1} is centered (in the optimal packing) appears strictly between the first candidate vertices v_p and $v_{p'}$ (with distance at least $2r$ from each of $\mathcal{C}(d_1), \mathcal{C}(d_2), \ldots, \mathcal{C}(d_j)$) respectively from $v_{\alpha_{up}}$ and $v_{\alpha_{low}}$. Otherwise, it should be one of the candidate vertices as the first vertices v_p and $v_{p'}$ are at the distance $2r$ from the centers of previously packed disks in clockwise and counter-clockwise directions, respectively. Now, we can perturb $\mathcal{C}(d_{j+1})$ to center d_{j+1} at the nearest candidate center without violating the packing property, and it also creates more space on $\partial\mathcal{P}$ (on the other side) for packing the following disks. Therefore, the claim follows by induction on j due to the above discussion (see Fig. 1). □

Lemma 1. *We can answer the decision question* DECISION$(\mathcal{P}, k, r)$ *in* $O(2^k n^2 \log n)$ *time.*

Proof. The correctness of our decision algorithm follows due to the following facts:

1. Fix some vertex v of $\mathcal{P}$ and a center of the disk d_1 at v. In the search space corresponding to the vertex v, along any (root-leaf) path (of the search tree) after the disk d_j is centered, by the claim above if $r \leq r_{max}$ there is always a candidate center vertex u in at least one direction along the boundary of $\mathcal{P}$ in order to pack the next disk d_{j+1}, for $j = 2, 3, \ldots, k-1$. We argued that the amortized time for finding this candidate vertex is $O(n + n \log j)$ (by accessing the array A_s (for a vertex s) computed in step 4 of Algorithm 1). Hence, the branching factor of every node of the search space is at most 2. Therefore, after the disk d_1 is centered at some vertex v of $\mathcal{P}$, the resulting search space for the remaining $k-1$ disks is a 2-way search tree, and its depth is $O(k)$.
2. Consider an element $2r \in C'$ such that $r \geq r_{max}$. Now, for this radius r let k' be the maximum number of disks that can be packed in the optimal packing OPT and $k \geq k'$. We can determine a vertex that is the center for the disk d_1^{OPT} in OPT, in $O(n)$ time by exploring the search space rooted corresponding to each vertex of the polygon $\mathcal{P}$. Then, by walking along $\partial \mathcal{P}$ from the point $\mathcal{C}(d_1^{OPT})$, we can charge each disk $d_{j'}^{OPT}$ with at least one disk d_j centered by Algorithm 1 if $(d_j \cap d_{j'}^{OPT}) \neq \emptyset$ for $j' = 1, 2, \ldots, k'$. Therefore, $\mathcal{C}(d_1^{OPT}) = \mathcal{C}(d_1)$, i.e., d_1^{OPT} gets charged with d_1 itself. Suppose there is some optimal disk $d_{j'}^{OPT}$ that does not get charged with any disk d_j centered by Algorithm 1. Then this will contradict with the termination of Algorithm 1. This implies that $k = k'$.
3. In the 2-way search tree, there are at most 2^j nodes at level j. Since we invest $O(n + n \log j)$ time at every node of the level j, the total time will be
$$\sum_{j=1}^{k-1} (2^j (n + n \log j)) = O(2^k (n + n \log k)).$$

If the radius $r \leq r_{max}$, then we can answer DECISION$(\mathcal{P}, k, r)$ correctly in time $n \cdot (2^k(n + n \log k)) = O(2^k(n^2 + n^2 \log k)) = O(2^k n^2 \log n)$ since we exhaustively search all n 2-way search trees and $k \leq n$. □

Theorem 1. *We have an exact fixed-parameter algorithm for the* DKCONP *problem in* $O(2^k n^2 \log^2 n)$ *time.*

Proof. Follows from Lemma 1 and by doing a binary search on the set C', since the total size of the search tree is bounded by a function of the parameter k alone, and every step takes polynomial time, and there are at most $\lceil 2 \log n \rceil$ calls to DECISION$(\mathcal{P}, k, r)$. □

4   An Exact Polynomial Time Algorithm

In this section, we discuss an exact polynomial time algorithm, based on dynamic programming, that solves the DKCONP problem in $O(n^4 k^2)$ time for any $k > 0$.

In order to accomplish this, we first let $S' \subseteq S$ be a subset of k points that are the center vertices of disks in OPT. Consider a Delaunay triangulation $\mathcal{DT}(S')$ of these optimal k centers. Obviously, all edges of $\mathcal{DT}(S')$ must lie within the convex hull $\mathcal{P}$ of S and also within the convex polygon Δ with S' as its vertices inscribed in $\mathcal{P}$. Additionally, the triangles and edges of a Delaunay triangulation of any set of points have some nice properties. As the Delaunay triangulation maximizes the minimum angle and follows the empty circle property [13], we have the following standard observation.

Observation 2. *The shortest diagonal or edge of Δ is always a Delaunay edge of $\mathcal{DT}(S')$.*

By Observation 2, another alternative way of viewing the DKCONP problem is to solve the problem of computing a subset $S''(\subseteq S)$ with k points such that the shortest edge of $\mathcal{DT}(S'')$ is as long as possible.

We know that the dual graph of a $\mathcal{DT}(S')$ is a Voronoi diagram $\mathcal{VD}(S')$. Since the k points that are in S' are in a convex position, $\mathcal{VD}(S')$ is a tree (see Fig. 2). This allows us to design a dynamic programming algorithm to solve the above optimization problem as follows:

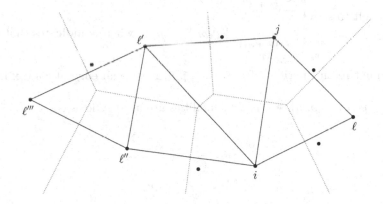

Fig. 2. OPT using Delaunay triangulation when $k = 6$

Let e_1 be an edge in $\mathcal{VD}(S')$ corresponding to the bisector of the pair $p_i, p_j \in S'$ (see Fig. 3). Now, assume that we know the three points p_i, p_j, p_ℓ are in the optimal solution S' such that the triangle $\Delta(p_i, p_j, p_\ell)$ formed by these points is a Delaunay triangle in $\mathcal{DT}(S')$. Let the segment (p_i, p_j) be oriented from p_i to p_j. Let $\mathcal{K} \in \mathbb{Z}^+$ be the budget (i.e., the number of facility centers remaining to be selected) and $\mathcal{E}_c(p_i, p_j, p_\ell)$ be a circle circumscribing the points p_i, p_j, p_ℓ. We define a subproblem $\phi(i, j, \ell; \mathcal{K})$, which returns the length of the smallest edge or diagonal of the optimal solution S'' (to DKCONP with $k = \mathcal{K} + 3$) in which the centers p_i, p_j, p_ℓ are already present, forming the Delaunay triangle in $\mathcal{DT}(S'')$ and $\mathcal{K}$ is the number of remaining centers to be selected. To solve $\phi(i, j, \ell; \mathcal{K})$, we need to find the best $p_{\ell'} \in S$ that lies to the left of line through $\overrightarrow{p_i p_j}$ (this region is the left half-plane of $\overrightarrow{p_i p_j}$, denoted as $L^-_{\overrightarrow{p_i p_j}}$), outside of $\mathcal{E}_c(i, j, \ell)$ such that it partitions $\mathcal{K}$ into two sub-parts $\mathcal{K}'$ and $\mathcal{K} - \mathcal{K}' - 1$ optimally, where $\mathcal{K}'$ is a positive integer. Now, recursively solve the subproblems $\phi(p_i, p_{\ell'}, p_j; \mathcal{K}')$ and $\phi(p_{\ell'}, p_j, p_i; \mathcal{K} - 1 - \mathcal{K}')$ lying to the left of $\overrightarrow{p_i p_{\ell'}}$ and to the left of $\overrightarrow{p_{\ell'} p_j}$ respectively (see Fig. 3). These two subproblems are invoked on only those points in S that lie to the left of the line through $\overrightarrow{p_i p_{\ell'}}$ and to the left of the line through $\overrightarrow{p_{\ell'} p_j}$, and that do not lie in the interior of $\mathcal{E}_c(p_{\ell'}, p_i, p_j)$. Hence we have the following recurrence:

$$\phi(p_i, p_j, p_\ell; \mathcal{K}) = \max_{\substack{\mathcal{K}' \leq \mathcal{K} - 1, \\ p_{\ell'} : p_{\ell'} \notin \bar{\mathcal{E}}_c(p_i, p_j, p_\ell), \\ p_{\ell'} \in L^-_{\overrightarrow{p_i p_j}}, \ p_{\ell'} \in S}} \left\{ \min \left\{ \begin{array}{c} |p_i p_{\ell'}|, |p_{\ell'} p_j|, \\ \phi(p_i, p_{\ell'}, p_j; \mathcal{K}'), \\ \phi(p_{\ell'}, p_j, p_i; \mathcal{K} - 1 - \mathcal{K}') \end{array} \right\} \right\}$$

The base cases are the following:

- $\phi(p_i, p_j, p_\ell; 1) = \min\limits_{\substack{p_{\ell'} : p_{\ell'} \notin \bar{\mathcal{E}}_c(p_i, p_j, p_\ell), \\ p_{\ell'} \in L^-_{\overrightarrow{p_i p_j}}, \ p_{\ell'} \in S}} \{|p_i p_{\ell'}|, |p_{\ell'} p_j|\}$, when we have only one facility left to select.

- $\phi(p_i, p_j, *; 1) = \min\limits_{\substack{p_{\ell'} : p_{\ell'} \notin \bar{\mathcal{E}}_c(p_i, p_j, p_\ell), \\ p_{\ell'} \in L^-_{\overrightarrow{p_i p_j}}, \ p_{\ell'} \in S}} \{|p_i p_{\ell'}|, |p_{\ell'} p_j|\}$, where $*$ indicates that there is no point p_ℓ on the right of $\overrightarrow{p_i p_j}$ (i.e., $\overline{p_i p_j}$ maybe an edge of convex hull of S').

- $\phi(p_i, p_j, p_\ell; 0) = |p_i p_j|$, when all k facilities are already located.

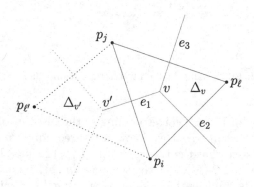

Fig. 3. Extending Voronoi diagram to next Voronoi center v' by picking vertex $p_{\ell'}$

In the above recurrence, there is no cyclic dependency between subproblems as the budget is partitioned into sub-parts for subproblems (i.e., the budget gets smaller and smaller for the subproblems). The optimal solution corresponds to a cell in the two-dimensional array of $\phi(p_i, p_j, *; k - 2)$ of the four-dimensional DP table.

Correctness of the Above Dynamic Programming Algorithm: The correctness follows from the fact that the dual to the Delaunay triangulation $\mathcal{DT}_{OPT}$ of an optimal solution is a tree $\mathcal{VD}_{OPT}$. Here, assume that we have one vertex v of $\mathcal{VD}_{OPT}$. Let Δ_v be the corresponding Delaunay triangle in $\mathcal{DT}_{OPT}$ (see Fig. 3). We know there are three edges e_1, e_2 and e_3 incident on v. We correctly determine the other endpoint v' of one of these three edges e_1, e_2 and e_3 by finding a $\Delta_{v'}$ that satisfies the empty circle property and maximizes the minimum Delaunay edge length with the best partitioning of the budget. From v', the Voronoi diagram will be extended recursively until the budget cannot be further divided. Since we reached v' from v, at v' we have two possible ways to find the next vertex of the tree $\mathcal{VD}_{OPT}$. The vertex v corresponds to any triple $p_i, p_j, p_{\ell'}$ that forms a Delaunay triangle, where p_i, p_j forms an edge of the convex hull of any feasible solution S''. As we are checking all possible combinations of optimal Delaunay triangulations that maximizes the minimum edge length of $\mathcal{DT}$, the above dynamic programming will return the optimal solution, and the optimal value can be obtained from one of the $O(n^2)$ cells $\phi(p_i, p_j, *; k - 2)$ in the DP table, where p_i, p_j corresponds to an edge of the convex hull of S'. To construct S', we will backtrack from the cell $\phi(p_i, p_j, *; k - 2)$ that has the maximum value.

Theorem 2. *We can solve the* DKCONP *problem in* $O(n^4 k^2)$ *time using dynamic programming.*

Proof. Note that there are $O(n^3 k)$ subproblems. In each subproblem, we have $O(n)$ choices to select $p_{\ell'}$ and $O(k)$ choices to partition $\mathcal{K}$. So, we spend $O(nk)$ time to combine optimal solutions to subproblems into an optimal solution to the bigger subproblem. Hence, the total running time of the dynamic programming algorithm for the DKCONP problem for any $k > 0$ is $O(n^4 k^2)$. □

5 Concluding Remarks

In this paper, we studied the k-dispersion problem on a convex polygon and proposed: (i) an exact fixed-parameter algorithm with runtime $O(2^k n^2 \log^2 n)$, (ii) an exact polynomial time algorithm with runtime $O(n^4 k^2)$ for any $k > 0$. For practical purposes, one may prefer (i) for small values of k and (ii) for large values of k as the polynomial dependency on n is smaller for the fixed-parameter algorithm than the polynomial time algorithm. Finally, we mention that the general Euclidean k-dispersion problem is still open from the point of polynomial time approximation algorithm with a better factor than $\frac{1}{2}$ and also from the point of designing an exact fixed-parameter algorithm.

References

1. Akagi, T., et al.: Exact algorithms for the max-min dispersion problem. In: Chen, J., Lu, P. (eds.) FAW 2018. LNCS, vol. 10823, pp. 263–272. Springer, Cham (2018). https://doi.org/10.1007/978-3-319-78455-7_20
2. Araki T., Nakano, S.I.: Max-min dispersion on a line. J. Comb. Optim., **44**, 1824–1830 (2020)
3. Avis, D., Toussaint, G.T., Bhattacharya, B.K.: On the multimodality of distances in convex polygons. Comput. Math. Appl. **8**(2), 153–156 (1982)
4. Baur, C., Fekete, S.P.: Approximation of geometric dispersion problems. Algorithmica **30**, 451–470 (2001)
5. Cormen, T.H., Leiserson, C.E., Rivest, R.L., Stein, C.: Introduction to Algorithms. MIT Press, Cambridge (2009)
6. Drezner, Z.: Facility location: A Survey of Applications and Methods. Springer Series in Operations, Springer-Verlag, New York (1995)
7. Drezner, Z., Hamacher, H.W.: Facility Location: Applications and Theory. Springer Berlin, Heidelberg (2004)
8. Erkut, E.: The discrete p-dispersion problem. Eur. J. Oper. Res. **46**(1), 48–60 (1990)
9. Fekete, S.P., Meijer, H.: Maximum dispersion and geometric maximum weight cliques. Algorithmica **38**, 501–511 (2004)
10. Horiyama, T., et al.: Max-Min 3-dispersion problems. In: Du, D.-Z., Duan, Z., Tian, C. (eds.) COCOON 2019. LNCS, vol. 11653, pp. 291–300. Springer, Cham (2019). https://doi.org/10.1007/978-3-030-26176-4_24
11. Kobayashi, Y., Nakano, S.I., Uchizawa, K., Uno, T., Yamaguchi, Y., Yamanaka, K.: Max-min 3-dispersion on a convex polygon. In: 37th European Workshop on Computational Geometry (2021)
12. Marx, D., Pilipczuk, M.: Optimal parameterized algorithms for planar facility location problems using Voronoi diagrams. In: Bansal, N., Finocchi, I. (eds.) ESA 2015. LNCS, vol. 9294, pp. 865–877. Springer, Heidelberg (2015). https://doi.org/10.1007/978-3-662-48350-3_72
13. Mark, D.B., Otfried, C., Marc, V.K, Mark, O.: Computational Geometry Algorithms and Applications, Springer, Heidelberg (2008). https://doi.org/10.1007/978-3-540-77974-2
14. Ravi, S.S., Rosenkrantz, D.J., Tayi, G.K.: Heuristic and special case algorithms for dispersion problems. Oper. Res. **42**(2), 299–310 (1994)
15. Shamos, M.I.: Computational Geometry. Yale University (1978)
16. Wang, D.W., Kuo, Y.S.: A study on two geometric location problems. Inf. Process. Lett. **28**(6), 281–286 (1988)

Arbitrary-Oriented Color Spanning Region for Line Segments

Sukanya Maji$^{(\boxtimes)}$ and Sanjib Sadhu

Department of CSE, National Institite of Technology, Durgapur, India
sm.20cs1102@phd.nitdgp.ac.in, sanjib.sadhu@cse.nitdgp.ac.in

Abstract. Given a set of colored geometric objects, a color spanning region of a desired shape is a region (of that shape) that contains at least one object of each color. Here, the objective is to optimize a specific parameter of the region as mentioned in the problem definition. In this paper, we study the optimal color spanning region recognition problem of different shapes for a given set $\mathcal{L}$ of n colored line segment objects in $\mathbb{R}^2$, where each segment is associated with any one of the m colors, namely $\{1, 2, \ldots, m\}$, where $3 \leq m < n$. These are (i) an arbitrary-oriented color spanning strip of minimum width, (ii) two congruent arbitrary-oriented minimum width color spanning strips which contain disjoint subset of the members in $\mathcal{L}$, (iii) two congruent arbitrary-oriented strips of minimum width, such that their union is color spanning, and (iv) an arbitrary-oriented color spanning rectangle of minimum area. The time complexities of the proposed algorithms for these problems are: (i) $O(n^2 \log n)$, (ii) $O(n^4 \log n)$, (iii) $O(n^4 m \log m)$, and (iv) $O(n^3 m)$. Better algorithm with reduced time complexities can be achieved for problems (ii) and (iii) if some restrictions are imposed on the relative orientation of the outputs. Each of these problems needs linear space.

Keywords: Color spanning region recognition · Geometric duality · Line sweep

1 Introduction

Given a set $\mathcal{L}$ of n line segments, each segment is attached with one of the m colors ($3 \leq m < n$), the objective of this paper is to study the problem of recognizing color spanning region of different shapes minimizing a specified parameter of the region depending on the problem requirement. Here the objects in $\mathcal{L}$ may be viewed as the facilities (e.g. hospitals, post-offices, schools etc.), available in a city, and the objective is to locate a region of minimum area where at least one facility of each type is available. The facilities may be points, line segments, convex polygons, etc. The desired region may be a strip, disk, rectangle, etc. The problem is well studied in the literature starting from the work of [1], and has found a lot of applications in facility location problem [1], pattern recognition [3], database queries [12], etc.

© The Author(s), under exclusive license to Springer Nature Switzerland AG 2023
A. Bagchi and R. Muthu (Eds.): CALDAM 2023, LNCS 13947, pp. 71–86, 2023.
https://doi.org/10.1007/978-3-031-25211-2_6

Related Work: The color spanning problem was studied by Abellanas et al. [1], where they computed a color spanning axis-parallel rectangle among a set of n colored points with m colors in $O(n(n-m)\log^2 m)$. Huttenlocher et al. [9] proposed algorithm for computing the smallest color spanning circle for a given set of n points with m colors in $O(mn\log n)$ time. The smallest color spanning strip and rectangle of arbitrary orientation for a given set of points can be computed in $O(n^2\log n)$ and $O(n^3\log m)$ time [6], respectively. The color spanning axis-parallel square and equilateral triangle can be determined in time $O(n\log^2 n)$ [10] and $O(n\log n)$ [8], respectively. Acharyya et al. [2] identified the smallest color spanning axis-parallel square, rectangle and circle for a colored point set around a given query point. Bae [4] computed the minimum width color spanning axis-parallel rectangular annulus for a set of points in $O((n-m)^3 n\log n)$.

Most of the research works on color spanning problem deals with the input facilities as a point set. However, in real application, it is not always reasonable to represent each facility by point only. This leads to studying the problem of recognizing a color spanning region of optimum size among a set of convex objects. For simplicity, we start research in this direction with colored line segments as the facilities. Note that, the method of solving color spanning region with point set facilities cannot be extended in a straightforward manner to handle this problem with line segment facilities. Huttenlocher et al. [9] computed the smallest color spanning axis-parallel square and disk with the line segments as facilities, in $O(n^2\log n)$ and $O(n^2\alpha(n)\log n)$ time, respectively.

Another related problem is the k-center problem, where a given set of geometric objects need to be covered by k congruent disks or squares of minimum size. The corresponding color spanning version is finding k congruent color spanning regions among a set of colored objects as facilities to place k demand points. Here, the concept is that the ith facility, denoted by r_i, can support at most $f(r_i)$ demand points (centers of the color spanning regions, each of equal size). We start studying this variation of the problem with $k = 2$ and $f(r_i) = 1$ for each facility in $\{r_1, r_2, \ldots, r_n\}$.

Depending on the problem instance, sometimes a single color spanning region may be more costly (measured in terms of width or area of the region) than the k congruent regions whose union is a color spanning. This motivates us to study further the union color spanning strips problem for a set of line segments $\mathcal{L}$. For simplicity, we have considered $k = 2$ strips in this paper.

Our Contributions: Given a set of n colored line segment objects with m different colors ($3 \le m < n$) in $\mathbb{R}^2$, we propose algorithms for computing color spanning (CS) arbitrary-oriented (i) a pair of strips of minimum width, and (ii) a rectangle of minimum area. The specific problems that are studied in this paper, are listed below in the Table 1 along with the time complexities of the proposed algorithms. The space complexity of all these problems is $O(n)$.

Table 1. The result of arbitrarily oriented color spanning object(s)

Problems on color spanning regions for a set $\mathcal{L}$ of line segments in $\mathbb{R}^2$	Segments covered $(\mathcal{L}', \mathcal{L}'' \subseteq \mathcal{L})$ by strip(s)/rectangle	Minimizes	Time complexity
A single strip	$\mathcal{L}'$ is color spanning (CS)	Strip width	$O(n^2 \log n)$
Two congruent strips	$\mathcal{L}'$ and $\mathcal{L}''$ are CS $(\mathcal{L}' \cap \mathcal{L}'' = \phi)$	Strip width	$O(n^4 \log n)$
Two congruent parallel strips	$\mathcal{L}'$ and $\mathcal{L}''$ are CS $(\mathcal{L}' \cap \mathcal{L}'' = \phi)$	Strip width	$O(n^3)$
Union color spanning by two strips	$\mathcal{L}' \cup \mathcal{L}''$ is CS $(\mathcal{L}' \cap \mathcal{L}'' = \phi)$	Strip width	$O(n^4 m \log m)$
Union color spanning by two parallel strips	$\mathcal{L}' \cup \mathcal{L}''$ is CS $(\mathcal{L}' \cap \mathcal{L}'' = \phi)$	Strip width	$O(n^3 \log m)$
A rectangle $(\mathcal{R})$	$\mathcal{L}'$ is CS	Area of $\mathcal{R}$	$O(n^3 m)$

2 Preliminaries and Notations

We use $\mathcal{L} = \{\ell_1, \ell_2, \ldots, \ell_n\}$ to denote the n input line segment facilities. The subset of the segments in $\mathcal{L}$ with color $i \in \{1, 2, \ldots, m\}$ is denoted by $\mathcal{L}_i$. We use $x(p)$ and $y(p)$ to denote the x- and y-coordinate of the point p, respectively. A line passing through any two points p and q is denoted by $\ell(p, q)$. A line segment ℓ in $\mathbb{R}^2$ is said to be *covered* by a region if every point on ℓ lies inside or on the boundary of that region. A segment with its two endpoints p and q is denoted by $[p, q]$.

Definition 1 (Color spanning). *A region R in $\mathbb{R}^2$ is said to be **color spanning** if it contains at least one member of $\mathcal{L}$ having color i for all $i = 1, 2, \ldots, m$.*

A strip $\mathcal{V}$ is an unbounded region enclosed by two parallel lines which are called the boundaries of $\mathcal{V}$. The width of a strip $\mathcal{V}$ is determined by the perpendicular distance between its two boundaries. We use CSS to denote any color spanning strip.

Definition 2 (Minimal and minimum-CSS). *A CSS is said to be a **minimal-CSS** if it cannot be shrunk further without violating the definition 1 of the color spanning region. There may exist more than one minimal-CSS for $\mathcal{L}$. The one having minimum width among all minimal-CSSs is said to be minimum width color spanning strip, and will be denoted by **minimum-CSS**.*

2.1 A Single Color Spanning Strip of Arbitrary Orientation

Problem 1 (Single color spanning strip). *Given a set $\mathcal{L} = \{\ell_1, \ell_2, \ldots, \ell_n\}$ of (possibly intersecting) line segments in $\mathbb{R}^2$; each segment $\ell_i \in \mathcal{L}$ is attached with one of m distinct colors ($3 \leq m < n$), compute a minimum-CSS $\mathcal{V}$ of arbitrary orientation.*

We use geometric duality [5] to solve this problem. Here, a point $p = (a, b)$ in the primal plane is represented by a line $p^* : y = ax - b$ in the dual plane, and a line $l : y = \mathtt{m}x + c$ in the primal plane is represented by a point $l^* = (\mathtt{m}, -c)$ in the dual plane. Note that, $\mathcal{V}$ may be vertical or non-vertical. We can compute the vertical strip $\mathcal{V}$ by sweeping a pair of vertical lines to locate all possible *minimal − CSSs'*. The one having minimum width is preserved as the *minimum − CSS* as the initialization of this algorithm. This needs $O(n \log n)$ time, maintaining an array of size m for storing the number of segments of each color i ($1 \le i \le m$) lying in the present position of the strip defined by the pair of sweep lines. We now concentrate on computing the smallest width non-vertical *CSS*. Since the point-line duality cannot handle any vertical line, if there exists any vertical line in $\mathcal{L}$, we rotate the entire set $\mathcal{L}$ by a small angle to make each segment non-vertical.

An arbitrary-oriented strip $\mathcal{V}$ is defined by its two boundaries, namely the upper boundary $ub(\mathcal{V})$ and the lower boundary $lb(\mathcal{V})$, that are mutually parallel lines; the point of intersection of $ub(\mathcal{V})$ with any vertical line lies above that of $lb(\mathcal{V})$ with the same vertical line. A strip $\mathcal{V}$ in the primal plane is mapped to a vertical line segment $\mathcal{V}^* = [lb^*(\mathcal{V}), ub^*(\mathcal{V})]$ in dual plane[1]. In duality transformation, a line segment $\ell_i = [p, q] \in \mathcal{L}$ in primal plane is mapped to a double wedge ℓ_i^* in dual plane [5], which is closure of the symmetric difference of the two half planes delimited by the lines p^* and q^*, and it does not contain any vertical line. The point of intersection of p^* and q^* is known as the *center-point* of the double wedge ℓ_i^* and is denoted by $cp(\ell_i^*)$. Let L_v be the vertical line passing through $cp(\ell_i^*)$; ℓ_{mid} be the line passing through ℓ_i^* with slope $\frac{1}{2}$(slope of p^* + slope of q^*). Each double wedge ℓ_i^* can be viewed as four rays, namely left-top $\ell t(\ell_i^*)$, left-bottom $\ell b(\ell_i^*)$, right-top $rt(\ell_i^*)$ and right-bottom $rb(\ell_i^*)$ emanating from $cp(\ell_i^*)$. The ray $\ell t(\ell_i^*)$ (resp. $\ell b(\ell_i^*)$) lies above (resp. below) ℓ_{mid} to the left of L_v, and the ray $rt(\ell_i^*)$ (resp. $rb(\ell_i^*)$) lies above (resp. below) ℓ_{mid} to the right of L_v (see Fig. 1). We refer to the union of $\ell t(\ell_i^*)$ and $rt(\ell_i^*)$ as $\mathcal{UT}(\ell_i^*)$ (*upper trace of ℓ_i^**), and the union of $\ell b(\ell_i^*)$ and $rb(\ell_i^*)$ as $\mathcal{LT}(\ell_i^*)$ (*lower trace of ℓ_i^**). While transforming a segment in primal plane to a double wedge using duality, the color associated with that segment remains same. The color of the double wedge (resp. line segment) ℓ^* (resp. ℓ) is denoted by $col(\ell^*)$ (resp. $col(\ell)$). The following result states the property of minimum width *CSS* for line segments.

Theorem 1. *A minimal-CSS $\mathcal{V}$ is defined by three segments, say $\ell_i, \ell_j, \ell_k \in \mathcal{L}$ lie inside $\mathcal{V}$, and one of its boundaries ($lb(\mathcal{V})$ or $ub(\mathcal{V})$) contains an endpoint of two segments $\in \{\ell_i, \ell_j, \ell_k\}$, and the other boundary contains an endpoint of a segment $\in \{\ell_i, \ell_j, \ell_k\}$. It may happen that both the boundaries of $\mathcal{V}$ may touch the two endpoints of a single segment $\in \{\ell_i, \ell_j, \ell_k\}$. The color of the segments defining $\mathcal{V}$ are different and none of their colors repeat inside the $\mathcal{V}$.*

---

[1] Due to the fact that that both the boundaries of $\mathcal{V}$ have same gradient.

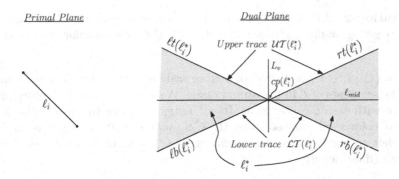

Fig. 1. Line segment ℓ_i in primal plane and its corresponding double wedge ℓ_i^* in dual plane

A segment s is said to be intersected by a double wedge ℓ^* if both $\mathcal{UT}(\ell^*)$ and $\mathcal{LT}(\ell^*)$ intersect with s. If a vertical segment in the dual plane (corresponding to a strip in the primal plane) is color spanning, it will be referred to as a $CS\_segment$. The dual $CS\_segment$ $\mathcal{V}^*$ of a $minimal\text{-}CSS$ $\mathcal{V}$ defined by three segments ℓ_i, ℓ_j and ℓ_k is shown in the Fig. 2. Due to the Theorem 1, we observe the following.

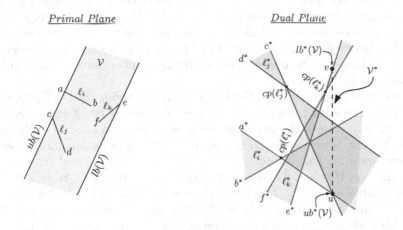

Fig. 2. Dual of strip $\mathcal{V}$ defined by three line segments ℓ_i, ℓ_j and ℓ_k.

Observation 1. *A strip $\mathcal{V}$ whose upper (resp. lower) boundary is defined by ℓ_i and ℓ_j, and lower (resp. upper) boundary is defined by ℓ_k in the primal plane, corresponds to the vertical segment $\mathcal{V}^* = [u, v]$ in the dual plane, where u is the point of intersection between the lower (resp. upper) traces of ℓ_i^* and ℓ_j^*, and v lies on the upper (resp. lower) trace of ℓ_k^* vertically above (resp. below) the point u (see Fig. 2). The width of this strip $\mathcal{V}$ is given by $\frac{|y(u)-y(v)|}{\sqrt{1+(x(u))^2}}$.*

Observation 2. *The dual $\mathcal{V}^*$ of a strip $\mathcal{V}$ is color spanning if at least one (left or right) wedge or the center-point of dual ℓ^* of ℓ of each color intersects with $\mathcal{V}^*$.*

Let $\mathcal{L}^* = \{\ell_i^* \mid \ell_i \in \mathcal{L}\}$ be the set of double wedges in the dual plane corresponding to the segments of $\mathcal{L}$ in the primal plane. We associate a vector $color[1..m]$ of length m with each $CS\_segment$. Its ith entry indicates the number of double wedges of color i are intersected by the $CS\_segment$. We sweep a vertical line λ from left to right among the members in $\mathcal{L}^*$ to identify a $CS\_segment$ of minimum (dual) length.

Data Structure: We use five pointers for the four rays of each double wedge $\ell^* \in \mathcal{L}^*$; the value of these pointers corresponding to a double wedge ℓ^* depend on the position (i.e. the x-coordinate) of the vertical sweep line λ. These five pointers of all the double wedges in $\mathcal{L}^*$ are initialized to $NULL$. We now describe the significance of these pointers of a double wedge ℓ^* at a particular position, say $x=\alpha$, of the sweep line λ.

Self: This pointer, associated with the ray $\ell t(\ell^*)$ (resp. $\ell b(\ell^*)$) points to $\ell b(\ell^*)$ (resp. $\ell t(\ell^*)$), and the same associated with $rt(\ell^*)$ (resp. $rb(\ell^*)$) points to $rb(\ell^*)$ (resp. $rt(\ell^*)$). From an upper trace of a double wedge, we can access its lower trace through this pointer, and vice versa.

CS_up: It is associated with the rays in the lower trace $\mathcal{LT}(\ell^*)$, and it points to the upper trace $\mathcal{UT}(t^*)$ of a double wedge t^* (say), vertically above it, such that a vertical segment at the present position ($x = \alpha$) of the sweep line λ lying between the $\mathcal{LT}(\ell^*)$ and $\mathcal{UT}(t^*)$ is color spanning (See Observation 2). Note that, for the upper trace of all the members in $\mathcal{L}^*$, this pointer is always set to $NULL$.

CS_dwn: It is associated with the two rays in the upper trace $\mathcal{UT}(\ell^*)$, which points to the lower trace $\mathcal{LT}(t^*)$ of a double wedge t^* (say), vertically below it, such that a vertical segment at the present position ($x = \alpha$) of the sweep line λ lying between the $\mathcal{LT}(t^*)$ and $\mathcal{UT}(\ell^*)$, is color spanning (See the Observation 2). This pointer is $NULL$ for the rays $\ell b(\ell^*)$ and $rb(\ell^*)$.

$Same\_col\_up$: It is associated with the two rays in the lower trace $\mathcal{LT}(\ell^*)$ of each double wedge ℓ^*. It points to the upper trace of a double wedge t^* ($\neq \ell^*$), that lies vertically above ℓ^* and is closest one to ℓ^* among all the double wedges having the color same as that of ℓ^* at the present position of the sweep line λ, provided such a double wedge t^* exists for ℓ^*; otherwise it is set to $NULL$. Also, this pointer is $NULL$ for the rays in $\mathcal{UT}(\ell^*)$, $\forall \ell^* \in \mathcal{L}^*$.

$Same\_col\_dwn$: It is associated with the two rays in the upper trace $\mathcal{UT}(\ell^*)$ of each double wedge ℓ^*. It points to the lower trace of a double wedge, say t^* ($\neq \ell^*$), if t^* lies vertically below ℓ^* and is closest one to ℓ^* among all double wedges having the color same as that of ℓ^* at the present position of the sweep line λ, provided such a double wedge t^* exists for ℓ^*; otherwise it is set to $NULL$. Also, this pointer is set to $NULL$ for the rays in $\mathcal{LT}(\ell^*)$, $\forall \ell^* \in \mathcal{L}^*$.

Algorithm: The event points of the vertical sweep line λ are the points of intersection of the dual of the endpoints of the input segments in $\mathcal{L}$ (see Observation 1) in sorted order with respect to their x-coordinates. These event points are created in $O(n^2)$ time [11], and are stored in an array A. We initialize the aforesaid pointers for each ray of the double wedges in $\mathcal{L}^*$ with their respective values at the first event position of the sweep line λ in the array A. During the sweep, we compute the $CS\_segment$ at each event point $e \in A$, and finally report the minimum length $CS\_segment$ observed.

During the sweep, the status of the sweep line λ is maintained as a list of the dual lines of $\mathcal{L}^*$ that appear on the sweep line λ in top to bottom order. The status of the sweep line is updated after processing each event point $e \in A$ as follows:

Case (i) The event point $e \in A$ corresponds to $cp(\ell_i^*)$, $\ell_i \in \mathcal{L}$:
We do the following updates:
Assign $Same\_col\_dwn(rt(\ell_i^*)) = Same\_col\_dwn(lt(\ell_i^*))$,
$Same\_col\_up(rb(\ell_i^*)) = Same\_col\_up(\ell b(\ell_i^*))$,
$CS\_up(rb(\ell_i^*)) = CS\_up(\ell b(\ell_i^*))$, and $CS\_dwn(rt(\ell_i^*)) = CS\_dwn(lt(\ell_i^*))$.

Case (ii) The event point $e \in A$ corresponds to the intersection of the upper (resp. lower) traces of two double wedges ℓ_i^* and ℓ_j^* of same color:
Let e be the point of intersection of upper traces of ℓ_i^* and ℓ_j^*, where $\mathcal{UT}(\ell_i^*)$ lies below $\mathcal{UT}(\ell_j^*)$ at the small distance $\epsilon > 0$ to the left of e. If at the left of e, the $Same\_col\_dwn$ of $\mathcal{UT}(\ell_i^*)$ points to the lower trace $\mathcal{LT}(\ell_k^*)$ of a double wedge ℓ_k^*, then at the event point e, the $Same\_col\_dwn$ of $\mathcal{UT}(\ell_i^*)$ needs to be updated to $\mathcal{LT}(\ell_j^*)$ provided $\mathcal{LT}(\ell_j^*)$ lies above $\mathcal{LT}(\ell_k^*)$.
Similarly, if the $Same\_col\_dwn$ pointer of $\mathcal{UT}(\ell_j^*)$ points to $\mathcal{LT}(\ell_i^*)$ just before the event e, then at the event point e, the $Same\_col\_dwn$ pointer of $\mathcal{UT}(\ell_j^*)$ will be updated to point to the old values (i.e. just before the event e) of $Same\_col\_dwn$ of $\mathcal{UT}(\ell_i^*)$; otherwise $Same\_col\_dwn$ pointer of $\mathcal{UT}(\ell_j^*)$ remains unaltered. However, the value of CS_dwn (resp. CS_up) pointer of the upper (resp. lower) trace remains unchanged at the event e.

Case (iii) The event point $e \in A$ of λ corresponds to the intersection of the upper (resp. lower) traces of the different colored double wedges ℓ_i^* and ℓ_j^*:
Suppose the upper trace of ℓ_i^* lies below that of ℓ_j^* at the $\epsilon > 0$ distance to the left of the event e. Without loss of generality, we assume that at $x = e$, the $\mathcal{UT}(\ell_i^*)$ and $\mathcal{UT}(\ell_j^*)$ are $lt(\ell_i^*)$ and $lt(\ell_j^*)$, respectively. At the event point e of λ, the CS_dwn pointers of $lt(\ell_i^*)$ and $lt(\ell_j^*)$ needs to be updated as follows.
Update of $CS\_dwn(lt(\ell_j^*))$: If at $\epsilon > 0$ distance to the left of the event e, the ℓ_i^* is essential[2] in the color spanning vertical segment $CS\_segment$ that spans from $lt(\ell_j^*)$ to $CS\_dwn(lt(\ell_j^*))$, then at the event e, we update the $CS\_dwn(lt(\ell_j^*))$ pointer to point to $Same\_col\_dwn(lt(\ell_i^*))$, otherwise the pointer $CS\_dwn(lt(\ell_j^*))$ remains unaltered.

---

[2] $color[col(\ell_i^*)]$ is 1 for an essential segment ℓ_i^* in the $CS\_segment$.

Update of $CS\_dwn(\ell t(\ell_i^*))$: At the event point e, the $CS\_dwn(\ell t(\ell_i^*))$ will be updated to point old value (i.e. just before the event e) of $CS\_dwn(\ell t(\ell_j^*))$, provided the lower trace of ℓ_i^* lies above that of double wedge pointed by old $CS\_dwn(\ell t(\ell_i^*))$ and the lower trace of the double wedge pointed by $CS\_dwn(\ell t(\ell_j^*))$ lies above that of $\ell t(\ell_i^*)$ before the event e.

Case (iv) The event point $e \in A$ of λ corresponds to the intersection of the lower (resp. upper) and upper (resp. lower) trace of the double wedges ℓ_i^* and ℓ_j^*, respectively:
In this case, only the status of the sweep line λ is changed.

All such aforesaid events take $O(1)$ time. For each types (i.e. aforesaid cases) of event e, if the pointer $Same\_col\_dwn$ (resp. $Same\_col\_up$) associated with a double wedge, say ℓ_i^*, is updated to point to a double wedge ℓ_k^*, then the pointer $Same\_col\_up$ (resp. $Same\_col\_dwn$) of the double wedge ℓ_k^* is also updated to point to ℓ_i^*. As the sweep line λ passes through each of its event point e, we update the CS_dwn (resp. CS_up) pointers of the ray involved in the upper (resp. lower) trace associated with the event point e. Also we need to update the CS_dwn (resp. CS_up) pointers of those rays whose CS_dwn (resp. CS_up) pointed to the rays associated with the event e. So this update may take linear amount of time to search for the rays whose CS_dwn (resp. CS_up) pointers need to be updated. This time can be expedited, if we use the idea of the following lemma and create two height balanced trees T_1 and T_2. These two trees are updated as the λ moves forward.

Lemma 1. *At a position, say $x = \alpha$, of the sweep line λ, if the CS_up (resp.CS_dwn) pointers of lower (resp. upper) traces of a pair of double wedges ℓ_i^* and ℓ_j^* point to the upper (resp. lower) trace of the same double wedge ℓ_k^* with $col(\ell_i) \neq col(\ell_j) \neq col(\ell_k)$, then the CS_up (resp. CS_dwn) pointers for the double wedges ℓ_{i+1}^*, ℓ_{i+2}^*, ..., ℓ_{j-1}^* lying between ℓ_i^* and ℓ_j^* also point to ℓ_k^*.*

In T_1 (resp. T_2), we store the triple (i, j, k) where the CS_up (resp. CS_dwn) pointer for the double wedges ℓ_i^*, ℓ_{i+1}^*, ..., ℓ_j^* points to the same double wedge ℓ_k^* (see the Lemma 1). The nodes in T_1 (resp. T_2) are mutually exclusive and exhaustive. These nodes are stored in T_1 (resp. T_2) with respect to the status of the λ (i.e. the ordered intersection of the double wedges with λ). We can create this T_1 (resp. T_2) in linear amount of time. We can update the CS_up pointers of the double wedges having the same value of CS_up pointers in $O(\log n)$ time using T_1 (or T_2). Hence each event needs $O(\log n)$ processing time and since there are total $O(n^2)$ events, we compute the minimal strips at each event point by checking the appropriate pointer (CS_up or CS_dwn) and report the one having minimum length among all the minimal strips obtained at each event points. The data structure of the sweep line λ at its current event point depends on the data structure of λ at its previous event point. As the sweep line λ moves forward through its event points, we need to compute $CS\_segment$ at the current event point e of λ using the data structure stored at e which can be obtained only from the information of the data structure at its previous event point. Hence as the sweep line moves through its event points e, we need to store the linear sized data structures of the previous event point of e instead of storing all the event points of λ altogether, and the same space can be reused as the λ moves forward to its next event point. Thus we obtain the following result.

Theorem 2. *The minimum width color spanning strip of arbitrary orientation for a given set of n colored line segments in $\mathbb{R}^2$ can be determined in $O(n^2 \log n)$ time and $O(n)$ space.*

2.2 Two Congruent Strips of Arbitrary Orientation

Problem 2 (Two congruent color spanning strips). *Given a set $\mathcal{L} = \{\ell_1, \ell_2, \ldots, \ell_n\}$ of (possibly intersecting) line segments in $\mathbb{R}^2$; each segment $\ell_i \in \mathcal{L}$ is attached with one of m distinct colors ($3 \leq m \leq n$), the objective is to compute arbitrary-oriented two congruent color spanning strips $\mathcal{V}_1$ and $\mathcal{V}_2$ of minimum width such that the set of segments covered by $\mathcal{V}_1$ and $\mathcal{V}_2$ are disjoint.*

In the context of Problem 2, note that if a segment lies inside $\mathcal{V}_1 \cap \mathcal{V}_2$, then it is suitably considered to lie completely inside one of $\mathcal{V}_1$ and $\mathcal{V}_2$. This problem is equivalent to compute a pair of *minimal-CSS* ($\mathcal{V}_1$, $\mathcal{V}_2$), so that the width of its larger strip is minimized among all possible pair of *minimal-CSS*s. We solve this problem by considering all possible *minimal-CSS* $\mathcal{V}_1$, and for each of them we choose a *minimum-CSS* $\mathcal{V}_2$ which covers the segments that are not covered by $\mathcal{V}_1$.

Fact 1. *If the two intersecting color spanning strips $\mathcal{V}_i$ and $\mathcal{V}_j$ are disjoint with respect to the segments ($\in \mathcal{L}$) covered by them, then there exists no double wedges in $\mathcal{L}^*$, that intersect with both the corresponding CS_segments $\mathcal{V}_i^*$ and $\mathcal{V}_j^*$, respectively.*

First we compute the minimal length $CS\_segment$ $\mathcal{V}_i^*$ at each event point e of a sweep line λ_1 using the procedure described in the Sect. 2.1. For each such $\mathcal{V}_i^*$, we compute all the $\mathcal{V}_j^*$ which are disjoint with $\mathcal{V}_i^*$ (see the Fact 1) using another sweep line λ_2 that lies to the right of λ_1. Among all these $\mathcal{V}_j^*$, we choose the one with minimum length. We check whether $\mathcal{V}_j^*$ is disjoint with $\mathcal{V}_i^*$ or not, as follows.

After computing the $CS\_segment$ $\mathcal{V}_i^*$ at the event point e of λ_1, we determine the set of double wedges, say $D_i^* \subseteq \mathcal{L}^*$, that completely intersect with $\mathcal{V}_i^*$ in the sorted order using the status of the sweep line λ_1. We consider the sweep line λ_2 at one of its event point, say e' which occurs to the right of e, and let, $\mathcal{V}_{i'}^*$ be the $CS\_segment$ at e' and the another endpoint of $\mathcal{V}_{i'}^*$ be p'. If e' is due to the intersection of any of its double wedge in D_i^*, then we only update D_i^* by swapping the corresponding two intersecting double wedges and move to the next event of λ_2. However, if e' is not due to the intersection of any of its double wedge in D_i^*, then we compute the double wedge $d \in D_i^*$ (resp. $d' \in D_i^*$) immediately below e' (resp. p'). This can be determined in $O(\log n)$ time from D_i^*. The strips $\mathcal{V}_{i'}^*$ and $\mathcal{V}_i^*$ will be disjoint in the following two cases:

(i) The d and d' exist, and they are same, (ii) both of the d and d' do not exist.

In all other cases, $\mathcal{V}_{i'}^*$ and $\mathcal{V}_i^*$ will be overlapping. As mentioned earlier in the Sect. 2.1, all the event points of the sweep lines need not to be stored altogether and the space complexity is also linear for the Problem 2. Since there are $O(n^2)$ such event points for both λ_1 and λ_2, we obtain the following result.

Theorem 3. *For a given set of n line segments in $\mathbb{R}^2$, we can compute two congruent disjoint (with respect to the segments covered) color spanning strips of the minimum width, if such a pair exists, otherwise we report that no such pair exists in $O(n^4 \log n)$ time and $O(n)$ space.*

Problem 3 (Restricted version of the Two congruent color spanning strip). *For the same inputs as in the Problem 2, compute two congruent, minimum width color spanning disjoint strips which are parallel to each other.*

The Theorem 1 leads to the following observation.

Observation 3. *If CSSs $\mathcal{V}_1$ and $\mathcal{V}_2$ are parallel to each other, then at least one boundary of one of the two strips $\mathcal{V}_1$ and $\mathcal{V}_2$ must contain an endpoint of two different colored segments in $\mathcal{L}$ that are covered by the corresponding strip. However, if $\mathcal{V}_1$ and $\mathcal{V}_2$ are not parallel, then one boundary of each of the strips must contain two endpoints of two different colored segments in $\mathcal{L}$. Note that, both the endpoints of the same segment may also define a boundary.*

If the two disjoint CSSs $\mathcal{V}_i$ and $\mathcal{V}_j$ are mutually parallel, then their corresponding dual $\mathcal{V}_i^*$ and $\mathcal{V}_j^*$, are two disjoint vertical segments, one lying vertically above the other.

Consider a minimal $CS\_segment$ $\mathcal{V}_i^*$ at an event point $e_i \in A$, determined by the procedure described in Sect. 2.1. Now, we will determine all possible minimal $CS\_segment$s $\mathcal{V}_j^*$ lying vertically above as well as below $\mathcal{V}_i^*$ in amortized $O(n)$ time. We explain the method of computing all the $CS\_segment$s below $\mathcal{V}_i^*$. Suppose ℓ_i^* be a double wedge lying immediately below $\mathcal{V}_i^*$. Take two pointers top and $bottom$, where top points to ℓ_i^* and $bottom$ points to a double wedge, say ℓ_j^*, below ℓ_i^*, such that the vertical segment from $\mathcal{UT}(\ell_i^*)$ to $\mathcal{LT}(\ell_j^*)$ at $x = x(e)$ is a $CS\_segment$, say $\mathcal{V}_j^*$. It can be determined from the sweep line status. The next $CS\_segment$ below $\mathcal{V}_j^*$, starting from double wedge that is just below ℓ_i^*, can be determined by shifting both the pointers top and $bottom$ downwards through the list of double wedges in the current sweep line status array. In this way, we compute all the $CS\_segment$s lying below $\mathcal{V}_i^*$ in linear amount of time. Similarly, we compute all possible minimal $CS\_segment$s that lie vertically above $\mathcal{V}_i^*$. Finally, we choose the one having minimum length as $\mathcal{V}_j^*$ that pairs with $\mathcal{V}_i^*$. The entire task is done in $O(n)$ time.

We repeat the above steps to compute all possible pairs $(\mathcal{V}_i^*, \mathcal{V}_j^*)$ at each event point $e_i \in A$ by sweeping the line λ. It may happen that there exists only one $CS\_segment$ in the entire floor. In that case, only one color spanning strip will be reported. Similar to the Problem 1 in the Sect. 2.1, we need not store all the events for this Problem 3 and hence, it needs $O(n)$ space. As there are $O(n^2)$ event points in the worst case, we have the following result.

Theorem 4. *For a given set of n colored line segments in $\mathbb{R}^2$, we can compute two congruent disjoint parallel color spanning strips of the minimum width in $O(n^3)$ time and $O(n)$ space.*

2.3 Two Congruent Strips of Arbitrary Orientation Whose Union is Color Spanning

Problem 4 (Union color spanning problem). *Given a set $\mathcal{L} = \{\ell_1, \ell_2, \ldots, \ell_n\}$ of (not necessarily disjoint) line segments in $\mathbb{R}^2$; each segment $\ell_i \in \mathcal{L}$ is associated with one of m distinct colors ($3 \le m \le n$), the objective is to compute arbitrary-oriented two congruent (i) disjoint (ii) non-disjoint strips $\mathcal{V}_1$ and $\mathcal{V}_2$ of minimum width, whose union is color spanning.*

We first compute two arbitrary-oriented disjoint strips $\mathcal{V}_1$ and $\mathcal{V}_2$. It is obvious that $\mathcal{V}_1$ and $\mathcal{V}_2$ are parallel. We use the line sweeping technique over the set of double wedges $\mathcal{L}^*$. For each color c_i ($1 \le i \le m$), we maintain two sorted lists of all the double wedges with respect to their lower traces and upper traces respectively, at each position $x = x(e)$, where e is the event point of the sweep line λ. These two lists are updated in constant time at each event point e as the line λ sweeps rightward. From these lists containing lower (resp. upper) traces of each color, we also compute a sorted array $first\_col\_\mathcal{LT}[1..m]$ (resp. $first\_col\_\mathcal{UT}[1..m]$) of size m at each event point e of λ, that keeps the first occurring lower (resp. upper) traces of each distinct colored double wedge that lies completely below (resp. above) the point e. These two arrays are sorted with respect to the point of intersections of the sweep line λ with the members of the array. Our algorithm executes the following tasks at each event point e of λ.

We take $\mathcal{V}_i^*$ with one of its endpoints starting at the event point e. Without loss of generality, we assume that the event point e is the intersection of two upper traces of two different colored double wedges. The other endpoint of $\mathcal{V}_i^*$ lie on the lower trace (lying vertically below e) of any one of the double wedges from the sorted array "$first\_col\_\mathcal{LT}$" and suppose this $\mathcal{V}_i^*$ intersects with the double wedges of k different colors. We use a color array $\mathcal{C}$ which keeps track of the color of wedges that completely intersect with $\mathcal{V}_i^*$ and it can be computed in $O(k)$ time. For each $\mathcal{V}_i^*$, we compute corresponding $\mathcal{V}_j^*$ which covers the double wedges of remaining $(m - k)$ colors. The two endpoints of the dual segment $\mathcal{V}_i^*$ (resp. $\mathcal{V}_j^*$) of the strip $\mathcal{V}_i$ (resp. $\mathcal{V}_j$) are denoted by top_1 (resp. top_2) and bot_1 (resp. bot_2). The color of the double wedges pointed by the top_1, top_2, bot_1 and bot_2 will be of different colors. Once we compute such a $\mathcal{V}_j^*$, we measure the length of $\mathcal{V}_j^*$. Next, we shift top_2 downward to the next upper trace below it and recompute $\mathcal{V}_j^*$. In this way we compute all possible $\mathcal{V}_j^*$ that lie below $\mathcal{V}_i^*$. Similarly we can compute all possible $\mathcal{V}_j^*$ lying above $\mathcal{V}_i^*$ in $O(n)$ time. We choose the $\mathcal{V}_j^*$ with minimum length. Now we increase the length of $\mathcal{V}_i^*$ by moving bot_1 pointer to next entry of the array "$first\_col\_\mathcal{LT}$" so that $\mathcal{V}_i^*$ covers now one extra color and we compute the corresponding minimum length $\mathcal{V}_j^*$. To obtain the optimal pair $\mathcal{V}_i^*$ and $\mathcal{V}_j^*$ at the event point e, we need not to compute all possible $\mathcal{V}_i^*$ due to the following lemma.

Lemma 2. *The function* $\max\{width(\mathcal{V}_i), width(\mathcal{V}_j)\}$ *is a convex function.*

Proof. If we increase the width of $\mathcal{V}_1$, the width of $\mathcal{V}_2$ either remains same or decreased, as the union of $\mathcal{V}_1$ and $\mathcal{V}_2$ is color spanning.   □

Since, the length of $\mathcal{V}_i^*$ is increased by shifting bot_1 through the members of the sorted array "$first\_col\_\mathcal{LT}$" (of size m), to minimize the $\max\{width(\mathcal{V}_i), width(\mathcal{V}_j)\}$ we need to iterate the above procedure at most $\log m$ times (see Lemma 2) at each event point e of λ. Finally we repeat the same procedure at each event point e of λ to find the optimal solution of the Problem 4. Similar to the Problem 2, we need linear space to solve the Problem 4. Thus we obtain the following result.

Theorem 5. *For the set $\mathcal{L}$ of n line segments in $\mathbb{R}^2$, we can compute two congruent disjoint parallel strips of the minimum width, whose union is color spanning in $O(n^3 \log m)$ time and $O(n)$ space.*

Now, we determine two arbitrary-oriented non-disjoint strips $\mathcal{V}_1$ and $\mathcal{V}_2$, whose union is color spanning. We apply almost the same technique as well as the data structures that are used to compute two disjoint strips (first part of the Problem 4). In this case, the two vertical $CS\_segments$ $\mathcal{V}_1^*$ and $\mathcal{V}_2^*$ will lie at two different event positions of two different sweep lines λ_1 and λ_2, respectively. Both $\mathcal{V}_1^*$ and $\mathcal{V}_2^*$ will be defined by three segments (see Observation 3). We first consider a $CS\_segment$ at an event point e of the sweep line λ_1 covering, say k colors (see the algorithm described for disjoint case). Then we compute the $\mathcal{V}_j^*$ (which covers the remaining $(m-k)$ colors) at each event e' of the sweep line λ_2 lying to the right of λ_1. We can compute the two sorted arrays (defined earlier) $first\_col\_\mathcal{LT}[1..m]$ (resp. $first\_col\_\mathcal{UT}[1..m]$) at the event point e' in $O(m)$ time. For a fixed length $CS\_segment$ $\mathcal{V}_i^*$ at e, we can determine minimum $\mathcal{V}_j^*$ at e' in $O(m)$ time. Now, the Lemma 2 says that in $O(m \log m)$ time, we can compute the optimum pair $(\mathcal{V}_i^*, \mathcal{V}_j^*)$ with one of their endpoints at e and e', respectively. Now, there are $O(n^2)$ different possible positions for each of the event points e and e'. Thus we obtain the following result.

Theorem 6. *For a given set of n line segments in $\mathbb{R}^2$, we can compute two congruent non-disjoint strips of the minimum width, whose union is color spanning in $O(n^4 m \log m)$ time and $O(n)$ space.*

2.4   Color Spanning Rectangle (CSR) of Arbitrary Orientation

Problem 5. *Given a set $\mathcal{L} = \{\ell_1, \ell_2, \ldots, \ell_n\}$ of (possibly intersecting) n line segments in $\mathbb{R}^2$; each segment $\ell_i \in \mathcal{L}$ is attached with one of m distinct colors ($3 \le m \le n$), the objective is to compute an arbitrary-oriented color spanning rectangle (CSR) $\mathcal{R}$ of minimum area.*

Fact 2. *A color spanning rectangle (CSR) $\mathcal{R}$ is the intersection of two color spanning strips, say $\mathcal{V}_1$ and $\mathcal{V}_2$, which are perpendicular to each other and boundaries of $\mathcal{V}_1$ and $\mathcal{V}_2$ pass through the opposite parallel sides of $\mathcal{R}$ (Fig. 3).*

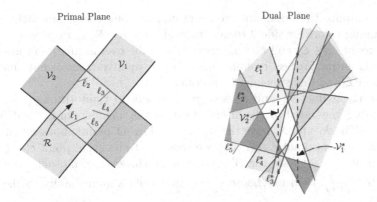

Fig. 3. Color spanning rectangle $\mathcal{R}$ in primal is represented by a pair $(\mathcal{V}_1^*, \mathcal{V}_2^*)$ in dual. (Color figure online)

Lemma 3. *One side of the minimum area CSR $\mathcal{R}$ for $\mathcal{L}$ must contain exactly two endpoints of any two segments (one endpoint of each) in $\mathcal{L}$ and each of the remaining sides of $\mathcal{R}$ must contain one endpoint of a segment in $\mathcal{L}$. Also the same colored endpoints can occur at most twice on the boundary of $\mathcal{R}$, and the color of the segments whose endpoint lie on the boundary of $\mathcal{R}$, will be distinct from remaining segments that lie completely inside $\mathcal{R}$.*

Proof. Suppose, $\mathcal{R}$ be the minimum area CSR and E be the set of segments enclosed by it. Since the area of $\mathcal{R}$ is minimum among all CSR of $\mathcal{L}$, the color of the segments in $\mathcal{L}$ whose endpoints lie on the boundary of $\mathcal{R}$ must be distinct from the others lying on or inside $\mathcal{R}$, otherwise we can rotate and/or shrink $\mathcal{R}$ to obtain another CSR with smaller area than $\mathcal{R}$, contradicting the assumption that $\mathcal{R}$ is minimum area CSR. Note that, the same colored endpoints may occur at most twice on the boundary of $\mathcal{R}$, if both the endpoints of a segment occur at the boundary of the $\mathcal{R}$. Let P be the convex hull of E. Two vertices of P must lie on a side of minimum area rectangle $\mathcal{R}$, if it encloses P [7]. $\qquad\square$

Since the adjacent sides of a rectangle are perpendicular to each other, we observe the following.

Observation 4. *A CSR which is the intersection of two perpendicular strips $\mathcal{V}_1$ and $\mathcal{V}_2$ in the primal plane (Fact 2), can be represented by two color spanning vertical segments (CS_ segment) $\mathcal{V}_1^*$ and $\mathcal{V}_2^*$ in the dual plane such that if the $\mathcal{V}_1^*$ is at $x = x_1$, then the $\mathcal{V}_2^*$ will be at $x = \frac{-1}{x_1}$, and the set of double wedges intersecting both $\mathcal{V}_1^*$ and $\mathcal{V}_2^*$ must be color spanning.*

Let $\mathcal{R}_{a,b}$ be the color spanning rectangle with one side defined by two points a and b of the two segments ℓ_a and ℓ_b, respectively. The four sides of $\mathcal{R}_{a,b}$ are numbered sequentially 1 to 4 in counter clockwise direction where the side 1 contains the points a and b. We first generate all possible CSR $\mathcal{R}_{a,b}$ with side 2 being defined by all possible segments with same (say blue) color only, and then we compute the minimum area rectangle among them. Similarly, we use the remaining colors for side 2 to generate all possible $\mathcal{R}_{a,b}$. Finally, we do this for all pairs of $\ell_a (\in \mathcal{L})$ and $\ell_b (\in \mathcal{L})$ to compute the overall minimum area CSR for $\mathcal{L}$. This entire process is done through the duality transformations of the segments in $\mathcal{L}$ and line sweeping technique.

Two vertical lines λ_1 and λ_2 sweep from left to right through the set of event points ξ generated by the intersections of the members in $\mathcal{L}^*$ as defined in Problem 1. The λ_1 and λ_2 have the same set of event points; however, if the λ_1 reaches at its event point $e \in \xi$, then we move λ_2 to its event point $e' \in \xi$ whose position is at $x = \frac{-1}{x(e)}$ (see the Observation 4). However, if no such event point exists at $x = \frac{-1}{x(e)}$, then we choose $e' \in \xi$ that occurs immediately to the left of $x = \frac{-1}{x(e)}$.

Consider a minimal width CSS $\mathcal{V}_1$ with its lower boundary passing through the endpoints a and b of the segments ℓ_a and ℓ_b, respectively. Let c, the endpoint of a segment ℓ_c, lies on the upper boundary of $\mathcal{V}_1$, so that $col(\ell_a) \neq col(\ell_b) \neq col(\ell_c)$ (see the Lemma 3). The double wedges ℓ_a^*, ℓ_b^* and ℓ_c^* represent the duals of ℓ_a, ℓ_b and ℓ_c, respectively. In dual, the point of intersection of $\mathcal{UT}(\ell_a^*)$ and $\mathcal{UT}(\ell_b^*)$ is the top endpoint of the $CS\_segment$ $\mathcal{V}_1^*$, and its bottom endpoint will lie vertically below its top endpoint and on the $\mathcal{LT}(\ell_c^*)$. Let $\mathcal{L}_1^* \subseteq \mathcal{L}^*$ be the set of double wedges intersecting with $\mathcal{V}_1^*$ and $\mathcal{L}_2^* \subset \mathcal{L}^*$ be the set of double wedges lying completely below $\mathcal{V}_1^*$. At each event point e of sweep line λ_1, we compute the $CS\_segments$ $\mathcal{V}_1^*$. For each such segment $\mathcal{V}_1^*$, we determine the corresponding $\mathcal{V}_2^*$ at $x = \frac{-1}{x(e)}$ which lies immediately after the event point, say e' of the sweep line λ_2. The order of the double wedges of $\mathcal{L}_1^*$ at $x = \frac{-1}{x(e)}$, can be obtained from the status of the sweep line λ_2 at $x = x(e')$ in linear time. The $\mathcal{V}_2^*$ to be determined, must intersect with ℓ_a^*, ℓ_b^* and ℓ_c^* in order to obtain a CSR made by the intersection of $\mathcal{V}_1$ and $\mathcal{V}_2$. We take two pointers top_1 (resp. top_2) and bot_1 (resp. bot_2) that are initialized to point to double wedges having top endpoint and bottom endpoint of $\mathcal{V}_1^*$ (resp. $\mathcal{V}_2^*$), respectively. At $x = \frac{-1}{x(e)}$, suppose $\mathcal{L}_{up}^* \subset \mathcal{L}_1^*$ be the set of double wedges with their upper trace lying above both $\mathcal{UT}(\ell_a^*)$ and $\mathcal{UT}(\ell_b^*)$. We choose a distinct colored (say blue) double wedge $w^* \in \mathcal{L}_{up}^*$ at $x = \frac{-1}{x(e)}$ which is closest to and above both the $\mathcal{UT}(\ell_a^*)$ and $\mathcal{UT}(\ell_b^*)$, and compute the $CS\_segment$ $\mathcal{V}_2^*$ (with top endpoint on $\mathcal{UT}(w^*)$) for the double wedges in $\mathcal{L}_1^*$ by maintaining a color array in linear time. Actually this w defines the side 2 of CSR (discussed above). The bottom endpoint of $\mathcal{V}_2^*$ will lie on the lower trace of a double wedge, say t^* and bot_2 will point to t^*. Note that w^* and t^* must be the essential[3] double wedges in $\mathcal{V}_2^*$. We determine the area of the rectangle whose sides are given by $\mathcal{V}_1^*$ and $\mathcal{V}_2^*$ in dual plane. Next we process each double wedge $d_i^* \in \mathcal{L}_2^*$ lying below ℓ_c^* at $x = x(e)$, as follows.

$col(d_i^*) = col(w^*)$: Here,
- if $\mathcal{UT}(d_i^*)$ lies below $\mathcal{UT}(\ell_a^*)$ or $\mathcal{UT}(\ell_b^*)$ and above $\mathcal{LT}(\ell_a^*)$ or $\mathcal{LT}(\ell_b^*)$ at $x = -\frac{1}{x(e)}$, then we stop, since no CSR $\mathcal{R}_{a,b}$ with $col(w^*)$ in side 2 is possible with such d_i^*.

---

[3] Essential color occurs exactly once inside the color spanning region.

- if $\mathcal{UT}(d_i^*)$ lies above both $\mathcal{UT}(\ell_a^*)$ and $\mathcal{UT}(\ell_b^*)$ at $x = \frac{-1}{x(e)}$, and below the double wedge pointed by top_2 pointer, then we reject all the double wedges lying above d_i^* (since same colored segment w as that of d_i cannot occur in CSR $\mathcal{R}_{a,b}$) by updating color array. We also update the top_2 pointer to point to d_i^*.
- if d_i^* lies below $\mathcal{LT}(\ell_a^*)$ and $\mathcal{LT}(\ell_b^*)$ at $x = -\frac{1}{x(e)}$, and above bot_2, then CSR $\mathcal{R}_{a,b}$ cannot include d_i^* and hence we move the bot_2 pointer to the double wedge lying immediately above d_i^*. Note that, now the vertical segment defined by the top_2 and bot_2 may not be the color spanning, and we should compute the CSR $\mathcal{R}_{a,b}$ whenever we get a $\mathcal{V}_2^*$.
- if d_i^* lies above top_2 at $x = -\frac{1}{x(e)}$, then we reject d_i^* (since $col(d_i^*)$ and $col(w^*)$ are same).

$col(d_i^*) \neq col(w^*)$: If d_i^* lies between the double wedges pointed by top_2 and bot_2, then we insert d_i^* in $\mathcal{V}_2^*$ and update the color array for $\mathcal{V}_2^*$. If the color of the double wedge pointed by bot_2 is same as that of d_i^* then we move bot_2 upwards to point to an essential wedge. If $\mathcal{V}_2^*$ becomes color spanning after insertion of d_i^*, then we update the bot_1 pointer to point to d_i^* and compute $\mathcal{V}_2^*$ and the corresponding CSR. Otherwise we reject d_i^*.

Since each segment is inserted and/or deleted in $\mathcal{V}_2^*$ at most once, the above procedure needs amortized linear time. We repeat this procedure for each distinct colored double wedge w^* to obtain the minimum area CSR $\mathcal{R}_{a,b}$. Finally, we execute this process at each event point of λ_1 to determine the overall minimum area CSR for $\mathcal{L}$. There are $O(n^2)$ event points for λ_1. Similar to the Problem 2, we also need the linear space to solve this problem. Since there are at most m distinct colors, and at each event point of λ_1, it takes amortized linear amount of time to compute all possible CSS $\mathcal{V}_2^*$ with each distinct color of the upper trace, we have the following result.

Theorem 7. *The minimum sized (area) color spanning rectangle of arbitrary orientation for a given set of n colored line segments in $\mathbb{R}^2$ can be determined in $O(n^3 m)$ time and $O(n)$ space.*

References

1. Abellanas, M., Hurtado, F., Icking, C., Klein, R., Langetepe, E., Ma, L., Palop, B., Sacristán, V.: Smallest color-spanning objects. Algorithms - ESA 2001. In: Proceedings of 9th Annual European Symposium, Aarhus, Denmark, 28–31 August 2001, vol. 2161, pp. 278–289 (2001)
2. Acharyya, A., Maheshwari, A., Nandy, S.C.: Color-spanning localized query. Theor. Comput. Sci. **861**, 85–101 (2021)
3. Asano, T., Bhattacharya, B.K., Keil, J.M., Yao, F.F.: Clustering algorithms based on minimum and maximum spanning trees. Proceedings of the Fourth Annual Symposium on Computational Geometry, Urbana-Champaign, IL, USA, 6–8 June 1988, pp. 252–257 (1988)
4. Bae, S.W.: An algorithm for computing a minimum-width color-spanning rectangular annulus. J. Korean Inst. Inf. Sci. Eng. **44**, 246–252 (2017)

5. de Berg, M., Cheong, O., van Kreveld, M.J., Overmars, M.H.: Computational Geometry: Algorithms and Applications, 3rd edn. Springer, Berlin (2008)
6. Das, S., Goswami, P.P., Nandy, S.C.: Smallest color-spanning object revisited. Int. J. Comput. Geom. Appl. **19**(5), 457–478 (2009)
7. Freeman, H., Shapira, R.: Determining the minimum-area encasing rectangle for an arbitrary closed curve. Commun. ACM **18**, 409–413 (1975)
8. Hasheminejad, J., Khanteimouri, P., Mohades, A.: Computing the smallest color-spanning equaliteral triangle. In: 31st European Workshop on Computational Geometry. pp. 32–35 (2015)
9. Huttenlocher, D.P., Kedem, K., Sharir, M.: The upper envelope of Voronoi surfaces and its applications. In: Proceedings of the Seventh Annual Symposium on Computational Geometry, North Conway, NH, USA, 10–12 June 1991. pp. 194–203. ACM (1991)
10. Khanteimouri, P., Mohades, A., Abam, M.A., Kazemi, M.R.: Computing the smallest color-spanning axis-parallel square. In: Cai, L., Cheng, S.-W., Lam, T.-W. (eds.) ISAAC 2013. LNCS, vol. 8283, pp. 634–643. Springer, Heidelberg (2013). https://doi.org/10.1007/978-3-642-45030-3_59
11. Lee, D.T., Ching, Y.: The power of geometric duality revisited. Inf. Process. Lett. **21**, 117–122 (1985)
12. Pruente, J.: Minimum diameter color-spanning sets revisited. Discret. Optim. **34**, 100550 (2019)

Game Theory

Rectilinear Voronoi Games with a Simple Rectilinear Obstacle in Plane

Arun Kumar Das[1]([✉]), Sandip Das[1], Anil Maheshwari[2], and Sarvottamananda[3]

[1] Indian Statistical Institute, Kolkata, India
arund426@gmail.com, sandipdas@isical.ac.in
[2] Carleton University, Ottawa, ON, Canada
anil@scs.carleton.ca
[3] Ramakrishna Mission Vivekananda Educational and Research Institute, Howrah, India
sarvottamananda@rkmvu.ac.in

Abstract. We study two player single round rectilinear Voronoi games in the plane for a finite set of clients where service paths are obstructed by a rectilinear polygon. The players wish to maximize the net number of their clients where a client is served by the nearest facility of players in $\mathbb{L}_1$ metric. We prove the tight bounds for the payoffs of both the players for the class of games with simple, convex and orthogonal convex polygons. We also generalize the results for $\mathbb{L}_\infty$ metric in the plane.

1 Introduction

Motivation. A *Voronoi game* is a *competitive facility location problem* where the goal is to maximize a service in the Voronoi cells of the players. Rectilinear versions of such problems naturally arise in several applications that deal with rectilinear paths, such as those related to city maps, electronic circuits, raster graphics, warehousing, architecture, civil engineering, network flows, etc. The motivation for the problem comes from real-life situations where an impassable zone restricts every player in a competitive facility location problem. We prove lower and upper bounds on the payoffs of such games in $\mathbb{L}_1$ and $\mathbb{L}_\infty$.

Previous Results. The concept of Voronoi games was introduced by Ahn et al. [1] for line segments and circles. Several variants of Voronoi games are available in the literature [2–15]. Ahn et al. [1], Cheong et al. [10] and, Fekete and Meijer [13] studied the versions where they tried to maximize the Voronoi cells themselves. Briefly, they proved that Bob is guaranteed at least half of the total payoff for their versions of Voronoi games. They also game suitable strategies for both the players. Durr and Thang [12], Teramoto et al. [15], Bandyopadhyay et al. [2] and Sun et al. [14] studied intractability of Voronoi games in graphs. Banik et al. [3] described a discrete vesion of the problem for line segments. Banik et al. [4,5] and, later, Berg et al. [9] solved the single round where two players can place a fixed number of facilities in a single round. Banik et al. [8] introduced Voronoi games in the interior of simple polygons and devised polynomial time optimal strategies for both Alice and Bob.

© The Author(s), under exclusive license to Springer Nature Switzerland AG 2023
A. Bagchi and R. Muthu (Eds.): CALDAM 2023, LNCS 13947, pp. 89–100, 2023.
https://doi.org/10.1007/978-3-031-25211-2_7

Banik et al. [6,7] described the version of the problem that we study in this paper. Banik et al. [6] and Das et al. [11] extended, generalized and improved the solutions of the problem. They also proved several tight lower and upper bounds on the payoffs of a similar nature as in this paper.

New Results. In this paper, we study the rectilinear Voronoi games with rectilinear polygonal obstacles for players similar to the games mentioned in [6,7,11]. A notable difference is that we restrict the players outside of a fixed polygon. We formally describe the rectilinear Voronoi game in Sect. 2.

We prove that the optimal payoff of Alice $\geq \lceil n/3 \rceil$ and $\leq n/2$ and that the optimal payoff of Bob $\geq n/2$ and $\leq \lfloor 2n/3 \rfloor$, where the net number of served clients is n. These bounds are tight. We also prove that these bounds hold irrespective of whether we fix the class of polygonal obstacles as simple polygons, convex polygons, or orthogonal convex polygons in contrast with the results of [6,7,11]. We then generalize these results for $\mathbb{L}_\infty$ metric.

Organization. We present some preliminary definitions, concepts and observations in Sect. 2. In Sect. 3, we prove the bounds for rectilinear Voronoi games for simple, convex and orthogonal convex polygonal obstacles. We show in Sect. 4 that the same bounds hold for extensions to $\mathbb{L}_\infty$.

2   Preliminaries

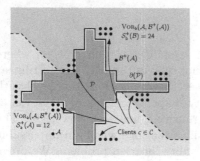

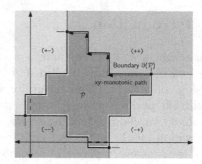

Fig. 1. A two player single round rectilinear Voronoi game with an orthogonal convex polygon obstacle in $\mathbb{L}_1$ metric in $\mathbb{R}^2$. Alice is at $\mathcal{A}$ and Bob wins with a payoff of 24 by playing at $\mathcal{B}^+(\mathcal{A})$.

Fig. 2. An illustration for orthogonal convex polygon with explanation of some notation for quadrants.

We present some definitions, notations and conventions first. A *Voronoi game* is a competitive game in which players compete to serve a set of clients by placing their facilities. A facility serves the clients in its Voronoi cell and shares the clients on its Voronoi cell boundary. The *payoff of each player* is determined by the net number of clients they serve. The *two players single round rectilinear Voronoi game with a polygon obstacle $\mathcal{P}$*, denoted by $\mathcal{G}_{\mathrm{P},\mathbb{L}_1}(\mathcal{C}, \mathcal{P})$, is a single round Voronoi game played between two players, conveniently named Alice and Bob. They place a single facility each in a region containing a *finite set of point* *clients* $C \subset \mathbb{R}^2$ with the *open simple polygonal obstacle* P in the $\mathbb{L}_1$ plane. Alice places her facility first, followed by Bob. The facility locations of Alice and Bob are denoted by $A \in \mathbb{R}^2$ and $B \in \mathbb{R}^2$, respectively. Effectively, $A \in \mathbb{R}^2 \setminus P$ and $B \in \mathbb{R}^2 \setminus P$, since $A \in P$ and $B \in P$ fetch exactly zero payoffs for Alice and Bob respectively. The *distance* from a facility f to a client c, denoted by $d_{\mathbb{L}_1}^P(f, c)$ is measured as the $\mathbb{L}_1$-length of any shortest path from f to c that avoids interior of P. The *payoffs of Alice and Bob*, denoted by $S_a(A, B)$ and $S_b(A, B)$, respectively, are the net count of clients they serve. See Fig. 1 for an example. We note that neither the shortest paths nor the best locations for Alice and Bob have to be unique. Moreover, we allow overlapping of clients and facilities, and in some cases, we also permit degenerate simple polygonal obstacles. Two problems arise naturally for these Voronoi games that we describe subsequently.

Problem 1. Let Alice and Bob play a two player single round rectilinear Voronoi game with a polygonal obstacle P. What is an optimal location of Alice that maximizes her minimum payoff? What is an optimal location of Bob that maximizes his payoff for a fixed Alice's facility location? ∎

An optimal location of Bob for a fixed location A for Alice's facility is denoted by $B^+(A)$ and the optimal payoff $S_b^+(A)$. The corresponding payoff of Alice is denoted by $S_a^+(A)$. Then,

$$S_a^+(A) = \min_{B \in \mathbb{R}^2} S_a(A, B) = S_a(A, B^+(A))$$

$$S_b^+(A) = \max_{B \in \mathbb{R}^2} S_b(A, B) = S_b(A, B^+(A))$$

We can compute $B^+(A)$ by solving any one of the above equations. An optimal locations of Alice and Bob are denoted by A^* and B^*, respectively, and the optimal payoffs by S_a^* and S_b^*, respectively. $B^* = B^+(A^*)$. Then,

$$S_a^* = \max_{A \in \mathbb{R}^2} \min_{B \in \mathbb{R}^2} S_a(A, B) = S_a^+(A^*) = S_a(A^*, B^*)$$

$$S_b^* = \min_{A \in \mathbb{R}^2} \max_{B \in \mathbb{R}^2} S_b(A, B) = S_b^+(A^*) = S_b(A^*, B^*)$$

We can compute A^* and then B^* by solving for $\min_{A \in \mathbb{R}^2} \max_{B \in \mathbb{R}^2} S_b(A, B) = \max_{B \in \mathbb{R}^2} S_b(A^*, B) = S_b(A^*, B^*)$.

Next, we propose the problem of determining the upper and lower bounds of Alice's and Bob's payoffs.

Problem 2. Let Alice and Bob play a two player single round rectilinear Voronoi game with polygonal obstacle P. What are the upper and lower bounds on the payoffs of Alice and Bob? ∎

The lower and upper bounds of Alice's payoffs are mathematically determined by expressions $\min_{\mathcal{G}_{P,\mathbb{L}_1}(C,P)} S_a^*$ and $\max_{\mathcal{G}_{P,\mathbb{L}_1}(C,P)} S_a^*$ respectively. Similarly, the lower and upper bounds of Bob's payoffs are $\min_{\mathcal{G}_{P,\mathbb{L}_1}(C,P)} S_b^*$ and $\max_{\mathcal{G}_{P,\mathbb{L}_1}(C,P)} S_b^*$ mathematically.

We can prove that the Voronoi game is a constant sum game. Hence

Theorem 1. $\mathcal{S}_a(\mathcal{A}, \mathcal{B}) + \mathcal{S}_b(\mathcal{A}, \mathcal{B}) = \mathcal{S}_a^+(\mathcal{A}) + \mathcal{S}_b^+(\mathcal{A}) = \mathcal{S}_a^* + \mathcal{S}_b^* = |\mathcal{C} \setminus \mathcal{P}|$

Proof. We note that $\mathcal{P}$ is open, and any client in the (strict) interior of $\mathcal{P}$ is not served. Other clients are either fully served by Alice or Bob or equally shared by them. Thus the net total of the payoffs in any Voronoi game is always equal to $|\mathcal{C} \setminus \mathcal{P}|$.   □

We study the class of the rectilinear Voronoi games when the obstacles are simple, convex or orthogonal convex polygons. We can also extend the Voronoi games and related problems described above to $\mathbb{L}_\infty$ metric. Instead of orthogonal convex polygons, we look at polygons that are oblique orthogonal convex polygons described later in Sect. 4.

We represent a simple polygon, and likewise, a simple polygonal region, $\mathcal{P}$ by its boundary $\partial(\mathcal{P})$ and assume that the polygon contains its open interiors. The boundary $\partial(\mathcal{P})$ is assumed to be represented by a non-crossing counter-clockwise sequence of edges such that the interior of $\mathcal{P}$ is on the left. The simple polygons are open and bounded. Likewise, both the interiors and exteriors are open. Though simply connected, they may possibly be degenerate. An *orthogonal convex polygon* is an open rectilinear polygon such that every horizontal and vertical line intersects the polygon no more than once in an interval. See Fig. 2 for an example. Let S be any finite or infinite bounded set of points. An *orthogonal convex hull* of S, possibly non-unique, is a minimal open orthogonal convex polygon that contains S and is denoted by OCHULL(S). The *smallest containing box* of S is denoted by BOX(S), i.e., BOX(S) = $\{ (q_x, q_y) \mid x_{\min}(S) < q_x < x_{\max}(S), y_{\min}(S) < q_y < y_{\max}(S) \}$ where $x_{\min}(S)$, $x_{\max}(S)$, $y_{\min}(S)$ and $y_{\max}(S)$ are respectively the left, right, bottom and top extremes of the set S.

We implicitly use the Voronoi regions of Alice and Bob in the discussion. The Voronoi regions of Alice and Bob are denoted by $\mathrm{VOR}_a(\mathcal{A}, \mathcal{B})$ and $\mathrm{VOR}_b(\mathcal{A}, \mathcal{B})$ respectively for their facility locations $\mathcal{A}$ and $\mathcal{B}$ respectively.

$$\mathrm{VOR}_a(\mathcal{A}, \mathcal{B}) = \{ p \in \mathbb{R}^2 \setminus \mathcal{P} \mid d_{\mathbb{L}_1}^{\mathcal{P}}(\mathcal{A}, p) \le d_{\mathbb{L}_1}^{\mathcal{P}}(\mathcal{B}, p) \},$$

$$\mathrm{VOR}_b(\mathcal{A}, \mathcal{B}) = \{ p \in \mathbb{R}^2 \setminus \mathcal{P} \mid d_{\mathbb{L}_1}^{\mathcal{P}}(\mathcal{A}, p) \ge d_{\mathbb{L}_1}^{\mathcal{P}}(\mathcal{B}, p) \}.$$

The clients $c \in \mathcal{C}$ for which $d_{\mathbb{L}_1}^{\mathcal{P}}(\mathcal{A}, c) = d_{\mathbb{L}_1}^{\mathcal{P}}(\mathcal{B}, c)$ are shared equally between Alice and Bob and contribute $1/2$ to each of the payoffs.

3   Bounds for Rectilinear Voronoi Games with Polygonal Obstacles

3.1   Unrestricted

Let S be a finite set of points in the polygonal region R. We define xy-*median* of S in R to be a point c_m, such that, any open horizontal and vertical chord of R that avoids c_m contain $\le \lfloor |S|/2 \rfloor$ points of S on the other side of c_m. We can

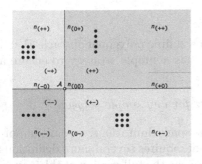
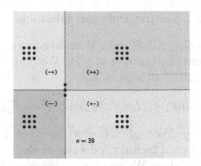

Fig. 3. Lemma 2: The lower bound of Alice's payoff for the unrestricted rectilinear Voronoi game in plane.

Fig. 4. Theorem 3: An unrestricted rectilinear Voronoi game proving tight bounds for non-overlapping clients.

argue that it is always possible to compute xy-*median* for any set of clients S in any bounded or unbounded simply connected region R. See Fig. 5.

Let $\mathcal{G}_{L_1}(\mathcal{C})$ be *an unrestricted rectilinear Voronoi game* with a finite set of clients $\mathcal{C}$ in plane, where the service paths are not restricted by any obstacle. Let $|\mathcal{C}|$ be n. We show that in an unrestricted rectilinear Voronoi game, Alice and Bob have an optimal strategy so that the other player does not have an advantage in their payoff. This is similar to the original result of [1].

Lemma 2. $\mathcal{S}_a^* \geq n/2$ and $\mathcal{S}_b^* \geq n/2$.

Proof. To prove Alice's bound, we put Alice's facility at the xy-median of $\mathcal{C}$. Then Alice is guaranteed a payoff of $n/2$. See Fig. 3 for the sketch of the proof. In the figure, $\lceil n/2 \rceil - (n_{(0+)} + n_{(0-)} + n_{(00)}) \leq (n_{(++)} + n_{(+-)} + n_{(+0)}) < \lceil n/2 \rceil$, etc. where $n_{(++)}$, etc., denotes the number of clients in the (++) quadrant, etc. We can show that $\mathcal{S}_a(\mathcal{A}, \mathcal{B}) > n/2$ by formulating it is an integer linear program while optimizing for the max-min payoff. To prove Bob's bound, we put Bob's facility overlapping Alice's facility. Bob is guaranteed a payoff of $n/2$ there. □

Moreover, as an immediate consequence of Lemma 2, we can show that the lower and upper bounds of $\mathcal{S}_a^* = \mathcal{S}_b^* = n/2$ are indeed same, invariably constant, and hence, tight for unrestricted case.

Theorem 3. *Let* $\mathcal{G}_{L_1}(\mathcal{C})$ *be an unrestricted rectilinear Voronoi game in* $\mathbb{R}^2$ *with* n *clients. Then* $\mathcal{S}_a^* = \mathcal{S}_b^* = n/2$.

Proof. We put Alice at an xy-median of $\mathcal{C}$. Bob is forced to place his facility at the same location to maximize his payoff. See Fig. 4 for an example unrestricted rectilinear Voronoi game. □

The implicit technique employed above is used several times later with some tight modifications for regions with obstacles.

3.2 Simple Polygon Obstacle

Let $\mathcal{G}_{\mathrm{P},\mathrm{L}_1}(\mathcal{C},\mathcal{P})$ be a Voronoi game in $\mathbb{R}^2$ with a simple polygonal obstacle $\mathcal{P}$ and clients $\mathcal{C}$. Let $n = |\mathcal{C} \setminus \mathcal{P}|$. We note that Bob has a simple strategy to ensure a payoff of at least $n/2$.

Lemma 4. *Let $\mathcal{A}$ be fixed. Then $\mathcal{S}_b^+(\mathcal{A}) \geq {}^n/_2$ for any simple polygon $\mathcal{P}$.*

Proof. We fix $\mathcal{B}^+(\mathcal{A})$ overlapping $\mathcal{A}$. Due to the space limitations, we omit important (and technical) details, as the formal proof requires several more definitions and claims. We request the interested reader to see the full version of this paper.
□

The strategy for Alice to ensure at least a minimum payoff is non-trivial. We show below a facility location where she can get a payoff of $\lceil n/3 \rceil$. With the aid of numerous figures, we sketch the main idea in our proof. We omit important (and technical) details, as the formal proof requires several more definitions and claims. It is impossible to provide all the necessary details within the page limit of the conference submission.

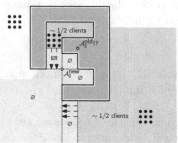
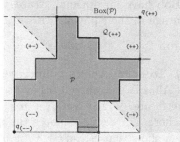

Fig. 5. The median horizontal and vertical chords for xy-median do not intersect in R for a set of points. We show the existence of another valid location for xy-median.

Fig. 6. The $\mathcal{Q}_{(++)}$ extended quadrant with respect to an orthogonal convex polygon $\mathcal{P}$. Naturally, $\mathcal{Q}_{(++)} \cup \mathcal{Q}_{(--)} = \mathbb{R}^2 \setminus \mathcal{P}$. Hence, $\mathcal{C} \setminus \mathcal{P} \subset \mathcal{Q}_{(++)} \cup \mathcal{Q}_{(--)}$.

We define the *extended quadrants relative to* $\mathcal{P}$ for the purpose of proofs below. Let us consider the Voronoi diagram with obstacle $\mathcal{P}$ in $\mathbb{L}_1$ of the two points $q_{(++)} = (x_{\max}(\mathcal{P}), y_{\max}(\mathcal{P}))$ and $q_{(--)} = (x_{\min}(\mathcal{P}), y_{\min}(\mathcal{P}))$. The closed (++) extended quadrant, denoted by $\mathcal{Q}_{(++)}$, is the set of points in the Voronoi cell of $q_{(++)}$, i.e., $\{ p \in \mathbb{R}^2 \setminus \mathcal{P} \mid d_{\mathbb{L}_1}^{\mathcal{P}}(q_{(++)}, p) \leq d_{\mathbb{L}_1}^{\mathcal{P}}(q_{(--)}, p) \}$. Likewise, we define $\mathcal{Q}_{(-+)}$, $\mathcal{Q}_{(--)}$ and $\mathcal{Q}_{(+-)}$ closed extended quadrants. See Fig. 6 for an illustration of the extended quadrant $\mathcal{Q}_{(++)}$.

Before proving a lower bound of $\lceil n/3 \rceil$, we show first that $\lceil n/4 \rceil$ is a weaker lower bound for $\mathcal{S}_a^*$.

Lemma 5. $S_a^* \geq \lceil n/4 \rceil$ *for any orthogonal convex polygon* $\mathcal{P}$.

Proof. We observe that at least one of any pair of diametrically opposite extended quadrants of $\mathcal{P}$, for example one of $\mathcal{Q}_{(++)}$ or $\mathcal{Q}_{(--)}$, will contain $\geq \lceil \frac{n}{2} \rceil$ clients because they cover $\mathbb{R}^2 \setminus \mathcal{P}$ and hence $\mathcal{C} \setminus \mathcal{P}$. We can show that $\mathcal{A}$ on a xy-median of the extended quadrant of the four that contains the most clients will get $\geq \lceil n/4 \rceil$ payoff. See Fig. 7.   □

Later, in Theorem 10, we show that $\lceil n/3 \rceil$ is the tight bound of S_a^* for even orthogonal convex polygonal obstacles.

Lemma 6. $S_a^* \geq \lceil n/3 \rceil$ *for any simple polygon* $\mathcal{P}$.

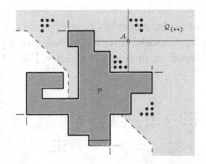
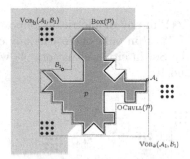

Fig. 7. A weak lower bound for Alice's payoff for rectilinear Voronoi game with a simple orthogonal polygonal obstacle. $\mathcal{Q}_{(++)}$ quadrant contains $\geq \lceil \frac{n}{2} \rceil$ clients. Alice gets at least half of the clients in $\mathcal{Q}_{(++)}$.

Fig. 8. Lower bound for Alice's payoff for rectilinear Voronoi game with a simple polygonal obstacle. Possible candidate location $\mathcal{A}_1$ in Lemma 6.

Proof. We consider two possible candidates for Alice's facility location for our claim. One of these will guarantee a payoff of $\lceil n/3 \rceil$. The first candidate location, denoted by $\mathcal{A}_1$, is a rightmost point of $\partial(\mathcal{P})$. See Fig. 8. If $S_a^+(\mathcal{A}_1) \geq \lceil n/3 \rceil$ then the proof is complete.

Otherwise, let $\mathcal{B}^+(\mathcal{A}_1) = \mathcal{B}_1$. We compute $\mathcal{V} = \text{VOR}_b(\mathcal{A}_1, \mathcal{B}_1)$. Naturally $\mathcal{C} \cap \mathcal{V}$ contains $> \lfloor 2n/3 \rfloor$, since, $S_a^+(\mathcal{A}_1) < \lceil n/3 \rceil \implies S_b^+(\mathcal{A}_1) > n - \lceil n/3 \rceil$. We fix $\mathcal{A}_2$ on the xy-median of the clients in $\mathcal{V}$, i.e., $\mathcal{C} \cap \mathcal{V}$. We give a proof sketch below that $S_a^+(\mathcal{A}_2) \geq \lceil n/3 \rceil$. Let $\mathcal{A}_2 = (a_x, a_y)$.

The following subcases arise.

Case 1. $\mathcal{A}_2 \in \mathbb{R}^2 \setminus \text{Box}(\mathcal{P})$.

If $\mathcal{A}_2$ is, without loss of generality, such that $a_x \geq x_{\min}(\mathcal{P})$ and $a_y \geq y_{\max}(\mathcal{P})$, then clearly there are at least $\lceil n/3 \rceil$ clients above and $\lceil n/3 \rceil$ clients right of $\mathcal{A}_2$. If $\mathcal{A}_2$ is, without loss of generality, such that $a_x \geq x_{\min}(\mathcal{P})$ and $y_{\min}(\mathcal{P}) < a_y > y_{\max}(\mathcal{P})$, then we can show that $\mathcal{B}_2$ either serves only shared clients of

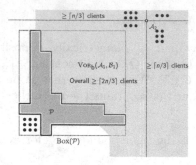

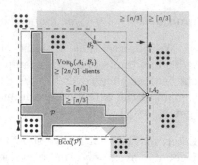

Fig. 9. Lemma 6: Case 1. (a) Both the median lines for $\mathcal{A}_2$ are unbounded.

Fig. 10. Lemma 6: Case 1. (b) One of the median lines for $\mathcal{A}_2$ is semi-bounded and the other one is unbounded.

$\mathcal{C} \cap \mathcal{V}$ in any diametrically opposite quadrants relative to $\mathcal{A}_2$ or serves clients of $\mathcal{C} \cap \mathcal{V}$ in only one of the any diametrically opposite quadrants relative to $\mathcal{A}_2$. So, in either situation, $\mathcal{A}_2$ will get a payoff of at least $\lceil n/3 \rceil$. See Figs. 9 and 10.

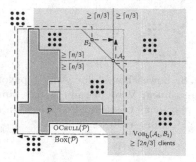

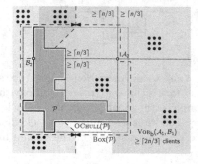

Fig. 11. Lemma 6: Case 2. (a) Both $\mathcal{A}_2$ and $\mathcal{B}_2$ in the same quadrant with respect to $\mathcal{P}$.

Fig. 12. Lemma 6: Case 2. (b) $\mathcal{A}_2$ and $\mathcal{B}_2$ in different quadrants with respect to $\mathcal{P}$.

Case 2. $\mathcal{A}_2 \in \text{Box}(\mathcal{P}) \setminus \text{OCHull}(\mathcal{P})$.

If $\mathcal{A}_2$ is, without loss of generality, in $\mathcal{Q}_{(++)}$, then we can show that $\mathcal{B}_2$ either serves only shared clients of $\mathcal{C} \cap \mathcal{V}$ in any diametrically opposite quadrants relative to $\mathcal{A}_2$ or serves clients of $\mathcal{C} \cap \mathcal{V}$ in only one of the diametrically opposite quadrants relative to $\mathcal{A}_2$. Thus, $\mathcal{A}_2$ is guaranteed a payoff of at least $\lceil n/3 \rceil$. See Figs. 11 and 12.

Case 3. $\mathcal{A}_2 \in \text{OCHull}(\mathcal{P}) \setminus \mathcal{P}$.

If $\mathcal{A}_2$ is sufficiently deep in a pocket such that the horizontal and vertical chords passing through $\mathcal{A}_2$ are also in the same pocket and, without loss of generality, the opening to the exterior is towards (++) quadrant with respect to $\mathcal{A}_2$ then, we can argue that since there are $\lceil n/3 \rceil$ clients in $\mathcal{C} \cap \mathcal{V}$ below the

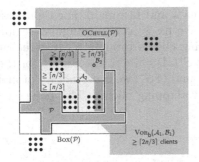

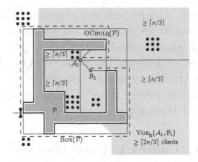

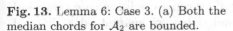

Fig. 13. Lemma 6: Case 3. (a) Both the median chords for $\mathcal{A}_2$ are bounded.

Fig. 14. Lemma 6: Case 3. (b) One of the median chord for $\mathcal{A}_2$ is bounded and the other semi-bounded.

horizontal chord and left of the vertical chord, Alice gets a payoff of at least $\lceil n/3 \rceil$. If $\mathcal{A}_2$ is shallow in a pocket, then too, we can show that $\mathcal{B}_2$ either serves only shared clients of $\mathcal{C} \cap \mathcal{V}$ in any diametrically opposite quadrants relative to $\mathcal{A}_2$ or serves clients of $\mathcal{C} \cap \mathcal{V}$ in only one of the diametrically opposite quadrants relative to $\mathcal{A}_2$. Thus $\mathcal{A}_2$ ensures a payoff of at least $\lceil n/3 \rceil$. See Figs. 13 and 14.

Case 4. $\mathcal{A}_2 \in \mathcal{P}$

This case does not arise.

Consequently, in all the cases, $\mathcal{S}_a^+(\mathcal{A}_2) \geq \lceil n/3 \rceil$. ∎

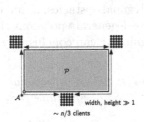

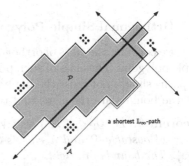

Fig. 15. The lower bound for Alice's payoff for rectilinear Voronoi game with a simple polygonal obstacle is tight.

Fig. 16. An oblique orthogonal convex polygon $\mathcal{P}$.

Theorem 7. *Let* $\mathcal{G}_{P,\mathbb{L}_1}(\mathcal{C}, \mathcal{P})$ *be a Voronoi game in* $\mathbb{R}^2$ *with a simple polygonal obstacle* $\mathcal{P}$ *and* n *clients* $\mathcal{C}$. *Then* $\lceil n/3 \rceil \leq \mathcal{S}_a^* \leq n/2$ *and* $n/2 \leq \mathcal{S}_b^* \leq \lfloor 2n/3 \rfloor$. *The bounds are tight.*

Proof. The bounds are consequences of Lemma 4 and Lemma 6. For tightness, we construct two Voronoi games as follows. We fix $\mathcal{P}$ as a rectangular region with three sets of about $n/3$ nearly overlapping clients totaling n clustered at three equidistant locations. We can show that $\mathcal{S}_a^+(\mathcal{A}) = \lceil n/3 \rceil$. See Fig. 15 for the construction. For the tightness of the upper bound of $\mathcal{S}_a^+$, we construct a Voronoi game as before with a single cluster of n overlapping clients. The polygonal obstacle does not matter.   □

3.3   Convex Polygon Obstacle

In [6], Alice was guaranteed a share of payoff for convex polygon case compared to the general case. In [11], both Alice and Bob were guaranteed a share of payoff for convex polygon case. However, unlike [6,11], in rectilinear Voronoi games with obstacles, there is no such advantage for either Alice or Bob, if we specialize to the class of convex polygonal obstacles. The proofs, on the other hand, are simplified. Also, we note that the convex polygons are special cases of orthogonal convex polygons though the opposite is not true. Thus we have the following theorem.

Theorem 8. *Let $\mathcal{G}_{P,L_1}(\mathcal{C}, \mathcal{P})$ be a Voronoi game in $\mathbb{R}^2$ with a convex polygonal obstacle $\mathcal{P}$ and n clients $\mathcal{C}$. Then $\lceil n/3 \rceil \leq \mathcal{S}_a^* \leq n/2$ and $n/2 \leq \mathcal{S}_b^* \leq \lfloor 2n/3 \rfloor$. The bounds are tight.*

Proof. The proof is similar to that of Theorem 7 though much simplified because of the convexity of the polygonal obstacle. The tightness's follow from the same constructions.   □

3.4   Orthogonal Simple Polygon Obstacle

Next, we consider the class of orthogonal simple polygonal obstacles, a subclass of simple polygons. The class of such polygons includes degenerate polygons though the paths should not cross the boundary edges. Again, we can conclusively show that the bounds are the same and tight. Hence,

Theorem 9. *Let $\mathcal{G}_{P,L_1}(\mathcal{C}, \mathcal{P})$ be a Voronoi game in $\mathbb{R}^2$ with a orthogonal simple polygonal obstacle $\mathcal{P}$ and n clients $\mathcal{C}$. Then $\lceil n/3 \rceil \leq \mathcal{S}_a^* \leq n/2$ and $n/2 \leq \mathcal{S}_b^* \leq \lfloor 2n/3 \rfloor$. The bounds are tight.*

3.5   Orthogonal Convex Polygon Obstacle

Lastly, we show the bounds for the class of orthogonal convex polygonal obstacles. As mentioned earlier, since the convex polygons are special cases of orthogonal convex polygons, hence the bounds here are valid for the earlier section too. Again this is unlike [11].

Theorem 10. *Let $\mathcal{G}_{P,L_1}(\mathcal{C}, \mathcal{P})$ be a Voronoi game in $\mathbb{R}^2$ with a orthogonal simple polygonal obstacle $\mathcal{P}$ and n clients $\mathcal{C}$. Then $\lceil n/3 \rceil \leq \mathcal{S}_a^* \leq n/2$ and $n/2 \leq \mathcal{S}_b^* \leq \lfloor 2n/3 \rfloor$. The bounds are tight.*

Proof. Case 3 of the proof of Lemma 6 does not arise because the orthogonal convex polygonal obstacle will not have any pockets. Also, for the tightness, we had deliberately constructed the Voronoi game so that the obstacle is at the same time simple, convex, orthogonal simple and orthogonal convex. Hence the same example game proves the tightness of each of these classes of obstacles. □

4   Bounds for $\mathbb{L}_\infty$ Metric in Plane

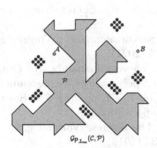
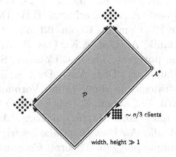

Fig. 17. A Voronoi game $\mathcal{G}_{P,L_\infty}(\mathcal{C},\mathcal{P})$ in $\mathbb{L}_\infty$ metric with the polygonal obstacle $\mathcal{P}$ in plane.

Fig. 18. Tightness of the lower bound for Alice's payoff for the Voronoi game $\mathcal{G}_{P,L_\infty}(\mathcal{C},\mathcal{P})$ in $\mathbb{L}_\infty$ metric with the convex polygonal obstacle $\mathcal{P}$ in plane.

Let $\mathcal{G}_{P,L_\infty}(\mathcal{C},\mathcal{P})$ be a Voronoi game in $\mathbb{L}_\infty$ metric with the polygonal obstacle $\mathcal{P}$ in plane with a set of clients $\mathcal{C}$. We note that the $\mathbb{L}_\infty$ metric is very similar to $\mathbb{L}_1$. Though it is not apparent, we can easily extend the bounds to these Voronoi games by modifying our proofs. Most of the arguments are valid if we choose an oblique pair of reference axes, i.e., if we choose the lines $x = y$ and $x + y = 0$ as the x-axis and the y-axis, respectively. Moreover, an *oblique orthogonal convex polygon* $\mathcal{P}$, which is an extension of orthogonal convex polygons, has a property that any lines parallel to the above two oblique axes will intersect the polygon $\mathcal{P}$ in at most one interval. See Figs. 16 and 17 for examples of an oblique orthogonal convex polygon and a Voronoi game $\mathcal{G}_{P,L_\infty}(\mathcal{C},\mathcal{P})$ respectively (Fig. 18).

Theorem 11. *Let $\mathcal{G}_{P,L_\infty}(\mathcal{C},\mathcal{P})$ be a Voronoi game in $\mathbb{R}^2$ with a simple polygonal obstacle and n clients. Then $\lceil n/3 \rceil \le \mathcal{S}_a^* \le n/2$ and $n/2 \le \mathcal{S}_b^* \le \lfloor 2n/3 \rfloor$.*

Also, there exist Voronoi games with a convex polygonal obstacle and n clients such that $\mathcal{S}_a^* = \lceil n/3 \rceil$ and $\mathcal{S}_b^* = n/2$.

Corollary 12. $\lceil n/3 \rceil \le \mathcal{S}_a^* \le n/2$ *and* $n/2 \le \mathcal{S}_b^* \le \lfloor 2n/3 \rfloor$ *for subclasses of Voronoi games in $\mathbb{L}_\infty$ with convex polygonal obstacles, oblique orthogonal polygonal obstacles and oblique orthogonal convex polygonal obstacles. The bounds are tight.*

References

1. Ahn, H.K., Cheng, S.W., Cheong, O., Golin, M., van Oostrum, R.: Competitive facility location: the Voronoi game. Theoret. Comput. Sci. **310**(1), 457–467 (2004)
2. Bandyapadhyay, S., Banik, A., Das, S., Sarkar, H.: Voronoi game on graphs. Theoret. Comput. Sci. **562**, 270–282 (2015)
3. Banik, A., Bhattacharya, B.B., Das, S.: Optimal strategies for the one-round discrete Voronoi game on a line. J. Comb. Optim. **26**(4), 655–669 (2013)
4. Banik, A., Bhattacharya, B.B., Das, S., Das, S.: The 1-dimensional discrete Voronoi game. Oper. Res. Lett. **47**(2), 115–121 (2019)
5. Banik, A., Bhattacharya, B.B., Das, S., Mukherjee, S.: The discrete Voronoi game in $\mathbb{R}^2$. Comput. Geom. **63**, 53–62 (2017)
6. Banik, A., Das, A.K., Das, S., Maheshwari, A.: Sarvottamananda: Voronoi game on polygons. Theoret. Comput. Sci. **882**, 125–142 (2021)
7. Banik, A., Das, A.K., Das, S., Maheshwari, A., Sarvottamananda, S.: Optimal strategies in single round Voronoi game on convex polygons with constraints. In: Wu, W., Zhang, Z. (eds.) COCOA 2020. LNCS, vol. 12577, pp. 515–529. Springer, Cham (2020). https://doi.org/10.1007/978-3-030-64843-5_35
8. Banik, A., Das, S., Maheshwari, A., Smid, M.: The discrete Voronoi game in a simple polygon. Theoret. Comput. Sci. **793**, 28–35 (2019)
9. de Berg, M., Kisfaludi-Bak, S., Mehr, M.: On one-round discrete Voronoi games. In: Lu, P., Zhang, G. (eds.) 30th International Symposium on Algorithms and Computation (ISAAC 2019). Leibniz International Proceedings in Informatics (LIPIcs), vol. 149, pp. 37:1–37:17. Schloss Dagstuhl-Leibniz-Zentrum fuer Informatik, Dagstuhl, Germany (2019)
10. Cheong, O., Har-Peled, S., Linial, N., Matousek, J.: The one-round Voronoi game. Discrete Comput. Geom. **31**(1), 125–138 (2004)
11. Das, A.K., Das, S., Maheshwari, A., Sarvattomananda, S.: Voronoi games using geodesics. In: Balachandran, N., Inkulu, R. (eds.) Algorithms and Discrete Applied Mathematics - 8th International Conference, CALDAM 2022, Puducherry, India, 10–12 February 2022, LNCS, vol. 13179, pp. 195–207. Springer, Cham (2022). https://doi.org/10.1007/978-3-030-95018-7_16
12. Dürr, C., Thang, N.K.: Nash equilibria in Voronoi games on graphs. In: Arge, L., Hoffmann, M., Welzl, E. (eds.) ESA 2007. LNCS, vol. 4698, pp. 17–28. Springer, Heidelberg (2007). https://doi.org/10.1007/978-3-540-75520-3_4
13. Fekete, S.P., Meijer, H.: The one-round Voronoi game replayed. Comput. Geom. **30**(2), 81–94 (2005)
14. Sun, X., Sun, Y., Xia, Z., Zhang, J.: The one-round multi-player discrete Voronoi game on grids and trees. Theoret. Comput. Sci. **838**, 143–159 (2020)
15. Teramoto, S., Demaine, E.D., Uehara, R.: The Voronoi game on graphs and its complexity. J. Graph Algorithms Appl. **15**(4), 485–501 (2011)

Diverse Fair Allocations: Complexity and Algorithms

Harshil Mittal[(✉)], Saraswati Nanoti, and Aditi Sethia

IIT Gandhinagar, Gujarat, India
{harshil.m,nanoti_saraswati,aditi.sethia}@iitgn.ac.in

Abstract. In this work, we initiate the study of diversity of solutions in the context of fair division of indivisible goods. In particular, we explore the notions of disjoint, distinct and symmetric allocations and study their complexity in terms of the fairness notions of *envy-freeness* and *equitability upto one item*. We show that for binary valuations, the above problems are polynomial time solvable. In contrast we show NP-hardness of disjoint and symmetric case, when the valuations are additive.

Keywords: Fair division · Diverse solutions · Disjoint allocations · Symmetric allocations

1 Introduction

Finding diverse optimal solutions is a computational task, where the aim is not only to find *one* solution, but to find *multiple* solutions that *look* sufficiently diverse from each other.

Given a computational task at hand, the decision question asks if there exists at least one solution and the search question aims at finding the solution by efficient computation. This caters to the setting where we are interested in exactly *one element* of the solution space. On the other extreme, there are counting problems, where the aim is to decide if there exist at least k many solutions, and if yes, the enumeration aims at listing all the possible solutions. This caters to the setting where we are interested in the *entire* solution space. A middle ground of these extremes is exactly where the concept of diverse solutions lie. Here, we are interested in a *subset* of solution space containing sufficiently dissimilar solutions.

Why do we need diverse solutions? Consider the classic problem finding a stable matching to pair up students for group projects. If for all the projects, the same stable matching solution is implemented, it leads to lack of collaborations, communication, and creates monotony. In such settings, a set of diverse solutions helps retain stability of the matchings while diversifying the experiences. Even in the instances where one solution suffices, the availability of diverse optimal solution allows the administrator in the real-world to assess the solutions based on their non-technical efficiencies such as such as gender diversity, environmental impacts, relocation costs, etc. and choose the one which is a better fit for the situation overall, even in terms of the parameters that are not part of the input. In recent years, many studies have been focussed at finding these diverse solutions in various contexts, for example, stable matchings [1],constrained programming [2,3], hitting sets [4], mixed integer programming [5], finding graph patterns like spanning trees, k-paths, isomorphisms [6].

In this paper, we consider diversity of solutions in the context of fair allocations of indivisible resources. Given a small list of optimal allocations, we can select one which is best for our purpose, perhaps by taking into account external factors. To the best of our knowledge, this direction has not been explored previously in this context. We say that two fair allocations are distinct if there is a good that reaches different owners under the respective allocations. A stronger notion of diversity is the pairwise disjointness of fair allocations, where every good is supposed to reach a different owner under the respective allocations. We also consider symmetric allocations for the restricted setting of two agents, where if the bundles are exchanged among the two agents, fairness is not compromised. That is, the agents not only value their own bundles, but also value the rivals bundle to such an extent that if they are to exchange the bundles, they still consider the allocations as fair. Disjoint allocation are always distinct but not vice-versa. Also, note that in the setting of two agents, a symmetric allocation is disjoint and vice versa.

We formulate the computational question as follows. For disjoint(distinct) version, one fair allocation Φ is part of the input and the goal is to check whether there exist an allocation disjoint(distinct) from Φ. While in context of symmetric allocations, there is no given base allocation and we ask whether there exist an allocation that is symmetric. Although the notions of symmetric and disjoint allocation coincide for 2 agents, but a yes-instance of symmetric version may not imply yes-instance of the disjoint version, although the other way round holds true.

The fairness notions that we consider in this work are *envy-freeness upto one good (EF1)* and *equitability upto one good (EQ1)*. Both EF1 and EQ1 always exists. [7] showed that the number of EF1 allocations is always exponential in the number of items when there are 2 agents. When there are more than 2 agents, we show that at least 2 distinct EF1 allocations can be found (Lemma 4). We present a stronger lowerbound for EQ1 where we show that atleast n distinct EQ1 allocations always exist for any number of agents (Lemma 5). For the restricted cases of binary and identical valuations respectively, we give bounds on the number of disjoint EF1 and EQ1 allocations (Lemma 1, 2, 3). Using a result from [8], we show that for binary valuations, deciding whether an EF1 allocation disjoint from a given one exists, can be done in polynomial time. On the other hand, for general additive valuations, we establish the hardness of the above problem (6, 1). Then in Sect. 5, we discuss the symmetric EF1 and EQ1 allocations for 2 agents and present polynomial time constructive algorithms for the cases when symmetric allocations exist. We defer the proof of the results marked with ⋆ to the Appendix 7. Table 1 contains a partial summary of our results.

Table 1. A partial summary of our results. The results highlighted in bold are existential.

	EF1		EQ1	
	Binary	Additive	Binary	Additive
Distinct	**P** (By implication)	**P** (Lemma 4)	**P** (By implication)	**P** (Lemma 5)
Disjoint	**P**(Lemma 6)	NP-Complete (Exact Set Cover) (Theorem 1)	?	?
Symmetric	**P**(Lemma 7)	?	P (Lemma 9)	NP-Complete (2-partition) (Theorem 2)

2 Preliminaries

For any $k \in \mathbb{N}$, let $[k] := \{1, \ldots, k\}$. A fair division instance consists of a set N of n *agents*, a set M of m *objects/goods*, and $V = (v_1, \ldots, v_n)$, a valuation profile that captures the preferences of every agent $i \in N$ over each subset of the goods in M via a valuation function $v_i : 2^M \rightarrow \mathbb{N} \cup \{0\}$. We will assume throughout that $v_i's$ satisfy *additivity* (unless otherwise stated), that is, for any agent $i \in N$ and any set of goods $S \subseteq M$, we have $v_i(S) := \sum_{j \in S} v_i(\{j\})$. The valuations are said to be normalized if $v_i(\emptyset) = 0$ and all the agents value the grand bundle (that is M), at a constant. They are monotone if for any $T \subseteq M$, $v_i(S) \leq v_i(T)$ for all $S \subseteq T$ and $i \in [n]$.

A *bundle* is any subset $S \subseteq M$ of the set of goods. An *allocation* $\Phi = (\Phi_1, \ldots, \Phi_n)$ is a partition of the set of goods into n bundles, one for each agent. We focus on complete allocations where $\bigcup_{i=1}^{n} \Phi_i = M$. Note that we will denote the bundle obtained by the agent a_i be either of the two notations Φ_i or $\Phi(a_i)$.

An allocation Φ is said to be *equitable* (*EQ*) if for every pair of agents $i, k \in N$, we have $v_i(\Phi_i) = v_k(\Phi_k)$, and *equitable up to one good* (EQ1) if for every pair of agents $i, k \in N$ such that $\Phi_k \neq \emptyset$, there exists some good $j \in \Phi_k$ such that $v_i(\Phi_i) \geq v_k(\Phi_k \setminus \{j\})$ [9,10].

An allocation Φ is said to be *envy-free* (EF) if for every pair of agents $i, k \in N$, we have $v_i(A_i) \geq v_i(A_k)$, and *envy-free up to one good* (EF1) if for every pair of agents $i, k \in N$ such that $\Phi_k \neq \emptyset$, there exists some good $j \in \Phi_k$ such that $v_i(\Phi_i) \geq v_i(\Phi_k \setminus \{j\})$ [11–13].

Two allocations Φ and Φ^* are said to be *disjoint* if for every agent a_i, the bundle she gets under Φ and Φ^* are disjoint, that is, $\Phi_i \cap \Phi_i^* = \emptyset$. Two allocations Φ and Φ^* are said to be *distinct* if for some agent a_i, the bundle she gets under Φ and Φ^* are distinct, that is, $\exists g \in [m]$ such that $\Phi^{-1}(g) \neq (\Phi^*)^{-1}(g)$. Given 2 agents, an EF1 (EQ1) allocation Φ is said to be *symmetric*, if the swapped allocation Φ^*, under which a_1 receives Φ_2 and a_2 receives Φ_1, is also EF1 (EQ1).

The computational questions that we address in this paper are as follows:

DISJOINT FAIR ALLOCATIONS
Input: A set $N = \{a_1, a_2, \ldots, a_n\}$ of agents and a set $M = \{g_1, g_2, \ldots, g_m\}$ of objects, a preference profile describing the preferences of all agents over the objects, a fair (EF1 or EQ1) allocation Φ
Question: Determine if there is an allocation Φ^* that is disjoint from Φ

DISTINCT FAIR ALLOCATIONS
Input: A set $N = \{a_1, a_2, \ldots, a_n\}$ of agents and a set $M = \{g_1, g_2, \ldots, g_m\}$ of objects, a preference profile describing the preferences of all agents over the objects, a fair (EF1 or EQ1) allocation Φ
Question: Determine if there is an allocation Φ^* distinct from Φ

SYMMETRIC FAIR ALLOCATIONS
Input: A set $N = \{a_1, a_2\}$ of two agents and a set $M = \{g_1, g_2, \ldots, g_m\}$ of objects, a preference profile describing the preferences of all agents over the objects
Question: Determine if there is a symmetric fair allocation Φ

3 Bounds on the Number of Disjoint and Distinct Allocations

Lemma 1 (Upper Bound on the Number of Pairwise Disjoint EF1 Allocations). *For any instance of the assignment problem that has $n \geq 2$ agents and $m \geq 1$ items with binary and additive valuations, it's not possible to have more than $\lfloor \frac{\alpha}{\lceil \frac{\alpha-n+1}{n} \rceil} \rfloor$ pairwise disjoint and complete EF1 allocations, where α is the maximum number of goods valued by any agent, assuming $\alpha \geq n$.*

Proof. For every agent $a_i \in N$, let $C(i)$ denote the set of items that are valued 1 by a_i. Then, α is the maximum of $|C(i)|$ over all agents a_i. We assume that $\alpha \geq n$. Let a_x be an agent for which $|C(x)| = \alpha$. Let σ be a complete EF1 allocation. As σ is EF1, $|\sigma(a_y) \cap C(x)| \leq |\sigma(a_x) \cap C(x)| + 1$ for all agents $a_y \neq a_x$. Also, as σ is complete, $\sum_{i=1}^{n} |\sigma(a_i) \cap C(x)| = |C(x)|$. Therefore, $|C(x)| \leq (n-1)(|\sigma(a_x) \cap C(x)| + 1) + |\sigma(a_x) \cap C(x)|$. That is, $|\sigma(a_x) \cap C(x)| \geq \lceil (|C(x)| - n + 1)/n \rceil$. Now, consider a collection of pairwise disjoint complete EF1 allocations, say $\{\sigma_1, \ldots, \sigma_p\}$. Due to pairwise disjointness, $|\sigma_1(a_x) \cap C(x)| + \ldots + |\sigma_p(a_x) \cap C(x)| \leq |C(x)|$. Also, for each $1 \leq j \leq p$, we know that $|\sigma_j(a_x) \cap C(x)| \geq \lceil (|C(x)| - n+1)/n \rceil$. Therefore, $p\lceil (|C(x)| - n+1)/n \rceil \leq |C(x)|$. So, as $|C(x)| \geq n$, it follows that $p \leq \lfloor \frac{\alpha}{\lceil \frac{\alpha-n+1}{n} \rceil} \rfloor$, as desired. $\square$

Lemma 2 (Counting EF1 Allocations for Identical Valuations). *For any instance of the assignment problem that has $n \geq 2$ agents and $m \geq 1$ items with identical valuations, if $m \geq n - 1$, then*

- *The number of complete EF1 allocations is at least the number of permutations of $[n]$.*
- *For every complete EF1 allocation Φ, the number of complete EF1 allocations disjoint from Φ is at least the number of derangements of $[n]$.*

Proof. To obtain the said number of EF1 allocations, we start with an arbitrary ordering on the n agents. All agents, one by one, pick an unallocated object that they value the most, according to the specified ordering. This process continues till there is any unallocated good available. We call the allocation obtained by the above Round Robin Procedure as Φ. It is easy to verify that Φ is EF1.

We now claim that any permutation of the allocation Φ corresponds to a complete EF1 allocation $\Phi^\star$, disjoint from Φ. To this end, suppose there is an agent a_i, who received $\Phi_i^\star$ under $\Phi^\star$, violates EF1 and envies the agent a_j even after removal of any good from the bundle of a_j ($=\Phi_j^\star$). That is, $u_i(\Phi_i^\star) < u_i(\Phi_j^\star \setminus \{g\}) \, \forall g \in \Phi_j^\star$. Since $\Phi^\star$ is a permutation of Φ, so $\Phi_i^\star$ and $\Phi_j^\star$ must be the allocated bundles under Φ. Suppose they are allocated to the agents a_p and a_q respectively. Then, $u_p(\Phi_i^\star) < u_p(\Phi_j^\star \setminus \{g\}) \, \forall g \in \Phi_j^\star$ as the valuations are identical and u_i and u_p are exactly the same functions. Therefore the agent a_p violates EF1 under the allocation Φ, which is a contradiction. This establishes that the number of complete EF1 allocations us at least the number of permutations of $[n]$ and the number of allocations disjoint from Φ are exactly the permutations that do not have any fixed point (which are exactly the number of derangements of $[n]$). □

Lemma 3 (($\star$) Counting Disjoint EQ1 Allocations for Identical Valuations). *For any instance of the assignment problem that has $n \geq 2$ agents and $m \geq 1$ items with identical valuations, if $m \geq n - 1$, then*

- *The number of complete EQ1 allocations is at least the number of permutations of $[n]$.*
- *For every complete EQ1 allocation Φ, the number of complete EQ1 allocations disjoint from Φ is at least the number of derangements of $[n]$.*

4   Computing Disjoint and Distinct Allocations

Lemma 4 (Existence of Two Complete and Distinct EF1 Allocations). *Every instance of the assignment problem that has $n \geq 2$ agents and $m \geq 1$ items with additive valuations admits two distinct allocations that are EF1 and complete.*

Proof. The proof proceeds by induction on m. If $1 \leq m \leq n$, we are done by simply taking any two distinct complete allocations that allocate at most 1 item to each agent. Assume that the claim is true for $1 \leq m \leq p - 1$, for some $p \geq n + 1$. Let's argue that it's true for $m = p$. Consider an instance with n agents (say $a_1, \ldots, a_n$) and p goods.
Case 1. Suppose there exist two agents (say a_i and a_j) who have a common most preferred item. Denote the most preferred good common to a_i and a_j by g. Now consider the following runs of Round Robin algorithm:

1. In Run 1, the agents are sorted such that a_i gets the first turn and further, a_i chooses g in this turn.
2. In Run 2, the agents are sorted such that a_j gets the first turn and further, a_j chooses g in this turn.

Clearly, the allocations returned by Run 1 and Run 2 are distinct and complete EF1 allocations, as desired.

Case 2. Now, suppose that for any pair of agents, the set of their most preferred items are disjoint. Let g_i be (one of) the most preferred item(s) of a_i. Note that all g_i's are distinct because of the case we are in. Project the instance by removing $g_1, \ldots, g_n$. As $p \geq n + 1$, at least one good remains. By the induction hypothesis, there exist two distinct EF1 allocations for the projected instance. Denote these allocations by Φ_1 and Φ_2. Now, obtain allocations $\Phi_1^\star$ and $\Phi_2^\star$ for the original instance by taking the allocations Φ_1 and Φ_2 (respectively), and then extending them by giving g_i to a_i for all $i \in [n]$. Clearly allocations $\Phi_1^\star$ and $\phi_2^\star$ are distinct (by the induction hypothesis). We now show that the allocation $\Phi_1^\star$ is EF1. Consider agents a_i and a_j. Suppose their bundles are $\Phi_{1i}^\star = X \cup \{g_i\}$ and $\Phi_{1j}^\star = Y \cup \{g_j\}$. Since Φ_1 is an EF1 allocation, we know that either there exists g such that $u_i(Y \setminus \{g\}) \leq u_i(X)$ or $u_i(Y) \leq u_i(X)$. In either case, since $u_i(g_i) \geq u_i(g_j)$, we have that a_i does not have envy towards a_j (upto one good) wrt the allocation $\Phi^\star$. Similarly, a_j does not envy a_i (upto one good) wrt the allocation $\phi_2^\star$. A similar argument shows that $\Phi^\star$ is also EF1. □

Lemma 5 (Existence of n Complete and Distinct EQ1 Allocations).
Every instance of the assignment problem that has $n \geq 2$ agents and $m \geq 1$ items with normalized monotone valuations admits n distinct allocations that are EQ1 and complete.

Proof. We proceed by induction on the number of goods. Let $a_1, a_2, \ldots a_n$ be the n agents. If $1 \leq m \leq n$, then the allocations under which every agent gets at most one good are all EQ1 allocations, and clearly there are at least n such allocations. For instance, let us fix an object g_1. We now describe n distinct EQ1 allocations $\Phi_1, \Phi_2, \Phi_3 \ldots, \Phi_n$. In Φ_i, allocate the item g_1 to agent a_i, and the remaining items arbitrarily among the remaining agents in such a way that every agent gets at most one item. We remark here that if $m = 1$, then there are exactly n such distinct EQ1 allocations.

Now we argue the case where $m > n$ using induction on m. Assume that the claim is true for $1 \leq m \leq p - 1$, for some $p \geq n + 1$. Now consider the case when $m = p$. Then by induction hypothesis, there exist n distinct EQ1 (partial) allocations $\Phi_1, \Phi_2, \Phi_3 \ldots, \Phi_n$ that allocate the object $\{g_1, g_2, \ldots, g_{p-1}\}$ among the n agents. Consider the unallocated good g_p under the above allocations. For each Φ_i, allocate g_p to the least happy agent a under Φ_i, resulting in a complete allocation. Notice that the completion does not violate equitability upto one good. Indeed, a is the least happy agent, so everyone else's utility is at least as much as that of a. So if $g_p's$ allocation to a makes any other agent unhappy, then also, she regains equitability after the removal of at most one good, that is, g_p itself. Hence the completed allocations are EQ1. This proves the result. □

Now we define a *matroid*, which we use in the following result. For any set $\mathcal{U}$ and a family $\mathcal{F} \subseteq 2^{\mathcal{U}}$ of subsets of $\mathcal{U}$, we say that the pair $(\mathcal{U}, \mathcal{F})$ is a *matroid on* $\mathcal{U}$ if the following conditions hold: i) The empty set belongs to $\mathcal{F}$. ii) For any set $A \in \mathcal{F}$, every subset of A belongs to $\mathcal{F}$. iii) For any sets $A, B \in \mathcal{F}$, if $|A| > |B|$, then there exists an element $u \in A \setminus B$ such that $B \cup \{u\}$ belongs to $\mathcal{F}$.

Lemma 6 (Disjoint EF1 Allocations for Binary Valuations). *The* DISJOINT ALLOCATION *problem is in* P *for binary and additive valuations.*

Proof. For every $1 \le i \le n$, let $\mathcal{M}_i$ denote the partition matroid on M whose categories are $M \setminus \Phi(a_i)$ and $\Phi(a_i)$, with capacities $|M \setminus \Phi(a_i)|$ and 0 respectively. That is, $\mathcal{M}_i$ is the matroid on M whose family consists of all subsets of $M \setminus \Phi(a_i)$. Note that an EF1 allocation σ is disjoint from the base allocation Φ if and only if σ is feasible under the matroid constraints specified by $\mathcal{M}_1, \ldots, \mathcal{M}_n$, i.e., for every $1 \le i \le n$, the bundle allotted to a_i under σ belongs to the family of $\mathcal{M}_i$. As shown in [8] (Theorem 4.1, Sect. 4), a feasible EF1 allocation always exists for the setting of partition matroids and binary additive valuations, and it can be found in polynomial time. □

Theorem 1 (Disjoint EF1 Allocations for General Valuations). *The* DISJOINT ALLOCATION *problem is NP-complete for general additive valuations.*

Proof. Let $\mathcal{I} := (U, \mathcal{F}, p)$ be an instance of EXACT COVER BY 3-SETS where, U is a universe of $3p$ elements, $\mathcal{F}$ is a family of 3-sized subsets of U such that $|\mathcal{F}| = t$. The problem is to decide if there exist p sets in $\mathcal{F}$, whose union is U. We create the disjoint allocation instance as follows.

- We create t set agents $\{a_1, a_2, \ldots a_t\}$ corresponding to the sets in $\mathcal{F}$, and one special agent s.
- We introduce $(t+1)$ many identity items for each agent in $\{a_1, a_2, \ldots a_t\}$. Precisely, for each $i \in t$, we introduce the items $\{g_i^j \mid j \in [t+1]\}$.
- We introduce $3p$ universe items $\{g_1, g_2, \ldots g_{3p}\}$ corresponding to the elements in the universe.
- We also introduce $(t-p)$ many pacifier items $\{q_1, q_2, \ldots q_{t-p}\}$.

A set agent a_i values all its identity items $\{g_i^1, g_i^2, \ldots g_i^{t+1}\}$ and all the pacifier items $\{q_1, q_2, \ldots q_{t-p}\}$ at 2 each. She values the three universe items in its corresponding set S_i at $\frac{2}{3}$ each. The special agent s values all the items at 0 each. We define one EF1 allocation Φ as follows.

$$\Phi(a) = \begin{cases} \{g_i^1, g_i^2, \ldots g_i^{t+1}\}, & \text{if } a = a_i \text{ for some } i \in [t] \\ \{q_1, q_2, \ldots q_{t-p}\} \cup \{g_1, g_2, \ldots g_{3p}\} & \text{if } a = s \end{cases}$$

Notice that Φ is indeed an EF1 allocation.

- The set agent a_i does not envy any other set agent a_j, as $u_{a_i}(\Phi(a_i)) = 2(t+1)$ and $u_{a_i}(\Phi(a_j)) = 0$
- The special agent s does not envy any a_i, as $u_s(\Phi(s)) = 0 = u_s(\Phi(a_i))$

– The set agent a_i do not envy the special agent s, as $u_{a_i}(\Phi(a_i)) = 2(t+1)$ and $u_{a_i}(\Phi(s)) = 2 + 2(t-p)$

This completes the construction of the reduced instance. We defer the equivalence argument to the Appendix 7. We demonstrate the reduction with the help of an example below (Fig. 1). □

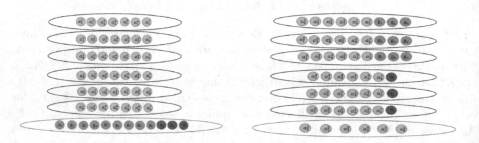

Fig. 1. Consider an instance of EXACT COVER BY 3-SETS where $U = \{1, 2, \ldots, 9\}$ and $\mathcal{F} = \{S_1, S_2, \ldots S_6\}$ where $S_1 = \{1, 2, 3\}$, $S_2 = \{4, 5, 6\}$, $S_3 = \{7, 8, 9\}$, $S_4 = \{2, 3, 4\}$, $S_5 = \{3, 4, 5\}$ and $S_6 = \{3, 5, 9\}$. Here $\{S_1, S_2, S_3\}$ forms an exact cover of U. In the reduced instance, we have 6 set agents and 7 set items corresponding to each set agent. Each set agent a_i for $i \in [1, 6]$ values items a_i^j for $j \in [1, 7]$ at 2 each and other set items at 0 each. There are 3 pacifier items q_1, q_2 and q_3 are valued at 2 by each set agent. There are 9 universe items. Each set agent values the universe items corresponding to her set at $\frac{2}{3}$ each and all other universe items at 0. The special agent values all the items at 0. The ovals denote the bundles of all the agents corresponding to the allocation Φ (LHS) and Φ' (RHS) from top to bottom respectively.

We now discuss a restricted setting of two agents, and the aim is to find whether there exists symmetric fair allocation.

5   Symmetric Allocations

Lemma 7 (Symmetric EF1 for Binary Valuations). *For $n = 2$ and additive 0/1 valuations, there always exists a symmetric complete EF1 allocation and can be found in polynomial time.*

Proof. Let a_1 and a_2 denote the two agents. Let X denote the set of items that are valued 1 by a_1 and 0 by a_2. Let Y denote the set of items that are valued 1 by a_2 and 0 by a_1. Let Z denote the set of items that are valued 1 by both a_1 and a_2. Consider the following complete allocation Φ: Allocate any $\lceil |X|/2 \rceil$ items of X to a_1, and the remaining $\lfloor |X|/2 \rfloor$ items of X to a_2. Allocate any $\lfloor |Z|/2 \rfloor$ items of Z to a_1, and the remaining $\lceil |Z|/2 \rceil$ items of Z to a_2. Allocate any $\lceil |Y|/2 \rceil$ items of Y to a_1, and the remaining $\lfloor |Y|/2 \rfloor$ items of Y to a_2. Here,

– a_1 values her bundle at $u_{11} := \lceil |X|/2 \rceil + \lfloor |Z|/2 \rfloor$
– a_1 values $a_2's$ bundle at $u_{12} := \lfloor |X|/2 \rfloor + \lceil |Z|/2 \rceil$

- a_2 values her bundle at $u_{22} := \lfloor |Y|/2 \rfloor + \lceil |Z|/2 \rceil$
- a_2 values a'_1s bundle at $u_{21} := \lceil |Y|/2 \rceil + \lfloor |Z|/2 \rfloor$

Note that u_{11} and u_{12} differ by at most 1. Also, u_{22} and u_{21} differ by at most 1. So when the bundles are swapped, then also, the resulting allocation respects EF1. Hence, Φ is a symmetric EF1 allocation.   □

Observe that for $n = 2$, $m = 3$, and additive valuations, there always exist a symmetric complete EF1 allocation. To see this, let a_1 and a_2 denote the two agents. Let $a(.)$ and $b(.)$ denote their valuation functions respectively. Let $\{g_1, g_2, g_3\}$ denote the three items. WLOG, assume $a(g_1) \geq a(g_2) \geq a(g_3)$. If $b(g_1) \geq b(g_2)$ or $b(g_1) \geq b(g_3)$), then the allocation where a_1 gets g_1 and a_2 gets $\{g_2, g_3\}$ is the required symmetric EF1 allocation. On the other hand, if $b(g_1) < b(g_2)$ and $b(g_1) < b(g_3)$, then, the allocation where a_1 gets g_2 and a_2 gets $\{g_1, g_3\}$ is the required allocation.

In contrast to the fact that symmetric EF1 allocations always exists for binary valuations, symmetric EQ1 allocations may not exist even with small number of goods in the binary setting.

Lemma 8 ((⋆) Symmetric EQ1 for Binary Valuations (non-existence)). *For $n = 2$, $m = 4$ and additive 0/1 valuations, there are instances where no complete EQ1 allocation is symmetric.*

Whenever an EQ1 allocation exists, it can be found in polynomial time, as suggested by the following lemma. The proof is based on a case analysis.

Lemma 9 (Symmetric EQ1 for Binary Valuations). *For $n = 2$ additive 0/1 valuations, a symmetric EQ1 allocation, if one exists, can be found in polynomial time.*

Proof (Idea). Consider two agents a_1 and a_2 and a set of goods $\{g_1.g_2, \ldots g_m\}$. Let X denote the set of items valued at 1 by a_1 and 0 by a_2. Let Y denote the set of items valued at 0 by a_1 and 1 by a_2. Let Z be the set of items valued at 1 by both a_1 and a_2. We argue the correctness of the claim by a case analysis depending on the cardinalities of the sets X, Y and Z. We will show that whenever $|X| \leq |Y| + 2$, a symmetric EQ1 allocation always exists, irrespective of the cardinality of Z. When $|X| \geq |Y| + 3$, we argue, by contradiction that such an allocation can not exist. For details of the proof, we defer the reader to the Appendix 7.   □

We now turn to the case of additive valuations for EF1. Let a_1 and a_2 be the two agents. Suppose that each of them orders the items in descending order of their valuations. We say that a pair of items forms an *inversion* if their relative order in a'_1s ordering is different from that in a'_2s ordering. Depending on the number of inversion in the input instance, we now identify the scenarios where a symmetric EF1 allocation always exists.

Lemma 10 ((⋆) Symmetric EF1 for Additive Valuations). *If there is atmost 1 inversion, then there always exists a symmetric complete EF1 allocation and can be found in polynomial time.*

Lemma 11 ((⋆) Symmetric EF1 for Additive Valuations). *If there are even number of goods and* $\binom{m}{2} - 1$ *inversions, then there always exists a symmetric EF1 allocation and can be found in polynomial time.*

Beyond the special cases above, we leave the complexity for finding symmetric EF1 allocations open. However, we show that finding symmetric EQ1 allocations is intractable for additive valuations. We note that the hard instances generated by our reduction do not have normalized utilities, and we leave open the issue of whether the problem is hard for this special case as well.

Theorem 2 (Symmetric EQ1 for Additive Valuations). *The problem of determining whether a symmetric EQ1 allocation exists is NP-complete for additive valuations.*

Proof. Given an allocation Φ, we can check whether it is EQ1 in polynomial time and we can again swap the bundles and check whether the allocation is EQ1. Thus, this problem belongs to the class NP.

For proving that it is NP-hard, we use a reduction from the 2-partition problem. Consider a typical instance of two partition problem where S is a set of positive integers and we need to find whether there exists a subset $A \subset S$ such that $\sum_{x \in A} x = \sum_{x \in S \setminus A} x$.

We create an instance of symmetric EQ1 allocation as follows: Create two agents a_1 and a_2. Let $S = \{x_1, x_2, \ldots, x_m\}$ for some $m \in \mathbb{N}$, let $\sum_{x \in S} x = s$. We create set goods $g_1, g_2, \ldots, g_m$ such that each g_i is valued at x_i by both a_1 and a_2. Create two dummy goods d_1 and d_2 such that both of them are valued at a_1 by 0 and at $s + 1$ by a_2. We show that S has a 2-partition if and only if this instance has a disjoint EQ1 allocation.

Forward Direction. Suppose there exists a subset A of S such that $\sum_{x \in A} x = \sum_{x \in S \setminus A} x = s/2$. Consider the allocation Φ where

$$\Phi(a) = \begin{cases} \{g_i \text{ such that } x_i \in A\} \cup \{d_1\} \text{ if } a = a_1 \\ \{g_i \text{ such that } x_i \in S \setminus A\} \cup \{d_2\} \text{ if } a = a_2 \end{cases}$$

In this allocation a_1 values her bundle at $s/2$ and a_2 values her bundle at $3s/2 + 1$. After removing d_2 from the bundle of a_2, she values it at $s/2$. Therefore, this allocation is EQ1. After swapping the bundles, a_1 values her bundle at $s/2$ and a_2 values her bundle at $3s/2 + 1$. After removing d_2 from the bundle of a_2, she values it at $s/2$. Therefore, again the allocation is EQ1. Thus Φ is a symmetric EQ1 allocation. This completes the proof of the forward direction.

Reverse Direction. Suppose we have a YES instance of disjoint EQ1 allocation. Let this allocation be Φ.

Suppose Φ allots both the dummy goods to a_2, then even after removing one good from the bundle of a_2, she will value her bundle at value at least $s + 1$. Agent a_1 cannot get a bundle of value $s + 1$ (according to a_1) even if she gets all the set goods. This contradicts the fact that Φ is an EQ1 allocation. Now if Φ allocates both the dummy goods to a_1, after swapping the bundles both the goods will go to a_2. So this allocation will not be an EQ1 allocation as shown above. This contradicts the fact that Φ is a symmetric EQ1 allocation.

Therefore Φ must allocate one dummy good to a_1 and one dummy good to a_2. Without loss of generality we assume that Φ allocates d_1 to a_1 and d_2 to a_2.

Let the set of all set goods allocated to a_1 by the allocation Φ be denoted by A and the set of all set goods allocated to a_2 by the allocation Φ be denoted by B. Consider $A' = \{x_i\}$ such that $g_i \in A$ for $i \in [1, m]$ and $B' = \{x_i\}$ such that $g_i \in B$ for $i \in [1, m]$. Since each set good is allocated to exactly one agent, it can be seen that A' and B' is a partition of S.

Observe that

$$\sum_{x_i \in A'} x_i = \sum_{g_i \in A} \Phi_{a_1}(g_i) = \sum_{g_i \in A} \Phi_{a_2}(g_i)$$

and

$$\sum_{x_i \in B'} x_i = \sum_{g_i \in B} \Phi_{a_2}(g_i) = \sum_{g_i \in B} \Phi_{a_1}(g_i).$$

Therefore to show that we have a YES instance of 2-partition, it is sufficient to show that

$$\sum_{g_i \in A} \Phi_{a_1}(g_i) = \sum_{g_i \in A} \Phi_{a_2}(g_i) - \sum_{g_i \in B} \Phi_{a_2}(g_i) - \sum_{g_i \in B} \Phi_{a_1}(g_i) \qquad (1)$$

Since in Φ the good d_2 is assigned to a_2, she will value her bundle at a value of at least $s + 1$. Even if a_1 gets all the set goods (i.e. $A = S$) she will value her bundle at a value s. Since we know that Φ is EQ1, we can now say that value of a_1's bundle according to a_1 is greater than or equal to the value of a_2's bundle after removing d_2 according to a_2. Thus value of goods in A according to a_1 is greater than or equal to the value of goods in B according to a_2. Thus we have the following:

$$\sum_{g_i \in A} \Phi_{a_2}(g_i) = \sum_{g_i \in A} \Phi_{a_1}(g_i) \geq \sum_{g_i \in B} \Phi_{a_2}(g_i) = \sum_{g_i \in B} \Phi_{a_1}(g_i) \qquad (2)$$

Now the allocation obtained after swapping the bundles in Φ allots d_1 to a_2. Thus she will value her bundle at a value of at least $s + 1$. Even if a_1 gets all the set goods (i.e. $B = S$) she will value her bundle at a value s. Since we know that Φ is EQ1, we can now say that value of a_1's bundle according to a_1 is greater than or equal to the value of a_2's bundle after removing d_2 according to a_2. Thus value of goods in B according to a_1 is greater than or equal to the value of goods in A according to a_2. Thus we have the following:

$$\sum_{g_i \in B} \Phi_{a_2}(g_i) = \sum_{g_i \in B} \Phi_{a_1}(g_i) \geq \sum_{g_i \in A} \Phi_{a_2}(g_i) = \sum_{g_i \in A} \Phi_{a_1}(g_i) \qquad (3)$$

Using 2 and 3, we get that 1 is satisfied and thus we have a YES instance of 2-partition. This completes the proof of the reverse direction. We refer Fig. 2 in the appendix for an example.   □

6   Conclusion and Future Directions

While we show that for any number of agents, there are at least 2 distinct EF1 allocations, we do not claim the result to be tight, and it would be interesting to get the upper-bounds in this context. The complexity of disjoint EQ1 allocations remains open. We settle the symmetric EQ1 question both in binary and additive setting, however complexity of symmetric EF1 is open. It would also be interesting to extend the scope of symmetric allocations to more than 2 agents.

Acknowledgement. The authors thank Neeldhara Misra for useful discussions.

7   Appendix

Proof (of Lemma 3). The proof idea is similar to that of Lemma 2. We start by an arbitrary EQ1 allocation obtained by the following procedure. The least happy agent comes first and picks an object she values the most. This procedure is repeated until all the objects are allocated. It is easy to verify that the final allocation Φ is indeed complete and EQ1. Now any permutation of Φ is also EQ1 by the similar argument as in Lemma 2 and this proves the result.   □

Proof (of Theorem 1). We argue here the equivalence of the reduction.

Forward Direction. Suppose $\mathcal{I}$ is a yes instance. Let $S = \{S_1, S_2, \ldots S_p\}$ be the exact cover. We construct a disjoint allocation Φ' as follows.

$$\Phi'(a) = \begin{cases} S_i \cup \{g_1^i, g_2^i, \ldots g_t^i\} \setminus \{a_i^i\} & \text{if } a = a_i \text{ such that } i \leq p \\ q_{i-p} \cup \{g_1^i, g_2^i, \ldots g_t^i\} \setminus \{g_i^i\} & \text{if } a = a_i \text{ such that } i > p \\ \{g_i^i, g_i^{t+1} \ \forall \ i \in [t]\} & \text{if } a = s \end{cases}$$

Notice that Φ' is disjoint from Φ. To see Φ' is indeed EF1, notice that

- The set agent $\{a_i : i \leq p\}$ does not envy any other set agent $\{a_j : j \leq p\}$, as $u_{a_i}(\Phi'(a_i)) = 2 = u_{a_i}(\Phi'(a_j))$
- The set agent $\{a_i : i \leq p\}$ envies $\{a_j : j > p\}$ upto one good, as $u_{a_i}(\Phi'(a_i)) = 2$ but $u_{a_i}(\Phi'(a_j)) = 4$
- The special agent s does not envy any a_i, as $u_s(\Phi'(s)) = 0 = u_s(\Phi'(a_i))$
- a_i envies s up to one good, as $u_{a_i}(\Phi'(a_i)) = 2 = u_{a_i}(\Phi'(s) \setminus \{g_i^{t+1}\}) \leq u_{a_i}(\Phi'(s)) = 4$

Reverse Direction. Suppose Φ' is an EF1 allocation, disjoint from Φ. Consider any set agent a_i. Under Φ', a_i can not get any of its $t + 1$ identity items. This implies that these $t + 1$ items must be allocated among the remaining t agents. By pigeonholing, one of the remaining agents, say x, must get at least two of the identity items of a_i. Then $u_{a_i}(\Phi'(x)) \geq 4$. As Φ' is EF1, it must happen that $u_{a_i}(\Phi'(a_i)) \geq 2$. The only way this can happen is either a_i gets a pacifier item q_i or gets all the items in its corresponding set S_i. Since this is true for every set agent $\{a_1, a_2, \ldots a_t\}$, and there are only $t - p$ of the pacifier items, therefore, p of the set agents, say $\{a_{i_1}, a_{i_2}, \ldots a_{i_p}\}$, must get their corresponding sets $\{S_{i_1}, S_{i_2}, \ldots S_{i_p}\}$. Since $|U| = 3p$, therefore $\{S_{i_1}, S_{i_2}, \ldots S_{i_p}\}$ must form an exact cover of U. Hence, $\mathcal{I}$ is a yes instance. $\square$

Proof (of Lemma 8). Let a_1 and a_2 be two agents. Let $a_1's$ valuation for the four goods $\{g_1, g_2, g_3, g_4\}$ be given by the vector $[1, 0, 0, 0]$ and $a_2's$ valuation be $[1, 1, 1, 1]$. Notice that under any EQ1 allocation, a_2 can not receive all the four goods. Also, if a_1 gets all the four goods under an allocation Φ, then the allocation $\Phi^\star$ obtained by swapping the bundles is not EQ1, therefore, Φ can not be symmetric. So both the agents must get at least one good under any symmetric EQ1 allocation. Now consider the following cases:

- Suppose under the EQ1 allocation Φ, a_1 gets g_1. Then she must also get one of the three remaining goods else, $u(a_2) = 3$ which will violate EQ1. Say, a_1 gets $\{g_1, g_2\}$, WLOG, and a_2 gets $\{g_3, g_4\}$. But notice that a_1 value $\{g_3, g_4\}$ at 0 and hence will violate EQ1 if the bundles are swapped. Therefore Φ is not symmetric.
  Also, if a_1 happened to receive $\{g_1, g_2, g_3\}$ under Φ, then although, Φ remains EQ1 as both the agents value their respective bundles at one, but if the bundles are swapped, then a_1 derives 0 value, while a_2 derives the value of 3, therefore violating EQ1.
- Suppose under the EQ1 allocation Φ, a_1 does not get g_1. Then she must get all the remaining 3 goods in order for Φ to be EQ1. But then under the allocation $\Phi^\star$ obtained by swapping the bundles, $u_{\Phi^\star}(a_1) = 1$ and $u_{\Phi^\star}(a_2) = 3$, violating EQ1. Therefore, Φ is not symmetric.

This concludes the argument. $\square$

Proof (of Lemma 9). Consider two agents a_1 and a_2 and a set of goods $\{g_1, g_2, \ldots g_m\}$. Let X denote the set of items valued at 1 by a_1 and 0 by a_2. Let Y denote the set of items valued at 0 by a_1 and 1 by a_2. Let Z be the set of items valued at 1 by both a_1 and a_2. We argue the correctness of the claim by a case analysis depending on the cardinalities of the sets X, Y and Z. Notice that without loss of generality we can assume that $|X| \geq |Y|$. (If that is not the case, then we can interchange the agents a_1 and a_2 so that we now have $|X| > |Y|$.)

Also let $z' = \frac{|Z|}{2}$. Let Z_1 denote a subset of Z of size $\lfloor z' \rfloor$ and Z_2 denote $Z \setminus Z_1$. Note that the size of Z_2 will be $\lceil z' \rceil$.

- $|X| \leq |Y| + 1$. We claim that in this case , a symmetric EQ1 allocation always exist, and can be computed as follows:
  Consider the following allocation Φ:

$$\Phi(a) = \begin{cases} X \cup Z_1 & a = a_1, \\ Y \cup Z_2 & a = a_2, \end{cases}$$

If $|Z|$ is even, the agent a_1 values her bundle at $|X| + \frac{|Z|}{2}$ and the agent a_2 values her bundle at $|Y| + \frac{|Z|}{2}$. Since $|X| \leq |Y| + 1$, the value of the bundle of a_1 (according to a_1) is at most 1 more than the value of the bundle of a_2 (according to a_2). Therefore, the value of the bundle of a_1 after removing one good is at most as much as the bundle of a_2. Thus this allocation is EQ1.

For showing that it is symmetric, we need to show that the allocation obtained after swapping the bundles of agents a_1 and a_2 is also EQ1.

In the new allocation, both a_1 and a_2 value their bundles at $\frac{|Z|}{2}$ each. Thus it is also EQ1.

Now consider the case where $|Z|$ is odd. Thus $|Z_1| = |Z_2| - 1$. The agent a_1 will value her bundle at $|X| + |Z_1|$ and the agent a_2 values her bundle at $|Y| + |Z_2|$. If $|X| = |Y|$, then the valuation of agent a_2 for her bundle will be one more than the valuation of agent a_1 for her bundle. But after removing one item from the bundle of a_2, both agents will have equal valuation for their respective bundle (according to themselves). If $|X| = |Y| + 1$, then both the agents will have an equal valuation for their bundle. Hence this is an EQ1 allocation.

After swapping the bundles, a_1 will value her bundle at $|Z_2|$ and a_2 will value her bundle at $|Z_1|$. Since we have $|Z_1| = |Z_2| - 1$, the value assigned by a_1 to her bundle is one more than the value assigned by a_2 to her bundle. Thus, after removing one good from the bundle of a_2, both agents will have same value for their bundle. Thus it is an EQ1 allocation even after swapping the bundles.

So we have shown that Φ is a symmetric EQ1 allocation.

- $|X| = |Y| + 2$. We claim that in this case, a symmetric EQ1 allocation always exists, and can be computed as follows: If $|Z|$ is even, consider the following allocation Φ:

$$\Phi(a) = \begin{cases} \{|Y| + 1 \text{ many goods from } X\} \cup Z_1 & a = a_1, \\ Y \cup Z_2 \cup \{1 \text{ good from } X\} & a = a_2, \end{cases}$$

The value assigned by a_1 to her bundle will be $|Y| + 1 + |Z_1|$ and the value assigned by a_2 to her bundle will be $|Y| + |Z_2|$. Since $|Z_1| = |Z_2|$ as $|Z|$ is even, the value assigned by a_1 to her bundle is one more than the value assigned by a_2 to her bundle. Therefore, after removing one good from the bundle of a_1, both the agents assign the same value to their respective bundles. Thus Φ is an EQ1 allocation.

After swapping the bundles of the two agents, the value assigned by a_1 to her bundle is $|Z_2| + 1$ and the value assigned by a_2 to her bundle is $|Z_1|$. Thus after removing one good from the bundle of a_1, both the agents assign the same value to their respective bundles. Thus we get an EQ1 allocation even after swapping the bundles.

Hence we have shown that Φ is a symmetric EQ1 allocation.

If $|Z|$ is odd, consider the following allocation Φ:

$$\Phi(a) = \begin{cases} X \cup Z_1 & a = a_1, \\ Y \cup Z_2 & a = a_2, \end{cases}$$

Since $|Z|$ is odd, we have $|Z_2| = |Z_1| - 1$. Now the value assigned by a_1 to her bundle is $|X| + |Z_1|$ which is equal to $|Y| + |Z_2| + 1$. The value assigned by a_2 to her bundle will be $|Y| + |Z_2|$. So, after removing one good from the bundle of a_1, both the agents will assign the same value to their respective bundles. Thus the allocation Φ is EQ1.

After exchanging the bundles, the agent a_1 will value her bundle at $|Z_2|$ and the agent a_2 will value her bundle at $|Z_1|$. Thus, after removing one good from the bundle of a_1, both the agents will assign the same value to their respective bundles. Thus the allocation obtained after the exchange is also EQ1.

Hence the allocation Φ is a symmetric EQ1 allocations.

$-$ $|X| \geq |Y| + 3$. We claim that in this case, a symmetric EQ1 allocation does not exist, as argued below. For the sake of contradiction, assume that there exists a symmetric EQ1 allocation, say Φ. Let x, y and z denote the number of items from X, Y and Z respectively allocated to a_1 under Φ. Note that under Φ,

- a_1 values her own bundle at $x + z$.
- a_2 values her own bundle at $|Y| - y + |Z| - z$.

Since Φ is EQ1, we have $x + z \leq (|Y| - y + |Z| - z) + 1$.
That is,

$$x + y + 2z - |Z| \leq |Y| + 1 \tag{4}$$

Also, under the swapped allocation, say Φ^*,

- a_1 values her own bundle at $|X| - x + |Z| - z$.
- a_2 values her own bundle at $y + z$.

Since Φ^* is EQ1, we have $|X| - x + |Z| - z \leq (y + z) + 1$.
That is,

$$x + y + 2z - |Z| \geq |X| - 1 \tag{5}$$

Using inequalities (1) and (2), we get $|X| \leq |Y| + 2$, a contradiction.

Proof (of Lemma 10). Notice that for any pair of items X and Y that do not appear consecutively in a_1's valuation, the relative order of X and Y is the same in a_1's and a_2's valuations. This is true because there's only one inversion.

$a_1 : g_m \geq g_{m-1} \geq g_{m-2} \geq \cdots \geq g_3 \geq g_2 \geq g_1$

$a_2 :$ any of the $(m-1)$ valuations that have 1 inversion

The following is a symmetric EF1 allocation:

a_1 gets $g_m, g_{m-3}, g_{m-4}, g_{m-7}, g_{m-8}, \cdots$

a_2 gets $g_{m-1}, g_{m-2}, g_{m-5}, g_{m-6}, g_{m-9}, \cdots$

This is because in a_2's (and also a_1's) valuation,

- $g_m \geq g_{m-2}$, $g_{m-3} \geq g_{m-5}$, $g_{m-4} \geq g_{m-6}$, $g_{m-7} \geq g_{m-9}, \cdots$ and
- $g_{m-1} \geq g_{m-3}$, $g_{m-2} \geq g_{m-4}$, $g_{m-5} \geq g_{m-7}$, $g_{m-6} \geq g_{m-8}, \cdots$ so on. $\square$

Proof (of Lemma 11). Notice that for any pair of items X and Y that do not appear consecutively in a_1's valuation, the relative orders of X and Y are different in a_1's and a_2's valuations - this is true because there are $\binom{m}{2} - 1$ inversions.

$a_1 : g_m \geq g_{m-1} \geq g_{m-2} \geq \ldots \geq g_3 \geq g_2 \geq g_1$

$a_2 :$ any of the $(m-1)$ valuations that have $\binom{m}{2} - 1$ inversions

The following is a symmetric EF1 allocation:

a_1 gets $g_m, g_{m-3}, g_{m-4}, g_{m-7}, g_{m-8}, \cdots$

a_2 gets $g_{m-1}, g_{m-2}, g_{m-5}, g_{m-6}, g_{m-9}, \cdots$

This is because

- In a_1's valuation,
  - $g_m \geq g_{m-2}, g_{m-3} \geq g_{m-5}, g_{m-4} \geq g_{m-6}, g_{m-7} \geq g_{m-9}, \cdots$ and
  - $g_{m-1} \geq g_{m-3}, g_{m-2} \geq g_{m-4}, g_{m-5} \geq g_{m-7}, g_{m-6} \geq g_{m-8}, \cdots$
- In a_2's valuation,
  - $g_{m-2} \geq g_m, g_{m-5} \geq g_{m-3}, g_{m-6} \geq g_{m-4}, g_{m-9} \geq g_{m-7}, \cdots$ and
  - $g_{m-3} \geq g_{m-1}, g_{m-4} \geq g_{m-2}, g_{m-7} \geq g_{m-5}, g_{m-8} \geq g_{m-6}, \cdots$

□

Proof. (Example for Theorem 2)

□

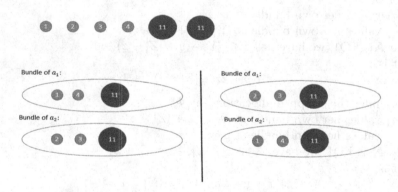

Fig. 2. Consider an instance of 2-partition with $S = \{1, 2, 3, 4\}$. We create set goods g_1, g_2, g_3 and g_4 such that both a_1 and a_2 value them at $1, 2, 3$ and 4 respectively. These are shown by the dark pink circles labelled $1, 2, 3$ and 4 respectively. We create two dummy goods d_1 and d_2 which are valued at 11 each by agent a_1 and at 0 each by agent a_2. These are denoted by green circles labelled 11. It can be verified that both the allocations shown (the allocation on the right is obtained by swapping the one on the left) are disjoint and EQ1. (Color figure online)

References

1. Ganesh, A., Vishwa Prakash, H., Nimbhorkar, P., Philip, G.: Disjoint stable matchings in linear time. In: International Workshop on Graph-Theoretic Concepts in Computer Science, pp. 94–105 (2021)
2. Hebrard, E., Hnich, B., O'Sullivan, B., Walsh, T.: Finding diverse and similar solutions in constraint programming. In: AAAI, pp. 372–377 (2005)
3. Petit, T., Trapp, A.C.: Finding diverse solutions of high quality to constraint optimization problems. In: Twenty-Fourth International Joint Conference on Artificial Intelligence (2015)
4. Baste, J., Jaffke, L., Masařík, T., Philip, G., Rote, G.: FPT algorithms for diverse collections of hitting sets. Algorithms 12(12), 254 (2019)
5. Danna, E., Woodruff, D.L.: How to select a small set of diverse solutions to mixed integer programming problems. Oper. Res. Lett. 37(4), 255–260 (2009)
6. Hanaka, T., Kobayashi, Y., Kurita, K., Otachi, Y.: Finding diverse trees, paths, and more. In: Proceedings of the AAAI Conference on Artificial Intelligence, pp. 3778–3786 (2021)
7. Suksompong, W.: On the number of almost envy-free allocations. Discret. Appl. Math. 284, 606–610 (2020)
8. Dror, A., Feldman, M., Segal-Halevi, E.: On fair division under heterogeneous matroid constraints. In: Proceedings of the Thirty-Fifth AAAI Conference on Artificial Intelligence, AAAI, pp. 5312–5320. AAAI Press (2021)
9. Dubins, L.E., Spanier, E.H.: How to cut a cake fairly. Am, Math. Monthly 68(1P1), 1–17 (1961) https://www.jstor.org/stable/2311357
10. Freeman, R., Sikdar, S., Vaish, R., Xia, L.: Equitable allocations of indivisible goods. In: Proceedings of the 28th International Joint Conference on Artificial Intelligence, pp. 280–286 (2019). https://www.ijcai.org/proceedings/2019/0040.pdf
11. Foley, D.: Resource Allocation and the Public Sector. Yale Economic Essays, New Haven (1967)
12. Lipton, R.J., Markakis, E., Mossel, E., Saberi, A.: On approximately fair allocations of indivisible goods. In: Proceedings of the 5th ACM Conference on Electronic Commerce, pp. 125–131 (2004). https://dl.acm.org/doi/10.1145/988772.988792
13. Budish, E.: The Combinatorial assignment problem: approximate competitive equilibrium from equal incomes. J. Polit. Econ.119(6), 1061–1103 (2011). https://www.journals.uchicago.edu/doi/abs/10.1086/664613

Graph Coloring

New Bounds and Constructions
for Neighbor-Locating Colorings
of Graphs

Dipayan Chakraborty[1] , Florent Foucaud[1] , Soumen Nandi[2], Sagnik Sen[3],
and D. K. Supraja[3]($\boxtimes$)

[1] Université Clermont-Auvergne, CNRS, Mines de Saint-Étienne,
Clermont-Auvergne-INP, LIMOS, 63000 Clermont-Ferrand, France
{dipayan.chakraborty,florent.foucaud}@uca.fr
[2] Netaji Subhas Open University, Kolkata, India
soumen2004@gmail.com
[3] Indian Institute of Technology Dharwad, Dharwad, India
sen007isi@gmail.com, dksupraja95@gmail.com

Abstract. A proper k-vertex-coloring of a graph G is a *neighbor-locating*
k-coloring if for each pair of vertices in the same color class, the sets of
colors found in their neighborhoods are different. The neighbor-locating
chromatic number $\chi_{NL}(G)$ is the minimum k for which G admits a
neighbor-locating k-coloring. A proper k-vertex-coloring of a graph G
is a *locating k-coloring* if for each pair of vertices x and y in the same
color-class, there exists a color class S_i such that $d(x, S_i) \neq d(y, S_i)$. The
locating chromatic number $\chi_L(G)$ is the minimum k for which G admits
a locating k-coloring. It follows that $\chi(G) \leq \chi_L(G) \leq \chi_{NL}(G)$ for any
graph G, where $\chi(G)$ is the usual chromatic number of G.

We show that for any three integers p, q, r with $2 \leq p \leq q \leq r$
(except when $2 = p = q < r$), there exists a connected graph $G_{p,q,r}$
with $\chi(G_{p,q,r}) = p$, $\chi_L(G_{p,q,r}) = q$ and $\chi_{NL}(G_{p,q,r}) = r$. We also show
that the locating chromatic number (resp., neighbor-locating chromatic
number) of an induced subgraph of a graph G can be arbitrarily larger
than that of G.

Alcon *et al.* showed that the number n of vertices of G is bounded
above by $k(2^{k-1} - 1)$, where $\chi_{NL}(G) = k$ and G is connected (this
bound is tight). When G has maximum degree Δ, they also showed that
a smaller upper-bound on n of order $k^{\Delta+1}$ holds. We generalize the latter by proving that if G has order n and at most $an + b$ edges, then
n is upper-bounded by a bound of the order of $k^{2a+1} + 2b$. Moreover,
we describe constructions of such graphs which are close to reaching the
bound.

Keywords: Coloring · Neighbor-locating coloring · Neighbor-locating
chromatic number · Identification problem · Location problem

© The Author(s), under exclusive license to Springer Nature Switzerland AG 2023
A. Bagchi and R. Muthu (Eds.): CALDAM 2023, LNCS 13947, pp. 121–133, 2023.
https://doi.org/10.1007/978-3-031-25211-2_9

1   Introduction

In the area of identification/location problems, one is given a discrete structure (such as a graph) and one wishes to identify its elements, that is, to be able to pairwise distinguish them from each other. This can be done by constructing, for example, dominating sets [15,22] or colorings [2,9,13] of the graph. The identification process may be based on distances [9,21] or on neighborhoods [2,22], and we may wish to distinguish all vertex pairs [15,21,22], only adjacent ones [13], or those with the same color [2,9]. This vast research area has many applications both in practical settings like fault-diagnosis in networks [15], biological testing [18], machine learning [11] and theoretical settings such as game analysis [12], isomorphism testing [4] or logical definability [16], to name a few.

Taking cues from the above research topics, recently, two variants of graph coloring were introduced, namely, *locating coloring* [9] and *neighbor-locating coloring* [2,5]. While the former concept has been well-studied since 2002 [5–10,19,20,23–25]), our focus of study is the latter, which was introduced in 2014 in [5] under the name of *adjacency locating coloring*, renamed in 2020 in [2] and studied in a few papers since then [1,3,14,17].

Throughout this article, we will use the standard terminologies and notations used in "Introduction to Graph Theory" by West [26].

Given a graph G, a *(proper) k-coloring* is a function $f : V(G) \to C$, where C is a set of k colors, such that $f(u) \neq f(v)$ whenever u is adjacent to v. The value $f(v)$ is called the *color* of v. The *chromatic number* of G, denoted by $\chi(G)$, is the minimum k for which G admits a k-coloring.

Given a k-coloring f of G, its ith color class is the collection S_i of vertices that have received the color i. The distance between a vertex x and a set S of vertices is given by $d(x, S) = \min\{d(x,y) : y \in S\}$, where $d(x,y)$ is the number of edges in a shortest path connecting x and y. Two vertices x and y are *metric-distinguished* with respect to f if either $f(x) \neq f(y)$ or $d(x, S_i) \neq d(y, S_i)$ for some color class S_i. A k-coloring f of G is a *locating k-coloring* if any two distinct vertices are metric-distinguished with respect to f. The *locating chromatic number* of G, denoted by $\chi_L(G)$, is the minimum k for which G admits a locating k-coloring.

Given a k-coloring f of G, suppose that a neighbor y of a vertex x belongs to the color class S_i. In such a scenario, we say that i is a *color-neighbor* of x (with respect to f). The set of all color-neighbors of x is denoted by $N_f(x)$. Two vertices x and y are *neighbor-distinguished* with respect to f if either $f(x) \neq f(y)$ or $N_f(x) \neq N_f(y)$. A k-coloring f is *neighbor-locating k-coloring* if each pair of distinct vertices are neighbor-distinguished. The *neighbor-locating chromatic number* of G, denoted by $\chi_{NL}(G)$, is the minimum k for which G admits a neighbor-locating k-coloring.

Observe that a neighbor-locating coloring is, in particular, a locating coloring. Thus, we have the following relation among the three parameters [2]:

$$\chi(G) \leq \chi_L(G) \leq \chi_{NL}(G).$$

Note that for complete graphs, all three parameters have the same value, that is, equality holds in the above relation. Nevertheless, the difference between the pairs of values of parameters χ, χ_{NL} and χ_L, χ_{NL}, respectively, can be arbitrarily large. Moreover, it was proved that for any pair p, q of integers with $3 \le p \le q$, there exists a connected graph G_1 with $\chi(G_1) = p$ and $\chi_{NL}(G_1) = q$ [2] and a connected graph G_2 with $\chi_L(G_2) = p$ and $\chi_{NL}(G_2) = q$ [17]. The latter of the two results positively settled a conjecture posed in [2]. We strengthen these results by showing that for any three integers p, q, r with $2 \le p \le q \le r$, there exists a connected graph $G_{p,q,r}$ with $\chi(G_{p,q,r}) = p$, $\chi_L(G_{p,q,r}) = q$ and $\chi_{NL}(G_{p,q,r}) = r$, except when $2 = p = q < r$.

One fundamental difference between coloring and locating coloring (resp., neighbor-locating coloring) is that the restriction of a coloring of G to an (induced) subgraph H is necessarily a coloring, whereas the analogous property is not true for locating coloring (resp., neighbor-locating coloring). Interestingly, we show that the locating chromatic number (resp., neighbor-locating chromatic number) of an induced subgraph H of G can be arbitrarily larger than that of G.

Alcon *et al.* [2] showed that the number n of vertices of G is bounded above by $k(2^{k-1} - 1)$, where $\chi_{NL}(G) = k$ and G has no isolated vertices, and this bound is tight. This exponential bound is reduced to a polynomial one when G has maximum degree Δ, indeed it was further shown in [2] that the upper-bound $n \le k \sum_{j=1}^{\Delta} \binom{k-1}{j}$ holds (for graphs with no isolated vertices and when $\Delta \le k - 1$). It was left open whether this bound is tight. The *cycle rank* c of a graph G, denoted by $c(G)$, is defined as $c(G) = |E(G)| - n(G) + 1$. Alcon *et al.* [3] gave the upper bound $n \le \frac{1}{2}(k^3 + k^2 - 2k) + 2(c - 1)$ for graphs of order n, neighbor-locating chromatic number k and cycle rank c. Further, they also obtained tight upper bounds on the order of trees and unicyclic graphs in terms of the neighbor-locating chromatic number [3], where a unicyclic graph is a connected graph having exactly one cycle.

As a connected graph with cycle rank c and order n has $n + c - 1$ edges and a graph of order n and maximum degree Δ has at most $\frac{\Delta}{2}n$ edges, the two latter bounds can be seen as two approaches for studying the neighbor-locating coloring for sparse graphs. We generalize this approach by studying graphs with given average degree, or in other words, graphs of order n having at most $an + b$ edges for some constants a, b (such graphs have average degree $2a + 2b/n$). For such graphs, we prove the upper bound $n \le 2b + k \sum_{i=1}^{2a}(2a + 1 - i)\binom{k-1}{i}$. Furthermore, we show that this bound is asymptotically tight, by a construction of graphs with $an + b$ edges (where $2a$ is any positive integer and $2b$ any integer) and neighbor-locating chromatic number $\Theta(k)$, whose order is $\Theta(k^{2a+1})$. Moreover, when $b = 0$, the graphs can be taken to have maximum degree $2a$. This implies that our bound and the one from [2] are roughly tight.

In Sect. 2, we study the connected graphs with prescribed values of chromatic number, locating chromatic number and neighbor-locating chromatic number. We also study the relation between the locating chromatic number (resp., neighbor-locating chromatic number) of a graph and its induced subgraphs. Finally, in Sect. 3 we study the density of graphs having bounded neighbor-locating chromatic number.

2 Gaps Among $\chi(G), \chi_L(G)$ and $\chi_{NL}(G)$

The first result we would like to prove involves three different parameters, namely, the chromatic number, the locating chromatic number, and the neighbor-locating chromatic number.

Theorem 1. *For all* $2 \leq p \leq q \leq r$, *except when* $p = q = 2$ *and* $r > 2$, *there exists a connected graph* $G_{p,q,r}$ *satisfying* $\chi(G_{p,q,r}) = p$, $\chi_L(G_{p,q,r}) = q$, *and* $\chi_{NL}(G_{p,q,r}) = r$.

Proof. First of all, let us assume that $p = q = r$. In this case, for $G_{p,q,r} = K_p$, it is trivial to note that $\chi(G_{p,q,r}) = \chi_L(G_{p,q,r}) = \chi_{NL}(G_{p,q,r}) = p$. This completes the case when $p = q = r$.

Second of all, let us handle the case when $p < q = r$. If $2 = p < q = r$, then take $G_{p,q,r} = K_{1,q-1}$. Therefore, we have $\chi(G_{p,q,r}) = 2$ as it is a bipartite graph, and it is known that $\chi_L(G_{p,q,r}) = \chi_{NL}(G_{p,q,r}) = q$ [2,9].

If $3 \leq p < q = r$, then we construct $G_{p,q,r}$ as follows: start with a complete graph K_p, on vertices $v_0, v_1, \cdots, v_{p-1}$, take $(q-1)$ new vertices $u_1, u_2, \cdots, u_{q-1}$, and make them adjacent to v_0. It is trivial to note that $\chi(G_{p,q,r}) = p$ in this case. Moreover, note that we need to assign q distinct colors to $v_0, u_1, u_2, \cdots, u_{q-1}$ under any locating or neighbor-locating coloring. On the other hand, $f(v_i) = i$ and $f(u_j) = j$ is a valid locating q-coloring as well as neighbor locating q-coloring of $G_{p,q,r}$. Thus we are done with the cases when $p < q = r$.

Thirdly, we are going to consider the case when $p = q < r$. If $3 = p = q < r$, then let $G_{p,q,r} = C_n$ where C_n is an odd cycle of suitable length, that is, a length which will imply $\chi_{NL}(C_n) = r$. It is known that such a cycle exists [1,5]. As we know that $\chi(G_{p,q,r}) = 3$, $\chi_L(G_{p,q,r}) = 3$ [9], and $\chi_{NL}(G_{p,q,r}) = r$ [1,5], we are done.

If $4 \leq p = q < r$, then we construct $G_{p,q,r}$ as follows: start with a complete graph K_p on vertices $v_0, v_1, \cdots, v_{p-1}$, and an odd cycle C_n on vertices $u_0, u_1, \cdots, u_{n-1}$, and identify the vertices v_0 and u_0. Moreover, we say that the length of the odd cycle C_n is a suitable length, that is, it is of a length which ensures $\chi_{NL}(C_n) = r$ and under any neighbor-locating r-coloring of C_n, every color is used at least twice. It is known that such a cycle exists [1,5]. Notice that $\chi(G_{p,q,r}) = p$ and $\chi_L(G_{p,q,r}) = q$. On the other hand, as the neighborhood of the vertices of the cycle C_n (subgraph of $G_{p,q,r}$) doesnot change if we consider it as an induced subgraph except for the vertex $v_0 = u_0$. Thus, we will need at least r colors to color C_n while it is contained inside $G_{p,q,r}$ as a subgraph. Hence $\chi_{NL}(G_{p,q,r}) = r$. Thus, we are done in this case also.

Finally, we are into the case when $p < q < r$. If $p = 2$, $q = 3$ and $r > 3$, then let $G_{p,q,r} = P_n$ where P_n is a path of suitable length, that is, a length which ensures $\chi_{NL}(G_{p,q,r}) = r$. It is known that such a path exists [3]. As we know that $\chi(G_{p,q,r}) = 2$, $\chi(G_{p,q,r}) = 3$ [9] and $\chi_{NL}(G_{p,q,r}) = r$ [1,5]. If $p = 2$ and $3 < q < r$, refer [17] for this case.

If $3 = p < q < r$, then we start with an odd cycle C_n on vertices $v_0, v_1, \cdots, v_{n-1}$ of a suitable length, where suitable means, a length that ensures

$\chi_{NL}(C_n) = r$ and under any neighbor-locating r-coloring of C_n, every vertex has two distinct color-neighbors. It is known that such a cycle exists [1,5]. Take $q - 1$ new vertices $u_1, u_2, \cdots, u_{q-1}$ and make all of them adjacent to v_0. This so obtained graph is $G_{p,q,r}$. It is trivial to note that $\chi(G_{p,q,r}) = 3$ in this case. Note that we need to assign q distinct colors to $v_0, u_1, u_2, \cdots, u_{q-1}$ under any locating or neighbor-locating coloring. One can show in a similar way like above that $\chi_L(G_{p,q,r}) = q$ and $\chi_{NL}(G_{p,q,r}) = r$.

If $4 \le p < q < r$, then we start with a path P_n of a suitable length, that is, it is of a length which ensures $\chi_{NL}(P_n) = r$ and under any neighbor-locating r-coloring of P_n, every color is used at least twice. It is known that such a path exists [1,5]. Let $P_n = u_0 u_1 \cdots u_{n-1}$. Now let us take a complete graph on p vertices $v_0, v_1, \cdots, v_{p-1}$. Identify the two graphs at u_0 and v_0 to obtain a new graph. Furthermore, take $(q - 2)$ independent vertices $w_1, w_2, \cdots, w_{q-2}$ and make them adjacent to u_{n-2}. This so obtained graph is $G_{p,q,r}$. One can show in a similar way like above that we have $\chi(G_{p,q,r}) = p, \chi_L(G_{p,q,r}) = q$, and $\chi_{NL}(G_{p,q,r}) = r$. □

Furthermore, we show that, unlike the case of chromatic number, an induced subgraph can have an arbitrarily higher locating chromatic number (resp., neighbor-locating chromatic number) than that of the graph.

Theorem 2. *For every $k \ge 0$, there exists a graph G_k having an induced subgraph H_k such that $\chi_L(H_k) - \chi_L(G_k) = k$ and $\chi_{NL}(H_k) - \chi_{NL}(G_k) = k$.*

Proof. The graph G_k is constructed as follows. We start with $2k$ independent vertices $a_1, a_2, \cdots, a_{2k}$ and k disjoint edges $b_1 b_1', b_2 b_2', \cdots, b_k b_k'$. After that we make all the above mentioned vertices adjacent to a special vertex v to obtain our graph G_k. Notice that v and the a_is must all receive distinct colors under any locating coloring or neighbor-locating coloring. On the other hand, the coloring f given by $f(v) = 0$, $f(a_i) = i$, $f(b_i) = 2i - 1$, and $f(b_i') = 2i$ is indeed a locating coloring as well as a neighbor-locating coloring of G_k. Hence we have $\chi_L(G_k) = \chi_{NL}(G_k) = (2k + 1)$.

Now take H_k as the subgraph induced by v, a_is and b_is. It is the graph $K_{1,3k}$, and we know that all vertices must get distinct colors under any locating coloring or neighbor-locating coloring. Hence we have $\chi_L(H_k) = \chi_{NL}(H_k) = (3k + 1)$.

This completes the proof. □

3   Bounds and Constructions for Sparse Graphs

In this section, we study the density of graphs having bounded neighbor-locating chromatic number.

3.1   Bounds

The first among those results provides an upper bound on the number of vertices of a graph in terms of its neighbor-locating chromatic number. This, in particular shows that the number of vertices of a graph G is bounded above by a polynomial function of $\chi_{NL}(G)$.

Theorem 3. *Let G be a connected graph on n vertices and m edges such that $m \leq an + b$, where $2a$ is a positive integer and $2b$ is an integer. If $\chi_{NL}(G) = k$, then*

$$n \leq 2b + k \sum_{i=1}^{2a} (2a + 1 - i)\binom{k-1}{i}.$$

In particular, any graph whose order attains the upper bound must be of maximum degree $2a + 1$ and with exactly $k\binom{k-1}{i}$ number of vertices of degree i.

Proof. Let D_i and d_i denote the set and the number of vertices in G having degree equal to i, respectively, and let D_i^+ and d_i^+ denote the set and the number of vertices in G having degree at least i, for all $i \geq 1$. Using the handshaking lemma, we know that

$$\sum_{v \in V(G)} deg(v) = 2|E(G)| = 2m \leq 2(an + b).$$

Notice that, as G is connected, and hence does not have any vertex of degree 0, it is possible to write

$$\sum_{v \in V(G)} deg(v) = \sum_{i=1}^{2a} i \cdot d_i + \sum_{v \in D_{2a+1}^+} deg(v).$$

Moreover, the number of vertices of G can be expressed as

$$n = (d_1 + d_2 + \cdots + d_{2a}) + d_{2a+1}^+ = d_{2a+1}^+ + \sum_{i=1}^{2a} d_i.$$

Therefore, combining the above equations and inequalities, we have

$$\sum_{i=1}^{2a} i \cdot d_i + \sum_{v \in D_{2a+1}^+} deg(v) \leq 2b + 2a\left(d_{2a+1}^+ + \sum_{i=1}^{2a} d_i\right)$$

which implies

$$d_{2a+1}^+ \leq \sum_{v \in D_{2a+1}^+} (deg(v) - 2a) \leq \left(\sum_{v \in D_{2a+1}^+} deg(v)\right) - 2ad_{2a+1}^+ \leq 2b + \sum_{i=1}^{2a}(2a-i)d_i$$

since there are exactly d_{2a+1}^+ terms in the summation $\sum_{v \in D_{2a+1}^+} (deg(v) - 2a)$ where each term is greater than or equal to 1, as $deg(v) \geq 2a+1$ for all $v \in D_{2a+1}^+$.

Let f be any neighbor-locating k-coloring of G. Consider an ordered pair $(f(u), N_f(u))$, where u is a vertex having degree at most s. Thus, u may receive one of the k available colors, while its color neighborhood may consist of at most s of the remaining $(k-1)$ colors. Thus, there are at most $k \sum_{i=1}^{s} \binom{k-1}{i}$ choices for the ordered pair $(f(u), N_f(u))$. As for any two vertices u, v of degree at most s, the following ordered pairs $(f(u), N_f(u))$ and $(f(v), N_f(v))$ must be distinct, we have

$$\sum_{i=1}^{s} d_i \leq k \sum_{i=1}^{s} \binom{k-1}{i}.$$

Using the above relation, we can show that

$$\sum_{i=1}^{2a}(2a+1-i)d_i = \sum_{s=1}^{2a}\left(\sum_{i=1}^{s} d_i\right) \leq \sum_{s=1}^{2a}\left(k\sum_{i=1}^{s}\binom{k-1}{i}\right) = k\sum_{i=1}^{2a}(2a+1-i)\binom{k-1}{i}.$$

As

$$\sum_{i=1}^{2a}(2a+1-i)d_i = \sum_{i=1}^{2a} d_i + \sum_{i=1}^{2a}(2a-i)d_i \text{ and } d_{2a+1}^+ \leq 2b + \sum_{i=1}^{2a}(2a-i)d_i,$$

we have

$$n - d_{2a+1}^+ + \sum_{i=1}^{2a} d_i \leq 2b + k\sum_{i=1}^{2a}(2a+1-i)\binom{k-1}{i}.$$

This completes the first part of the proof.

For the proof of the second part of the Theorem, we notice that if the order of a graph G^* attains the upper bound, then equality holds in all of the above inequations. In particular, we must have $d_{2a+1}^+ = \sum_{v \in D_{2a+1}^+}(deg(v) - 2a)$ which implies that G^* cannot have a vertex of degree more than $2a+1$. Moreover, we also have the following equality.

$$\sum_{i=1}^{s} d_i = k\sum_{i=1}^{s}\binom{k-1}{i} \text{ for } s = 1, 2, \ldots, 2a+1.$$

This proves that G^* has exactly $k\binom{k-1}{i}$ vertices of degree i.   □

Next we are going to present some immediate corollaries of Theorem 3. A *cactus* is a connected graph in which no two cycles share a common edge.

Corollary 1. *Let G be a cactus on n vertices and m edges. If $\chi_{NL}(G) = k$, then*

$$n \leq \frac{k^4 + 11k^2 - 12k - 6}{6}.$$

Moreover, if the cactus has exactly t cycles, then we have

$$n \leq 2(t-1) + \frac{k^3 + k^2 - 2k}{2}.$$

Proof. Observe that G has at most $\frac{3(n-1)}{2}$ edges. So, by substituting $a = \frac{3}{2}$ and $b = -\frac{3}{2}$ in the bound for n established in Theorem 3, we have

$$n \le 2b + k \sum_{i=1}^{2a}(2a+1-i)\binom{k-1}{i} = -3 + k\sum_{i=1}^{3}(4-i)\binom{k-1}{i}$$

$$= -3 + 3k\binom{k-1}{1} + 2k\binom{k-1}{2} + k\binom{k-1}{3}$$

$$= \frac{k^4 + 11k^2 - 12k - 6}{6}.$$

Note that, if the cactus G has exactly t cycles, then G has exactly $(n + t - 1)$ edges. Hence, replacing $a = 1$ and $b = (t - 1)$ in the bound for n established in Theorem 3, we obtain the required bound for the cactus. $\square$

A graph is *t-degenerate* if its every subgraph has a vertex of degree at most t.

Corollary 2. *Let G be a t-degenerate graph on n vertices and m edges. If $\chi_{NL}(G) = k$, then*

$$n \le k \sum_{i=1}^{2t}(2t+1-i)\binom{k-1}{i} - t(t+1).$$

Proof. Observe that the number of edges in a t-degenerate graph is $m \le tn - \frac{t(t+1)}{2}$. Substituting $a = t$ and $b = -\frac{t(t+1)}{2}$ in the bound for n established in Theorem 3, we obtain the required bound. $\square$

A planar graph is 5-degenerate, thus using the above corollary, we know that for a planar graph G one can obtain an upper bound of $|V(G)|$. However, since $|E(G)| \le 3|V(G)| - 6$, we are able to obtain a better bound.

Corollary 3. *Let G be a planar graph on n vertices and m edges. If $\chi_{NL}(G) = k$, then*

$$n \le k \sum_{i=1}^{6}(7-i)\binom{k-1}{i} - 12.$$

Proof. Note that the number of edges in a planar graph is at most $3n - 6$. Substituting $a = 3$ and $b = -6$ in the bound for n established in Theorem 3, we get the required bound. $\square$

3.2 Tightness

Next we show the asymptotic tightness of Theorem 3. To that end, we will prove the following result.

Theorem 4. *Let $2a$ be a positive integer and let $2b$ be an integer. Then, there exists a graph G on n vertices and m edges satisfying $m \le an + b$ such that $n = \Theta(k^{2a+1})$ and $\chi_{NL}(G) = \Theta(k)$. Moreover, when $b = 0$, G can be taken to be of maximum degree $2a$.*

The proof of this theorem is contained within a number of observations and lemmas. Also, the proof is constructive, and the constructions depend on particular partial colorings. Therefore, we are going to present a series of graph constructions, their particular colorings, and their structural properties. We are also going to present the supporting observations and lemmas in the following.

Lemma 1. *Let us consider a* $(p \times q)$ *matrix whose* ij^{th} *entry is* $m_{i,j}$, *where* $p < q$. *Let* M *be a complete graph whose vertices are the entries of the matrix. Then there exists a matching of* M *satisfying the following conditions:*

(i) The endpoints of an edge of the matching are from different columns.
(ii) Let e_1 *and* e_2 *be two edges of the matching. If one endpoint of* e_1 *and* e_2 *are from the* i^{th} *columns, then the other endpoints of them must belong to distinct columns.*
(iii) The matching saturates all but at most one vertex of M *per column.*

Proof. Consider the permutation $\sigma = (1\ 2\ \cdots q)$. The matching consists of edges of the type $m_{(2i-1),j}m_{2i,\sigma^i(j)}$ for all $i \in \{1, 2, \cdots, \lfloor \frac{p}{2} \rfloor\}$ and $j \in \{1, 2, \cdots, q\}$. We will show that this matching satisfies all listed conditions.

Observe that, a typical edge of the matching is of the form $m_{(2i-1),j}m_{2i,\sigma^i(j)}$. As the second co-ordinates of the subscript of the endpoints of the said edge is different, condition (i) from the statement is verified.

Suppose that there are two edges of the type $m_{(2i-1),j}m_{2i,\sigma^i(j)}$ and $m_{(2i'-1),j'}m_{2i',\sigma^{i'}(j')}$. If $m_{(2i-1),j}$ and $m_{(2i'-1),j'}$ are from the same column, that is, $j = j'$, then we must have $i \neq i'$ as they are different vertices. Thus $\sigma^i(j) \neq \sigma^{i'}(j) = \sigma^{i'}(j')$ as $i \neq i'$. If $m_{(2i-1),j}$ and $m_{2i',\sigma^{i'}(j')}$ are from the same column, then we have $j = \sigma^{i'}(j')$. Moreover, if we have $j' = \sigma^i(j)$, then it will imply that

$$j = \sigma^{i'}(\sigma^i(j)) = \sigma^{i+i'}(j).$$

This is only possible if $q|(i + i')$, which is not possible as $i, i' \in \{1, 2, \cdots, \lfloor \frac{p}{2} \rfloor\}$. Therefore, we have verified condition (ii) of the statement.

Notice that, the matching saturates all the vertices of M when p is even, whereas it saturates all except the vertices in the pth row of the matrix when p is odd. This verifies condition (iii) of the statement.   □

Corollary 4. *Let* G *be a graph with an independent set* M *of size* $(p \times q)$, *where* $M = \{m_{ij} : 1 \leq i \leq p, 1 \leq j \leq q\}$ *and* $p < q$. *Moreover, let* ϕ *be a* $(k' + q)$-*coloring of* G *satisfying the following conditions:*

1. $k' + 1 \leq \phi(x) \leq k' + q$ if and only if $x \in M$,
2. x and y are neighbor-distinguished unless both belong to M,
3. $\phi(m_{ij}) = k' + j$.

Then it is possible to find spanning supergraph G' *of* G *by adding a matching between the vertices of* M *which will make* ϕ *a neighbor-locating* $(k' + q)$-*coloring of* G'.

Proof. First of all build a matrix whose ij^{th} entry is the vertex m_{ij}. After that, build a complete graph whose vertices are entries of this matrix. Now using Lemma 1, we can find a matching of this complete graph that satisfies the three conditions mentioned in the statement of Lemma 1. We construct G' by including exactly the edges corresponding to the edges of the matching, between the vertices of M. We want to show that after adding these edges and obtaining G', indeed ϕ is a neighbor-locating $(k' + q)$-coloring of G'.

Notice that by the definition of ϕ, $(k' + q)$ colors are used. So it is enough to show that the vertices of G' are neighbor-distinguished with respect to ϕ. To be exact, it is enough to show that two vertices x, y from M are neighbor-distinguished with respect to ϕ in G'. If $\phi(x) = \phi(y)$, then they must have different color-neighborhood inside M according to the conditions of the matching. This is enough to make x, y neighbor-distinguished. □

Now we are ready to present our iterative construction. However, given the involved nature of it, we need some specific nomenclatures to describe it. For convenience, we will list down some points to describe the whole construction.

(i) An *i-triplet* is a 3-tuple of the type (G_i, ϕ_i, X_i) where G_i is a graph, ϕ_i is a neighbor-locating (ik)-coloring of G_i, X_i is a set of $(i+1)$-tuples of vertices of G_i, each having non-repeating elements. Also, two $(i+1)$-tuples from X_i do not have any entries in common.

(ii) Let us describe the 1-triplet (G_1, ϕ_1, X_1) explicitly. Here G_1 is the path $P_t = v_1 v_2 \cdots v_t$ on t vertices where $t = 4 \left\lfloor \frac{k(k-1)(k-2)+4}{8} \right\rfloor$. As

$$\frac{(k-1)^2(k-2)}{2} < 4 \left\lfloor \frac{k(k-1)(k-2)+4}{8} \right\rfloor \leq \frac{k^2(k-1)}{2},$$

we must have $\chi_{NL}(P_t) = k$ (see [2]). Let ϕ_1 be any neighbor-locating k-coloring of G_1 and

$$X_1 = \{(v_{i-1}, v_{i+1}) : i \equiv 2, 3 \pmod 4\}.$$

(iii) Suppose an *i-triplet* (G_i, ϕ_i, X_i) is given. We will (partially) describe a way to construct an $(i+1)$-triplet from it. To do so, first we will construct an intermediate graph G'_{i+1} as follows: for each $(i+1)$-tuple $(x_1, x_2, \cdots, x_{i+1}) \in X_i$ we will add a *new vertex* x_{i+2} adjacent to each vertex from the $(i+1)$-tuple. Moreover, $(x_1, x_2, \cdots, x_{i+1}, x_{i+2})$ is designated as an $(i+2)$-tuple in G'_{i+1}. After that, we will take k copies of G'_{i+1} and call this so-obtained graph as G''_{i+1}. Furthermore, we will extend ϕ_i to a function ϕ_{i+1} by assigning the color $(ik + j)$ to the new vertices from the jth copy of G'_{i+1}. The copies of the $(i+2)$-tuples are the $(i+2)$-tuples of G''_{i+1}.

(iv) Consider the $(i+1)$-triplet $(G''_{i+1}, \phi_{i+1}, X''_{i+1})$ where X''_{i+1} denotes the set of all $(i+2)$-tuples of G''_{i+1}. The *color of an $(i+2)$-tuple* $(x_1, x_2, \cdots, x_{i+2})$ is the set

$$C((x_1, x_2, \cdots, x_{i+2})) = \{\phi_i(x_1), \phi_i(x_2), \cdots, \phi_i(x_{i+2})\}.$$

Let us partition the set of new vertices based on the colors used on the elements (all but the last one) of the $(i + 2)$-tuple of which it is (uniquely) part of. To be explicit, the last elements of two $(i + 2)$-tuples are in the same partition if and only if they have the same color. Let this partition be denoted by $X_{i1}, X_{i2}, \cdots, X_{is_i}$, for some integer s_i.

(v) First fix a partition X_{ir} of X_i. Next construct a matrix with its ℓth column having vertices from X_{ir} as its entries if they are also from the ℓth copy of G'_{i+1} in G''_{i+1}. Thus the matrix is a $(p \times q)$ matrix where $p = |X_{ir}|$ and $q = k$. We are going to show that, $p < q$. However, for convenience, we will defer it to a later part (Lemma 2).

(vi) Let us delete all the new vertices from G''_{i+1} except for the ones in X_{ir}. This graph has the exact same properties of the graph G from Corollary 4 where X_{ir} plays the role of the independent set M. Thus it is possible to add a matching and extend the coloring (like in Corollary 4). We do that for each value of r and add the corresponding matching to our graph G''_{i+1}. After adding all such matchings, the graph we obtain is G_{i+1}.

Lemma 2.   *We have* $|X_{ir}| < k$, *where* X_{ir} *is as in Item(v) of the above list.*

Proof. It is easy to calculate that the set of 2-tuples having the same color in G_1 is strictly less than k. After that we are done by induction.   □

Lemma 3. *The function* ϕ_{i+1} *is a neighbor-locating coloring of* G_{i+1}.

Proof. The function ϕ_{i+1} is constructed from ϕ_i, alongside constructing the triplet G_{i+1} from G_i. While constructing, we use the same steps from that of Corollary 4. Thus, the newly colored vertices become neighbor-distinguished in G_{i+1} under ϕ_{i+1}.   □

The above two lemmas validate the correctness of the iterative construction of G_is. However, it remains showing how G_is help us prove our result. To do so, let us prove certain properties of G_is.

Lemma 4. *The graph* G_i *is not regular and has maximum degree* $(i + 1)$.

Proof. As we have started with a path, our G_1 has maximum degree 2 and is not regular. In the iteration step for constructing the graph G_{i+1} from G_i, the degree of an old vertex (or its copy) can increase at most by 1, while a new vertex of G_{i+1} is adjacent to exactly $(i + 1)$ old vertices and at most one new vertex. Hence, a new vertex in G_{i+1} can have degree at most $(i + 2)$. Therefore, the proof is done by induction.   □

Finally, we are ready to prove Theorem 4.

Proof of Theorem 4. Given a and b, to build the example that will prove the theorem, one can consider $G = G_{2a+1}$.   □

Acknowledgements. This work is partially supported by the following projects: "MA/IFCAM/18/39", "SRG/2020/001575", "MTR/2021/000858" and "NBHM/RP-8 (2020)/Fresh". Research by the first and second authors is partially sponsored by a public grant overseen by the French National Research Agency as part of the "Investissements d'Avenir" through the IMobS3 Laboratory of Excellence (ANR-10-LABX-0016), the IDEX-ISITE initiative CAP 20–25 (ANR-16-IDEX-0001) and the ANR project GRALMECO (ANR-21-CE48-0004).

References

1. Alcon, L., Gutierrez, M., Hernando, C., Mora, M., Pelayo, I.M.: The neighbor-locating-chromatic number of pseudotrees. arXiv preprint arXiv:1903.11937 (2019)
2. Alcon, L., Gutierrez, M., Hernando, C., Mora, M., Pelayo, I.M.: Neighbor-locating colorings in graphs. Theoret. Comput. Sci. **806**, 144–155 (2020)
3. Alcon, L., Gutierrez, M., Hernando, C., Mora, M., Pelayo, I.M.: The neighbor-locating-chromatic number of trees and unicyclic graphs. Discussiones Mathematicae Graph Theory (2021)
4. Babai, L.: On the complexity of canonical labeling of strongly regular graphs. SIAM J. Comput. **9**(1), 212–216 (1980)
5. Behtoei, A., Anbarloei, M.: The locating chromatic number of the join of graphs. Bull. Iran. Math. Soc. **40**(6), 1491–1504 (2014)
6. Behtoei, A., Mahdi, A.: A bound for the locating chromatic numbers of trees. Trans. Comb. **4**(1), 31–41 (2015)
7. Behtoei, A., Omoomi, B.: On the locating chromatic number of Kneser graphs. Discret. Appl. Math. **159**(18), 2214–2221 (2011)
8. Behtoei, A., Omoomi, B.: On the locating chromatic number of the cartesian product of graphs. Ars Combin. **126**, 221–235 (2016)
9. Chartrand, G., Erwin, D., Henning, M., Slater, P., Zhang, P.: The locating-chromatic number of a graph. Bull. Inst. Combin. Appl. **36**, 89–101 (2002)
10. Chartrand, G., Erwin, D., Henning, M.A., Slater, P.J., Zhang, P.: Graphs of order n with locating-chromatic number n-1. Discret. Math. **269**(1–3), 65–79 (2003)
11. Chlebus, B.S., Nguyen, S.H.: On finding optimal discretizations for two attributes. In: Polkowski, L., Skowron, A. (eds.) RSCTC 1998. LNCS (LNAI), vol. 1424, pp. 537–544. Springer, Heidelberg (1998). https://doi.org/10.1007/3-540-69115-4_74
12. Chvátal, V.: Mastermind. Combinatorica **3**(3), 325–329 (1983)
13. Esperet, L., Gravier, S., Montassier, M., Ochem, P., Parreau, A.: Locally identifying coloring of graphs. Electron. J. Comb. **19**(2), 40 (2012)
14. Hernando, C., Mora, M., Pelayo, I.M., Alcón, L., Gutierrez, M.: Neighbor-locating coloring: graph operations and extremal cardinalities. Electron. Notes Discret. Math. **68**, 131–136 (2018)
15. Karpovsky, M.G., Chakrabarty, K., Levitin, L.B.: On a new class of codes for identifying vertices in graphs. IEEE Trans. Inf. Theory **44**(2), 599–611 (1998)
16. Kim, J.H., Pikhurko, O., Spencer, J.H., Verbitsky, O.: How complex are random graphs in first order logic? Random Struct. Algorithms **26**(1–2), 119–145 (2005)
17. Mojdeh, D.A.: On the conjectures of neighbor locating coloring of graphs. Theoret. Comput. Sci. **922**, 300–307 (2011)
18. Moret, B.M.E., Shapiro, H.D.: On minimizing a set of tests. SIAM J. Sci. Stat. Comput. **6**(4), 983–1003 (1985)
19. Purwasih, I., Baskoro, E.T., Assiyatun, H., Suprijanto, D., Baca, M.: The locating-chromatic number for Halin graphs. Commun. Combin. Optim. **2**(1), 1–9 (2017)

20. Purwasih, I.A., Baskoro, E.T., Assiyatun, H., Suprijanto, D.: The bounds on the locating-chromatic number for a subdivision of a graph on one edge. Proc. Comput. Sci. **74**, 84–88 (2015)
21. Slater, P.J.: Leaves of trees. In: Proceedings of the 6th Southeastern Conference on Combinatorics, Graph Theory, and Computing. Congressus Numerantium, vol. 14, pp. 549–559 (1975)
22. Slater, P.J.: Dominating and reference sets in a graph. J. Math. Phys. Sci. **22**(4), 445–455 (1988)
23. Syofyan, D.K., Baskoro, E.T., Assiyatun, H.: The locating-chromatic number of binary trees. Procedia Comput. Sci. **74**, 79–83 (2015)
24. Welyyanti, D., Baskoro, E.T., Simajuntak, R., Uttunggadewa, S.: On the locating-chromatic number for graphs with two homogenous components. J. Phys. Conf. Ser. **893**, 012040 (2017)
25. Welyyanti, D., Baskoro, E.T., Simanjuntak, R., Uttunggadewa, S.: On locating-chromatic number for graphs with dominant vertices. Procedia Comput. Sci. **74**, 89–92 (2015)
26. West, D.B.: Introduction to Graph Theory, vol. 2. Prentice Hall, Upper Saddle River (2001)

5-List Coloring Toroidal 6-Regular Triangulations in Linear Time

Niranjan Balachandran[iD] and Brahadeesh Sankarnarayanan[(✉)][iD]

Department of Mathematics, Indian Institute of Technology Bombay,
Mumbai 400076, Maharashtra, India
{niranj,bs}@math.iitb.ac.in

Abstract. We give an explicit procedure for 5-list coloring a large class of toroidal 6-regular triangulations in linear time. We also show that these graphs are not 3-choosable.

Keywords: List coloring · Toroidal graph · Triangulation · Regular graph · Linear time algorithm

1 Introduction

We will be concerned with the following coloring variant known as *list coloring*, defined independently by Vizing [34] and by Erdős, Rubin, and Taylor [18]. A *list assignment* $\mathcal{L}$ on a graph $G = (V, E)$ is a collection of sets of the form $\mathcal{L} = \{L_v \subset \mathbb{N} : v \in V(G)\}$, where one thinks of each L_v as a *list* of colors available for coloring the vertex $v \in V(G)$. A graph G is $\mathcal{L}$-*choosable* if there exists a function $\mathrm{color} \colon V(G) \to \mathbb{N}$ such that $\mathrm{color}(v) \in L_v$ for every $v \in V(G)$ and $\mathrm{color}(v) \neq \mathrm{color}(w)$ whenever $vw \in E(G)$. A graph G is called k-*choosable* if it is $\mathcal{L}$-choosable for every k-list assignment $\mathcal{L}$ (i.e., an assignment of lists of size at least k, also called k-lists). The least integer k for which G is k-choosable is the *choice number*, or *list chromatic number*, of G and is denoted $\chi_\ell(G)$. If $\chi_\ell(G) = k$, we also say that G is k-*list chromatic*. Notice that the usual notion of graph coloring is equivalent to $\mathcal{L}$-coloring when all the lists assigned by $\mathcal{L}$ are identical. This also shows that $\chi(G) \leq \chi_\ell(G)$ for all graphs G, and in general the inequality can be strict [18,34].

1.1 Motivation

k-Choosability is Computationally Hard. It is well-known that computing the chromatic number is an NP-hard problem [25]. The restricted problem of finding a 4-coloring of a 3-chromatic graph is also NP-hard [24]. Even the problem of 3-colorability of 4-regular planar graphs is known to be NP-complete [14].

Naturally, list coloring is also a computationally hard problem, but much more: for instance, it is well-known [22] that the problem of deciding whether a

A. Bagchi and R. Muthu (Eds.): CALDAM 2023, LNCS 13947, pp. 134–146, 2023.
https://doi.org/10.1007/978-3-031-25211-2_10

given planar graph is 4-choosable is NP-hard—even if the 4-lists are all chosen from $\{1, 2, 3, 4, 5\}$ [13]—and so is deciding whether a given planar triangle-free graph is 3-choosable [22]. But, contrast the latter with the fact that every planar triangle-free graph is 3-colorable by Grötzsch's theorem [21], and that a 3-coloring can be found in linear time [15]. In other words, restrictions on graph parameters—such as the girth, as in Grötzsch's theorem—that allow for efficient coloring algorithms need to be strengthened further in order to get list coloring algorithms of a similar flavor.

Note that even proving nontrivial bounds for the choice number is far tougher than the corresponding problem for the chromatic number. Some of the notable instances of such bounds being determined include Brooks's theorem for choosability [18, 34], Thomassen's remarkable proof that every planar graph is 5-choosable [31], and Galvin's solution to the famous Dinitz problem [20]. Other interesting examples include the fact that planar bipartite graphs are 3-choosable [5] and that any 4-regular graph decomposable into a Hamiltonian circuit and vertex-disjoint triangles is 3-choosable [19]. However, there is a fundamental difference between the former and latter examples, as we elaborate below.

$\mathcal{L}$-**Coloring is Algorithmically Hard.** Consider the problem: given a list assignment $\mathcal{L}$ on a graph G, can one efficiently determine whether or not G is $\mathcal{L}$-choosable, and in the case when G is $\mathcal{L}$-choosable can one also efficiently specify a proper coloring from these lists? The theorems of Brooks, Thomassen and Galvin mentioned earlier are some of the few instances where such algorithms are known for a large class of graphs. In the other examples that we mentioned, the proof uses the combinatorial nullstellensatz [4], in particular a powerful application found by Alon and Tarsi [5]. Hence, it does not allow one to extract an efficient algorithmic solution to the problem of $\mathcal{L}$-coloring when the list assignment $\mathcal{L}$ is specified, except in certain special cases. That there is no known efficient algorithm that produces a 3-list coloring from a given list assignment in these examples illustrates the difficulty of the problem of efficiently finding a proper $\mathcal{L}$-coloring even for graphs of small maximum degree. Even just for planar bipartite graphs, an algorithmic determination of a list coloring largely remains open [13].

Hence, efficient $\mathcal{L}$-coloring algorithms for large classes of graphs are interesting. We also place our work within the context of recent results on efficient list coloring algorithms for similar classes of graphs in Sect. 1.3 below.

1.2   Our Work

As noted earlier, in order to find good bounds for the choice number it is natural to place restrictions on certain graph parameters. We shall focus on a certain class of graphs G having bounded *degeneracy number* $d(G)$, defined as $d(G) := \max_{H \leq G}\{\delta(H)\}$, where the maximum is taken over all induced subgraphs H of G, and $\delta(H)$ is the minimum degree of H. A simple inductive argument [3] shows that $\chi_\ell(G) \leq d(G) + 1$ for every simple graph G. This improves the rudimentary upper bound $\chi_\ell(G) \leq \Delta(G) + 1$, where $\Delta(G)$ is the maximum degree of G. A

natural choice for a large collection of graphs with bounded degeneracy number is
the class of graphs that are embeddable in a fixed surface, where by a *surface* we
mean a compact connected 2-manifold, and a graph is *embeddable* in a surface if,
informally speaking, it can be drawn on the surface without any crossing edges.
In this paper, we will be concerned only with toroidal graphs, that is, graphs
that are embeddable on the torus S_1.

Let $G = (V, E)$ be a toroidal graph, and let F be the set of its faces in
an embedding into S_1. The graphs satisfying degree$(v) = d$ for all $v \in V$ and
degree$(f) = m$ for all $f \in F$, for some $d, m \geq 1$, have been of interest [6,7]
especially in the study of vertex-transitive graphs on the torus [30]. A simple
calculation using Euler's formula shows that the only possible values of (d, m)
are $(3, 6)$, $(4, 4)$ and $(6, 3)$. Our focus will be on the graphs of the last kind,
namely the 6-regular triangulations on the torus. Since triangulations have the
maximum possible number of edges in any graph with a fixed number of vertices
and embeddable on a given surface, one might additionally expect this class of
graphs to present a greater obstacle to an efficient solution to the list coloring
problem as compared to the others.

The main result of this paper, Theorem 1, is a linear time algorithm for 5-
list coloring a large class of these toroidal 6-regular triangulations. Our result is
nearly tight for this class in the sense that the list size is at most one more than
the choice number for any graph in this family. In fact, in Corollary 1 we find
an infinite family of 5-chromatic-choosable graphs for which a list coloring can
be specified in linear time.

Here, $T(r, s, t)$ is a triangulation obtained from an $r \times s$ toroidal grid, $r, s \geq 1$
(see Definition 1 for a precise statement):

Theorem 1. *Let G be a simple 6-regular toroidal triangulation. Then, G is 5-
choosable under any of the following conditions:*

(1) G is isomorphic to $T(r, s, t)$ for $r \geq 4$;
(2) G is isomorphic to $T(1, s, 2)$ for $s \geq 9$, $s \neq 11$;
(3) G is isomorphic to $T(2, s, t)$ for s and t both even;
(4) G is 3-chromatic.

*Moreover, the 5-list colorings can be given in linear time. Furthermore, none of
these graphs are 3-choosable. Hence, $\chi_\ell(G) \in \{4, 5\}$ if any of the cases (1) to (4)
hold for G.*

We are currently unable to comment on the choosability of the excluded graphs,
but we note that they consist only of nine nonisomorphic 5-chromatic graphs,
as well as a subcollection of triangulations of the specific form $T(1, s, t)$ that
are 4-chromatic. For any tuple (r, s, t), there is a simple formula describing each
tuple (r', s', t') such that $T(r, s, t)$ is isomorphic to $T(r', s', t')$ (see [6,29]), and
there are at most 6 such tuples for any (r, s, t). It is also not difficult to see
that the loopless multigraphs $T(r, s, t)$ are all 5-choosable. So, in this sense,
Theorem 1 covers the 5-choosability of "most" 6-regular toroidal triangulations.
Furthermore, among those graphs covered in Theorem 1, the 5-chromatic ones
are precisely those isomorphic to $T(1, s, 2)$ for $s \not\equiv 0 \pmod 4$. Thus, we have:

Corollary 1. *If G is isomorphic to $T(1, s, 2)$ for $s \not\equiv 0 \pmod 4$, $s \geq 9$, $s \neq 11$, then G is 5-chromatic-choosable, i.e. $\chi(G) = \chi_\ell(G) = 5$. Moreover, a 5-list coloring can be found in linear time.*

To the best of our knowledge, the method of proof that we employ is novel, in that we develop a framework that allows us to systematically compare the lists on vertices that are not too far apart, and that allows us to compute the list coloring in an efficient manner. By using the differential information between lists on nearby vertices, we reduce the *list configurations* that need to be considered. This kind of "list calculus" differs from other list coloring algorithms in the literature, which instead reduce the possible *graph configurations* by exploiting general structure results on the family of graphs under consideration (minimum girth, edge-width, etc.), while the specific lists on the graphs remain nebulous. Our method of proof could prove fruitful in other areas where a structure theorem—such as Theorem 2 in our case—allows one to shift attention towards the configuration of the lists themselves. We also emphasize that our linear-time algorithm for 5-list coloring these graphs is nearly best possible, since any fixed vertex needs to be "scanned" very few times.

1.3 Related Work

Colorability vs. Choosability. Note that it follows from Brooks's theorem for choosability that any 6-regular toroidal triangulation not isomorphic to K_7 is 6-choosable. Albertson and Hutchinson [2] showed that there is a unique simple graph in this family that is 6-chromatic, which has 11 vertices, and Thomassen [32] later classified all the 5-colorable toroidal graphs. But a precise characterization of all the 5-chromatic 6-regular toroidal triangulations was completed only recently [12, 29, 35]. Our results are the first in this line to attempt to characterize the list colorability of the 6-regular triangulations on the torus.

Choosability of Grids. The problem of determining the choice number of 4-regular toroidal $m \times n$ grids, for $m, n \geq 3$, has been raised by Cai, Wang and Zhu [10]. These graphs are a special case of those satisfying $(d, m) = (4, 4)$. It is easy to show by induction that these grids are all 3-colorable, and the above authors conjecture that they are also 3-choosable. Recent work by Li, Shao, Petrov and Gordeev [26] has nearly determined the choice number of these grids as follows: if mn is even, then the choice number is 3, else it is either 3 or 4. Contrasting this with Theorem 1, we note that both nearly determine the choice number in the sense that the true value of the choice number is either equal to, or one less than, the computed value for each member of the family. However, their result does not a priori give an efficient algorithm for $\mathcal{L}$-coloring the toroidal grids since their proof uses the combinatorial nullstellensatz, whereas our result actually gives a linear time algorithm for $\mathcal{L}$-coloring the toroidal triangulations.

Recent Algorithmic Advances for List Colorings. Dvořák and Kawarabayashi [16] have shown that there exists a polynomial time algorithm for 5-list coloring graphs embedded on a fixed surface. Postle and Thomas [28] have proved that for any surface Σ and every $k \in \{3, 4, 5\}$ there exists a linear time algorithm for determining whether or not an input graph G embedded in Σ and having girth at least $8 - k$ is k-choosable. In particular, when $\Sigma = S_1$ and $k = 5$, this implies that there is a linear time algorithm for determining whether or not any of the 6-regular triangulations under consideration in this paper are 5-choosable. This work was later extended by Postle in [27], wherein he showed that for each fixed surface Σ there exists a linear time algorithm to find a k-list coloring of a graph G with girth at least $8 - k$ for $k \in \{3, 4, 5\}$. Again, when $\Sigma = S_1$ and $k = 5$, this says that there is a linear time algorithm to find a 5-list coloring of a 6-regular triangulation on the torus.

Our results in this paper are stronger than those mentioned above for the class of 6-regular toroidal triangulations. Firstly, the high degree of the polynomial time algorithm in [16] makes it impractical to implement, though the authors suggest that it should likely be possible to reduce the bound enough to make the algorithm practical at least for planar graphs. Secondly, the linear time algorithm in [28] is contingent upon an enumeration of the 6-*list critical* graphs on the torus. Indeed, the authors show that there are only finitely many 6-list critical graphs on the torus, but a full list of these graphs is not explicitly known, and their bound on the maximum number of vertices any 6-list critical graph on the torus can have is far too large to be amenable to a straightforward enumerative check[1]. Also, their linear time algorithm does not specify an $\mathcal{L}$-coloring in the case when the graph is $\mathcal{L}$-choosable for a given list assignment $\mathcal{L}$. Thirdly, the linear time algorithm in [27] first requires a brute-force computation of the list colorings for any such list assignment on graphs of "small" order. However, the bound on the sizes of these small graphs is far too large to be computationally feasible, which makes the algorithm itself of mostly theoretical interest, as noted in a recent work by Dvořák and Postle [17].

This is in contrast with the results in this paper, wherein the 5-choosable graphs identified in Theorem 1 can also be given 5-list colorings in linear time, unlike as in [28]. Furthermore, the non-3-choosability of the 3-chromatic graphs $T(r, s, t)$ is not covered by the results in [28] since these graphs have girth equal to 3, whereas their algorithm for 3-list coloring is applicable only for graphs having girth at least 5. Lastly, our proof of Theorem 1 supplies an implementable algorithm for 5-list coloring all the toroidal graphs under consideration without the need for running a brute-force check on any of them, in contrast with [27].

Structure of This Paper. In the rest of this paper, we sketch the proof of Theorem 1. We relegate the full details to the arXiv version [8] due to space constraints.

---

[1] It is worth contrasting this with the corresponding colorability problem: while Thomassen [33] has shown that for every fixed surface there are only finitely many 6-*critical* graphs that embed on that surface, explicit lists of these 6-critical graphs are known only for the projective plane [1], the torus [32] and the Klein bottle [11,23].

2   Proof of Theorem 1

Definition 1. *For integers $r \geq 1$, $s \geq 1$ and $0 \leq t \leq s-1$, take $V = \{(i,j) : 1 \leq i \leq r, 1 \leq j \leq s\}$ to be the vertex set of the graph $T(r,s,t)$ equipped with the following edges. If $r = 1$, then $(1,j)$ is adjacent to $(1,j \pm 1)$, $(1,j \pm t)$ and $(1, j \pm (t+1))$. If $r > 1$, then: for each $1 < i < r$, (i,j) is adjacent to $(i, j \pm 1)$, $(i \pm 1, j)$ and $(i \pm 1, j \mp 1)$; $(1,j)$ is adjacent to $(1, j \pm 1)$, $(2,j)$, $(2, j-1)$, $(r, j+t+1)$ and $(r, j+t)$; (r,j) is adjacent to $(r, j \pm 1)$, $(r-1, j+1)$, $(r-1, j)$, $(1, j-t)$ and $(1, j-t-1)$.*

Here, addition in the first coordinate is taken modulo r and in the second coordinate is taken modulo s. It is clear that each $T(r,s,t)$ is a 6-regular triangulation of the torus. Conversely, we have

Theorem 2 (Altshuler [7], 1973). *Every 6-regular triangulation on the torus is isomorphic to $T(r,s,t)$ for some integers $r \geq 1$, $s \geq 1$, and $0 \leq t < s$.*

By the "column" C_i we shall mean the set $C_i := \{v \in V : v = (i,j) \text{ for some } 1 \leq j \leq s\}$. Define the *cylindrical triangulation* $C(r,s)$ to be the graph obtained from $T(r+1, s, 0)$ by deleting the column C_{r+1}. Next, we compile some well-known results (see [18], for instance) on the colorability of paths and cycles:

Lemma 1. *1. An even cycle is 2-list chromatic.*
2. An odd cycle is not 2-colorable, and hence not 2-choosable. However, if $\mathcal{L}$ is a list assignment of 2-lists on an odd cycle such that not all the lists are identical, then the cycle is $\mathcal{L}$-choosable.
3. If $\mathcal{L}$ is a list assignment on an odd cycle having one 1-list, one 3-list, and all the rest as 2-lists, then the cycle is $\mathcal{L}$-choosable.
4. If $\mathcal{L}$ is a list assignment on a path graph having one 1-list, and all the rest as 2-lists, then the path is $\mathcal{L}$-choosable.
   Moreover, the $\mathcal{L}$-colorings can all be found in linear time.

The following lemma due to S. Sinha (during an undergraduate research internship with the first author) is in a similar spirit to Thomassen's list coloring of a near-triangulation of the plane [31], and is a key ingredient in the proof of case (1) in Theorem 1. Note that $L_{(i,j)}$ denotes the list assigned on the vertex $(i,j) \in V$ by the list assignment $\mathcal{L}$.

Lemma 2. *For $r \geq 3$, $s \geq 3$, let $G = C(r,s)$ be a cylindrical triangulation. Suppose that $\mathcal{L}$ is a list assignment on G such that:*

(1) there exists $1 \leq j \leq s$ such that the exterior vertices $(1,j)$ and $(1, j-1)$ have lists of size equal to 4;
(2) every other exterior vertex has a list of size equal to 3;
(3) every interior vertex has a list of size equal to 5.

Then, G is $\mathcal{L}$-choosable. Moreover, an $\mathcal{L}$-coloring can be found in linear time.

Proof (sketch). By inductively coloring the rightmost column using Lemma 1, it suffices to consider the case $r = 3$. Color the vertex $(2, s)$ with $c \in L_{(2,s)} \setminus L_{(1,s)}$, which exists since C_2 has 5-lists. This reduces the sizes of the lists on each of the neighbors of $(2, s)$ by 1, except for $L_{(1,s)}$, which still has size equal to 4. Now, use Lemma 1 to color C_3, and then color the remaining vertices in the order as indicated in Fig. 1. The numbers indicate the reduced list sizes at that step of the coloring algorithm (note that the edges between the top and bottom rows are not shown in this and all subsequent figures). At the final step we are left to color a triangle with lists of sizes 1, 2, and 3, which is easily done.

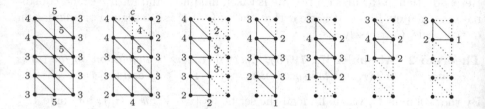

Fig. 1. Illustration of the sizes of the lists on the vertices at each step for $G = C(3, 5)$

2.1   Reductions for the Proof of Case (1)

Suppose that $T(r, s, t)$, for $r \geq 4$, $s \geq 3$, has a 5-list assignment $\mathcal{L}$. We assume that not all the lists are identical, since these graphs are all 5-colorable in linear time by the results in [12, 29, 35].

Reduction 1. *For every vertex, its list is contained in the union of its lists on its two left neighbors (as well as of its right neighbors).*

Proof (sketch). If not, choose a color for v that is not in the union of the lists of those two neighboring vertices. Use Lemma 1 to color the entire column of v, and notice that Lemma 2 is now applicable.

Next, focus on a pair of adjacent vertices on the same column that have distinct lists. Applying Lemmas 1 and 2 as before to this pair and their neighbors on an adjacent column (say, the left one), we can cut down to

Reduction 2. *Whenever (i, j) and $(i, j - 1)$ have distinct lists assigned by $\mathcal{L}$, one of the following three configurations holds (and also one of a similar set of configurations obtained by analysing the vertices adjacent on the right column):*

(a) $L_{(i,j)} = L_{(i-1,j+1)}$ *and* $L_{(i,j-1)} = L_{(i-1,j-1)}$;
(b) $L_{(i,j)} = L_{(i-1,j+1)} = L_{(i-1,j)}$ *and* $L_{(i,j-1)} \neq L_{(i-1,j-1)}$;
(c) $L_{(i,j)} \neq L_{(i-1,j+1)}$ *and* $L_{(i,j-1)} = L_{(i-1,j)} = L_{(i-1,j-1)}$.

For the third reduction, focus on a pair of adjacent vertices on adjacent columns that have the same lists. For the fourth reduction, focus on a face in which the two adjacent vertices lying on the same column have the same lists. Using Reduction 2 on the former, and Lemmas 1 and 2 on the latter, we get

Reduction 3. *Whenever u and v are adjacent vertices on distinct columns with $L_u = L_v$, there is a vertex w adjacent to both u and v such that $L_w = L_u = L_v$.*

Reduction 4. *Whenever u, v and w are mutually adjacent vertices having identical lists, with v and w lying on the same column, at least one of the vertical neighbors of u has a list identical to L_u.*

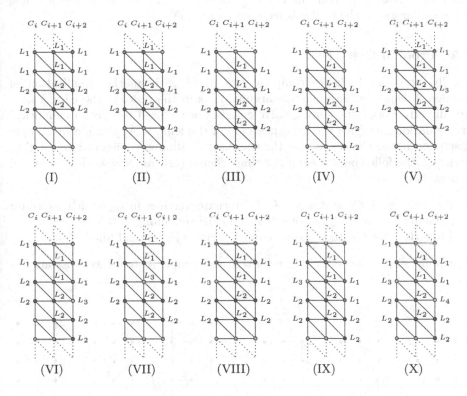

Fig. 2. Illustration of the ten configurations arrived at after reductions

What remains is to exploit the structure of 6-regular triangulations given by Theorem 2 with the rigidity imposed on the list assignment by Reductions 1 to 4. For a list L, define the *list-class* of L in G, denoted $G[L]$, to be the induced subgraph of G on those vertices v such that $L_v = L$. Let $L \in \mathcal{L}$ and let H be a (maximal connected) component of $G[L]$. If $V(H)$ is a singleton, we call H an *isolated component*, else we call H a *nonisolated component*.

Lemma 3. *Suppose that $\mathcal{L}$ obeys Reductions 1 to 4.*

(1) Let H be an isolated component, $V(H) = \{(i,j)\}$. Then, there are distinct lists $L', L'' \in \mathcal{L}$ such that $L_{(i-1,j+1)} = L_{(i,j+1)} = L_{(i+1,j+1)} = L_{(i+1,j)} = L'$ and $L_{(i-1,j)} = L_{(i-1,j-1)} = L_{(i,j-1)} = L_{(i+1,j-1)} = L''$.

(2) Let H be a nonisolated component of a list-class $G[L]$, with $v \in V(H)$. Then, at least one vertical neighbor of v also belongs to $V(H)$.

Putting together all of the above, we get a complete description of the list assignment $\mathcal{L}$ from only the information of lists assigned on every four or five consecutive vertices in any one column: the lists propagate across columns in any of precisely ten ways, as shown in Fig. 2. The lists labelled L_3 and L_4 in Fig. 2 belong to isolated components. For the full details for how one arrives at these ten configurations, the reader is requested to see [8].

2.2   Proof of Case (1)

Ideally, one would like to complete the proof with another application of Lemma 2. However, an induction argument as in the proof of the lemma does not directly work here, since a naive coloring of C_1 need not give a cylindrical triangulation satisfying the hypothesis (1) of the lemma. Applying a little more discretion in our choices, we use the small set of allowed configurations for $\mathcal{L}$ to arrive at the following two-step coloring scheme (assume $r = 4$ without loss of generality):

1. Properly color C_1 and a set J of alternate vertices in C_3 such that (after reducing the lists) C_2 has one 4-list and the remaining as 3-lists.
2. Properly color C_4, then the remaining vertices in C_3, and finally C_2.

Assuming step 1 is successfully achieved, we complete step 2 as illustrated in Fig. 3.

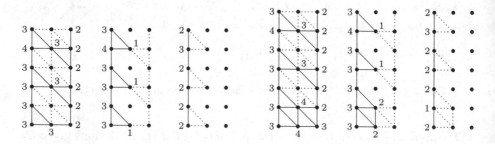

Fig. 3. Sizes of the lists on the columns C_2, C_3 and C_4 in step 2 when $s = 6$ and 7, respectively

Step 1 crucially uses the reduction into the ten cases illustrated in Fig. 2. Indeed, for each of the ten configurations that could appear on the column C_1, we describe an explicit procedure for coloring C_1, as well as for picking out the set J and a coloring for it, so that step 1 is completed. This is a three stage process; for the full details, the reader is again referred to [8].

2.3 Proofs of Cases (2) to (4)

Notice that Lemma 2 is not applicable on $C(r, s)$ for $r \leq 2$, so cases (2) and (3) of Theorem 1 require a different line of attack. So, we shall instead use the narrow length of the $r \times s$ grid to place restrictions on the list assignment $\mathcal{L}$. The analysis is therefore shorter in these cases compared to case (1) as discussed above, but we omit the proof here due to space constraints and instead refer the reader to [8].

For case (4), the 5-choosability of the 3-chromatic 6-regular toroidal triangulations was settled in a previous work [9], but a small modification is required to get a linear time algorithm, for which we apply a lemma of Bondy, Bopanna, and Siegel [5] instead of the theorem of Alon and Tarsi [5]. The necessary changes are explicated in [8].

2.4 Proof of Non-3-Choosability of the Graphs in Cases (1) to (4)

Note that $T(r, s, t)$ is 3-chromatic if and only if $s \equiv 0 \equiv r - t \pmod 3$. Let $L_1 := \{1, 2, 3\}$, $L_2 := \{2, 3, 4\}$, and $L_3 := \{1, 3, 4\}$. Let $\mathcal{L}$ be the list-assignment that assigns these lists to the columns of $T(r, s, t)$ ($r \geq 4$, $s \geq 3$) as follows: C_1 and C_2 are assigned L_1, C_3 is assigned L_2, and $C_4, \ldots, C_r$ are assigned L_3. Let the vertices $(1, 1)$ and $(1, 2)$ be properly colored using $\mathcal{L}$ in any manner. This uniquely determines a proper coloring of the induced subgraph on $C_1 \cup C_2$.

Now, there is a unique way to extend this coloring properly to the induced subgraph on $C_2 \cup C_3$ as follows: simply extend the coloring from C_2 to C_3 using the same lists used on C_2, namely $L_1 = \{1, 2, 3\}$; then, recolor all the vertices in C_3 that have the color 1 with the color 4. In this manner, one can see that the coloring is extended uniquely to the rest of C_3, with 4 occuring in those places where 1 would have occured had C_3 also been colored using $L_1 = \{1, 2, 3\}$.

Next, repeat the same process to extend the coloring on C_3 to a proper coloring on the induced subgraph on $C_3 \cup C_4 \cup \cdots \cup C_r$ as follows: color the vertices in $C_4 \cup \cdots \cup C_r$ using the colors used on C_3, namely $L_2 = \{2, 3, 4\}$, and then recolor those vertices in $C_4 \cup \cdots \cup C_r$ that have the color 2 with the color 1.

Now, we note that this coloring cannot be proper on all of $T(r, s, t)$ because this process of successive relabelling has mapped the tuple $(1, 2, 3)$ to $(2, 1, 3)$. Thus, for this to be a proper coloring of $T(r, s, t)$, the original coloring on C_1 must arise as the unique extension of the coloring on C_r to the induced subgraph on $C_r \cup C_1$; but, $(2, 1, 3)$ is not a cyclic permutation of $(1, 2, 3)$, so this cannot happen for any t.

The rest of the cases (i.e., $r \leq 3$) are handled similarly, and we direct the reader to [8] for the full details.

Acknowledgements. Research of Brahadeesh Sankarnarayanan is supported by the National Board for Higher Mathematics (NBHM), Department of Atomic Energy (DAE), Govt. of India.

References

1. Albertson, M.O., Hutchinson, J.P.: The three excluded cases of Dirac's map-color theorem. Ann. New York Acad. Sci. **319**(1), 7–17 (1979). https://doi.org/10.1111/j.1749-6632.1979.tb32768.x. MR0556001, Zbl 0489.05023

2. Albertson, M.O., Hutchinson, J.P.: On six-chromatic toroidal graphs. Proc. Lond. Math. Soc., Third Ser. **41**(Part 3), 533–556 (1980). https://doi.org/10.1112/plms/s3-41.3.533. MR0591654, Zbl 0394.05018

3. Alon, N.: Restricted colorings of graphs. In: Walker, K. (ed.) Surveys in Combinatorics, 1993. London Mathematical Society Lecture Note series, vol. 187, pp. 1–34. Cambridge University Press, Cambridge (1993). https://doi.org/10.1017/CBO9780511662089.002. MR1239230, Zbl 0791.05034

4. Alon, N.: Combinatorial Nullstellensatz. Comb. Probab. Comput. **8**(1–2), 7–29 (1999). https://doi.org/10.1017/S0963548398003411. MR1684621, Zbl 0920.05026

5. Alon, N., Tarsi, M.: Colorings and orientations of graphs. Combinatorica **12**(2), 125–134 (1992). https://doi.org/10.1007/BF01204715. MR1179249, Zbl 0756.05049

6. Altshuler, A.: Hamiltonian circuits in some maps on the torus. Discrete Math. **1**(4), 299–314 (1972). https://doi.org/10.1016/0012-365X(72)90037-4. MR0297597, Zbl 0226.05109

7. Altshuler, A.: Construction and enumeration of regular maps on the torus. Discrete Math. **4**(3), 201–217 (1973). https://doi.org/10.1016/S0012-365X(73)80002-0. MR0321797, Zbl 0253.05117

8. Balachandran, N., Sankarnarayanan, B.: 5-list-coloring toroidal 6-regular triangulations in linear time. arXiv:2106.01634 [math.CO] (2021). https://arxiv.org/abs/2106.01634

9. Balachandran, N., Sankarnarayanan, B.: The choice number versus the chromatic number for graphs embeddable on orientable surfaces. Electron. J. Comb. **28**(4), #P4.50, 20 (2021). https://doi.org/10.37236/10263. MR4394670, Zbl 1486.05101

10. Cai, L., Wang, W.F., Zhu, X.: Choosability of toroidal graphs without short cycles. J. Graph Theory **65**(1), 1–15 (2010). https://doi.org/10.1002/jgt.20460. MR2682511, Zbl 1205.05083

11. Chenette, N., Postle, L., Streib, N., Thomas, R., Yerger, C.: Five-coloring graphs on the Klein bottle. J. Comb. Theory, B **102**(5), 1067–1098 (2012). https://doi.org/10.1016/j.jctb.2012.05.001. MR2959391, Zbl 1251.05054

12. Collins, K.L., Hutchinson, J.P.: Four-coloring six-regular graphs on the torus. In: Hansen, P., Marcotte, O. (eds.) Graph Colouring and Applications. CRM Proceedings on Lecture Notes, vol. 23, pp. 21–34. American Mathematical Society, Rhode Island (1999). https://doi.org/10.1090/crmp/023/02. MR1723634, Zbl 0944.05044

13. Dabrowski, K.K., Dross, F., Johnson, M., Paulusma, D.: Filling the complexity gaps for colouring planar and bounded degree graphs. J. Graph Theory **92**(4), 377–393 (2019). https://doi.org/10.1002/jgt.22459. MR4030964, Zbl 1443.05064

14. Dailey, D.P.: Uniqueness of colorability and colorability of planar 4-regular graphs are NP-complete. Discrete Math. **30**(3), 289–293 (1980). https://doi.org/10.1016/0012-365X(80)90236-8. MR0573644, Zbl 0448.05030

15. Dvořák, Z., Kawarabayashi, K.-i., Thomas, R.: Three-coloring triangle-free planar graphs in linear time. ACM Trans. Algorithms **7**(4), 14, Article no. 41 (2011). https://doi.org/10.1145/2000807.2000809. MR2836980, Zbl 1295.05231

16. Dvořák, Z., Kawarabayashi, K.-i.: List-coloring embedded graphs. In: Khanna, S. (ed.) Proceedings of the Twenty-Fourth Annual ACM-SIAM Symposium on Discrete Algorithms, SODA, New Orleans, LA, 6–8 January 2013, pp. 1004–1012 (2013). https://doi.org/10.1137/1.9781611973105.72. MR3202964, Zbl 1423.05063
17. Dvořák, Z., Postle, L.: On decidability of hyperbolicity. Combinatorica (2022). https://doi.org/10.1007/s00493-022-4891-8. Advance online publication
18. Erdös, P., Rubin, A.L., Taylor, H.: Choosability in graphs. In: Chinn, P.Z., McCarthy, D. (eds.) Proceedings of the West Coast Conference on Combinatorics, Graph Theory and Computing (Humboldt State University, Arcata, California, 5–7 September 1979). Congr. Numer., vol. 26, pp. 125–157. Utilitas Mathematica Publishing Inc., Manitoba (1980). MR0593902, Zbl 0469.05032
19. Fleischner, H., Stiebitz, M.: A solution to a colouring problem of P. Erdős. Discrete Math. **101**(1–3), 39–48 (1992). https://doi.org/10.1016/0012-365X(92)90588-7. MR1172363, Zbl 0759.05037
20. Galvin, F.: The list chromatic index of a bipartite multigraph. J. Comb. Theory, B **63**(1), 153–158 (1995). https://doi.org/10.1006/jctb.1995.1011. MR1309363, Zbl 0826.05026
21. Grötzsch, H.C.: Zur Theorie der diskreten Gebilde. 7. Mitteilung: Ein Dreifarbensatz für dreikreisfreie Netze auf der Kugel. Wiss. Z. Martin-Luther-Univ. Halle-Wittenberg, Math.-Natur. Reihe **8**(1), 109–120 (1958). MR0116320, Zbl 0089.39506
22. Gutner, S.: The complexity of planar graph choosability. Discrete Math. **159**(1–3), 119–130 (1996). https://doi.org/10.1016/0012-365X(95)00104-5. MR1415287, Zbl 0865.05066
23. Kawarabayashi, K.-i., Král', D., Kynčl, J., Lidický, B.: 6-critical graphs on the Klein bottle. SIAM J. Discrete Math. **23**(1), 372–383 (2009). https://doi.org/10.1137/070706835. MR2476836, Zbl 1215.05067
24. Khanna, S., Linial, N., Safra, S.: On the hardness of approximating the chromatic number. Combinatorica **20**(3), 393–415 (2000). https://doi.org/10.1007/s004930070013. MR1774844, Zbl 0964.68065
25. Kleinberg, J., Tardos, É.: Algorithm Design, 1st edn. Pearson Educ. Ltd., Harlow, Essex (2014)
26. Li, Z., Shao, Z., Petrov, F., Gordeev, A.: The Alon-Tarsi number of a toroidal grid. arXiv:1912.12466 [math.CO] (2019). https://doi.org/10.48550/arXiv.1912.12466
27. Postle, L.: Linear-time and efficient distributed algorithms for list coloring graphs on surfaces. In: Zuckerman, D. (ed.) Proceedings of the 2019 IEEE 60th Annual Symposium on Foundations of Computer Science, FOCS, Baltimore, MD, 9–12 November 2019, pp. 929–941. IEEE, Los Alamitos, CA (2019). https://doi.org/10.1109/FOCS.2019.00060. MR4228207
28. Postle, L., Thomas, R.: Hyperbolic families and coloring graphs on surfaces. Trans. Am. Math. Soc., B **5**, 167–221 (2018). https://doi.org/10.1090/btran/26. MR3882883, Zbl 1401.05126
29. Sankarnarayanan, B.: Note on 4-coloring 6-regular triangulations on the torus. Ann. Comb. **26**(3), 559–569 (2022). https://doi.org/10.1007/s00026-022-00573-8. MR4473892, Zbl 07583948
30. Thomassen, C.: Tilings of the torus and the Klein bottle and vertex-transitive graphs on a fixed surface. Trans. Am. Math. Soc. **323**(2), 605–635 (1991). https://doi.org/10.2307/2001547. MR1040045, Zbl 0722.05031
31. Thomassen, C.: Every planar graph is 5-choosable. J. Comb. Theory, B **62**(1), 180–181 (1994). https://doi.org/10.1006/jctb.1994.1062. MR1290638, Zbl 0805.05023
32. Thomassen, C.: Five-coloring graphs on the torus. J. Comb. Theory, B **62**(1), 11–33 (1994). https://doi.org/10.1006/jctb.1994.1052. MR1290628, Zbl 0805.05022

33. Thomassen, C.: Color-critical graphs on a fixed surface. J. Comb. Theory, B **70**(1), 67–100 (1997). https://doi.org/10.1006/jctb.1996.1722. MR1441260, Zbl 0883.05051
34. Vizing, V.G.: Coloring the vertices of a graph in prescribed colors (in Russian). Metody Diskretn. Anal. **29**, 3–10 (1976). MR0498216, Zbl 0362.05060
35. Yeh, H.G., Zhu, X.: 4-colorable 6-regular toroidal graphs. Discrete Math. **273**(1–3), 261–274 (2003). https://doi.org/10.1016/S0012-365X(03)00242-5. MR2025955, Zbl 1034.05024

On Locally Identifying Coloring of Graphs

Sriram Bhyravarapu[1(✉)], Swati Kumari[2], and I. Vinod Reddy[2]

[1] The Institute of Mathematical Sciences, HBNI, Chennai, India
sriramb@imsc.res.in
[2] Department of Electrical Engineering and Computer Science, IIT, Bhilai, India
{swatik,vinod}@iitbhilai.ac.in

Abstract. A proper coloring of a graph G is said to be *locally identifying coloring* (lid-coloring for short), if for every pair of adjacent vertices u and v with distinct closed neighborhood, the sets of colors assigned in the closed neighborhood of u and v are distinct. The minimum number of colors required in any lid-coloring of a graph G is called the *lid-chromatic number* of G, denoted by $\chi_{lid}(G)$.

In this paper, we give a characterization of graphs whose lid-chromatic number equals the number of vertices. Next, we study lid-coloring on several restricted graph classes. We show that for any block graph G, $\chi_{lid}(G) \leq 2\chi(G)$, where $\chi(G)$ denotes the chromatic number of G. We show that the lid-chromatic number of a biconvex bipartite graph can be computed in polynomial time. Finally, we find the lid-chromatic number of the Cartesian and Lexicographic products of paths and cycles.

1 Introduction

All graphs considered in this paper are finite, undirected and simple (without loops and multiple edges). For a graph $G = (V, E)$, we use $V(G)$ and $E(G)$ to denote the vertex set and edge set of G respectively. The open neighborhood of a vertex v, denoted $N(v)$, is the set of vertices adjacent to v and the set $N[v] = N(v) \cup \{v\}$ denote the closed neighborhood of v. Let $f : V(G) \to \mathbb{N}$ be a vertex coloring of G. For a subset $X \subseteq V(G)$, $f(X) = \{f(v) \mid v \in X\}$ denotes the set of colors that appear in X.

A vertex coloring f of a graph G is called *locally identifying coloring* (lid-coloring for short) if (i) f is a proper coloring of G (no two adjacent vertices have the same color) and (ii) for each pair of adjacent vertices u, v with $N[u] \neq N[v]$, we have $f(N[u]) \neq f(N[v])$. The smallest integer k for which G admits a lid-coloring is called the *lid-chromatic number* of G, denoted by $\chi_{\text{lid}}(G)$. Note that the lid-chromatic number of a graph G is the maximum of the lid-chromatic numbers of its connected components. Therefore, throughout this paper we restrict ourselves to connected graphs.

Locally identifying coloring was introduced by Esperet et al. [2]. They showed that the decision version of the lid-coloring is NP-complete on bipartite graphs but polynomial time solvable on trees. They also proved that $\chi_{lid}(G) \leq 2\chi(G)$ for interval graphs, split graphs and cographs and conjectured the same bound for chordal graphs. They also showed that for an outerplanar graph G, $\chi_{lid}(G) \leq 20$. Foucaud et al. [4] showed that any graph G with maximum degree Δ has a locally identifying coloring with at most $2\Delta^2 - 3\Delta + 3$ colors, by answering positively a question raised in the paper [2]. Goncalves [6] showed that for any planar graph G, $\chi_{lid}(G) \leq 1280$. They also proved that for any graph class of bounded expansion, the lid-chromatic number is bounded. Martins and Sampaio [7] obtained a linear time algorithms to calculate the lid-chromatic number for some classes of graphs having few P_4's.

Our Contributions: It is easy to see that $\chi(G) = n$ if and only if $G = K_n$. In Sect. 3, we investigate the graphs for which the lid-chromatic number equals n and give a complete characterization of graphs having lid-chromatic number n.

Esperet et al. [2] conjectured that, for any chordal graph G, $\chi_{lid}(G) \leq 2\chi(G)$. They verified the conjecture for subclasses of chordal graphs such as interval graphs, split graphs and cographs. In Sect. 4, we show that the conjecture holds for block graphs, which are a subclass of chordal graphs.

Esperet et al. [2] showed that, if G is bipartite then $\chi_{lid}(G) \leq 4$. They also proved that deciding whether a bipartite graph G has $\chi_{lid}(G) \in \{3, 4\}$ is NP-complete. As lid-coloring is NP-complete on bipartite graphs, we focus on biconvex bipartite graphs, which is a subclass of bipartite graphs. In Sect. 5, we show that lid-chromatic number of biconvex bipartite graphs can be computed in polynomial time.

Finally, we investigate lid-coloring on Cartesian product and Lexicographic product of graphs. Proper coloring of various graph products has been well studied [3,5,8,9]. For example, the chromatic number of the Cartesian product of two graphs G and H [8] is equal to $\max\{\chi(G), \chi(H)\}$. The chromatic number of a lexicographic product of two graphs G and H is equal to the b-fold chromatic number [5] of G, where $b = \chi(G)$. In Sects. 6 and 7, we give exact values of the lid-chromatic number of Cartesian and Lexicographic products of paths and cycles.

2   Preliminaries

We denote the set $\{1, 2, \ldots, k\}$ with $[k]$. Let $c : V(G) \to [k]$ be a coloring of the vertices of G using k colors. Let $S \subseteq V(G)$, then $c(S) = \{c(v) \mid v \in S\}$. We say an edge $uv \in E(G)$ *respects lid-coloring* if either (i) $N[u] = N[v]$, or (ii) $c(N[u]) \neq c(N[v])$. We associate a distinguishing color for each edge satisfying the condition (ii) in the above sentence. We define a *distinguishing color* for the edge uv to be a color from $(c(N[u]) \setminus c(N[v])) \cup (c(N[v]) \setminus c(N[u]))$. That is, a color that is seen in the closed neighborhood of u and not in the closed neighborhood of v or vice-versa. The minimum degree of the graph G is denoted by $\delta(G)$. For more details on graph theoretic notation or terminology, we refer the reader to the textbook [1]. The lid-chromatic numbers of paths and cycles are stated below.

Due to space constraints, the proofs of the results marked $(\star)$ are presented in the full version of the paper.

Lemma 1 ([4]). *For a positive integer n, where $n \geq 2$, we have*

$$\chi_{lid}(P_n) = \begin{cases} 2 & \text{if } n = 2; \\ 3 & \text{if } n \text{ is odd} \\ 4 & \text{if } n \text{ is even and } n \neq 2; \end{cases}$$

Lemma 2 ([4]). *For a positive integer n, where $n \geq 4$, we have*

$$\chi_{lid}(C_n) = \begin{cases} 3 & \text{if } n \equiv 0 \bmod 4; \\ 5 & \text{if } n = 5 \text{ or } 7; \\ 4 & \text{otherwise}; \end{cases}$$

3 Graphs with $\chi_{lid}(G) = |V(G)|$

Let $G = (V, E)$ be a graph on n vertices. It is known that $\chi(G) = n$ if and only if $G = K_n$. In this section, we investigate graphs for which the lid-chromatic number equals n. We first show the characteristics of graphs G whose lid-chromatic number is at most $n - 1$. Using this, we conclude with the structure of graphs that require n colors.

Theorem 1. *Let $G = (V, E)$ be a graph on n vertices. Then $\chi_{lid}(G) \leq n - 1$ if and only if there exist two non-adjacent vertices $x, y \in V(G)$ such that for every edge $uv \in E(G)$ at least one of the following conditions is satisfied.*

(1) *if either u or v belong to $\{x, y\}$*
    (a) $N[u] = N[v]$, *or*
    (b) $N[u] \setminus \{x, y\} \neq N[v] \setminus \{x, y\}$.
(2) *Both u and v does not belong to $\{x, y\}$.*
    (a) $N[u] = N[v]$, *or*
    (b) $N[u] \setminus \{x, y\} \neq N[v] \setminus \{x, y\}$, *or*
    (c) $N[u] \setminus \{x, y\} = N[v] \setminus \{x, y\}$ *and*
        (i) *if* $N(u) \cap \{x, y\} \neq \emptyset$, *then* $N(v) \cap \{x, y\} = \emptyset$, *or*
        (ii) *if* $N(u) \cap \{x, y\} = \emptyset$, *then* $N(v) \cap \{x, y\} \neq \emptyset$.

Proof. ($\Longrightarrow$) There exists a lid-coloring of G using at most $n - 1$ colors. We need to show that there exist two non-adjacent vertices $x, y \in V(G)$ such that for every edge $uv \in E(G)$ at least one of the given two conditions is satisfied. Without loss of generality, let $\chi_{lid}(G) = n - 1$. Otherwise, if $\chi_{lid}(G) < n - 1$, we can always replace the repeated colors in the lid-coloring of G with new colors to get a lid-coloring of G that uses $n - 1$ colors. Let $c : V(G) \to [n - 1]$ be a lid-coloring of G. There exists two non-adjacent vertices x and y in G such that $c(x) = c(y)$.

Consider an arbitrary edge uv of G.

Case 1: Either u or v belong to $\{x, y\}$.

Without loss of generality, let $u = x$. Suppose the edge xv does not satisfy the conditions 1(a) and 1(b). That is, we have

$$N[x] \neq N[v], \text{ and} \tag{1}$$

$$N[x] \setminus \{x, y\} = N[v] \setminus \{x, y\}. \tag{2}$$

From Eqs. (1), (2) and the fact that $c(x) = c(y)$, it is clear that $c(N[x]) = c(N[v])$, which is a contradiction to the fact that c is a lid-coloring of G.

Case 2: Both u and v does not belong to $\{x, y\}$.

Suppose that the edge uv does not satisfy any of the three conditions 2(a), 2(b) and 2(c), we have

$$N[u] \neq N[v], \tag{3}$$

$$N[u] \setminus \{x, y\} = N[v] \setminus \{x, y\}, \text{ and} \tag{4}$$

$$N(u) \cap \{x, y\} \neq \emptyset, N(v) \cap \{x, y\} \neq \emptyset. \tag{5}$$

If $N(u) \cap \{x, y\} = N(v) \cap \{x, y\}$, then using (4) we get $N[u] = N[v]$, which is a contradiction to (3). Therefore, using (3),(4) and (5), we get that

$$N(u) \cap \{x, y\} \neq N(v) \cap \{x, y\}. \tag{6}$$

Using Eq. (4), we have $c(N[u] \setminus \{x, y\}) = c(N[v] \setminus \{x, y\})$. From Eq. (5), we know that u and v have some neighbor in $\{x, y\}$. Since, $c(x) = c(y)$ the set of colors assigned to vertices in $N[u]$ and $N[v]$ are the same in G even though $N[u] \neq N[v]$. This is a contradiction to the assumption that c is a lid-coloring of G.

($\Longleftarrow$) Let G be a graph with two non-adjacent vertices x and y such that for every edge $uv \in E(G)$, at least one of the given conditions is satisfied. We need to show that $\chi_{\text{lid}}(G) \leq n - 1$.

Let $f : V(G) \rightarrow [n-1]$ be a coloring of G constructed as follows. Assign $f(x) = f(y) = 1$. Each of the vertices in $V(G) \setminus \{x, y\}$ are assigned distinct colors from $[n-1] \setminus \{1\}$. We now argue that f is a lid-coloring of G.

Since every vertex in $V(G) \setminus \{x, y\}$ is assigned a distinct color and $xy \notin E(G)$, we have that f is a proper coloring. Consider an arbitrary edge $uv \in E(G)$.

Case 1: Either u or v belongs to $\{x, y\}$.

Without loss of generality we assume that $u = x$. If $N[x] = N[v]$ then there is nothing to prove.

If $N[x] \setminus \{x, y\} \neq N[v] \setminus \{x, y\}$, then as no two vertices in $V(G) \setminus \{x, y\}$ are colored with the same color by f. Therefore $f(N[x]) \neq f(N[v])$.

Case 2: Both u and v does not belong to $\{x, y\}$.

We assume that the edge uv does not satisfy condition 2(a), otherwise there is nothing to prove.

If $N[u] \setminus \{x, y\} \neq N[v] \setminus \{x, y\}$, then as no two vertices in $V(G) \setminus \{x, y\}$ are colored with the same color by f. Therefore, $f(N[u]) \neq f(N[v])$.

Lastly, for any edge $uv \in E(G)$, if $N[u] \setminus \{x, y\} = N[v] \setminus \{x, y\}$, then set of colors (say S) used in $N[u] \setminus \{x, y\}$ is same as the set of colors used in $N[v] \setminus \{x, y\}$. If $N(u) \cap \{x, y\} = \emptyset$, then set of colors in $N[u]$ is S. Since $N(v) \cap \{x, y\} \neq \emptyset$, set of colors used in $N[v]$ is $S \cup \{1\}$. Therefore, set of colors used in $N[u]$ is not same as set of colors used in $N[v]$. Similarly, if $N(u) \cap \{x, y\} \neq \emptyset$, then set of colors in $N[u]$ is $S \cup \{1\}$. Since, $N(v) \cap \{x, y\} = \emptyset$, set of colors in $N[v]$ is S. Therefore, set of colors used in $N[u]$ is not same as set of colors used in $N[v]$.

Hence, f is a lid-coloring that uses $n - 1$ colors.   □

Given a graph G, if G does not satisfy the conditions mentioned in Theorem 1, then the lid-chromatic number of G is equal to $|V(G)| = n$.

Thus, as a consequence of Theorem 1, we have the following corollary.

Corollary 1. *Given a graph G, we can decide in polynomial time if $\chi_{lid}(G) = n$, where n is the number of vertices of the graph G.*

4   Block Graphs

This section is devoted to block graphs. We prove that every block graph has lid-chromatic number at most $2\omega(G)$, where $\omega(G)$ is the size of a largest clique in G.

Definition 1 (Block Graphs [1]). *A vertex u of a connected graph G is called a cut vertex if $G-u$ is disconnected. A block of a graph G is a maximal connected induced subgraph of G that has no cut vertex. A block graph is a graph in which every block is a clique.*

To prove the result, we use the notion of block decomposition of graphs. Let $G = (V, E)$ be a block graph with q blocks $B_1, \ldots, B_q$. The block decomposition of G is a tree denoted by $T_B = (V_B, E_B)$ where $V_B = \{B_1, \ldots, B_q\}$ and $E_B = \{B_i B_j \mid B_i \cap B_j \neq \emptyset\}$. We root the tree T_B at a node B_R having at least two cut vertices. For example, the block decomposition of P_4 (the path on four vertices) contains three blocks $\{B_1, B_2, B_3\}$ where every block is a K_2 with B_2 being adjacent to both B_1 and B_3. The level of a block B denoted by $\ell(B)$ is the distance of B from the root block B_R in T_B. Clearly $\ell(B_R) = 0$. We also define level of a vertex v as $\ell(v) = \min\{\ell(B) \mid v \in V(B)\}$. For each block B at level $p \geq 1$, we call the vertex $v \in B$ as the *distinguishing ancestor vertex* of B, denoted by $dav(B)$, if and only if $\ell(v) < \ell(B)$. We call B' as the *parent block* of B if B' is the parent of B in T_B.

For a vertex $v \in V(G)$, let

$$D_1(v) = \{B \mid \ell(B) = \ell(v) + 1 \text{ and } dav(B) = v\}, \text{ and}$$

$$D_2(v) = \{B \mid \ell(B) = \ell(v) + 2, \text{ and } dist_G(dav(B), v) = 1\},$$

where $dist_G(w, w')$ is the length of a shortest path from w to w' in G.

Notice that if v is not a cut-vertex in G, then $D_1(v) = D_2(v) = \emptyset$. An illustration of a block decomposition is shown in Fig. 1.

We first present the overall idea of the algorithm. We consider the blocks level by level (starting with level 0) and in each level we process the blocks from left to the right in some arbitrary order. First, all the vertices in the root block are assigned colors. When a non-root block B is considered for coloring as in the above ordering, all vertices at level at most $\ell(B) - 1$ are assigned colors.

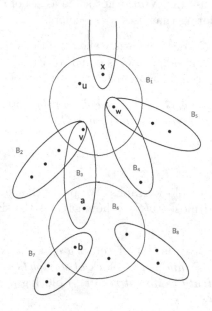

Fig. 1. An illustration of a block decomposition of G where each B_i represents a block. Considering $\ell(B_1) = p$, we have $\ell(B_2) = \ell(B_3) = \ell(B_4) = \ell(B_5) = p+1$, $\ell(B_6) = p+2$ and $\ell(B_7) = \ell(B_8) = p+3$. Also $dav(B_1) = x$, $dav(B_2) = dav(B_3) = v$, and $dav(B_6) = a$. For the vertex x, we have $D_1(x) = \{B_1\}$ and $D_2(x) = \{B_2, B_3, B_4, B_5\}$. For the vertex a, we have $D_1(a) = \{B_6\}$ and for the vertex b, we have $D_1(b) = \{B_7\}$.

We color each block B in two phases. In its first phase, exactly two vertices of B are colored while coloring its parent block. After the first phase, we find a set of forbidden colors for B. In its second phase, we arbitrarily assign colors to the remaining uncolored vertices of B from the set $[2k]$ excluding its forbidden colors.

Theorem 2. *If G is a block graph, then $\chi_{lid}(G) \leq 2\omega(G)$.*

Proof. Given a block graph G and its block decomposition, we show that $\chi_{lid}(G) \leq 2k$, where $k = \omega(G)$. To prove the result, we present a lid-coloring $c : V(G) \to [2k]$ of G using at most $2k$ colors.

Coloring Procedure: Let $N(B)$ be the set of colors that are forbidden to be used in the block B. Initially $N(B) = \emptyset$, for each block B. We consider the blocks level by level and from left to right in each level.

At each block B, for each cut vertex $v \in V(B) \setminus \{dav(B)\}$, we identify two colors $W(v)$ and $A(v)$. For each edge vw, where $w \in V(B) \setminus \{dav(B)\}$, the color $W(v)$ serves as a distinguishing color for the edge vw. Similarly, $A(v)$ serves as a distinguishing color for the edges vw where $w \notin V(B)$. The intuition is that the edge vw respects lid-coloring because of the colors $W(v)$ and $A(v)$ depending on whether or not $w \in V(B)$.

Coloring the Root Block B_R: Let $\{v_1, v_2, \ldots, v_{k'}\}$, where $k' \leq k$, be the vertices of B_R. Each vertex in $V(B_R)$ is assigned a distinct color from $\{1, 2, \ldots, 2k\}$. WLOG, let two vertices v_1, v_2 from B_R be assigned the colors c_1 and c_2 respectively. For each cut vertex v of B_R, we do the following:

1. (a) If $v \neq v_1$ set $A(v) = c_1$, else set $A(v) = c_2$.
   (b) Forbid the color $A(v)$ to be used in all the blocks from $D_1(v) \cup D_2(v)$.
   That is, for all $B \in D_1(v) \cup D_2(v)$, update $N(B) = N(B) \cup \{A(v)\}$.
2. Choose $W(v)$ to be a color from $[2k] \setminus (c(B_R) \cup N(B'))$ such that for any two cut vertices u, u' of B_R, $W(u) \neq W(u')$, where B' is some block in $D_1(v)$. Here $c(B_R)$ represents the colors assigned to the vertices in $V(B_R)$, i.e., $c(B_R) = \{c(w) \mid w \in V(B_R)\}$, and $N(B')$ is the set of colors forbidden to be used in $V(B')$.
   Let $S(B_R)$ be the set of associated colors for the cut vertices of B_R. That is, $S(B_R) = \{W(v) \mid v$ is a cut vertex of $B_R\}$. We now color a vertex in each block of $D_1(v)$ in the following manner. For each $B' \in D_1(v)$, choose an arbitrary uncolored vertex $v' \in V(B')$ and assign $c(v') = W(v)$. Then update $N(B') = N(B') \cup (S(B_R) \setminus \{c(v')\})$.

Notice that after processing all the cut vertices of B_R, exactly two vertices in each of the blocks at level 1 are colored.

Coloring a Non-Root Block B: In the first phase coloring of B, exactly two vertices of B are colored. Let the two colored vertices be $dav(B)$ and v'. In the second phase coloring of B, we arbitrarily assign colors to the uncolored vertices in $V(B)$ from $[2k] \setminus (N(B) \cup \{c(dav(B)), c(v')\})$.

For each cut vertex $v \in V(B) \setminus \{dav(B)\}$, we do the following:

1. (a) Set $A(v) = c(dav(B))$.
   (b) Forbid the color $A(v)$ to be used in all blocks from $D_1(v) \cup D_2(v)$. That is, for all $B \in D_1(v) \cup D_2(v)$, update $N(B) = N(B) \cup \{A(v)\}$.
2. This step is similar to the Step 2 of the root block case where each instance of B_R is to be replaced by B. Also whenever we talk about the cut vertices of B_R, we replace it with the the cut vertices in $V(B) \setminus \{dav(B)\}$.

We recursively apply the above coloring procedure on the blocks in order and complete the coloring. We now show that the coloring obtained is a lid-coloring of G. Before we prove the correctness, we will look at the following claim.

Claim. For any non-root block B, we have $|N(B)| \leq k + 2$.

Proof. Recall that, the set $N(B)$ represents the set of colors that are forbidden in B. Let B' and B'' be the parent and the grand-parent of B in T_B respectively. From the description of the algorithm, the colors of the vertices $dav(B)$ and $dav(B')$ are forbidden in B. In the worst case, B' contains at most k cut vertices.

For each cut vertex $v \in V(B')$, we assigned a color $W(v)$ which is forbidden to be used in B'. Therefore by combining all, at most $k + 2$ colors are forbidden in B.

Correctness: We first need to show why it is possible to color the uncolored vertices of each block B in its second phase of coloring. From the above claim, we have that $|N(B)| \le k + 2$. Since B has exactly two colored vertices after its first phase of coloring, the number of uncolored vertices in B is at most $k - 2$. We have the budget to color the remaining with $[2k] \setminus N(B)$.

It is easy to see that the coloring yields a proper coloring. We now show that each edge respects lid-coloring. Let uv be an edge. We have the following cases.

- $\ell(u) = \ell(v)$ Let $u, v \in V(B)$. If none of them are cut vertices of B, then $N[u] = N[v]$. If exactly one of them is a cut vertex, say u. Then by the coloring procedure, there exists a color $W(u) \in c(N[u])$ and $W(u) \notin c(N[v])$. Else both u and v are cut vertices. Then $W(u) \in c(N[u])$ and $W(u) \notin c(N[v])$. Also $W(v) \notin c(N[u])$ and $W(v) \in c(N[v])$.
- $\ell(u) \ne \ell(v)$
  Let $u, v \in V(B)$ for some block B. Then it follows that exactly one of v or u is $dav(B)$. WLOG let $dav(B) = u$ and B^* be the parent block of B where $u \in V(B^*)$. While coloring the block B^*, we had chosen the color of a vertex in B^* to be the $A(u)$ and forbid the color $A(u)$ in $D_1(u) \cup D_2(u)$. This forces none of the neighbors of v in $D_1(v)$ to be assigned the color $A(u)$. Hence the edge uv respects lid-coloring.

The edge uv respects lid-coloring in all the above cases. This completes the proof of Theorem 2. □

Remark: Consider the graph P_4. From Lemma 1, we have that $\chi_{lid}(P_4) = 4$. It is easy to see that P_4 is a block graph and the bound in Theorem 2 is tight.

5 Biconvex Bipartite Graphs

Definition 2 (Biconvex Bipartite Graph). *A bipartite graph $G = (X \cup Y, E)$ is called* convex bipartite *graph over the vertex set X if X can be enumerated such that for all $y \in Y$ the vertices adjacent to y are consecutive with respect to the ordering on X. If G is convex over both X and Y, it is said to be* biconvex bipartite *graph.*

Theorem 3. *If $G = (X \cup Y, E)$ is a connected biconvex bipartite graph having at least three vertices, then*

$$\chi_{lid}(G) = \begin{cases} 4, & \text{if } Z \cap X \ne \emptyset \text{ and } Z \cap Y \ne \emptyset \\ 3, & \text{otherwise} \end{cases}$$

where $Z = \{u \in X \cup Y \mid deg(u) = 1\}$.

Proof. Let $G = (X \cup Y, E)$ be a biconvex bipartite graph. Let $\sigma = x_1, x_2, \ldots, x_p$ be an enumeration of vertices of X and $\pi = y_1, y_2, \ldots, y_q$ be an enumeration of vertices of Y. Let Z represent the set of degree one vertices in G. We will divide the proof into the following cases depending on the existence of degree one vertices in X and Y.

Case 1: $|Z| = 0$. That is $\delta(G) \geq 2$.

Let $c : V(G) \to \{1, 2, 3\}$ be a coloring of G defined as follows: $c(x_i) = 1$ for each $i \in [p]$, $c(y_j) = 2$ when j is even and $c(y_j) = 3$ otherwise for each $j \in [q]$. Clearly c is a proper coloring. Next, we show that c is a lid-coloring of G.

Consider two adjacent vertices $u \in X$ and $v \in Y$. Clearly $N[u] \neq N[v]$, otherwise $G = K_2$, contradicting the assumption that G is connected and has at least three vertices. Hence, we have $N[u] \neq N[v]$ for any pair of adjacent vertices of G. As $deg(u) \geq 2$, we have $c(N[u]) = \{1, 2, 3\}$ and $c(N[v]) = \{1, c(v)\}$. That is $c(N[u]) \neq c(N[v])$ for any pair of adjacent vertices u and v of G. Therefore, c is a lid coloring of G.

Case 2: Either $Z \cap X \neq \emptyset$ or $Z \cap Y \neq \emptyset$ but not both.

Without loss of generality we assume that $Z \cap X = \emptyset$ and $Z \cap Y \neq \emptyset$. It is easy to see that the coloring $c : V(G) \to \{1, 2, 3\}$ defined in Case 1 is also a lid-coloring of G in this case.

Case 3: Both $Z \cap X \neq \emptyset$ and $Z \cap Y \neq \emptyset$.

Suppose $\chi_{lid}(G) = 3$ and let $f : V(G) \to [3]$ be a lid-coloring of G. Let $x \in Z \cap X$ and y be the neighbor of x. Without loss of generality, assume that $f(x) = 1$, $f(y) = 2$ and $f(N[x]) = \{1, 2\}$. Then $f(N[y]) = \{1, 2, 3\}$, since G is connected and has at least three vertices. If $z \in N(y)$, then $f(N[z]) \neq \{1, 2, 3\}$, otherwise $f(N[y]) = f(N[z])$ and $N[y] \neq N[z]$. Thus, $f(N[z]) = \{2, f(z)\}$ for every $z \in N(y)$. Note that z is at even distance from x and y is at odd distance from x. Thus for any vertex w, $|f(N[w])| = 3$ if w is at odd distance from x and $|f(N[w])| = 2$ if w is at even distance from x. Every vertex v in the set $Z \cap Y$ is at odd distance from x, that is $|f(N[v])| = 3$, which is a contradiction to the fact that the degree of v is one in G. Therefore $\chi_{lid}(G) \geq 4$. It was shown in [2] that $\chi_{lid}(G) \leq 4$ when G is bipartite. Hence, we conclude that $\chi_{lid}(G) = 4$. □

Corollary 2. *Given a biconvex bipartite graph G, the lid-chromatic number of G can be computed in polynomial time.*

Proof. Given a biconvex bipartite graph $G = (X \cup Y, E)$, from Theorem 3 we only need to determine whether both X and Y contain degree one vertices or not which can be done in polynomial time. Based on this we decide the lid-chromatic number of G. □

6   Cartesian Product

Definition 3 (Cartesian product). *Cartesian product $G \square H$ of graphs G and H is a graph such that $V(G \square H) = V(G) \times V(H)$, where $\times$ represents cartesian* *product and two vertices (u_1, v_1) and (u_2, v_2) in $G \square H$ are adjacent if and only if either $u_1 = u_2$ and v_1 is adjacent to v_2 in H, or $v_1 = v_2$ and u_1 is adjacent to u_2 in G.*

Theorem 4 ([2]). *If G and H are bipartite graphs without isolated vertices, then $\chi_{lid}(G \square H) = 3$.*

As a corollary, we obtain the lid-chromatic number of Cartesian product of paths.

Corollary 3. *For every pair of integers m and n, where $2 \leq m \leq n$, we have $\chi_{lid}(P_m \square P_n) = 3$.*

Taking the work forward, we study lid-coloring of Cartesian product of a path and a cycle, and Cartesian product of two cycles.

6.1   Cartesian Product of a Cycle and a Path

Theorem 5. *For every pair of integers m and n, where $m \geq 3$, $n \geq 2$, we have*

$$\chi_{lid}(C_m \square P_n) = \begin{cases} 5 & m = 3 \text{ and } n \geq 2 \\ 4 & m \text{ is odd, } m > 3 \text{ and } n \geq 2 \\ 3 & m \text{ is even and } n \geq 2 \end{cases}$$

Proof. Let $G = C_m \square P_n$ and let $V(C_3) = \{u_1, u_2, u_3\}$, $V(P_n) = \{v_1, v_2, \cdots, v_n\}$ and $V(C_3 \square P_n) = \{(u_1, v_i), (u_2, v_i), (u_3, v_i) \mid i \in [n]\}$.

Case 1: When $m = 3$ and $n \geq 2$. A 5-lid-coloring of $C_3 \square P_n$ is given in the Fig. 2a. Hence, we have $\chi_{lid}(C_3 \square P_n) \leq 5$.

Next, we show that $\chi_{lid}(C_3 \square P_n) \geq 5$. Let $X = \{(u_1, v_1), (u_2, v_1), (u_3, v_1)\}$. Clearly the graph $G[X] \cong C_3$. Hence $\chi_{lid}(G[X]) = 3$. Observe that (a) every pair of vertices in X have distinct closed neighborhoods and (b) all the three colors used in X in any lid-coloring of $C_3 \square P_n$ appear in the neighborhood of any vertex of X. In order to maintain distinct colors in their closed neighborhood, we must use at least two new colors in $\{(u_1, v_2), (u_2, v_2), (u_3, v_2)\}$. Therefore, any lid-coloring of $C_3 \square P_n$ must use at least 5 colors. Hence, we have $\chi_{lid}(C_3 \square P_n) = 5$.

Case 2: $m > 3$ and m is odd, $n \geq 2$.

Let $V(C_m) = \{u_1, u_2, \cdots, u_m\}$ and $V(P_n) = \{v_1, v_2, \cdots, v_n\}$, and $V(C_m \square P_n) = \{(u_i, v_j) \mid i \in [m], j \in [n]\}$. A 4-lid-coloring of $C_m \square P_n$ is given in the Fig. 2b. Hence, we have $\chi_{lid}(C_m \square P_n) \leq 4$.

Next, we show that $\chi_{lid}(C_m \square P_n) \geq 4$. Let $X = \{(u_i, v_1) \mid i \in [m]\}$. As the graph $G[X]$ induced by the vertices of X is an odd cycle, we have $\chi_{lid}(G) \geq \chi(G[X]) \geq 3$. Observe that in any proper coloring $f : V(G[X]) \to \{1, 2, 3\}$ there exists two adjacent vertices (u_i, v_1) and (u_j, v_1) such that $f(N[(u_i, v_1)]) = f(N[(u_j, v_1)]) = \{1, 2, 3\}$ in $G[X]$. As $N[(u_i, v_1)] \neq N[(u_j, v_1)]$, in any lid-coloring of G at least one new color must be used to color a vertex of either $N[(u_i, v_1)]$ or $N[(u_j, v_1)]$. Hence $\chi_{lid}(C_m \square P_n) \geq 4$. Altogether we have $\chi_{lid}(C_m \square P_n) = 4$.

Case 3: m is even and $n \geq 2$,

In this case both C_m and P_n are bipartite and hence from Theorem 4 we have $\chi_{lid}(C_m \square P_n) = 3$. ☐

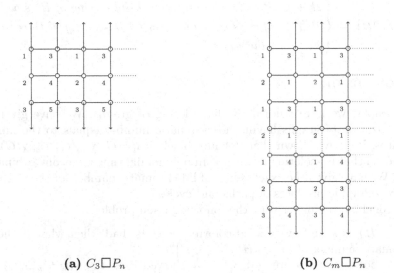

(a) $C_3 \square P_n$ **(b)** $C_m \square P_n$

Fig. 2. (a) A 5-lid-coloring of $C_3 \square P_n$ for $n \geq 2$, and (b) A 4-lid-coloring of $C_m \square P_n$, when m is odd, $m > 3$ and $n \geq 2$.

6.2 Cartesian Product of Two Cycles

Theorem 6 ($\star$). *For every pair of integers m and n, where $m \geq 3$, $n \geq 3$, we have*

$$\chi_{lid}(C_m \square C_n) = \begin{cases} 5 & m = 3 \text{ and } n \geq 3 \\ 3 & m \text{ is even and } n \text{ is even} \\ 4 & m \text{ is odd, } m > 3 \text{ and } n \text{ is even} \\ \leq 5 & m \text{ is odd, } m > 3 \text{ and } n \text{ is odd, } n > 3 \end{cases}$$

7 Lexicographic Product

Definition 4 (Lexicographic product). *Lexicographic product $G[H]$ of graphs G and H is a graph such that $V(G[H]) = V(G) \times V(H)$, where $\times$ represents cartesian product and two vertices (u_1, v_1) and (u_2, v_2) in $G[H]$ are adjacent if and only if either u_1 is adjacent to u_2 in G, or $u_1 = u_2$ and v_1 is adjacent to v_2 in H.*

Definition 5. *A lid coloring $f : V(H) \to [k]$ of a graph H is called 'bad' if there exists a vertex v in H such that $f(N[v]) = [k]$. Otherwise, we call f as a 'good' lid-coloring of H.*

Theorem 7 (⋆). *Let H be a graph on n vertices with $\chi_{lid}(H) = k$. For any integer $m \geq 3$, we have*

$$\chi_{lid}(P_m[H]) = \begin{cases} 2k+1 & \text{if } m \text{ is odd and every } k\text{-lid-colring of } H \text{ is bad ;} \\ 2k+2 & \text{if } m \text{ is even and every } k\text{-lid-colring of } H \text{ is bad ;} \\ 2k & \text{otherwise;} \end{cases}$$

8 Conclusion

In this paper, we have studied the lid-coloring of graphs. We have given the characterization of graphs having lid-chromatic number equals to the number of vertices. We have shown that, for any block graph G, $\chi_{lid}(G) \leq 2\chi(G)$. We have proved that lid-coloring is solvable in polynomial time on biconvex bipartite graphs. We have given the exact values of lid-chromatic number for the Cartesian and Lexicographic products of paths and cycles.

We conclude the paper with the following open problems.

1. If $\chi_{lid}(H) = k$ and every k-lid-coloring of H is 'bad' then what is the lid-chromatic number of (a) $P_2[H]$ (b) $C_3[H]$?
2. When both m and n are odd, we have showed that $4 \leq \chi_{lid}(C_m \square C_n) \leq 5$. We do not know the exact value of the lid-chromatic number of $C_m \square C_n$.

Acknowledgments. We would like to thank anonymous referees for their helpful comments. The first author and the third author acknowledges SERB-DST for supporting this research via grants PDF/2021/003452 and SRG/2020/001162 respectively for funding to support this research.

References

1. Diestel, R.: Graph Theory. Graduate Texts in Mathematics (2005)
2. Esperet, L., Gravier, S., Montassier, M., Ochem, P., Parreau, A.: Locally identifying coloring of graphs. Electron. J. Comb. **19**(2), 40 (2012)
3. Esperet, L., Wood, D.R.: Colouring strong products. arXiv preprint arXiv:2205.04953 (2022)
4. Foucaud, F., Honkala, I., Laihonen, T., Parreau, A., Perarnau, G.: Locally identifying colourings for graphs with given maximum degree. Discrete Math. **312**(10), 1832–1837 (2012)
5. Geller, D., Stahl, S.: The chromatic number and other functions of the lexicographic product. J. Comb. Theory, Ser. B **19**(1), 87–95 (1975)
6. Gonçalves, D., Parreau, A., Pinlou, A.: Locally identifying coloring in bounded expansion classes of graphs. Discrete Appl. Math. **161**(18), 2946–2951 (2013)
7. Martins, N., Sampaio, R.: Locally identifying coloring of graphs with few P4s. Theor. Comput. Sci. **707**, 69–76 (2018)
8. Sabidussi, G.: Graphs with given group and given graph-theoretical properties. Canadian J. Math. **9**, 515–525 (1957)
9. Shitov, Y.: Counterexamples to hedetniemi's conjecture. Ann. Math. **190**(2), 663–667 (2019)

On Structural Parameterizations of Star Coloring

Sriram Bhyravarapu[1]([✉]) and I. Vinod Reddy[2]

[1] The Institute of Mathematical Sciences, HBNI, Chennai, India
sriramb@imsc.res.in
[2] Department of Electrical Engineering and Computer Science,
IIT Bhilai, Bhilai, India
vinod@iitbhilai.ac.in

Abstract. A *star coloring* of a graph G is a proper vertex coloring such that every path on four vertices uses at least three distinct colors. The minimum number of colors required for such a star coloring of G is called star chromatic number, denoted by $\chi_s(G)$. Given a graph G and a positive integer k, the STAR COLORING PROBLEM asks whether G has a star coloring using at most k colors. This problem is NP-complete even on restricted graph classes such as bipartite graphs.

In this paper, we initiate a study of STAR COLORING from the parameterized complexity perspective. We show that STAR COLORING is fixed-parameter tractable when parameterized by (a) neighborhood diversity, (b) twin-cover, and (c) the combined parameters clique-width and the number of colors.

1 Introduction

A coloring $f : V(G) \to \{1, 2, \ldots, k\}$ of a graph $G = (V, E)$ is a *star coloring* if (i) $f(u) \neq f(v)$ for every edge $uv \in E(G)$, and (ii) every path on four vertices uses at least three distinct colors. The *star chromatic number* of G, denoted by $\chi_s(G)$, is the smallest integer k such that G is star colorable using k colors. Given a graph G and a positive integer k, the STAR COLORING problem asks whether G has a star coloring using at most k colors. The name star coloring is due to the fact that the subgraph induced by any two color classes (subset of vertices assigned the same color) is a disjoint union of stars.

STAR COLORING [14] is used in the computation of the Hessian matrix. A Hessian matrix is a square matrix of second order partial derivatives of a scalar-valued function. Hessian matrices are used in large-scale optimization problems, parametric sensitivity analysis [3], image processing, computer vision [22], and control of dynamical systems in real time [3]. Typically, Hessian matrices that arise in a large-scale application are sparse. The computation of a sparse Hessian matrix using the automatic differentiation technique requires a seed matrix. Coleman and Moré [5] showed that the computation of a seed matrix can be formulated using a star coloring of the adjacency graph of a Hessian matrix.

STAR COLORING was first introduced by Grünbaum in [15]. The computational complexity of the problem is studied on several graph classes. The problem is polynomial time solvable on cographs [23] and line graphs of trees [24]. It is NP-complete to decide if there exists a star coloring of bipartite graphs [4] using at most k colors, for any $k \geq 3$. It has also been shown that STAR COLORING is NP-complete on planar bipartite graphs [1] and line graphs of subcubic graphs [20] when $k = 3$. Recently, Shalu and Cyriac [26] showed that k-STAR COLORING is NP-complete for graphs of degree at most four, where $k \in \{4, 5\}$.

To the best of our knowledge, the problem has not been studied in the framework of parameterized complexity. In this paper, we initiate the study of STAR COLORING from the viewpoint of parameterized complexity. In parameterized complexity, the running time of an algorithm is measured as a function of input and a secondary measure called a parameter. A parameterized problem is said to be fixed-parameter tractable (FPT) with respect to a parameter k, if the problem can be solved in $f(k)n^{O(1)}$ time, where f is a computable function independent of the input size n and k is a parameter associated with the input instance. For more details on parameterized complexity, we refer the reader to the texts [9]. As STAR COLORING is NP-complete even when $k = 3$, the problem is para-NP complete when parameterized by the number colors k. This motivates us to study the problem with respect to structural graph parameters, which measure the structural properties of the input graph. The parameter tree-width [25] introduced by Robertson and Seymour is one of the most investigated structural graph parameters for graph problems.

The STAR COLORING problem is expressible in monadic second order logic (MSO) [17]. Using the meta theorem of Courcelle [6], the problem is FPT when parameterized by the tree-width of the input graph. Clique-width [8] is another graph parameter which is a generalization of tree-width. If a graph has bounded tree-width, then it has bounded clique-width, however, the converse may not always be true (e.g., complete graphs). Courcelle's meta theorem can also be extended to graphs of bounded clique-width. It was shown in [7] that all problems expressible in MSO logic that does not use edge set quantifications (called as MS_1-logic) are FPT when parameterized by the clique-width. However, the STAR COLORING problem cannot be expressed in MS_1 logic [10,17]. Motivated by this, we study the parameterized complexity of the problem with respect to the combined parameters clique-width and the number of colors and show that STAR COLORING is FPT.

Next, we consider the parameters neighborhood diversity [19] and twin-cover [13]. These parameters are weaker than clique-width in the sense that graphs of bounded neighborhood diversity (resp. twin-cover) have bounded clique-width, however, the converse may not always be true. Moreover, these two parameters are not comparable with the parameter tree-width and they generalize the parameter vertex cover [13] (see Fig. 1). We show that STAR COLORING is FPT with respect to neighborhood diversity or twin-cover.

2   Preliminaries

For $k \in \mathbb{N}$, we use $[k]$ to denote the set $\{1, 2, \ldots, k\}$. If $f : A \rightarrow B$ is a function and $C \subseteq A$, $f|_C$ denotes the restriction of f to C, that is $f|_C : C \rightarrow B$ such

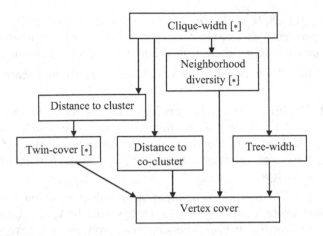

Fig. 1. Hasse diagram of some structural graph parameters. An edge from a parameter k_1 to a parameter k_2 means that there is a function f such that for all graphs G, we have $k_1(G) \leq f(k_2(G))$. The parameters considered in this paper are indicated by $*$.

that for all $x \in C$, $f|_C(x) = f(x)$. All graphs we consider in this paper are undirected, connected, finite and simple. For a graph $G = (V, E)$, we denote the vertex set and edge set of G by $V(G)$ and $E(G)$ respectively. We use n to denote the number of vertices and m to denote the number of edges of the graph. For simplicity, an edge between vertices x and y is denoted as xy. For a subset $X \subseteq V(G)$, the graph $G[X]$ denotes the subgraph of G induced by vertices of X. If $f : V(G) \to [k]$ is a coloring of G using k colors, then we use $f^{-1}(i)$ to denote the subset of vertices of G which are assigned the color i. For a subset $U \subseteq V(G)$, we use $f(U)$ to denote the set of colors used to color the vertices of U, i.e., $f(U) = \bigcup_{u \in U} f(u)$.

For a vertex set $X \subseteq V(G)$, we denote $G - X$, the graph obtained from G by deleting all vertices of X and their incident edges. The open neighborhood of a vertex v, denoted $N(v)$, is the set of vertices adjacent to v and the set $N[v] = N(v) \cup \{v\}$ denotes the closed neighborhood of v. The neighbourhood of a vertex set $S \subseteq V(G)$ is $N(S) = (\bigcup_{v \in S} N(v)) \setminus S$. For a fixed coloring of G, we say a path is *bi-colored* if there exists a proper coloring of the path using two colors.

The proofs of the theorems marked with $(\star)$ are presented in the full version of this paper [2].

3   Neighborhood Diversity

In this section, we show that STAR COLORING is FPT when parameterized by neighborhood diversity. The key idea is to reduce star coloring on graphs of bounded neighborhood diversity to the integer linear programming problem (ILP). The latter is FPT when parameterized by the number of variables.

Theorem 1 ([11,18,21]**).** *The q-variable* INTEGER LINEAR PROGRAMMING FEASIBILITY *problem can be solved using $O(q^{2.5q+o(q)}n)$ arithmetic operations and space polynomial in n, where n is the number of bits of the input.*

We now define the parameter neighborhood diversity and state some of its properties.

Definition 1 (Neighborhood Diversity [19]**).** *Let $G = (V, E)$ be a graph. Two vertices $u, v \in V(G)$ are said to have the* same type *if and only if $N(u) \setminus \{v\} = N(v) \setminus \{u\}$. A graph G has neighborhood diversity at most t if there exists a partition of $V(G)$ into at most t sets $V_1, V_2, \ldots, V_t$ such that all vertices in each set have same type.*

Observe that each V_i either forms a clique or an independent set in G. Also, for any two distinct types V_i and V_j, either each vertex in V_i is adjacent to each vertex in V_j, or no vertex in V_i is adjacent to any vertex in V_j. We call a set V_i as a *clique type* (resp, independent type) if $G[V_i]$ is a clique (resp, independent set). It is known that a smallest sized partition of $V(G)$ into clique types and independent types can be found in polynomial time [19]. Hence, we assume that the types $V_1, V_2, \ldots, V_t$ of the graph G are given as input.

We now present the main result of the section.

Theorem 2. STAR COLORING *can be solved in $O(q^{2.5q+o(q)}n)$ time, where $q = 2^t$ and t is the neighborhood diversity of the graph.*

Let $G = (V, E)$ be a graph with the types $V_1, V_2, \ldots, V_t$. For each $A \subseteq \{1, 2, \ldots, t\}$, we denote a *subset type* of G by $T_A = \{V_i \mid i \in A\}$. We denote the set of all types adjacent to type V_i by $adj(V_i)$. That is, $V_j \in adj(V_i)$ if every vertex in V_j is adjacent to every vertex of V_i. Given a graph G and its types, we construct the ILP instance in the following manner.

Construction of the ILP Instance: For each $A \subseteq [t]$, let n_A be the variable that denotes the number of colors assigned to vertices in every type of T_A and not used in any of the types from $\{V_1, V_2, \ldots, V_t\} \setminus T_A$. For example, if $A = \{1, 3, 4\}$ (i.e., $T_A = \{V_1, V_3, V_4\}$) and $n_A = 2$, then there are two colors, say c_1 and c_2, such that both c_1 and c_2 are assigned to vertices in each of the types V_1, V_3 and V_4 and not assigned to any of the vertices in types $\{V_1, V_2, \ldots, V_t\} \setminus \{V_1, V_3, V_4\}$. This is the critical part of the proof where we look at how many colors are used exclusively in each type of T_A rather than what colors are used. Since we have a variable n_A for each $A \subseteq [t]$, the number of variables is 2^t.

We now describe the constraints for ILP with a short description explaining the significance or the information being captured by the constraints.

(C0) Discard all subset types T_A containing two types V_i, V_j where $V_j \in adj(V_i)$. To ensure that no two adjacent vertices are assigned the same color, we introduce this constraint that only considers T_A in which no two types in T_A are adjacent.

(C1) The sum of all the variables is at most k. That is $\sum_{A \subseteq [t]} n_A \leq k$.

We introduce this constraint to ensure that the number of colors used in any coloring is at most k.

(C2) For each clique type V_i, $i \in [t]$, the sum of the variables n_A for which $V_i \in T_A$ is equal to the number of vertices in V_i. That is, $\sum_{A:V_i \in T_A} n_A = |V_i|$.

To ensure that no two vertices in the clique type V_i are assigned the same color, we introduce this constraint.

(C3) For each independent type V_i, where $i \in [t]$, the sum of the variables n_A for which $V_i \in T_A$ is at least one and at most the minimum of k and the number of vertices in V_i. That is, $1 \leq \sum_{A:V_i \in T_A} n_A \leq \min\{k, |V_i|\}$.

To ensure that the number of colors used for coloring an independent type V_i is at least one and at most the minimum of k and $|V_i|$, we introduce this constraint.

(C4) For each combination of four distinct types, say $V_{i_1}, V_{i_2}, V_{i_3}$ and V_{i_4}, where $i_1, i_2, i_3, i_4 \in [t]$, with $V_{i_1}, V_{i_3} \in adj(V_{i_2})$ and $V_{i_4} \in adj(V_{i_3})$, we have the following constraint:

If the sum of the variables n_A for which $V_{i_1}, V_{i_3} \in T_A$ is at least one, then sum of variables n_B for which $V_{i_2}, V_{i_4} \in T_B$ should be equal to zero. That is,

$$\sum_{\substack{A:V_{i_1},V_{i_3} \in T_A \text{ where} \\ V_{i_1},V_{i_3} \in adj(V_{i_2}) \text{ and } V_{i_4} \in adj(V_{i_3})}} n_A \geq 1 \implies \sum_{B:V_{i_2},V_{i_4} \in T_B} n_B = 0.$$

This constraint ensures that if there exists a vertex in V_{i_1} and a vertex in V_{i_3} that are assigned the same color, then the sets of colors used to color the vertices of V_{i_2} and V_{i_4} are disjoint.

(C5) For every combination of three distinct types, say $V_{i_1}, V_{i_2}, V_{i_3}$, where $i_1, i_2, i_3 \in [t]$, with V_{i_1} being an independent type and $V_{i_2}, V_{i_3} \in adj(V_{i_1})$, we have the following constraint:

If the sum of the variables n_A for which $V_{i_1} \in T_A$ is strictly less than the number of vertices in V_{i_1}, then the sum of variables n_B for which $V_{i_2}, V_{i_3} \in T_B$ is equal to zero.

$$\sum_{\substack{A:V_{i_1} \in T_A, \text{ where } V_{i_2},V_{i_3} \in adj(V_{i_1}) \text{ and} \\ V_{i_1} \text{ is an independent type}}} n_A < |V_{i_1}| \implies \sum_{B:V_{i_2},V_{i_3} \in T_B} n_B = 0.$$

This constraint ensures that if there exist two vertices in V_{i_1} that are assigned the same color, then every vertex in V_{i_2} is assigned a color different from every vertex in V_{i_3}.

(C6) For every combination of two distinct independent types V_{i_1}, V_{i_2}, where $i_1, i_2 \in [t]$ with $V_{i_1} \in adj(V_{i_2})$, if the sum of the variables n_A for which $V_{i_1} \in T_A$ is less than the number of vertices in V_{i_1}, then the sum of variables n_B for which $V_{i_2} \in T_B$ is equal to the number of vertices in V_{i_2}, and vice-versa. The former constraint is illustrated below while the latter constraint can be constructed by swapping V_{i_1} and V_{i_2} in the former constraint.

$$\sum_{\substack{A:V_{i_1}\in T_A \text{ where } V_{i_1}\in adj(V_{i_2}) \\ \text{and } V_{i_1},V_{i_2} \text{ are independent types}}} n_A < |V_{i_1}| \implies \sum_{B:V_{i_2}\in T_B} n_B = |V_{i_2}|.$$

This constraint ensures that if there exist two vertices in V_{i_1} that are assigned the same color then all vertices in V_{i_2} are assigned distinct colors. We can say similar things for the latter constraint.

(C7) For each $A \subseteq [t]$, $n_A \geq 0$.

The number of colors used exclusively in all the types of T_A is at least 0.

The construction of the ILP instance is complete. We use Theorem 1 to obtain a feasible assignment for ILP. Using this, we find a star coloring of G. We now show that G has a star coloring using at most k colors if and only if there exists a feasible assignment to ILP.

Lemma 1. *If there exists a feasible assignment to ILP then G has a star coloring using at most k colors.*

Proof. Using a feasible assignment returned by the ILP, we construct a star coloring $f : V(G) \to [k]$ of G. Let $A_1, A_2, \ldots A_{2^t}$ be the subsets of $[t]$ in some fixed order. For each A_i, we associate the set of colors $c(A_i) = \{\sum_{j=0}^{i-1} n_{A_j} + 1, \sum_{j=0}^{i-1} n_{A_j} + 2, \ldots, \sum_{j=0}^{i-1} n_{A_j} + n_{A_i}\}$, where $n_{A_0} = 0$.

Now, for each V_j, we associate the set of colors $c(V_j) = \cup_{j\in A_i} c(A_i)$. If V_j is a clique type, then from constraint (C2), $|c(V_j)| = |V_j|$ for every j. Therefore, we color the vertices of V_j with distinct colors from the set $c(V_j)$. If V_j is an independent type, then from constraint (C3), $1 \leq |c(V_j)| \leq \min\{k, |V_j|\}$. In this case, we greedily color the vertices of V_j with colors from the set $c(V_j)$ such that each color is used at least once in V_j. This finishes the description of the coloring f of G.

We now argue that f is a star coloring of G. To show that f is a proper coloring, we need to show that every vertex is assigned a color and adjacent vertices do not receive the same color. The coloring process described above ensures that every vertex is colored. Also, f is a proper coloring because of the constraints (C0) and (C2). The former constraint ensures that subset types T_A considered do not contain a pair of adjacent types in it while the latter constraint ensures that no two vertices in a clique type are assigned the same color. Thus f is a proper coloring.

We now show that there is no bi-colored path of length 3. Suppose that there exists a path $u_1 - u_2 - u_3 - u_4$ on four vertices such that $f(u_1) = f(u_3)$ and $f(u_2) = f(u_4)$.

- **Vertices u_1, u_2, u_3, u_4 belong to four distinct types.**
  WLOG, let u_1, u_2, u_3, u_4 belong to V_1, V_2, V_3, V_4 respectively. From the definition of neighborhood diversity, we have $V_1, V_3 \in adj(V_2)$ and $V_4 \in adj(V_3)$. As $f(u_1) = f(u_3)$ and $f(u_2) = f(u_4)$, there exists two sets $A \subseteq [t]$ and $B \subseteq [t]$ such that $V_1, V_3 \in T_A$, $V_2, V_4 \in T_B$, $n_A \geq 1$ and $n_B \geq 1$. This cannot happen because of the constraint (C4).
- **Vertices u_1, u_2, u_3, u_4 belong to three distinct types.**
  WLOG, let u_1, u_2, u_3, u_4 belong to V_1, V_2, V_1, V_3 respectively. Since $f(u_1) = f(u_3)$, it is the case that $\sum_{A:V_1 \in T_A} n_A < |V_1|$ implying V_1 is an independent type. Since V_1 is an independent type with two vertices assigned the same color and $f(u_2) = f(u_4)$, there exists $B \subseteq [t]$ such that $V_2, V_3 \in T_B$ and $n_B \geq 1$. This cannot happen because of the constraint (C5).
- **Vertices u_1, u_2, u_3, u_4 belong to two distinct types.**
  WLOG, let $u_1, u_3 \in V_1$ and $u_2, u_4 \in V_2$. Similar arguments as in the above case can be applied to show that V_1 and V_2 are independent types and this case cannot arise due to constraint (C6).

Thus f is a star coloring of G using at most k colors. □

Lemma 2. *If G has a star coloring using at most k colors then there exists a feasible assignment to ILP.*

Proof. Let $f : V(G) \to [k]$ be a star coloring of G using k colors. For each $A \subseteq [t]$, we set

$$n_A = |\bigcap_{V_i \in T_A} f(V_i) - \bigcup_{V_i \notin T_A} f(V_i)|.$$

That is n_A is the number of colors that appear in each of the types in T_A and does not appear in any of the types from $\{V_1, \ldots, V_t\} \setminus T_A$. We now show that such an assignment satisfies the constraints (C0)–(C7).

1. Since f is a proper coloring of G, no two vertices in two adjacent types are assigned the same color. Hence the constraint (C0) is satisfied.
2. Using the fact that f is a star coloring that uses k colors and from the definition of n_A, where each color is counted towards only exactly one variable, we see that the constraint (C1) is satisfied. For each of the remaining variables n_A for which no color is associated with it, we have that $n_A = 0$. Hence the constraint (C7) is satisfied.
3. When V_i is a clique type, we have that $|f(V_i)| = |V_i|$. The expression $\sum_{A:V_i \in T_A} n_A$ denotes the number of colors used in V_i in the coloring f, which equals $|V_i|$. Hence the constraint (C2) is satisfied.
4. When V_i is an independent type, the number of colors used in V_i is at most the minimum of k and $|V_i|$. In addition, we need at least one color to color the vertices of V_i. Hence $1 \leq |f(V_i)| \leq \min\{k, |V_i|\}$. Since $\sum_{A:V_i \in T_A} n_A = |f(V_i)|$, the constraint (C3) is satisfied.

5. Since f is a star coloring, there is no bi-colored P_4. Thus for every combination of four types, say V_1, V_2, V_3 and V_4, if there exists a color assigned to a vertex in V_1 and a vertex in V_3 with $V_1, V_3 \in adj(V_2)$ and $V_4 \in adj(V_3)$, then all the vertices in $V_2 \cup V_4$ should be assigned distinct colors. That is, there is no $B \subseteq [t]$ for which $V_2, V_4 \in T_B$ and $n_B \geq 1$. Hence the constraint (C4) is satisfied.

Similarly, we can show that constraints (C5) and (C6) are also satisfied.   □

The running time of the algorithm depends on the time taken to construct an ILP instance and obtain a feasible assignment for the ILP using Theorem 1. The former takes polynomial time while the latter takes $O(q^{2.5q+o(q)}n)$ time where $q = 2^t$ is the number of variables. This completes the proof of Theorem 2.

4   Twin Cover

In this section, we show that STAR COLORING is FPT when parameterized by twin cover. Ganian [13] introduced the notion of twin-cover which is a generalization of vertex cover. Note that the parameters neighborhood diversity and twin-cover are not comparable (see Sect. 3.4 in [13]). We now define the parameter twin-cover and state some of its properties.

Definition 2 (Twin Cover [13]**).** *Two vertices u and v of a graph G are said to be twins if $N(u) \setminus \{v\} = N(v) \setminus \{u\}$ and true twins if $N[u] = N[v]$. A twin-cover of a graph G is a set $X \subseteq V(G)$ of vertices such that for every edge $uv \in E(G)$ either $u \in X$ or $v \in X$, or u and v are true twins.*

Remark 1. If $X \subseteq V(G)$ is a twin-cover of G then (i) $G - X$ is disjoint union of cliques, and (ii) for each clique K in $G - X$ and each pair of vertices u, v in K, $N(u) \cap X = N(v) \cap X$.

Theorem 3. STAR COLORING *can be solved in $O(q^{2.5q+o(q)}n)$ time where $q = 2^{2^t}$ and t is the size of a twin-cover of the graph.*

Overview of the Algorithm: Given an instance (G, k, t) of STAR COLORING, and a twin cover $X \subseteq V(G)$ of size t in G, the goal is to check if there exists a star coloring of G using at most k colors. The algorithm consists of the following four steps.

1. We guess the coloring $f : X \to [t']$ of X in a star coloring of G (where $t' \leq t$). Then construct an auxiliary graph G' from G where the neighborhood diversity of G' is bounded by a function of t.
2. We show that G has a star coloring g, using at most k colors, such that $g|_X = f$ if and only if G' has a star coloring h, using at most k colors, such that $h|_X = f$.
3. We construct a graph $\mathcal{B}$, which is a subgraph of G' such that G' has a star coloring h, using at most k colors, such that $h|_X = f$ if and only if $\mathcal{B}$ has a proper coloring using at most $k - t'$ colors, where $t' = |f(X)|$.

4. We show that the neighborhood diversity of $\mathcal{B}$ is bounded by a function of t. Then we use the FPT algorithm parameterized by neighborhood diversity from [13] to check whether $\mathcal{B}$ has a proper coloring using at most $k - t'$ colors and decide if there exists a star coloring of G' using at most k colors.

Given a graph G, there exists an algorithm to compute a twin-cover of size at most t (if one exists) in $O(1.2738^t + tn)$ time [13]. Hence we assume that we are given a twin-cover $X = \{v_1, v_2, \ldots, v_t\} \subseteq V(G)$ of size t.

Let (G, k, t) be an instance of STAR COLORING and $X = \{v_1, v_2, \ldots, v_t\} \subseteq V(G)$ be a twin-cover of size t in G. That is, $G[V \setminus X]$ is a disjoint union of cliques. By the definition of twin cover, all vertices in a clique K from $G[V \setminus X]$ has the same neighborhood in X. Similar to the proof of Theorem 2, we define subset types. For each $A \subseteq [t]$, let $T_A = \{v_i \mid i \in A\} \subseteq X$ denote a subset type of G. For every subset type T_A, we denote a *clique type* of G by $K_A = \{K \mid K \text{ is a clique in } G[V \setminus X] \text{ and } N(K) \cap X = T_A\}$.

Step 1 of the algorithm is to initially guess the colors of the vertices in X in a star coloring of G. Let $f : X \to [t']$ be such a coloring, where $t' \leq t$. The rest of the proof is to check if f could be extended to a coloring $g : V(G) \to [k]$ such that $g|_X = f$. Let $X_i = f^{-1}(i) \subseteq X$ be the set of vertices from X that are assigned the color $i \in [t']$ in f. We now construct an auxiliary graph G' from G by repeated application of the Claims 1, 2 and the Reduction Rule 1.

Claim 1. *Let K_A be a clique type with $|K_A| \geq 2$ and there exist two vertices in $X_i \cap T_A$ for some $i \in [t']$. Let $G^\star$ be the graph obtained from G by adding additional edges between every pair of non-adjacent vertices in $\bigcup_{K \in K_A} V(K)$. Then (G, k, t) is a yes-instance of STAR COLORING if and only if $(G^\star, k, t)$ is a yes-instance of STAR COLORING.*

Proof. Let $K, K' \in K_A$ be two cliques and $u, v \in X_i \cap T_A$ (i.e., $f(u) = f(v) = i$). For the forward direction, let g be a star coloring of (G, k, t). Since $g|_X = f$ and g is a star coloring, no two vertices in $\bigcup_{K \in K_A} V(K)$ are assigned the same color. Suppose not, there exists two vertices $w, w' \in \bigcup_{K \in K_A} V(K)$ such that $g(w) = g(w')$, then $w - u - w' - v$ is a bi-colored P_4. Observe that g is also a star coloring of $(G^\star, k, t)$.

For the reverse direction, let h be a star coloring of $(G^\star, k, t)$ that uses at most k colors. Since G is a subgraph of $G^\star$, we have that h is also a star coloring of (G, k, t). □

Notice that a clique type K_A satisfying the assumptions of Claim 1 will now have $|K_A| = 1$. We now look at the clique types K_A such that $|K_A| \geq 2$ and apply the following reduction rule. Let $K \in K_A$ be an arbitrarily chosen clique with maximum number of vertices.

Reduction Rule 1. *Let K_A be a clique type with $|K_A| \geq 2$ and $|X_i \cap T_A| \leq 1$, for all $i \in [t']$. Also, let $K \in K_A$ be an arbitrarily chosen clique with maximum* *number of vertices over all cliques in K_A. Then (G, k, t) is a yes-instance of* STAR COLORING *if and only if* $(G - \bigcup_{K' \in K_A \setminus \{K\}} V(K'), k, t)$ *is a yes-instance of* STAR COLORING.

Lemma 3 ($\star$). *Reduction Rule 1 is safe.*

We repeatedly apply Reduction Rule 1 on the clique types K_A for which $|K_A| \geq 2$ after the application of Claim 1. Thereby ensuring $|K_A| = 1$ for all clique types K_A for which $|K_A| \geq 2$ in G. Thus for all clique types K_A, we have that $|K_A| \leq 1$. Notice that after the application of Claim 1 and Reduction Rule 1, the resulting graph has bounded neighborhood diversity. However, a proper coloring of the resulting graph may not yield a star coloring. The following claim help us to reduce our problem to proper coloring parameterized by neighborhood diversity.

Claim 2. *Let K_A and K_B, with $A \neq B$, be two clique types such that there exists two vertices $u, v \in X_i$ such that $u \in T_A \cap T_B$ and $v \in T_B$, for some $i \in [t]$. Let $G^\star$ be the graph obtained from G by adding additional edges between every pair of non-adjacent vertices in $V(K_A) \cup V(K_B)$. Then (G, k, t) is a yes-instance of* STAR COLORING *if and only if $(G^\star, k, t)$ is a yes-instance of* STAR COLORING.

Proof. For the forward direction, let g be a star coloring of (G, k, t). This implies that no two vertices in $V(K_A) \cup V(K_B)$ are assigned the same color because of the vertices u and v. Hence g is also a star coloring of $(G^\star, k, t)$.

The reverse direction is trivial. Since G is a subgraph of $G^\star$, the star coloring of $(G^\star, k, t)$ is also a star coloring of (G, k, t). $\qquad\square$

We are now ready to explain the steps of our algorithm in detail.

Step 1: Given an instance (G, k, t) of STAR COLORING, we construct an auxiliary graph. The graph constructed after repeated application of Claim 1, Reduction Rule 1 and Claim 2, is the auxiliary graph G'. We now argue that the neighborhood diversity of G' is bounded by a function of t. Consider the partition $\{V(K_A) \mid A \subseteq [t]\} \cup \{\{v_i\} \mid v_i \in X\}$ of $V(G')$. Notice that each clique type K_A of G' is a clique type (see Sect. 3 for more details). That is, all the vertices in K_A have the same neighborhood in X. This is true because initially all vertices in K_A have the same neighborhood (by definition of twin cover) and during the process of adding edges (Claims 1 and 2), either all the vertices in K_A are made adjacent to all the vertices in a type K_B ($A \neq B$) or none of them are adjacent to any vertex in K_B. Thus the number of such types is at most 2^t. Including the vertices of X, we have that the neighborhood diversity of G' is at most $2^t + t$.

Step 2: We need to show that (G, k, t) is a yes-instance of STAR COLORING if and only if (G', k, t) is a yes-instance of STAR COLORING. This is accomplished by the correctness of the Claims 1, 2 and the Reduction rule 1.

Step 3: The next step of the algorithm is to find a set of colors from $[t']$ that can be assigned to the vertices in $V \setminus X$. Towards this, for each $A \subseteq [t]$, we guess a subset of colors $D_A \subseteq [t']$ of size at most $|V(K_A)|$, that can be assigned to the vertices in the clique type K_A in a star coloring of G (extending the coloring f of X) that uses at most k colors. For the guess D_A, we arbitrarily (as it does not matter which vertices are assigned a specific color) assign colors from D_A to vertices in K_A such that $|D_A|$ vertices in K_A are colored distinctly. Given the guess D_A for each K_A, we can check in $2^{O(t)}$ time if the color set D_A associated with K_A is indeed a proper coloring (considering the coloring f of X and its neighboring types). In a valid guess, some vertices $Q \subseteq V(G') \setminus X$ are assigned colors from $[t']$. The uncolored vertices of G' should be given a color from $[k] \setminus [t']$. Let $g : X \cup Q \to [t']$ be a coloring such that $g(v) = f(v)$ if $v \in X$, and $g(v) = \ell$, where ℓ is the assigned color as per the above greedy assignment, if $v \in Q$.

We now extend this partial coloring g of $Q \cup X$ to a full coloring of G', where the vertices in $V(G') \setminus (Q \cup X)$ are assigned colors from $[k] \setminus [t']$. Let $\mathcal{B}$ be the subgraph of G' obtained by deleting the vertices $Q \cup X$ from G'. Notice that $\mathcal{B}$ has neighborhood diversity at most 2^t.

Claim 3. *There exists a star coloring of G' extending g using at most k colors if and only if there exists a proper coloring of $\mathcal{B}$ using at most $k - t'$ colors.*

Proof. Let $h : V(G') \to [k]$ be a star coloring of G' such that $h|_{Q \cup X} = g$. Clearly $|h(\mathcal{B})| = |h(V \setminus (Q \cup X))| \leq k - t'$. That is, h restricted to the vertices of $\mathcal{B}$ is a proper coloring of $\mathcal{B}$ which uses at most $k - t'$ colors.

For the reverse direction, let $c : V(\mathcal{B}) \to [k - t']$ be a proper coloring. We construct a coloring $h : V(G') \to [k]$ using the coloring c as follows: $h(v) = c(v)$ if $v \in \mathcal{B}$, and $h(v) = g(v)$ otherwise. We show that h is a star coloring of G'. Suppose not, without loss of generality, let $u_1 - u_2 - u_3 - u_4$ be a bi-colored P_4, with $u_1 \in K_{A_1}$, $u_2 \in X$, $u_3 \in K_{A_2}$ and $u_4 \in X$ (notice that this is the only way a bi-colored P_4 exists), for some $A_1, A_2 \subseteq [t]$. That is, $c(u_1) = c(u_3)$ and $c(u_2) = c(u_4)$. Also, $A_1 \neq A_2$ because of the proper coloring. If this were the case, we would have applied Claim 2 as K_{A_1} and K_{A_2} satisfy the assumptions along with coloring of the vertices u_2 and u_4 in X. As a consequence, each vertex in K_{A_1} would have been adjacent to each vertex in K_{A_2}. $\square$

Step 4: It is known that proper coloring is FPT parameterized by neighborhood diversity [12]. The algorithm in [12] uses integer linear programming with 2^{2k} variables, where k is the neighborhood diversity of the graph. Since $\mathcal{B}$ has neighborhood diversity at most 2^t, we have that the number of variables $q \leq 2^{2^t}$. We use the algorithm to test whether $\mathcal{B}$ has a proper coloring using at most $k - t'$ colors.

Running Time: Step 1 of the algorithm takes $O(t^t)$ time to guess a coloring of X. Reduction Rule 1, Claims 1 and 2 can be applied in $2^{O(t)} n^{O(1)}$ time. Step 2 can be processed in $2^{O(t)} n^{O(1)}$ time. Step 3 involves guessing the colors that the clique types can take from the colors used in X and this takes $O(2^{2^t})$ time. Constructing $\mathcal{B}$ takes polynomial time. Step 4 is applying the FPT algorithm parameterized by neighborhood diversity from [12] on $\mathcal{B}$ which takes $O(q^{2.5q + o(q)} n)$ time where

$q \leq 2^{2^t}$. The latter dominates the running time and hence running time of the algorithm is $O(2^{2^t} q^{2.5q+o(q)} n^{O(1)})$, where $q \leq 2^{2^t}$.

This completes the proof of Theorem 3.

5   Clique-Width

Theorem 4 (⋆). *Given a graph G, its nice w-expression and an integer k, we can decide if there exists a star coloring of G using k colors in $O((3^{w^3 k^2 + w^2 k^2})^2 n^{O(1)})$ time.*

6   Conclusion

In this paper, we study the parameterized complexity of STAR COLORING with respect to several structural graph parameters. We show that STAR COLORING is FPT when parameterized by (a) neighborhood diversity, (b) twin cover, and (c) the combined parameter clique-width and the number of colors.

We conclude the paper with the following open problems for further research.

1. What is the parameterized complexity of STAR COLORING when parameterized by distance to cluster or distance to co-cluster?
2. It is known that graph coloring admits a polynomial kernel when parameterized by distance to clique [16]. Does STAR COLORING also admit a polynomial kernel parameterized by distance to clique?

Acknowledgments. We would like to thank anonymous referees for their helpful comments. The first author and the second author acknowledges SERB-DST for supporting this research via grants PDF/2021/003452 and SRG/2020/001162 respectively for funding to support this research.

References

1. Albertson, M., Chappell, G., Kierstead, H., Kündgen, A., Ramamurthi, R.: Coloring with no 2-colored P_4's. Electr. J. Comb. **11**, 03 (2004)
2. Bhyravarapu, S., Reddy, I.V.: On structural parameterizations of star coloring. arXiv preprint arXiv:2211.12226 (2022)
3. Büskens, C., Maurer, M.: Sensitivity analysis and real-time optimization of parametric nonlinear programming problems. In: Grotschel, M., Krumke, S.O., Rambau, J. (eds.) Online Optimization of Large Scale Systems, pp. 3–16. Springer, Heidelberg (2001)
4. Coleman, T.F., Moré, J.J.: Estimation of sparse Jacobian matrices and graph coloring blems. SIAM J. Numer. Anal. **20**(1), 187–209 (1983)
5. Coleman, T.F., Moré, J.J.: Estimation of sparse hessian matrices and graph coloring problems. Math. Program. **28**(3), 243–270 (1984). https://doi.org/10.1007/BF02612334
6. Courcelle, B.: The monadic second-order logic of graphs III: tree-decompositions, minors and complexity issues. RAIRO-Theor. Inf. Appl. **26**(3), 257–286 (1992)

7. Courcelle, B., Makowsky, J.A., Rotics, U.: Linear time solvable optimization problems on graphs of bounded clique-width. Theory Comput. Syst. **33**(2), 125–150 (2000). https://doi.org/10.1007/s002249910009
8. Courcelle, B., Olariu, S.: Upper bounds to the clique width of graphs. Discret. Appl. Math. **101**(1–3), 77–114 (2000)
9. Cygan, M., et al.: Parameterized Algorithms, vol. 4. Springer, Cham (2015)
10. Fomin, F.V., Golovach, P.A., Lokshtanov, D., Saurabh, S.: Intractability of clique-width parameterizations. SIAM J. Comput. **39**(5), 1941–1956 (2010)
11. Frank, A., Tardos, É.: An application of simultaneous diophantine approximation in combinatorial optimization. Combinatorica **7**(1), 49–65 (1987). https://doi.org/10.1007/BF02579200
12. Ganian, R.: Using neighborhood diversity to solve hard problems. CoRR, abs/1201.3091 (2012)
13. Ganian, R.: Improving vertex cover as a graph parameter. Discrete Math. Theor. Comput. Sci. **17**(2), 77–100 (2015)
14. Gebremedhin, A.H., Tarafdar, A., Pothen, A., Walther, A.: Efficient computation of sparse hessians using coloring and automatic differentiation. INFORMS J. Comput. **21**(2), 209–223 (2009)
15. Grünbaum, B.: Acyclic colorings of planar graphs. Israel J. Math. **14**(4), 390–408 (1973)
16. Gutin, G., Majumdar, D., Ordyniak, S., Wahlström, M.: Parameterized pre-coloring extension and list coloring problems. SIAM J. Discrete Math. **35**(1), 575–596 (2021)
17. Harshita, K., Mishra, S., Sadagopan, N.: FO and MSO approach to some graph problems: approximation and poly time results. arXiv preprint arXiv:1711.02889 (2017)
18. Kannan, R.: Minkowski's convex body theorem and integer programming. Math. Oper. Res. **12**(3), 415–440 (1987)
19. Lampis, M.: Algorithmic meta-theorems for restrictions of treewidth. Algorithmica **64**(1), 19–37 (2012)
20. Lei, H., Shi, Y., Song, Z.-X.: Star chromatic index of subcubic multigraphs. J. Graph Theory **88**(4), 566–576 (2018)
21. Lenstra, H.W.: Integer programming with a fixed number of variables. Math. Oper. Res. **8**(4), 538–548 (1983)
22. Lorenz, C., Carlsen, I.-C., Buzug, T.M., Fassnacht, C., Weese, J.: A multi-scale line filter with automatic scale selection based on the Hessian matrix for medical image segmentation. In: ter Haar Romeny, B., Florack, L., Koenderink, J., Viergever, M. (eds.) Scale-Space 1997. LNCS, vol. 1252, pp. 152–163. Springer, Heidelberg (1997). https://doi.org/10.1007/3-540-63167-4_47
23. Lyons, A.: Acyclic and star colorings of cographs. Discrete Appl. Math. **159**(16), 1842–1850 (2011)
24. Omoomi, B., Roshanbin, E., Dastjerdi, M.V.: A polynomial time algorithm to find the star chromatic index of trees. Electron. J. Comb. **28**(1), 1 (2021)
25. Robertson, N., Seymour, P.D.: Graph minors. I. excluding a forest. J. Comb. Theory Ser. B **35**(1), 39–61 (1983)
26. Shalu, M.A., Antony, C.: The complexity of star colouring in bounded degree graphs and regular graphs. In: Balachandran, N., Inkulu, R. (eds.) CALDAM 2022. LNCS, vol. 13179, pp. 78–90. Springer, Cham (2022). https://doi.org/10.1007/978-3-030-95018-7_7

Perfectness of G-generalized Join of Graphs

T. Kavaskar$^{(\boxtimes)}$

Department of Mathematics, Central University of Tamil Nadu,
Thiruvarur 610005, India
t_kavaskar@cutn.ac.in

Abstract. In this paper, we prove that the G-generalized join of complete or totally disconnected graphs is perfect if and only if G is perfect. As a result, we deduce some results proved in (Saeid et al. Rocky Mountain J. Math. 48(3) (2018), 729–751) and (Nilesh et al. arXiv (2022), arXiv:2205.04916). We also characterize rings, posets and reduced semigroups whose zero-divisor graphs and ideal based zero-divisor graphs are perfect. As a consequence, we characterize distributive lattices with 0, reduced semirings and boolean rings whose zero divisor graphs are perfect, which are proved in (Patil et al. in Discrete Math. 340: 740–745, 2017).

Keywords: Perfect graphs · G-generalized join of graphs · Zero-Divisor Graphs

1 Introduction

All the graphs considered in the paper are finite, simple and undirected. Let $G = (V(G), E(G))$ be a graph. For $v \in V(G)$ and $S \subseteq V(G)$, let $N_G(v)$ denote the open neighborhood of v in G and $\langle S \rangle$ denote the subgraph induced by S. Let $\overline{G}$ denote the complement of a graph G. A *proper k-coloring* of a graph G is a function from $V(G)$ into a set of k colors such that no two adjacent vertices receive the same color. The *chromatic number* of a graph G, denoted by $\chi(G)$, is the least positive integer k such that there exists a proper k-coloring of G. A *clique* in a graph G is a complete subgraph of G. The *clique number* of G is the largest size of a clique in G and it is denoted by $\omega(G)$. Let G be a graph with $V(G) = \{u_1, u_2, \ldots, u_n\}$ and $H_1, H_2, \ldots, H_n$ be pairwise disjoint graphs. The G-*generalized join* graph, denoted by $G[H_1, H_2, \ldots, H_n]$, of $H_1, H_2, \ldots, H_n$ is the graph obtained by replacing each vertex u_i of G by H_i and joining each vertex of H_i to each vertex of H_j by an edge if u_i is adjacent to u_j in G. If $H_i \cong H$, for $1 \le i \le n$, then $G[H_1, H_2, \ldots, H_n]$ becomes the standard lexicographic product $G[H]$.

For a graph G, we define a relation $\sim_G$ on $V(G)$ as follows: For any $x, y \in V(G)$, define $x \sim_G y$ if and only if $N_G(x) = N_G(y)$. Clearly, $\sim_G$ is an equivalence relation on $V(G)$. Let $[x]$ be the equivalence class which contains x and S be the set of all equivalence classes of this relation $\sim_G$. Based on this equivalence classes we define the reduced graph G_r of a graph G as follows. The *reduced graph* G_r of G (defined

A. Bagchi and R. Muthu (Eds.): CALDAM 2023, LNCS 13947, pp. 172–183, 2023.
https://doi.org/10.1007/978-3-031-25211-2_13

in [13]) is the graph with vertex set $V(G_r) = S$ and two distinct vertices $[x]$ and $[y]$ are adjacent in G_r if and only if x and y are adjacent in G.

Note that if $V(G_r) = \{[x_1], [x_2], \ldots, [x_k]\}$, then G is the G_r-generalized join of $\langle[x_1]\rangle, \langle[x_2]\rangle, \ldots, \langle[x_k]\rangle$, that is, $G = G_r[\langle[x_1]\rangle, \langle[x_2]\rangle, \ldots, \langle[x_k]\rangle]$ and each $[x_i]$ is an independent subset of G (that is, $\langle[x_i]\rangle$ has no edge). Clearly, G_r is isomorphic to an induced subgraph of G. It is easy to observe the following observation.

Observation 1. If G_r is the reduced graph of G with $\omega(G_r) = \chi(G_r)$, then $\chi(G) = \omega(G_r)$.

Let G be a graph with $\omega(G) = k$, and let $\Delta_k(G)$ be the set of all the vertices of a graph G which lie in some clique of size k of G. A connected graph G is called a *generalized complete k-partite* graph (see [13]) if the vertex set $V(G)$ of G is a disjoint union of A and H satisfying the following conditions:

(1) $A = \Delta_k(G)$ and the subgraph induced by A is a complete k-partite graph with parts, say, $A_i, i = 1, 2, \ldots, k$.

(2) For any $h \in H$ and $i \in \{1, 2, \ldots, k\}$, h is adjacent to some vertex of A_i if and only if h is adjacent to any vertex of A_i.
Set $W(h) = \{1 \le i \le k \mid N(h) \cap A_i \neq \emptyset\}$ for any $h \in H$.

(3) For any $h_1, h_2 \in H$, h_1 is adjacent to h_2 if and only if $W(h_1) \cup W(h_2) = \{1, 2, \ldots, k\}$.

A graph G is called a *compact* graph (see [13]) if G contains no isolated vertices and for each pair x, y of non-adjacent vertices of G, there is a vertex z in G with $N(x) \cup N(y) \subseteq N(z)$. A graph G is said to be k-*compact* if it is compact and $\omega(G) = k$.

Throughout this paper, rings are finite non-zero commutative rings with unity. Let R be a ring. A non-zero element x of R is said to be a *zero-divisor* if there exists a non-zero element y of R such that $xy = 0$. A non-zero element u of R is *unit* in R if there exists v in R such that $uv = 1$. For $x \in R$, the *annihilator* of x is the set $Ann(x) = \{y \in R \mid xy = 0\}$. A ring R is said to be *local* if it has unique maximal ideal M. The *nilradical* of a ring R is the set $J = \{x \in R : x^t = 0$, for some positive integer $t\}$. The *index of nilpotency* of J is the least positive integer m for which $J^m = \{0\}$, where $J^m = JJ \ldots J$ (m-times). A ring R is said to be *reduced* if $J = \{0\}$. A ring is said to be *indecomposable* if it can not be written as a direct product of two rings. Let $\mathbb{Z}_n$ be the ring of integer modulo n.

For any ring R, in [6], Beck associated a simple graph with R whose vertices are the elements of R and any two distinct vertices x and y are adjacent if and only if $xy = 0$ in R. Beck conjectured that (see [6]) the chromatic number and clique number of this graph are the same and this was disproved by Anderson and Naseer in [2] (also, see [10]). It can be observed that for the graph associated with the ring, the vertex 0 is adjacent to every other vertex. Anderson and Livingston in [5] slightly modified the definition of the graph associated with a ring by considering the zero-divisors as the vertices and any two distinct vertices x and y are adjacent if and only if $xy = 0$ in R. They called this *zero-divisor graph* of the ring R and it is denoted by $\Gamma(R)$. Zero-divisor graphs have been extensively studied in the past. This can be seen in [1,3,4,11,20].

The following definitions and results can be found in [4, 20]. For $x, y \in R$, define $x \sim_R y$ if and only if $Ann(x) = Ann(y)$. It is proved in [4] that the relation $\sim_R$ is an equivalence relation on R. For $x \in R$, let $D_x = \{r \in R \mid x \sim_R r\}$ be the equivalence class of x. Let $R_E = \{D_{x_1}, D_{x_2}, \ldots, D_{x_k}\}$ be the set of all equivalence classes of the relation $\sim_R$. The *compressed zero-divisor graph* $\Gamma_E(R)$ of R (defined in [20]) is a simple graph with vertex set $R_E \backslash \{D_0, D_1\}$ and two distinct vertices D_x and D_y are adjacent if and only if $xy = 0$. The following result can be found in [18].

Theorem 1 [18]. *If R is a ring, then*

(i) $\Gamma(R) \cong \Gamma_E(R)[\langle D_{x_1} \rangle, \langle D_{x_2} \rangle, \ldots, \langle D_{x_{k-2}} \rangle]$, *where* $D_{x_i} \neq D_0, D_1$, *for* $1 \leq i \leq k - 2$,
(ii) $\langle D_{x_i} \rangle$ *is complete if and only if* $x_i^2 = 0$, *and*
(iii) $\langle D_{x_i} \rangle$ *is totally disconnected (that is, $\langle D_{x_i} \rangle$ has no edge) if and only if* $x_i^2 \neq 0$.

The following result is proved in [3].

Theorem 2 [3]. *If R is a non-zero reduced ring, then there exists a positive integer k such that the compressed zero-divisor graph $\Gamma_E(R) \cong \Gamma(\mathbb{Z}_2^k)$, where $\mathbb{Z}_2^k = \mathbb{Z}_2 \times \mathbb{Z}_2 \times \ldots \times \mathbb{Z}_2$ (k-times).*

In [9], Hala and Jukl introduced the concept of the zero-divisor graph of a poset. Let $(P, \leq)$ be a finite poset with the least element 0. For any $a, b \in P$, denote $L(a, b) = \{c \in P \mid c \leq a \text{ and } c \leq b\}$. A non-zero element $a \in P$ is said to be a *zero-divisor* if $L(a, b) = \{0\}$ for some $0 \neq b \in P$. We say a non-zero element $a \in P$ is an *atom (primitive)* if for any $0 \neq b \in P$, $b \leq a$ implies $a = b$. The *zero-divisor graph $\Gamma(P)$ of a poset P* is a graph whose vertex set $V(\Gamma(P))$ consists of the zero-divisors of P, in which a is adjacent to b if and only if $L(a, b) = \{0\}$. It is shown in [9] that for any poset P, the clique number and the chromatic number of $\Gamma(P)$ are the same.

By a *semigroup*, we mean a finite commutative semigroup with the zero element 0. A semigroup S is said to be *reduced* if for any $a \in S$ and any positive integer n, $a^n = 0$ implies $a = 0$. A semigroup S is said to be *idempotent* (it is a so-called semilattice, see [13]) if for each $a \in S$, $a^2 = a$.

We define a *zero-divisor graph of a semigroup* in a similar manner in the definition of zero-divisor graph of a ring.

Let $R = \mathbb{Z}_2^k$. Clearly, it is a Boolean ring and it becomes a poset by defining $a \leq b$ iff $ab = a$ for any $a, b \in R$. Note that, the zero-divisor graphs of R as a ring (or a semigroup) and as a poset coincide. Let H be a subgraph of $\Gamma(\mathbb{Z}_2^k)$. We say that H is *minimal* (see [13]) if H is an induced subgraph of $\Gamma(\mathbb{Z}_2^k)$ which contains all the atoms of the poset $\mathbb{Z}_2^k$, and we say H is *minimal closed* (see [13]) if H is minimal and $V(H) \cup \{0\}$ is a sub-semigroup of $\mathbb{Z}_2^k$. The following results can be found in [13].

Theorem 3 [13]. *Let G be a simple graph with $\omega(G) = k$. Then the following statements are equivalent:*

(i) G is the zero-divisor graph of a poset.

(ii) G is a k-compact graph.

(iii) G is a generalized complete k-partite graph.

(iv) The reduced graph G_r of G is isomorphic to a minimal subgraph of $\Gamma(\mathbb{Z}_2^k)$.

Theorem 4 [13]. *Let G be a simple graph with $\omega(G) = k$. Then the following statements are equivalent:*

(i) G is the zero-divisor graph of a reduced semigroup with 0.

(ii) G is a generalized complete k-partite graph such that for any non-adjacent vertices $a, b \in V(G)$, there is a vertex $c \in V(G)$ with $W(c) = W(a) \cup W(b)$.

(iii) The reduced graph G_r of G is isomorphic to a minimal closed subgraph of $\Gamma(\mathbb{Z}_2^k)$.

(iv) G is the zero-divisor graph of a semilattice (or equivalently, idempotent semigroup) with 0.

A graph G is *perfect* if $\omega(H) = \chi(H)$ for every induced subgraph H of G.
The following result was proved by Lovasz, see [12].

Theorem 5 [12]. *The complement of every perfect graph is perfect.*

In [7], Berge conjectured the following and it was proved by Chudnovsky et al., see [8].

Theorem 6 (Strong Perfect Graph Theorem [8]**).** *A graph G is perfect if and only if it does not contain an induced subgraph which is either an odd cycle of length at least 5 or the complement of such a cycle.*

The paper mainly deals with the results on perfect graph using the Strong Perfect Graph Theorem. As a result, we deduced many known results in the literature. This is precisely as follows.

In Sect. 2, we prove that the G-generalized join of complete graphs and totally disconnected graphs is perfect if and only if G is perfect. As a consequence, we deduce the results proved in [14] and [17] and prove that the lexicographic product of a perfect graph and a complete graph and the lexicographic product of a perfect graph and a complement of a complete graph are perfect.

In Sect. 3, we characterize rings, posets and reduced semigroups whose zero-divisor graphs and ideal based zero-divisors are perfect. As a result, we characterize distributive lattices with 0, reduced semirings and boolean rings whose zero divisor graphs are perfect, which are proved in [15]. Further, we completely characterize rings the ideal based zero-divisor graph of the ring $\mathbb{Z}_n$ is perfect.

2   When a G-generalized Join of Complete and Totally Disconnected Graphs is Perfect

In this section, we prove the following result on perfect graphs.

Theorem 7. *If G is a graph with vertex set $V(G) = \{v_1, v_2, \ldots, v_n\}$ and $H_1, H_2, \ldots, H_n$ are graphs such that each H_i is either complete or a totally disconnected graph, then G is perfect if and only if $G[H_1, H_2, \ldots, H_n]$ is perfect.*

Proof. Let $G' = G[H_1, H_2, \ldots, H_n]$. It is enough to prove if G is perfect, then G' is perfect. Suppose G' is not perfect, then by Theorem 6, G' contains either an odd cycle of length at least 5 as an induced subgraph or the complement of an odd cycle of length at least 5 as an induced subgraph.

Case 1. G' contains an odd cycle C_{2k+1} as an induced subgraph, where $k \geq 2$.

Let $V(C_{2k+1}) = \{x_0, x_1, \ldots, x_{2k}\}$ such that x_i is adjacent to x_{i+1} (where the addition in subscript is taken modulo $2k+1$) and x_i is not adjacent to x_j, where $j \neq i-1, i+1$. Suppose there exists $1 \leq t \leq n$ such that $|V(C_{2k+1}) \cap V(H_t)| \geq 2$.

First, if there exists $0 \leq i \leq 2k$ such that $x_i, x_{i+1} \in V(H_t)$. Then H_t is complete and hence $x_{i-1} \notin V(H_t)$ (otherwise, C_{2k+1} would not be induced in G'). Thus there exists $1 \leq s \leq n$ with $s \neq t$ such that $x_{i-1} \in V(H_s)$ and hence x_{i-1} is adjacent to x_{i+1}, which is a contradiction.

Next, if there exist $0 \leq i, j \leq 2n$ such that $j \neq i-1, i, i+1$ and $x_i, x_j \in V(H_t)$. Then H_t has no edge in G' and $x_{i+1}, x_{i-1} \notin V(H_t)$. Suppose if $j \neq i+2$, then there exists $1 \leq s \leq n$ such that $s \neq t$ and $x_{i+1} \in V(H_s)$ and hence x_j is adjacent to x_{i+1}, (because of $x_i x_{i+1} \in E(C_{2k+1})$) which is a contradiction. Therefore, if $j = i + 2$, then there exists $1 \leq s \leq n$ such that $s \neq t$ and $x_{i-1} \in V(H_s)$ and therefore x_{i-1} is adjacent to x_j, which is again a contradiction.

Hence $|V(C_{2k+1}) \cap V(H_i)| = 1$, for $0 \leq i \leq 2k$ which implies that G contains an odd cycle of length at least 5 as an induced subgraph, which is a contradiction.

Case 2. G' contains a complement of an odd cycle of length at least 5 as an induced subgraph.

Let $\overline{C_{2k+1}}$ be the complement of the odd cycle C_{2k+1} as an induced subgraph of G', where $k \geq 2$ with $V(\overline{C_{2k+1}}) = \{x_0, x_1, \ldots, x_{2k}\}$ such that x_i is not adjacent to x_j for $j = i-1, i+1$ and x_i is adjacent to x_j, for $j \neq i-1, i, i+1$ (where the addition in subscripts is taken modulo $2k + 1$). Suppose there exists $1 \leq t \leq n$ such that $|V(C_{2k+1}) \cap V(H_t)| \geq 2$.

First, if there exists $0 \leq i \leq 2k$ such that $x_i, x_{i+1} \in V(H_t)$. Then H_t has no edge and $x_{i-1} \notin H_t$ and hence there exists $1 \leq s \leq n$ with $s \neq t$ such that $x_{i-1} \in V(H_s)$. But x_{i+1} is adjacent to x_{i-1} and hence x_i is adjacent to x_{i-1}, which is a contradiction.

Next, if there exist $0 \leq i, j \leq 2n$ such that $j \neq i-1, i, i+1$ and $x_i, x_j \in V(H_t)$. Then H_t is complete and $x_{i-1}, x_{i+1} \notin V(H_t)$. Suppose if $j \neq i+2$, then there exists $1 \leq s \leq n$ such that $s \neq t$ and $x_{i+1} \in V(H_s)$. But x_j is adjacent to x_{i+1} and therefore x_i is adjacent to x_{i+1}, which is impossible. Hence, if $j = i + 2$, then there exists $1 \leq s \leq n$ such that $s \neq t$ and $x_{i-1} \in V(H_s)$ and therefore x_{i-1} is adjacent x_i, which is again a contradiction.

Thus $|V(\overline{C_{2k+1}}) \cap V(H_i)| = 1$, for $0 \leq i \leq 2k$, which implies that G contains a complement of an odd cycle of length at least 5 as an induced, which is a contradiction.

The following corollary is an immediate consequence of Theorem 7.

Corollary 1. *If G is perfect and n is a positive integer, then $G[K_n]$ and $G[K_n^c]$ are perfect.*

Proof. As $G[K_n] \cong G[K_n, K_n, \ldots, K_n]$ and $G[K_n^c] \cong G[K_n^c, K_n^c, \ldots, K_n^c]$, the result follows from Theorem 7.

The following result proved in [17] is deduced from Theorem 7.

Corollary 2 (Corollary 3.2, [17]). *A graph G is perfect if and only if it's reduced graph G_r is perfect.*

The following relation is defined on a graph G in [14]. For $x, y \in V(G)$, define $x \approx y$ if and only if either $x = y$ or $xy \in E(G)$ and $N(x) \setminus \{y\} = N(y) \setminus \{x\}$. Clearly, it is an equivalence relation. Let $[x]$ be the equivalence class of x, and $S = \{[x_1], [x_2], \ldots, [x_r]\}$ be the set of all equivalence classes of the relation $\approx$. Based on these equivalence classes of the relation $\approx$, we defined (This can be seen in [14]) the graph G_{red} with vertex set $V(G_{red}) = S$ and two distinct vertices $[x]$ and $[y]$ are adjacent in G_{red} if and only if x and y are adjacent in G. Clearly, for any graph G, $G = G_{red}[\langle[x_1]\rangle, \langle[x_2]\rangle, \ldots, \langle[x_r]\rangle]$ and $\langle[x_i]\rangle$ is complete, for $1 \le i \le r$.

By Theorem 7, we deduce the following result proved in [14].

Corollary 3 (Theorem 4.4, [14]). *A graph is perfect if and only if G_{red} is perfect.*

3   Perfect Zero-Divisor Graph of a Ring

In this section, we ask the following interesting question. When does the zero-divisor graph of a ring R perfect? To answer this question, we provide a necessary and sufficient condition for which the zero-divisor graph of a ring is perfect.

Theorem 8. *If R is a ring, then $\Gamma(R)$ is perfect if and only if its compressed zero-divisor graph $\Gamma_E(R)$ of R is perfect.*

Proof. The result follows from Theorems 1 and 7.

Let $R_1, R_2, \ldots, R_k$ be rings. For $x_j \in R_1 \times R_2 \times \ldots \times R_k$, there exists a unique $x_j(i) \in R_i$, for $1 \le i \le k$, such that $x_j = (x_j(1), x_j(2), \ldots, x_j(k))$.

Note that there are several rings satisfying Beck's conjecture; see [2,4,6,9,10, 20]. One of them is a finite reduced ring. Using Observation 1, we give a shorter proof of this result as follows.

Corollary 4 [6,20]. *If R is a non-zero reduced ring, then $\chi(\Gamma(R)) = \omega(\Gamma(R))$.*

Proof. By Observation 1 and Theorem 2, it is enough to prove $\omega(\Gamma(\mathbb{Z}_2^k)) = \chi(\Gamma(\mathbb{Z}_2^k))$. Clearly $\{e_i \mid 1 \le i \le k\}$, where $e_i = (0, \ldots, 0, 1, 0, \ldots, 0)$, induces a clique. Color first e_i by i, for $1 \le i \le k$.

For any $x = (x(1), x(2), \ldots, x(k)) \in V(\Gamma(\mathbb{Z}_2^k)) \setminus \{e_i \mid 1 \le i \le k\}$, there exists a least j with $1 \le j \le k$, such that $x(i) = 0$ for $1 \le i \le j - 1$ and $x(j) = 1$. Color x by j, then the resulting coloring is a proper k-coloring of $\Gamma(\mathbb{Z}_2^k)$.

The following result gives a necessary condition for a product of rings whose zero-divisor graphs are perfect.

Theorem 9. *Let $R = R_1 \times R_2 \times \ldots \times R_k$, where R_i's are indecomposable rings. If $\Gamma(R)$ is perfect, then $k \leq 4$.*

Proof. Suppose $k \geq 5$. Then the set of vertices $\{(1,1,0,0,0,0,\ldots,0), (0,0,1,1,0,\ldots,0), (1,0,0,0,1,0,\ldots,0), (0,1,0,1,0,0,\ldots,0), (0,0,1,0,1,0,\ldots,0)\}$ forms an induced cycle of length 5. By Theorem 6, we get a contradiction.

Next, let us prove the following result.

Theorem 10. *If $R = \mathbb{Z}_2^4$ $(= \mathbb{Z}_2 \times \mathbb{Z}_2 \times \mathbb{Z}_2 \times \mathbb{Z}_2)$, then $\Gamma(R)$ is perfect.*

Proof. Suppose $\Gamma(R)$ is not perfect. Then, by Theorem 6, we consider the following cases.

Case 1. $\Gamma(R)$ contains an odd cycle of length at least 5 as an induced subgraph.

Let C_{2r+1} be an induced cycle in $\Gamma(R)$ of length $2r + 1$ with the vertex set $\{x_0, x_1, \ldots, x_{2r}\}$, where $r \geq 2$. If exactly one co-ordinate of x_i is non-zero, for $0 \leq i \leq 2r$, then $2r + 1 \leq 4$, a contradiction. Therefore there exists an x_i containing at least two non-zero co-ordinates. WLOG, $x_i = (1, 1, x_i(3), x_i(4))$, for some i, $0 \leq i \leq 2r$. Then the 1st two coordinates of x_{i-1}, x_{i+1} are zeros, that is, $x_{i-1}(1) = x_{i-1}(2) = x_{i+1}(1) = x_{i+1}(2) = 0$. Since x_{i-1} and x_{i+1} are not adjacent, either the third coordinate or forth coordinate of x_{i-1} and x_{i+1} are non-zero. WLOG, $x_{i-1}(3) = x_{i+1}(3) = 1$. If $x_{i-1}(4) = 1$, then $x_{i+1}(4) = 0$, as $x_{i-1} \neq x_{i+1}$ and hence $x_{i-1} = (0, 0, 1, 1)$ and $x_{i+1} = (0, 0, 1, 0)$. Since x_{i-2} is adjacent to x_{i-1}, we have $x_{i-2} = (x_{i-2}(1), x_{i-2}(2), 0, 0)$. Thus x_{i-2} is adjacent to x_{i+1}, which is a contradiction. Hence $x_{i-1}(4) = 0$, which implies that $x_{i+1}(4) = 1$ and thus $x_{i+1} = (0, 0, 1, 1)$ and $x_{i-1} = (0, 0, 1, 0)$. Since x_{i+2} is adjacent x_{i+1}, we have $x_{i+2} = (x_{i+2}(1), x_{i+2}(2), 0, 0)$ and hence x_{i+2} is adjacent to x_{i-1}, which is a contradiction.

Case 2. $\Gamma(R)$ contains the complement of an odd cycle of length at least 5 as an induced subgraph.

Let $\overline{C_{2r+1}}$ be an induced subgraph of $\Gamma(R)$ with vertex set $\{x_0, x_1, \ldots, x_{2r}\}$, where $r \geq 2$. If no x_i contains exactly two coordinates that are non-zeros, then there exists j, $1 \leq j \leq k$ such that x_j contains exactly three that coordinates that are non-zero (otherwise $2r + 1 \leq 4$), which is impossible. Thus there exists i, $1 \leq i \leq k$ such that x_i contains exactly two coordinates that are non-zeros. WLOG, $x_i = (1, 1, x_i(3), x_i(4))$. Since x_i is adjacent to $2r - 2$ vertices in $\overline{C_{2r+1}}$, namely $x_{i+2}, x_{i+3}, \ldots, x_{i+2r-1}$ (where the addition in subscripts taken modulo $2r + 1$), we have the 1st two coordinates of $x_{i+2}, x_{i+3}, \ldots, x_{i+2r-1}$ are zero's and hence $x_{i+2}, x_{i+3}, \ldots, x_{i+2r-1} \in \{(0, 0, 1, 1), (0, 0, 1, 0), (0, 0, 0, 1)\}$. Thus $2r - 2 \leq 3$, which implies $2r + 1 \leq 6$. As it is an odd number and $r \geq 2$, we have $2r + 1 = 5$. Therefore $\overline{C_5} \cong C_5$. By Case 1, which is impossible.

The following result in [14] is a consequence of Theorems 9 and 10.

Corollary 5 [14]. *If $R = \mathbb{Z}_2^k$, then $\Gamma(R)$ is perfect if and only if $k \leq 4$.*

Proof. By Theorems 9 and 10, it is enough to prove that $\Gamma(R)$ is perfect if $k \leq 3$. In this case we have $|V(\Gamma(R))| \leq 6$, and hence $\Gamma(R)$ does not contain a cycle of length 5 as an induced subgraph of $\Gamma(R)$ and, thus the result follows.

It is well-known that any finite non-zero reduced commutative ring R is isomorphic to a finite direct product of finite fields, say $\mathbb{F}_{p_1^{\alpha_1}}, \mathbb{F}_{p_2^{\alpha_2}}, \ldots, \mathbb{F}_{p_\ell^{\alpha_\ell}}$, where p_i's are prime numbers and α_i's are positive integers, that is $R \cong \mathbb{F}_{p_1^{\alpha_1}} \times \mathbb{F}_{p_2^{\alpha_2}} \times \ldots \times \mathbb{F}_{p_\ell^{\alpha_\ell}}$.

By Theorem 2, the compressed zero-divisor graph of a reduced ring R is isomorphic to the zero-divisor graph of $\mathbb{Z}_2^k$, for some $k \geq 1$, that is $\Gamma_E(R) \cong \Gamma(\mathbb{Z}_2^k)$. So, the following result is a consequence of Theorem 9 and Corollary 5.

Theorem 11. *If $R \cong \mathbb{F}_{p_1^{\alpha_1}} \times \mathbb{F}_{p_2^{\alpha_2}} \times \ldots \times \mathbb{F}_{p_\ell^{\alpha_\ell}}$ is a non-zero reduced ring, where $\mathbb{F}_{p_i^{\alpha_i}}$'s are finite fields, then $\Gamma(R)$ is perfect if and only if $\ell \leq 4$.*

Proof. The first part is clear from Theorem 9. For the second part, let us assume that $\ell \leq 4$. Then $\omega(\Gamma(R)) \leq 4$. By the above discussion, $\Gamma_E(R) \cong \Gamma(\mathbb{Z}_2^k)$ for some $k \geq 1$. Suppose $k \geq 5$, then $\Gamma(\mathbb{Z}_2^k)$ contains a clique $\langle\{e_i : 1 \leq i \leq k\}\rangle$ of size at least 5 (where e_i's are defined in Corollary 4) and hence $\omega(\Gamma(R)) \geq 5$, which is impossible. Thus $k \leq 4$ and therefore, by Corollary 5 $\Gamma(\mathbb{Z}_2^k)$ is perfect, and hence $\Gamma(R)$ is perfect by Theorem 8.

The following result in [15] is an immediate consequence of Corollary 5, because every finite Boolean ring R is isomorphic to $\mathbb{Z}_2^k$, for some $k \geq 1$.

Corollary 6 [15]. *Let R be a finite Boolean ring. Then the following are equivalent,*

(1) $\Gamma(R)$ is perfect.
(2) $\Gamma(R)$ does not contain K_5 as a subgraph.
(3) $|R| \leq 2^4$.

3.1 Perfect Ideal Based Zero-Divisor Graph of Rings

In this subsection, we characterize rings whose ideal based zero-divisor graphs are perfect. In particular, under what values of n, the ideal based zero divisor graph of the ring $\mathbb{Z}_n$ of integers modulo n is perfect.

The following observation is observed in [16] and [21].

(i) If I is an ideal of R and $x_1 + I, x_2 + I, \ldots, x_k + I$ are the distinct cosets of I, which are zero-divisors of the quotient ring $\frac{R}{I}$, then $\Gamma_I(R)$ is a $\Gamma(\frac{R}{I})$-generalized join of $\langle x_1 + I \rangle, \langle x_2 + I \rangle, \ldots, \langle x_k + I \rangle$, that is,

$$\Gamma_I(R) = \Gamma\left(\frac{R}{I}\right)[\langle x_1 + I \rangle, \langle x_2 + I \rangle, \ldots, \langle x_k + I \rangle],$$

(ii) $\langle x_i + I \rangle$ is a complete subgraph of $\Gamma_I(R)$ if and only if $x_i^2 \in I$,

(iii) $\langle x_i + I \rangle$ is a totally disconnected subgraph of $\Gamma_I(R)$ if and only if $x_i^2 \notin I$.

Hence, by Theorems 7 and 8, we have

Theorem 12. *Let I be an ideal of R, then the following are equivalent,*

(i) $\Gamma_I(R)$ *is perfect;*
(ii) $\Gamma(\frac{R}{I})$ *is perfect;*
(iii) $\Gamma_E(\frac{R}{I})$ *is perfect.*

We recall the following result proved in [19].

Theorem 13 [19]. *The zero divisor graph $\Gamma(\mathbb{Z}_n)$ of a ring $\mathbb{Z}_n$ is perfect if and only if $n = p^a, p^a q^b, p^a qr$, or $pqrs$, where p, q, r and s are distinct primes and a and b are positive integers.*

It is well known that if I is an ideal of $\mathbb{Z}_n$ generated by m, then $\frac{\mathbb{Z}_n}{I} \cong \mathbb{Z}_m$. So, we have

Corollary 7. *If I is an ideal of $\mathbb{Z}_n$ generated by m, then $\Gamma_I(\mathbb{Z}_n)$ is perfect if and only if $m = p^a, p^a q^b, p^a qr$, or $pqrs$, where p, q, r and s are distinct primes and a and b are positive integers.*

Proof. By Theorems 12 and 13, $\Gamma_I(\mathbb{Z}_n)$ is perfect if and only if $\Gamma(\mathbb{Z}_m)$ is perfect if and only if $m = p^a, p^a q^b, p^a qr$, or $pqrs$.

3.2   Zero-Divisor Graph of Rings, Reduced Semigroups and Posets

In [13], it is shown that the chromatic number is equal to the clique number of zero-divisor graphs of poset, reduced semiring with 0 and reduced semigroup with 0. So it is interesting to consider the following problem.

Problem. Characterize the posets, reduced rings and reduced semigroups whose zero-divisor graphs are perfect.

Now we characterize posets whose zero-divisor graphs are perfect using Theorem 3.

Theorem 14. *Let G be a zero-divisor graph of a poset with 0 and $\omega(G) = k$. Then the following are equivalent,*

(i) G *is perfect.*
(ii) *The reduced graph G_r of G is perfect.*
(iii) *The reduced graph H_r of H (where H is in the Definition of generalized complete k-partite graph) is perfect.*

Proof. (i) $\Leftrightarrow$ (ii) It follows from Corollary 2.
(ii) $\Rightarrow$ (iii) It follows from the definition of perfect.

(iii) $\Rightarrow$ (ii) Suppose G_r is not perfect graph, then by the Theorem 6, G_r contains an odd cycle of length at least 5 as an induced subgraph or the complement of an odd cycle of length at least 5 as an induced subgraph.

Let $e_1, e_2, \ldots, e_k$ be the atoms of G.

Case 1. G_r contains an odd cycle of length at least 5 as an induced subgraph.

Let C_{2s+1} be an odd cycle of G_r as an induced subgraph with vertex set $V(C_{2s+1}) = \{a_0, a_1 \ldots, a_{2s}\}$, where $s \geq 2$. Then $V(C_{2s+1})$ is not a subset of $V(H_r)$. As the atoms forms a clique, we have $|V(C_{2s+1}) \cap \{e_1, e_2, \ldots, e_k\}| \leq 2$. First if $|V(C_{2s+1}) \cap \{e_1, e_2, \ldots, e_k\}| = 2$, then there exist $i, j \in \{1, 2, \ldots, k\}$ and $\ell \in \{0, 1, 2, \ldots, 2s\}$ such that $a_\ell = e_i$ and $a_{\ell+1} = e_j$. Since $a_{\ell+2}$ and $a_{\ell+3}$ are not adjacent to $a_\ell = e_i$, we have $i \notin W(a_{\ell+2}) \cup W(a_{\ell+3})$ and hence $W(a_{\ell+2}) \cup W(a_{\ell+3}) \neq \{1, 2, \ldots, k\}$, which is a contradiction to the definition of generalized complete k-partite graph. Next if $|V(C_{2s+1}) \cap \{e_1, e_2, \ldots, e_k\}| = 1$, then we get a contradiction in a similar way as above. Thus $V(C_{2s+1}) \cap \{e_1, e_2, \ldots, e_k\} = \emptyset$ and hence C_{2s+1} is an induced odd cycle of H_r, which is a contradiction.

Case 2. G_r contains the complement of an odd cycle of length at least 5 as an induced subgraph.

Let $\overline{C_{2s+1}}$ be the complement of the odd cycle C_{2s+1} in G_r with vertex set $V(\overline{C_{2s+1}}) = \{a_0, a_1, \ldots, a_{2s}\}$, where $s \geq 2$. If $V(\overline{C_{2s+1}}) \cap \{e_1, e_2, \ldots, e_k\} \neq \emptyset$, then there exists $i \in \{1, 2, \ldots, k\}$ such that $e_i = a_\ell$, for some $\ell \in \{0, 1, 2, \ldots, 2s\}$. Then $a_{\ell-1}, a_{\ell+1} \notin \{e_1, e_2, \ldots, e_k\}$ and they are not adjacent to e_i and hence $i \notin W(a_{\ell-1}) \cup W(a_{\ell+1})$, which is impossible. Thus, $V(\overline{C_{2s+1}}) \cap \{e_1, e_2, \ldots, e_k\} = \emptyset$ and therefore $\overline{C_{2s+1}}$ lies in H_r, which is a contradiction.

Next, we present equivalent conditions for a zero-divisor graph of a reduced semigroup to be perfect using Theorem 4.

Theorem 15. *Let G be a zero-divisor graph of a reduced semigroup with $\omega(G) = k$. Then the following are equivalent,*

(i) G is perfect.
(ii) The reduced graph G_r of G is perfect.
(iii) The reduced graph H_r of H (where H is given in the definition of generalized complete k-partite graph) is perfect.

Proof. The proof is similar to that of Theorem 14.

A lattice $L = (L, \wedge, \vee)$ with 0 is *distributive* if for $x, y, z \in L$, $x \wedge (y \vee z) = (x \wedge y) \vee (x \wedge z)$. As every lattice is a poset, we have the following result proved in [15].

Corollary 8 [15]. *Let L be a distributive lattice with 0. Then the following are equivalent,*

(i) $\Gamma(L)$ is perfect.
(ii) $\Gamma(L)$ contains no induced cycle of length 5.
(iii) $\omega(\Gamma(L)) \leq 4$, (equivalently, the number of atoms of $\Gamma(L)$ is at most 4).

Proof. (i) $\Rightarrow$ (ii) It is trivial from the definition of perfect graph.

(ii) $\Rightarrow$ (iii) If $\langle\{a_1, a_2, \ldots, a_s\}\rangle$ is a clique in $\Gamma(L)$, where $s \geq 5$, then the subgraph induced by $\{a_1 \vee a_2, a_3 \vee a_4, a_1 \vee a_5, a_2 \vee a_3, a_4 \vee a_5\}$ is an induced cycle of length 5 (as L is distributive) which is a contradiction.

(iii) $\Rightarrow$ (i) Suppose $\Gamma(L)$ is not perfect. Then by Theorem 14, the reduced subgraph H_r of H, defined in Theorem 14, is not perfect. By Theorem 6, H_r contains an odd cycle of length at least 5 as an induced subgraph or its complement of an odd cycle of length at least 5 as an induced subgraph. If H_r contains an induced odd cycle C_{2s+1} with vertex set $V(C_{2s+1}) = \{a_1, a_2, \ldots, a_{2s+1}\}$, where $s \geq 2$. Then $a_i \wedge a_{i+1} = 0$, for $1 \leq i \leq 2s$, $a_{2s+1} \wedge a_1 = 0$ and $a_i \wedge a_j \neq 0$, for $j \neq i - 1, i, i + 1$ and hence the subgraph induced by $\{a_1 \wedge a_3, a_1 \wedge a_4, a_2 \wedge a_4, a_2 \wedge a_5, a_3 \wedge a_{2s+1}\}$ is a clique in $\Gamma(L)$ of size 5, which is a contradiction. Similarly if H_r contains the complement $\overline{C_{2s+1}}$ of an induced odd cycle C_{2s+1} with vertex set $V(\overline{C_{2s+1}}) = \{a_1, a_2, \ldots, a_{2s+1}\}$, where $s \geq 2$, then the subgraph induced by $\{a_1 \wedge a_2, a_2 \wedge a_3, a_3 \wedge a_4, a_4 \wedge a_5, a_5 \wedge a_1\}$ is a clique in $\Gamma(L)$ of size 5, which is again a contradiction.

As every semiring is a semigroup and by Theorem 15, we have the following result proved in [15].

Corollary 9 [15]. *Let R be a reduced semiring with 0. Then the following are equivalent,*

(i) $\Gamma(R)$ is perfect.
(ii) $\Gamma(R)$ contains no induced cycle of length 5.
(iii) $\omega(\Gamma(R)) \leq 4$, (equivalently, the number of atoms of $\Gamma(R)$ is at most 4).

Proof. The proof is similar to that of Corollary 8 by replacing $\vee$ and $\wedge$ by addition and multiplication, respectively.

Acknowledgment. This research was supported by the University Grant Commissions Start-Up Grant, Government of India grant No. F. 30-464/2019 (BSR) dated 27.03.2019.

References

1. Akbari, S., Mohammadian, A.: On the zero-divisor graph of a commutative ring. J. Algebra. **274**, 847–855 (2004)
2. Anderson, D.D., Naseer, M.: Beck's coloring of commutative ring. J. Algebra. **159**, 500–514 (1993)
3. Anderson, D.F., LaGrange, J.D.: Commutative Boolean monoids, reduced rings, and the compressed zero-divisor graph. J. Pure Appl. Algebra. **216**, 1626–1636 (2012)
4. Anderson, D.F., Levy, R., Shapiro, J.: Zero-divisor graphs, von Neumann regular rings, and Boolean algebras. J. Pure Appl. Algebra **180**, 221–241 (2003)
5. Anderson, D.F., Livingston, P.S.: The zero-divisor graph of a commutative ring. J. Algebra **217**, 434–447 (1999)
6. Beck, I.: Coloring of commutative rings. J. Algebra **116**, 208–226 (1988)

7. Berge, C.: Perfect graphs six papers on graph theory. Indian Statistical Institute, Calcutta, pp. 1–21 (1963)
8. Chudnovsky, M., Robertson, N., Seymour, P., Thomas, R.R.: The strong perfect graph theorem. Ann. Math. **164**, 51–229 (2006)
9. Halas, R., Jukl, M.: On Beck's coloring of Posets. Discret. Math. **309**, 4584–4589 (2009)
10. Kavaskar, T.: Beck's coloring of finite product of commutative ring with unity. Graph Combin. **38**(3), 1–9 (2022)
11. LaGrange, J.D.: On realizing zero-divisor graphs. Commun. Algebra **36**, 4509–4520 (2008)
12. Lovász, L.: Normal hypergraphs and the perfect graph conjecture. Discret. Math. **2**, 253–267 (1972)
13. Lu, D., Wu, T.: The zero-divisor graphs of posets and an application to semigroups. Graphs Combin. **26**, 793–804 (2010)
14. Nilesh, K., Joshi, V.: Coloring of zero-divisor graphs of posets and applications to graphs associated with algebraic structures. arXiv (2022). arXiv:2205.04916
15. Patil, A., Waphare, B.N., Joshi, V.: Perfect zero-divisor graphs. Discret. Math. **340**, 740–745 (2017)
16. Redmond, S.: An ideal-based zero-divisor graph of a commutative ring. Comm. Algebra **32**(9), 4425–4443 (2003)
17. Bagheri, S., Nabaei, F., Rezaeii, R., Samei, K.: Reduction graph and its application on algebraic graphs. Rocky Mt. J. Math. **48**(3), 729–751 (2018)
18. Selvakumar, K., Gangaeswari, P., Arunkumar, G.: The Wiener index of the zero-divisor graph of a finite commutative ring with unity. Discret. Appl. Math. **311**, 72–84 (2022)
19. Smith, B.: Perfect zero-divisor graphs of $\mathbb{Z}_n$. Rose-Hulman Undergrad. Math. J. **17**(2), 113–132 (2016)
20. Spiroff, S., Wickham, C.: A zero-divisor graph determined by equivalence classes of zero-divisors. Commun. Algebra **39**, 2338–2348 (2011)
21. Balamoorthy, S., Kavaskar, T., Vinothkumar, K.: Wiener index of ideal-based zero-divisor graph of a commutative ring with unity. Communicated

On Coupon Coloring of Cayley Graphs

Reji Thankachan[ID] and Pavithra Rajamani[(✉)][ID]

Department of Mathematics, Government College Chittur, University of Calicut,
Palakkad, Kerala 678104, India
pavithrajeni94@gmail.com

Abstract. A k-coupon coloring of a graph G without isolated vertices is an assignment of colors from $[k] = \{1, 2, \ldots, k\}$ to the vertices of G such that the neighborhood of every vertex of G contains vertices of all colors from $[k]$. The maximum k for which a k-coupon coloring exists is called the coupon coloring number of G. The Cayley graph $Cay(\Gamma, C)$ of a group Γ is a graph with vertex set Γ and edge set $E(Cay(\Gamma, C)) = \{gh : hg^{-1} \in C\}$, where C is a subset of Γ that is closed under taking inverses and does not contain the identity. Let R be a commutative ring with unity. Then $Cay(R^+, Z^*(R))$ is denoted by $\mathbb{CAY}(R)$, where R^+ is the additive group and $Z^*(R)$ is the non-zero zero-divisors of R. For a natural number n, the generalized Cayley graph, Γ_R^n is a simple graph with vertex set $R^n \setminus \{0\}$ and two distinct vertices X and Y are adjacent if and only if there is a lower triangular matrix A over R whose entries on the main diagonal are non-zero and such that $AX^T = Y^T$ or $AY^T = X^T$, where B^T is the transpose of the matrix B. In this paper, we have studied the coupon coloring of $\mathbb{CAY}(R)$ and generalized Cayley graph Γ_R^n.

Keywords: Coupon coloring · Cayley graph · Generalized Cayley graph

1 Introduction

The concept of coupon coloring number was introduced by Chen et al. in [6]. Let G be a graph without isolated vertices. A k-vertex coloring, or simply a k-coloring of G is a mapping c from the vertex set of G to $[k] = \{1, 2, \ldots, k\}$. A k-coupon coloring of a graph G is an assignment of colors from $[k] = \{1, 2, \ldots, k\}$ to the vertices of G such that the neighborhood of every vertex of G contains vertices of all colors from $[k]$. The maximum k for which a k-coupon coloring exists is called the coupon coloring number of G and it is denote by $\chi_c(G)$.

Colors can be imagined as coupons of different types. The idea of coupon coloring is to receive all different tokens from the neighbors of each vertex. The concept can be applied to different practical problems. Suppose an information or message is separated into mutually exhaustive parts and is assigned to different members of a group. Each member is assigned only one part of these decomposed parts. Members can be considered as vertices of a graph. Then the conditions of coupon coloring ensures that each member gathers the whole piece of information

A. Bagchi and R. Muthu (Eds.): CALDAM 2023, LNCS 13947, pp. 184–191, 2023.
https://doi.org/10.1007/978-3-031-25211-2_14

or message from her neighbors and the maximum k in coupon coloring is then associated with maximizing the length of the original information. This concept can be applied into problems in network science and one such application in multi-robot network is given in [2].

Let $G = (V, E)$ be a graph. $D \subseteq V$ is a dominating set if every vertex in $V \setminus D$ is adjacent to at least one vertex in D. Let $G = (V, E)$ be a graph without isolated vertices. $D' \subseteq V$ is a total dominating set (TDS) if every vertex of G is adjacent to at least one vertex in D'. The minimum cardinality among all the total dominating sets in G is called the total domination number, $\gamma_t(G)$. The coupon coloring number is also referred to as the total domatic number, introduced in [4], which is the maximum number of disjoint total dominating sets. Coupon coloring was studied by many authors [7,9,10]. In [10] Y Shi et al. determined coupon coloring number of complete graphs, complete k-partite graphs, wheels, cycles, unicyclic graphs and bicyclic graphs.

In this paper, we have studied the coupon coloring of Cayley graphs and generalized Cayley graphs of finite commutative rings. Every finite commutative ring can be written as a product of some finite local rings. The proof of this result can be found in [5]. First, we found the exact coupon coloring number of $\mathbb{CAY}(R)$. Further, we have studied the coupon coloring of Γ_R^n. We present some bounds for the coupon coloring number of Γ_R^n in terms of $|Z(R)|$ and $|U(R)|$.

2   Preliminaries

All graphs considered in this paper are simple, finite and undirected. As usual K_n denotes the complete graph with n vertices. Let G be a graph without isolated vertices. In a k-coupon coloring c of G, the neighborhood of a vertex must contain vertices of all colors from $[k]$. Clearly, coupon coloring is an improper coloring and $\chi_c(G) \leq \delta(G)$.

Let Γ be a group and let C be a subset of Γ that is closed under taking inverses and does not contain the identity. Then the Cayley graph $Cay(\Gamma, C)$ is a graph with vertex set Γ and edge set

$$E(Cay(\Gamma, C)) = \{gh : hg^{-1} \in C\}.$$

Let $\mathbb{Z}_n$ denote the additive group of integers modulo n. If C is a subset of $\mathbb{Z}_n \setminus \{0\}$, then construct the graph $Cay(\mathbb{Z}_n, C)$ as follows. The vertices of $Cay(\mathbb{Z}_n, C)$ are elements of $\mathbb{Z}_n$ and (i, j) is an edge of $Cay(\mathbb{Z}_n, C)$ if and only if $j - i \in C$. The graph $Cay(\mathbb{Z}_n, C)$ is called a circulant graph of order n, and C is called its connection set. Let R be a commutative ring with unity, $Z(R)$ be the set of zero-divisors of R and $U(R)$ be the set of units of R. Then the Cayley graph of R with respect to its non-zero zero-divisors is the graph $Cay(R^+, Z^*(R))$ denoted by $\mathbb{CAY}(R)$, where $Z^*(R) = Z(R) \setminus \{0\}$. This is the Cayley graph whose vertices are all elements of the additive group R^+ and in which two distinct vertices x and y are joined by an edge if and only if $x - y \in Z^*(R)$.

Let R be a commutative ring with identity element. For a natural number n, Afkhami et al. [3] defined the generalized Cayley graph Γ_R^n in 2012. It is a simple graph with vertex set $R^n \setminus \{0\}$ and two distinct vertices X and Y are adjacent if and only if there is a lower triangular matrix A over R whose entries on the main diagonal are non-zero and such that $AX^T = Y^T$ or $AY^T = X^T$, where B^T is the transpose of the matrix B.

The following results will be useful for the upcoming sections.

Theorem 1. *[10] Let G be a complete graph with n vertices. Then*

$$\chi_c(G) = \left\lfloor \frac{n}{2} \right\rfloor.$$

Theorem 2. *[1] Let R be a ring. Then the following statements hold:*

1. $\mathbb{CAY}(R)$ *has no edge if and only if R is an integral domain.*
2. *If (R, M) is an Artinian local ring, then $\mathbb{CAY}(R)$ is a disjoint union of $\left|\frac{R}{M}\right|$ copies of the complete graph $K_{|M|}$.*
3. $\mathbb{CAY}(R)$ *cannot be a complete graph.*
4. $\mathbb{CAY}(R)$ *is a regular graph of degree $|Z(R) - 1|$ with isomorphic components.*

Theorem 3. *[3] Let R be a commutative ring with unity. If $X = (x_1, x_2, \ldots, x_n)$, $x_1 \in U(R)$ and $Y = (y_1, y_2, \ldots, y_n)$, $y_1 \neq 0$ are two vertices of Γ_R^n, then X and Y are adjacent in Γ_R^n.*

3   Coupon Coloring Number of $\mathbb{CAY}(R)$

Coupon coloring is defined for graphs without isolated vertices. So, we consider $\mathbb{CAY}(R)$, where R is a finite commutative ring which is not an integral domain. If R is a finite local ring, then by Theorem 2, $\mathbb{CAY}(R)$ is the disjoint union of $\left|\frac{R}{M}\right|$ copies of the complete graph $K_{|M|}$. So, $\chi_c(\mathbb{CAY}(R)) = \left\lfloor \frac{|M|}{2} \right\rfloor$. The following theorem gives the exact coupon coloring number of $\mathbb{CAY}(R)$, when R is a finite commutative ring.

Theorem 4. *Let R be a finite commutative ring and $R \cong R_1 \times R_2 \times \cdots \times R_n$, $n > 1$ be the local ring decomposition. If $k = \min\{|R_i/m_i| : i = 1, 2, \ldots, n\}$, m_i be the maximal ideal of R_i, then*

$$\chi_c(\mathbb{CAY}(R)) = \frac{|R|}{k}.$$

Proof. Assume that $k = |R_1/m_1|$. Let

$$R_1/m_1 = \{r_j + m_1 : j = 1, 2, \ldots, k\}$$

Define D_t for each $(n-1)$-tuple $y^{(t)} = (y_2^{(t)}, y_3^{(t)}, \ldots y_n^{(t)}) \in R_2 \times \cdots \times R_n$,

$$D_t = \{(y_j, y_2^{(t)}, y_3^{(t)}, \ldots, y_n^{(t)}) : y_j's \text{ are distinct elements from the } k \text{ cosets of } m_1\}$$

So $|D_t| = k$ and there are $\frac{|R|}{k}$ such sets which are pairwise disjoint. Define $c : R \to \left[\frac{|R|}{k}\right]$ by $c(x) = t$ if $x \in D_t$. We claim that c is a coupon coloring of $\mathbb{CAY}(R)$. For let $x \in R$. Then $x = (x_1, x_2 \ldots, x_n)$ and $x_1 \in r_l + m_1$, for some $l \in \{1, 2, \ldots, k\}$. So there exists $y = (y_1, y_2^{(t)}, y_3^{(t)}, \ldots, y_n^{(t)}) \in D_t$ such that $y_1 \in r_l + m_1$. Then $x_1 - y_1 \in m_1$ and x is adjacent to y. If $x = y \in D_t$, then x is adjacent to all other elements of D_t, since $x - y' = (x_1 - y_j, 0, 0, \ldots, 0)$ for all $y' = (y_j, y_2^{(t)}, y_3^{(t)}, \ldots, y_n^{(t)}) \in D_t$. Hence the claim. So,

$$\chi_c(\mathbb{CAY}(R)) \geq \frac{|R|}{k}.$$

Let c be a k-coupon coloring of $\mathbb{CAY}(R)$ such that the color 1 is given to at most $k - 1$ vertices. Suppose that S is the set of vertices with color 1. Since m_i has at least k cosets, for each $i = 1, 2, \ldots, n$, there exists a coset $r_{l_i} + m_i$ of m_i such that none of the elements of S has i^{th} co-ordinate from $r_{l_i} + m_i$. That is, if $T_i = \{(x_1, x_2, \ldots, x_n) \in R : x_i \in r_{l_i} + m_i\}$, then $S \cap T_i = \phi$, for all $i = 1, 2, \ldots, n$. Consider the vertex $z = (z_1, z_2, \ldots, z_n), z_i \in r_{l_i} + m_i$ for all $i = 1, 2, \ldots, n$. Since c is a k-coupon coloring of $\mathbb{CAY}(R)$, there exists $y \in S$ such that $y - z \in Z^*(R)$. That is, $y_j - z_j \in Z(R_j)$ for some j. Then $y_j \in r_{l_j} + m_j$ and so $y \in T_j$, a contradiction. Hence in a k-coupon coloring of $\mathbb{CAY}(R)$ each color must be given to at least k vertices and

$$\chi_c(\mathbb{CAY}(R)) \leq \frac{|R|}{k}.$$

Therefore, $\chi_c(\mathbb{CAY}(R)) = \frac{|R|}{k}$. □

4   Coupon Coloring Number of Γ_R^n

In this section, we have studied the coupon coloring of generalized Cayley graphs of finite commutative rings. If $R \not\cong \mathbb{Z}_2$ is a field, then Γ_R^n is a union of n complete graphs [3]. These n complete graphs are the graphs induced by C_i, for $i = 1, 2, \ldots, n$, where C_i is the set of all vertices whose first non-zero components are in the ith place. Therefore, $\chi_c(\Gamma_R^n) = \min \{\chi_c(K_{|C_i|}) : i = 1, 2, \ldots, n\} = \min \left\{ \left\lfloor \frac{|C_i|}{2} \right\rfloor : i = 1, 2, \ldots, n \right\} = \left\lfloor \frac{|C_n|}{2} \right\rfloor = \left\lfloor \frac{|R|-1}{2} \right\rfloor$, since $|C_i| = (|R| - 1)|R|^{n-i}$.

For the coupon coloring number, we have to find the maximum number of disjoint total dominating sets of Γ_R^n. If R is not an integral domain, Selvakumar [8] proved that $\{X = (x_1, x_2, \ldots, x_n), Y = (y_1, y_2, \ldots, y_n) : x_1 \in U(R), y_1 \in Z^*(R), y_2 \in U(R)\}$ is a total dominating set of Γ_R^n. Hereafter assume that R is a finite commutative ring which is not an integral domain. So, there exists at least one non-zero zero-divisor in R.

Theorem 5. *Let R be a finite commutative ring and $n > 1$ be a positive integer. Then*

$$\chi_c(\Gamma_R^n) \geq (|Z(R)| - 1)|U(R)||R|^{n-2}.$$

Proof. Let $X = (x_1, x_2, \ldots, x_n), Y = (y_1, y_2, \ldots, y_n)$, and let $k = (|Z(R)| - 1)|U(R)||R|^{n-2}$. Consider the two element sets of the form

$$A = \left\{ X, Y : x_1 \in U(R), x_2 \in Z^*(R), y_1 \in Z^*(R), y_2 \in U(R) \right\}$$

Note that, for each pair $(a, b) \in U(R) \times Z^*(R)$,

$$\left\{ X = (a, b, x_3, x_4, \ldots, x_n), Y = (b, a, x_3, x_4, \ldots, x_n) \right\}$$

are disjoint sets of the form A, for all $(x_3, x_4, \ldots, x_n) \in R^{n-2}$. So, there are $|R|^{n-2}$ disjoint sets of the form A, for each $(a, b) \in U(R) \times Z^*(R)$. Thus we can have k sets of the form A. Name these k disjoint sets as $A_1, A_2, \ldots, A_k$. Define $c : V(\Gamma_R^n) \to [k]$ by

$$c(X) = \begin{cases} i & \text{if } X \in A_i \\ 1 & \text{otherwise.} \end{cases}$$

Then c is a coupon coloring on Γ_R^n, since each A_i is a total dominating set of Γ_R^n. Hence,

$$\chi_c(\Gamma_R^n) \geq k.$$

$\square$

Lemma 1. *Let R be a finite commutative ring and $n > 1$ be a positive integer. In any k-coupon coloring of Γ_R^n, each color should be given to at least a vertex of the form $X = (x_1, x_2, \ldots, x_n)$ with $x_1 \in Z(R)$.*

Proof. Let H be the set of all vertices with first co-ordinate is a zero divisor. Suppose that there is a k-coupon coloring with none of the vertices of H has color 1. Then the vertex $Z = (0, z_2, \ldots, z_n)$ has no neighbor with color 1. $\square$

Suppose that Γ_R^n has a k-coupon coloring. Denote H_t as the set of all vertices with color $t \in \{1, 2, \ldots, k\}$. Then H_t must be a total dominating set of Γ_R^n for all $t = 1, 2, \ldots, k$.

Lemma 2. *Suppose that Γ_R^n has a k-coupon coloring. If $X = (0, x_2, \ldots, x_n) \in H_t$, then there is a $Y = (y_1, y_2, \ldots, y_n) \neq X$ with $y_1 \in Z(R)$ in H_t.*

Proof. Suppose that $X = (0, x_2, \ldots, x_n) \in H_t$ and $y_1 \notin Z(R)$ for all $Y = (y_1, y_2, \ldots, y_n) \neq X$ in H_t. That is, $y_1 \in U(R)$ for all $Y = (y_1, y_2, \ldots, y_n) \neq X$ in H_t. Then any vertex $Z = (z_1, z_2, \ldots, z_n)$ with $z_1 \neq 0$ in Γ_R^n is adjacent to a vertex Y with color t. But the vertex X has no neighbor with the color t, a contradiction. $\square$

Theorem 6. *Let R be a finite commutative ring and $n > 1$ be a positive integer. Then*

$$\chi_c(\Gamma_R^n) \leq (|Z(R)| - 1)|R|^{n-1} + \left\lfloor \frac{|R|^{n-1}}{2} \right\rfloor.$$

Proof. Suppose that Γ_R^n has a k-coupon coloring. From Lemma 1, there is an $X = (x_1, x_2, \ldots, x_n) \in H_t$, with $x_1 \in Z(R)$ in every H_t. So we can have at most $|Z(R)||R|^{n-1}$ such H_t's. By Lemma 2, if $X = (0, x_2, \ldots, x_n) \in H_t$, then there is a $Y = (y_1, y_2, \ldots, y_n) \neq X$ with $y_1 \in Z(R)$ in H_t. Thus to get the maximum number of color classes H_t, each H_t must contain either a vertex with non-zero zero-divisor in first co-ordinate or two vertices with 0 in their first co-ordinate, but not both. Hence we can have at most $(|Z(R)| - 1)|R|^{n-1} + \left\lfloor \frac{|R|^{n-1}}{2} \right\rfloor$ distinct color classes.   □

Theorem 7. *Suppose that $Z(R)$ is an ideal of the finite commutative ring R and $n > 1$ is a positive integer. If Γ_R^n has a k-coupon coloring, then there exist a vertex $X = (x_1, x_2, \ldots, x_n), x_1 \in Z^*(R), x_2 \in U(R)$ or a vertex $X' = (0, x_2', \ldots, x_n')$ in every H_t.*

Proof. By Lemma 1, H_t must have a vertex with zero-divisor in its first co-ordinate. Assume that $y_1 \neq 0$ for every $Y = (y_1, y_2, \ldots, y_n)$ in H_t. Suppose that the second co-ordinate is also a zero-divisor for every $Y = (y_1, y_2, \ldots, y_n)$ in H_t with $y_1 \in Z^*(R)$. Consider $Z = (z_1, z_2, \ldots, z_n), z_1 = 0, z_2 \in U(R)$. Then there must be an $X = (x_1, x_2, \ldots, x_n) \in H_t$ with

$$
\begin{pmatrix} a & 0 & \cdots & 0 \\ b & c & \cdots & 0 \\ & & \vdots & \\ a_{n1} & a_{n2} & \cdots & a_{nn} \end{pmatrix} \begin{pmatrix} x_1 \\ x_2 \\ \vdots \\ x_n \end{pmatrix} = \begin{pmatrix} 0 \\ z_2 \\ \vdots \\ z_n \end{pmatrix} \tag{1}
$$

or

$$
\begin{pmatrix} a' & 0 & \cdots & 0 \\ b' & c' & \cdots & 0 \\ & & \vdots & \\ a_{n1}' & a_{n2}' & \cdots & a_{nn}' \end{pmatrix} \begin{pmatrix} 0 \\ z_2 \\ \vdots \\ z_n \end{pmatrix} = \begin{pmatrix} x_1 \\ x_2 \\ \vdots \\ x_n \end{pmatrix}. \tag{2}
$$

Equation (2) is not possible, since x_1 is non-zero. Since $a \neq 0$, $x_1 \in Z^*(R)$. Then there exists $a \in Z^*(R)$ such that $ax_1 = 0$. But $bx_1 + cx_2 = z_2$ cannot be a unit, since $x_1, x_2 \in Z(R)$ and $Z(R)$ is an ideal of R, $z_2 = bx_1 + cx_2 \in Z(R)$, a contradiction.   □

Next corollary follows from the proof of Theorem 6 and Theorem 7.

Corollary 1. *Let R be a finite commutative ring and $n > 1$ be a positive integer. If $Z(R)$ is an ideal of R, then*

$$
\chi_c(\Gamma_R^n) \leq (|Z(R)| - 1)|U(R)||R|^{n-2} + \left\lfloor \frac{|R|^{n-1}}{2} \right\rfloor.
$$

Example: Consider a finite commutative ring R with $Z(R)$ is an ideal of R and $n = 2$. Let $r = (|Z(R)| - 1)|U(R)|$, $s = \left\lfloor \frac{|U(R)|}{2} \right\rfloor$.

$$
A = \left\{ X = (x_1, x_2), Y = (y_1, y_2) : x_1, y_2 \in Z^*(R), y_1, x_2 \in U(R) \right\}
$$

$$B = \left\{ X = (0, x_2), Y = (0, y_2), Z = (x_2, 0), V = (y_2, 0) : x_2, y_2 \in U(R), x_2 \neq y_2 \right\}$$

Note that, there can be r disjoint sets of the form A and name them as A_i, $i = 1, 2, \dots, r$. Similarly, there can be s disjoint sets of the form B and name them as B_j for $j = 1, 2, \dots, s$. Clearly, A_i is a total dominating set for all $i = 1, 2, \dots, r$. We claim that B_j is also a total dominating set of Γ_R^n. For, let $(a, b) \in V(\Gamma_R^n)$. If $a \neq 0$, then $(a, b) \neq Z$ is adjacent to Z in B_j and Z is adjacent to V. If $a = 0$, then $b \neq 0$ and

$$\begin{pmatrix} b & 0 \\ 0 & x_2^{-1}b \end{pmatrix} \begin{pmatrix} 0 \\ x_2 \end{pmatrix} = \begin{pmatrix} 0 \\ b \end{pmatrix}.$$

So, if $b \neq x_2$, (a, b) is adjacent to $(0, x_2)$ and if $b = x_2$, then (a, b) is adjacent to $(0, y_2)$,

$$\begin{pmatrix} x_2 & 0 \\ 0 & x_2^{-1}y_2 \end{pmatrix} \begin{pmatrix} 0 \\ x_2 \end{pmatrix} = \begin{pmatrix} 0 \\ y_2 \end{pmatrix}.$$

Thus B_j is a total dominating set for all j. Note that $A_i \cap B_j = \phi$ for all i and j. So there are $r + s$ disjoint total dominating sets and

$$\chi_c(\Gamma_R^2) \geq (|Z(R)| - 1)|U(R)| + \left\lfloor \frac{|U(R)|}{2} \right\rfloor.$$

By the above corollary,

$$(|Z(R)| - 1)|U(R)| + \left\lfloor \frac{|U(R)|}{2} \right\rfloor \leq \chi_c(\Gamma_R^2) \leq (|Z(R)| - 1)|U(R)| + \left\lfloor \frac{|R|}{2} \right\rfloor,$$

since $n = 2$.

If $R = \mathbb{Z}_4$, then $Z(R) = \{0, 2\}$, $U(R) = \{1, 3\}$ and $3 \leq \chi_c(\Gamma_{\mathbb{Z}_4}^2) \leq 4$. Here, $A_1 = \{(2, 1), (1, 2)\}$, $A_2 = \{(2, 3), (3, 2)\}$, $B_1 = \{(0, 1), (0, 3), (1, 0), (3, 0)\}$. Note that, $C = \{(0, 2), (2, 2), (1, 1)\}$ is also a total dominating set. Therefore, $\chi_c(\Gamma_{\mathbb{Z}_4}^2) = 4$. Hence, the bound in Corollary 1 is sharp.

Acknowledgments. The authors would like to thank the reviewers for their invaluable suggestions and comments which greatly contributed in the improvement of the paper.

The second author is thankful to University Grants Commission, India, for their financial support through Junior Research Fellowship (JUNE18-411992).

References

1. Aalipour, G., Akbari, S.: On the Cayley graph of a commutative ring with respect to its zero-divisors. Comm. Algebra **44**(4), 1443–1459 (2016)
2. Abbas, W., Egerstedt, M., Liu, C.H., Thomas, R., Whalen, P.: Deploying robots with two sensors in $k_{1,6}$-free graphs. J. Graph Theory **82**(3), 236–252 (2016)
3. Afkhami, M., Khashyarmanesh, K., Nafar, K.: Generalized Cayley graphs associated to commutative rings. Linear Algebra Appl. **437**(3), 1040–1049 (2012)

4. Aram, H., Sheikholeslami, S.M., Volkmann, L.: On the total domatic number of regular graphs. Trans. Combinat. **1**(1), 45–51 (2012)
5. Bini, G., Flamini, F.: Finite commutative rings and their applications, vol. 680. Springer Science & Business Media (2002). https://doi.org/10.1007/978-1-4615-0957-8
6. Chen, B., Kim, J.H., Tait, M., Verstraete, J.: On coupon colorings of graphs. Discret. Appl. Math. **193**, 94–101 (2015)
7. Chen, H., Jin, Z.: Coupon coloring of cographs. Appl. Math. Comput. **308**, 90–95 (2017)
8. Krishnan, S.: Domination in generalized Cayley graph of commutative rings. J. Math. Study **54**, 427–434 (2021)
9. Nagy, Z.L.: Coupon-coloring and total domination in Hamiltonian planar triangulations. Graphs Combinat. **34**(6), 1385–1394 (2018)
10. Shi, Y., Wei, M., Yue, J., Zhao, Y.: Coupon coloring of some special graphs. J. Comb. Optim. **33**(1), 156–164 (2017)

Coloring of a Superclass of $2K_2$-free graphs

Athmakoori Prashant$^{(\boxtimes)}$ⓘ and S. Francis Rajⓘ

Department of Mathematics, Pondicherry University, Puducherry 605014, India
11994prashant@gmail.com, francisraj_s@pondiuni.ac.in

Abstract. The class of $2K_2$-free graphs has been well studied in various contexts in the past. In this paper, we study the chromatic number of {$butterfly, hammer$}-free graphs, a superclass of $2K_2$-free graphs and show that a connected {$butterfly, hammer$}-free graph G with $\omega(G) \neq 2$ admits $\binom{\omega+1}{2}$ as a χ-binding function which is also the best available χ-binding function for $2K_2$-free graphs. In addition, we show that if $H \in \{C_4 + K_p, P_4 + K_p\}$, then any {$butterfly, hammer, H$}-free graph G with no components of clique size two admits a linear χ-binding function. Furthermore, we also establish that any connected {$butterfly, hammer, H$}-free graph G where $H \in \{(K_1 \cup K_2) + K_p, 2K_1 + K_p\}$, is perfect for $\omega(G) \geq 2p$.

Keywords: Chromatic number · χ-binding function · $2K_2$-free graphs

2000 AMS Subject Classification: 05C15 · 05C75

1 Introduction

All graphs considered in this paper are simple, finite and undirected. Let G be a graph with vertex set $V(G)$ and edge set $E(G)$. For any positive integer k, a *proper k-coloring* or simply *k-coloring* of a graph G is a mapping $c : V(G) \rightarrow \{1, 2, \ldots, k\}$ such that for any two adjacent vertices $u, v \in V(G)$, $c(u) \neq c(v)$. If a graph G admits a proper k-coloring, then G is said to be *k-colorable*. The *chromatic number*, $\chi(G)$, of a graph G is the smallest k such that G is k-colorable. In this paper, K_n, P_n and C_n denote the complete graph, the path and the cycle on n vertices respectively. For $T, S \subseteq V(G)$, let $N_S(T) = N(T) \cap S$ (where $N(T)$ denotes the set of all neighbors of T in G), let $\langle T \rangle$ denote the subgraph induced by T in G and let $[T, S]$ denote the set of all edges in G with one end in T and the other end in S. If every vertex in T is adjacent with every vertex in S, then $[T, S]$ is said to be *complete*. For any graph G, we write $H \sqsubseteq G$ if H is an induced subgraph of G. A set $S \subseteq V(G)$ is said to be an *independent set (clique)* if any two vertices of S are non-adjacent (adjacent). The *clique number* of a graph G,

The first author's research was supported by the Council of Scientific and Industrial Research, Government of India, File No: 09/559(0133)/2019-EMR-I.

is the size of a maximum clique in G and is denoted by $\omega(G)$. When there is no ambiguity, $\omega(G)$ will be denoted by ω. A subset S of $V(G)$ is known as a dominating set if every vertex in $V \backslash S$ has a neighbor in S.

Let $\mathcal{F}$ be a family of graphs. We say that G is $\mathcal{F}$-free if it contains no induced subgraph which is isomorphic to a graph in $\mathcal{F}$. For a fixed graph H, let us denote the family of H-free graphs by $\mathcal{G}(H)$. For two vertex-disjoint graphs G_1 and G_2, the *join* of G_1 and G_2, denoted by $G_1 + G_2$, is the graph whose vertex set $V(G_1 + G_2) = V(G_1) \cup V(G_2)$ and the edge set $E(G_1 + G_2) = E(G_1) \cup E(G_2) \cup \{xy : x \in V(G_1), y \in V(G_2)\}$.

A graph G is said to be *perfect* if $\chi(H) = \omega(H)$, for every induced subgraph H of G. A graph G is said to be a *multisplit graph*, if $V(G)$ can be partitioned into two subsets V_1 and V_2 such that V_1 induces a complete multipartite graph and V_2 is an independent set in G. In particular, if V_1 induces a complete graph, then G is said to be a *split graph*. A hereditary graph class $\mathcal{G}$ is said to be χ-*bounded* [11] if there is a function f (called a χ-binding function) such that $\chi(G) \leq f(\omega(G))$, for every $G \in \mathcal{G}$. We say that the χ-binding function f is *special linear* if $f(x) = x + c$, where c is a constant. There has been extensive research done on χ-binding functions for various graph classes. See for instance, [4,7,8,10,11,17,19,20]. Let us recall a famous result by P.Erdős.

Theorem 1 ([9]). *For any positive integers $k, l \geq 3$, there exists a graph G with girth at least l and $\chi(G) \geq k$ (girth of G is the length of the shortest cycle in G).*

As a consequence of Theorem 1, Gyárfás in [11] observed that there exists no χ-binding function for $\mathcal{G}(H)$ whenever H contains a cycle. He further went on to conjecture that $\mathcal{G}(H)$ is χ-bounded for every fixed forest H. One of the earlier works in this direction, was done by S.Wagon in [20] where he showed that the class of $2K_2$-free graphs admits $\binom{\omega+1}{2}$ as a χ-binding function. There has been extensive studies on the class of $2K_2$-free graphs ; see for instance [2–4,7,8,10,12,16,20]. He further extended this result to show that the class of pK_2-free graphs admit a χ-binding function of $O(\omega^{2p-2})$, when $p \in \mathbb{N}$. In [5], S. Dhanalakshmi et al. established the following structural characterization for $2K_2$-free graphs.

Theorem 2 ([5]). *A connected graph is $2K_2$-free if and only if it is $\{P_5, butterfly, hammer\}$-free.*

Theorem 2 brings to attention three specific superclasses of $2K_2$-free graphs namely, $\{P_5, hammer\}$-free graphs, $\{P_5, butterfly\}$-free graphs and $\{butterfly, hammer\}$-free graphs. In [3] Brause et al. showed that the class of $\{P_5, hammer\}$-free graphs admits the same χ-binding function as the class of $2K_2$-free graphs, namely $\binom{\omega+1}{2}$. I. Schiermeyer in [18] established a cubic χ-binding function for $\{P_5, butterfly\}$-free graphs which was recently improved to $\frac{3}{2}(\omega^2 - \omega)$ by Wei Dong et al. in [6]. Finally, it remains to study the χ-binding function for $\{butterfly, hammer\}$-free graphs.

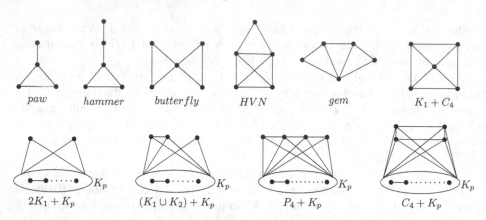

Fig. 1. Some special graphs

We begin this paper by showing that if G is a connected {$butterfly$, $hammer$}-free graph with $\omega(G) \neq 2$, then G is χ-bounded and the χ-binding function is same as the class of $2K_2$-free graphs. As a result the connected {$butterfly, hammer$}-free graphs that are χ-unbounded should be of clique size 2. In addition, we were able to show that if G is a connected {$butterfly, hammer$}-free graph G with $\omega(G) \neq 2$ then $V(G)$ can be partitioned into four sets where three of them are independent and one induce a $2K_2$-free graph. As a consequence we see that for any graph H, if the class of {$2K_2, H$}-free graphs admits f as a χ-binding function then the class of connected {$butterfly, hammer, H$}-free graphs G with $\omega(G) \neq 2$ will admit $f+3$ as a χ-binding function. In [12] T. Karthick and S. Mishra posed a problem seeking tight\linear chromatic bounds for {$2K_2, H$}-free graphs for various H, where H is a graph on $t \geq 6$ vertices. We partially answer this question by providing linear χ-bounds for {$butterfly, hammer, H$}-free graphs G which contain no components of clique size two and where $H \in \{C_4+K_p, P_4+K_p, (K_1\cup K_2)+K_p, 2K_1+K_p\}$. Some graphs that are considered as forbidden induced subgraphs in this paper are given in Fig. 1. Notations and terminologies not mentioned here are as in [21].

2   Preliminaries

Throughout this paper, we use a particular partition of the vertex set of a graph G which was initially defined by S. Wagon in [20] and later improved by A. P. Bharathi and S. A. Choudum in [1] as follows. Let $A = \{v_1, v_2, \ldots, v_\omega\}$ be a maximum clique of G. The *lexicographic ordering* on the set $L = \{(i,j) : 1 \leq i < j \leq \omega\}$ is defined in the following way. For two distinct elements $(i_1, j_1), (i_2, j_2) \in L$, we say that (i_1, j_1) precedes (i_2, j_2), denoted by $(i_1, j_1) <_L (i_2, j_2)$ if either $i_1 < i_2$ or $i_1 = i_2$ and $j_1 < j_2$. For every $(i,j) \in L$, let $C_{i,j} = \{v \in V(G)\backslash A : v \notin N(v_i) \text{ and } v \notin N(v_j)\} \backslash \left\{ \bigcup_{(i',j')<_L(i,j)} C_{i',j'} \right\}$. That is for $(i,j) \in L$, $C_{i,j}$ consists of all the vertices v in $V(G)\backslash A$ such that i and j are the least distinct positive integers for which vv_i and vv_j are not adjacent in G. Clearly $C_{i,j} \cap C_{i',j'} = \emptyset$ if $(i,j) \neq (i',j')$.

For $1 \leq k \leq \omega$, let us define $I_k = \{v \in V(G)\backslash A : v \in N(v_i),$ for every $i \in \{1,2,\ldots,\omega\}\backslash\{k\}\}$. Clearly we see that $[A\backslash\{v_k\}, I_k]$ is complete, for every $k \in \{1,2,\ldots,\omega\}$. Since A is a maximum clique, for $1 \leq k \leq \omega$, I_k is an independent set and for any $x \in I_k$, $xv_k \notin E(G)$. Also for any $(i,j) \in L$ and $k \in \{1,2,\ldots,\omega\}$, from the defintion of $C_{i,j}$ and I_k, we see that $I_k \cap C_{i,j} = \emptyset$. In addition, each vertex in $V(G)\backslash A$ is non-adjacent to at least one vertex in A. Hence those vertices will be contained either in I_k for some $k \in \{1,2,\ldots,\omega\}$, or in $C_{i,j}$ for some $(i,j) \in L$. Thus $V(G) = A \cup (\overset{\omega}{\underset{k=1}{\cup}} I_k) \cup (\underset{(i,j)\in L}{\cup} C_{i,j})$. Throughout this paper we shall use the partition $V(G) = V_1 \cup V_2$, where $V_1 = \underset{1\leq k\leq\omega}{\cup} (\{v_k\} \cup I_k) = \underset{1\leq k\leq\omega}{\cup} U_k$ and $V_2 = \underset{(i,j)\in L}{\cup} C_{i,j}$.

Let us observe the following.

Fact 3 *For every $(i,j) \in L$, if $a \in C_{i,j}$, then $N_A(a) \supseteq \{v_1, v_2, \ldots, v_j\}\backslash\{v_i, v_j\}$.*

Proof. If $a \in C_{i,j}$, then i and j are the least distinct positive integers for which vv_i and vv_j are not adjacent in G which implies $vv_k \in E(G)$ for every $k \in \{1,2,\ldots,j\}\backslash\{i,j\}$.

3   {*butterfly, hammer*}-free graphs

In [4], Chung et al. proved the existence of a dominating maximum clique in a connected $2K_2$-free graph G, when $\omega(G) \geq 3$. We begin Sect. 3 by showing that if G is a connected graph in $\mathcal{G}(hammer)$, a superclass of $2K_2$-free graphs, with $\omega(G) \neq 2$, then it contains a dominating ω-partite graph with clique number ω.

Theorem 4. *If G is a connected hammer-free graph with $\omega(G) \neq 2$, then G contains a dominating ω-partite graph with clique number ω.*

Proof. Let G be a connected *hammer*-free graph. Clearly, $\langle V_1 \rangle$ is an ω-partite graph such that $\omega(\langle V_1 \rangle) = \omega(G)$. We shall show that V_1 is a dominating set of G. For $\omega(G) = 1$, there is nothing to prove. So let us assume that $\omega(G) \geq 3$. Suppose there exists $x \in V_2$ such that $[x, V_1] = \emptyset$, then there exists $y \in V_2$ such that $[y, V_1] \neq \emptyset$ and x and y are connected by a path. Let $y' \in V_2$ such that $[y', V_1] \neq \emptyset$ and $[x', V_1] = \emptyset$ for every $x' \in V_2$ such that $d(x,x') \leq d(x,y')$. Let $P_{xy'}$ denote the shortest path between x and y' and z be the preceeding vertex of y' in $P_{xy'}$. Let us assume that $y' \in C_{m,n}$, for some $(m,n) \in L$. If $N_A(y') \neq \emptyset$, say $v_s \in N_A(y')$, then $\langle\{z,y',v_s,v_m,v_n\}\rangle \cong hammer$, a contradiction. Hence as $[y', V_1] \neq \emptyset$ and $N_A(y') = \emptyset$, there exists a vertex $a \in I_s$, where $s \in \{1,2,\ldots,\omega\}$ such that $ay' \in E(G)$. Also since $\omega(G) \geq 3$, there exists two integers $r,q \in \{1,2,\ldots,\omega\}\backslash\{s\}$ such that $\langle\{z,y',a,v_r,v_q\}\rangle \cong hammer$, a contradiction. Hence every vertex in V_2 has a neighbor in V_1.

One can notice that by Theorem 1, for any natural number $k \geq 3$, there exists a graph G with girth greater than 3 (which implies $\omega = 2$) and $\chi(G) \geq k$. This can also be observed by using the Mycielskian construction given in [13]. Since $\omega(hammer) = \omega(butterfly) = 3$, these graphs belong to the class of $\{butterfly, hammer\}$-free graphs. Hence, while considering the χ-binding functions of $\{butterfly, hammer\}$- free graphs we shall assume that $\omega \neq 2$. Let us now find a structural characterization and a χ-binding function for connected $\{butterfly, hammer\}$-free graphs.

Theorem 5. *If G is a connected $\{butterfly, hammer\}$-free graph with $\omega(G) \neq 2$, then*

(i) *There exists a partition of $V(G)$ such that $V(G) = X_1 \cup X_2 \cup X_3 \cup X_4 \cup X_5$ where $\langle X_1 \rangle$ is a dominating ω-partite graph with $\omega(\langle X_1 \rangle) = \omega(G)$, $\langle X_2 \rangle$, $\langle X_3 \rangle$ and $\langle X_4 \rangle$ are independent sets and $(\langle X_1 \cup X_5 \rangle)$ is $2K_2$-free.*

(ii) $\chi(G) \leq \binom{\omega(G)+1}{2}$.

Proof. Let G be a connected $\{butterfly, hammer\}$-free graph with $\omega(G) \neq 2$. For $\omega(G) = 1$, there is nothing to prove. So let us assume that $\omega(G) \geq 3$. By Theorem 4, we see that V_1 dominates G. We shall now show that $C_{i,j}$'s are independent.

Claim 1: $C_{i,j}$ is an independent set for every $(i, j) \in L$.

By Fact 3, it is not difficult to see that if ab is an edge in $\langle C_{i,j} \rangle$, then $(i, j) = (1, 2)$. Since V_1 dominates G, $[a, V_1] \neq \emptyset$ and $[b, V_1] \neq \emptyset$. Now if $[\{a, b\}, A] \neq \emptyset$, say v_s is adjacent to either a or b then $\langle \{a, b, v_s, v_1, v_2\} \rangle$ will induce a *butterfly* or a *hammer*, a contradiction. Similarly, we can attain a contradiction when $[\{a, b\}, I_s] \neq \emptyset$ for some $s \in \{1, 2, \ldots, \omega\}$. Hence $C_{1,2}$ is also independent.

Next, we shall show that $\langle V_1 \cup (\underset{j \geq 4 \; i \leq j-1}{\cup} C_{i,j}) \rangle$ is $2K_2$-free.

On the contrary, let us assume that there exists an induced $2K_2$ in $\langle V_1 \cup (\underset{j \geq 4 \; i \leq j-1}{\cup} C_{i,j}) \rangle$, say a, b, c, d with edges $ab, cd \in E(G)$. We begin by considering $\omega(G) = 3$. Here, $V_1 \cup (\underset{j \geq 4 \; i \leq j-1}{\cup} C_{i,j}) = V_1$. One can observe that $\{a, b, c, d\} \cap A = \emptyset$ (on the contrary, if one of the vertices of $\{a, b, c, d\}$ is in A, say $a = v_r$, then $c, d \in I_r$, a contradiction to $cd \in E(G)$). Hence there exists an integer $k \in \{1, 2, 3\}$ such that I_k contains exactly two non-adjacent vertices of $\{a, b, c, d\}$. Depending on whether the remaining two vertices of $\{a, b, c, d\}$ belong to the same I_i or not, we see that there exists an $r \in \{1, 2, 3\}$ such that $\langle \{v_r, a, b, c, d\} \rangle \cong$ *butterfly* or *hammer*, a contradiction.

Next, let us consider $\omega(G) \geq 4$. Let us break our proof into three cases.

Case 1: $|\{a, b, c, d\} \cap (\underset{j \geq 4 \; i \leq j-1}{\cup} C_{i,j})| \leq 1$.

Without loss of generality, let us assume that $b, c, d \in V_1$. Since $\omega(G) \geq 4$, there exists $q \in \{1, 2, \ldots, \omega\}$ such that $U_q \cap \{b, c, d\} = \emptyset$. Hence $\langle \{a, b, c, d, v_q\} \rangle \cong$ *butterfly* or *hammer*, a contradiction.

Case 2: $|\{a, b, c, d\} \cap (\underset{j \geq 4 \; i \leq j-1}{\cup} C_{i,j})| = 2$.

Let $x, y \in \{a, b, c, d\}$ such that $x, y \in V_1$. Let $r_1, r_2 \in \{1, 2, \dots, \omega\}$ (not necessarily distinct) such that $x \in U_{r_1}$ and $y \in U_{r_2}$. Clearly, there exists an integer $q_1 \in \{1, 2, 3\} \backslash \{r_1, r_2\}$ such that $[v_{q_1}, \{x, y\}]$ is complete. As a consequence, $|[v_{q_1}, \{a, b, c, d\} \backslash \{x, y\}]| = 0$, which implies $\{\{a, b, c, d\} \backslash \{x, y\}\} \subseteq \underset{j \geq 4}{\cup} C_{q_1, j}$. If there exists an integer $q_2 \in \{1, 2, 3\} \backslash \{r_1, r_2, q_1\}$, then $\langle \{v_{q_2}, a, b, c, d\} \rangle \cong butterfly$, a contradiction. Hence $\{r_1, r_2, q_1\} = \{1, 2, 3\}$ and thereby $\langle \{v_{r_2}, a, b, c, d\} \rangle \cong butterfly$ or $hammer$, a contradiction.

Case 3: $|\{a, b, c, d\} \cap (\underset{j \geq 4}{\cup} \underset{i \leq j-1}{\cup} C_{i,j})| \geq 3$.

Similar to Case 2, but with a little more involved arguments we can show that Case 3 is also not possible.

Therefore $V_1 \cup (\underset{j \geq 4}{\cup} \underset{i \leq j-1}{\cup} C_{i,j})$ is $2K_2$-free.

Let $\{1, 2, \dots, \binom{\omega+1}{2}\}$ be the set of colors. For establishing a $\binom{\omega+1}{2}$ coloring of G, let us start by assigning the color k to the vertices in $(v_k \cup I_k)$, for every $k \in \{1, 2, \dots, \omega\}$. Next, for $(i, j) \in L$, color each $C_{i,j}$ with a new color and hence G can be colored with at most $\omega + \sum_{j=2}^{\omega(G)} (j - 1) = \binom{\omega(G)+1}{2}$ colors. Clearly, this is a proper coloring of G.

One can observe that if G is a $2K_2$-free graph then it contains at most one edge containing component and hence any k-coloring for the non-trivial component of G will yield a k-coloring for G. Now as a consequence of Theorem 5 we get Corollary 1 and Corollary 2. Note that Corollary 1 is a result due to S. Wagon in [20].

Corollary 1 ([20]). *If G is a $2K_2$-free graph, then $\chi(G) \leq \binom{\omega(G)+1}{2}$.*

Corollary 2. *For any graph H, if the class of $\{2K_2, H\}$-free graphs admits the χ-binding function f, then a connected $\{butterfly, hammer, H\}$-free graph G such that $\omega(G) \neq 2$ would admit $f + 3$ as a χ-binding function.*

3.1 $\{butterfly, hammer, P_4 + K_p\}$-free graphs

Without much difficulty, one can make the following observations on $(P_4 + K_p)$-free graphs.

Lemma 1. *Let p be a positive integer and G be a $(P_4 + K_p)$-free graph with $\omega(G) \geq p + 2$. Then*

(i) *For $k, \ell \in \{1, 2, \dots, \omega(G)\}$, $[I_k, I_\ell]$ is complete. Thus $\langle V_1 \rangle$ is a complete multipartite graph with $U_k = \{v_k\} \cup I_k$, $1 \leq k \leq \omega(G)$ as its partitions.*

(ii) *For $(i, j) \in L$ with $j \geq p + 2$, if $a \in C_{i,j}$ such that $av_\ell \notin E(G)$, for some $\ell \in \{1, 2, \dots, \omega\}$, then $[a, I_\ell] = \emptyset$.*

Now, let us establish a linear χ-binding function for $\{butterfly, hammer, P_4 + K_p\}$-free graphs. When $p = 0$, $P_4 + K_p \cong P_4$ and since P_4-free graphs are perfect, G is ω-colorable. So let us consider $p \geq 1$.

Theorem 6. *Let p be a positive integer. If G is a {butterfly, hammer, $P_4 + K_p$}-free graph such that no component has clique size two, then*

$$\chi(G) \leq \begin{cases} \omega(G) + \frac{p(p+1)}{2} & \text{for } 1 \leq \omega(G) \leq p+1 \\ \omega(G) + \frac{p(p+1)}{2} - 1 & \text{for } \omega(G) \geq p+2. \end{cases}$$

Proof. Let G be a connected {butterfly, hammer, $P_4 + K_p$}-free graph such that $\omega(G) \neq 2$. Clearly, the vertices of V_1 can be colored with $\omega(G)$ colors by assigning the color k to the vertices in U_k, for $1 \leq k \leq \omega$. Now, by Claim 1 of Theorem 5, $C_{i,j}$ is an independent set for every $(i,j) \in L$. In order to color the vertices of V_2, let us break the proof into two cases depending upon the value of $\omega(G)$.

Case 1 $1 \leq \omega(G) \leq p+1$

For $\omega(G) = 1$, there is nothing to prove. Let $\omega(G) \geq 3$. Since each $C_{i,j}$ is an independent set, each $C_{i,j}$ can be properly colored by assigning a new color. Therefore V_2 can be colored with at most $\sum_{j=2}^{p+1} j - 1 = \frac{p(p+1)}{2}$ colors. Hence $\chi(G) \leq \omega(G) + \frac{p(p+1)}{2}$.

Case 2 $\omega(G) \geq p+2$

Let $\{1, 2, \ldots, \omega + \frac{p(p+1)}{2} - 1\}$ be the set of colors. If one observes closely, it can be seen that for $j \geq p+3$, $(\bigcup_{i=1}^{j-1} C_{i,j})$ is an independent set. Hence by using (ii) of Lemma 1, for $j \geq p+3$ we can assign the color j to each vertex in $(\bigcup_{i=1}^{j-1} C_{i,j})$. In addition, we can assign the color $p+2$ to the vertices in $C_{1,p+2}$ and color k to the vertices in $C_{k,p+2}$, $2 \leq k \leq p+1$. Let us next color the vertices of $C_{1,p+1}$. If $a \in C_{1,p+1}$ such that $[a, I_1] = \emptyset$, then assign the color 1 to a. If $[a, I_1] \neq \emptyset$ then we shall show that it can be assigned the color $p+1$. We begin by showing that for $r \geq p+2$, $[a, U_r] = \emptyset$. Let $b \in I_1$ such that $ab \in E(G)$. If there exists an $r \geq p+2$ such that $[a, U_r] \neq \emptyset$, then let $u_r \in U_r$ such that $au_r \in E(G)$. Hence by (i) of Lemma 1 and Fact 3 we can see that if $p = 1$, then $\langle\{a, b, v_{p+1}, v_1, u_r\}\rangle \cong P_4 + K_p$, a contradiction and if $p \geq 2$, then $\langle\{a, b, v_{p+1}, v_1, u_r, v_2, \ldots, v_p\}\rangle \cong P_4 + K_p$, a contradiction. Hence $[a, U_r] = \emptyset$ for $r \geq p+2$ and this in turns implies that $[a, C_{p+1,p+2}] = \emptyset$ and $[a, I_{p+1}] = \emptyset$. Therefore, when $[a, I_1] \neq \emptyset$, then a can be colored with $p+1$. Finally, the vertices of the remaining $C_{i,j}$ can be colored by assigning a new color to each $C_{i,j}$ and hence G can be colored with at most $\omega + \sum_{j=2}^{p+1}(j-1) - 1 = \omega + \frac{p(p+1)}{2} - 1$ colors.

Clearly this is a proper coloring of G.

Without much difficulty one can see that the same can be extended for the disconnected case.

By using Corollary 1 and Theorem 6 we get a χ-binding function for {$2K_2, P_4 + K_p$}-free graphs and partially answer the question raised by T. Karthick and S. Mishra in [12].

Corollary 3. *If p is a positive integer and G is a $\{2K_2, P_4 + K_p\}$-free graph,*
then $\chi(G) \leq \begin{cases} \omega(G) + \frac{p(p+1)}{2} & \text{for } 1 \leq \omega(G) \leq p+1 \\ \omega(G) + \frac{p(p+1)}{2} - 1 & \text{for } \omega(G) \geq p+2. \end{cases}$

As a simple consequence of Corollary 3 we get Corollary 4 and Corollary 5, results in [3] and [16] respectively.

Corollary 4. *[3] Let G be a $\{2K_2, gem\}$-free graph, then $\chi(G) \leq \max\{3, \omega\}$.*

Corollary 5. *[16] If G is a $\{2K_2, K_2 + P_4\}$-free graph with $\omega(G) \geq 4$, then $\chi(G) \leq \omega(G) + 2$.*

3.2 $\{butterfly, hammer, C_4 + K_p\}$-free graphs

While considering $\{butterfly, hammer, C_4 + K_p\}$-free graphs, as in Theorem 6, with similar type of cases but with a slightly different argument one can establish a linear χ-binding function.

Theorem 7. *Let p be a positive integer. If G is a $\{butterfly, hammer, C_4 + K_p\}$-free graph such that no component has clique size two, then $\chi(G) \leq \omega(G) + \frac{p(p+1)}{2}$.*

One can easily observe that Corollary 6 and Corollary 7 follows from Corollary 1 and Theorem 7.

Corollary 6. *Let p be a positive integer. If G is a $\{2K_2, C_4 + K_p\}$-free graph, then $\chi(G) \leq \omega(G) + \frac{p(p+1)}{2}$.*

Corollary 7. *[16] If G is a $\{2K_2, K_1 + C_4\}$-free graph, then $\chi(G) \leq \omega(G) + 1$.*

We see that here also Corollary 6 partially answers the question raised by T. Karthick and S. Mishra in [12]. The question of whether this bound is optimal remains open.

3.3 $\{butterfly, hammer, (K_1 \cup K_2) + K_p\}$-free graphs

Let us begin by recalling some of the results in [14,15].

Theorem 8. *[14] Let G be a connected graph. Then G is paw-free if and only if G is either K_3-free or complete multipartite.*

Proposition 1. *[15] Let G be a $((K_1 \cup K_2) + K_p)$-free graph with $\omega(G) \geq p+2$, $p \geq 1$. Then G satisfies the following.*

  (i) For $k, \ell \in \{1, 2, \ldots, \omega(G)\}$, $[I_k, I_\ell]$ is complete. Thus, $\langle V_1 \rangle$ is a complete multipartite graph with $U_k = \{v_k\} \cup I_k$, $1 \leq k \leq \omega(G)$ as its partitions.
  (ii) For $j \geq p+2$ and $1 \leq i < j$, $C_{i,j} = \emptyset$.

(iii) For $x \in V_2$, *x has neighbors in at most* $(p-1)$ U_ℓ's *where* $\ell \in \{1, 2, \ldots, \omega(G)\}$.

By using Proposition 1, let us show that connected $\{butterfly, hammer, (K_1 \cup K_2) + K_p\}$-free graphs with $\omega \geq 2p$ are multisplit graphs.

Theorem 9. *Let p be an integer greater than 1 and G be a connected $\{butterfly, hammer, (K_1 \cup K_2) + K_p\}$-free graph. If $\omega(G) \geq 2p$, then G is a multisplit graph.*

Proof. Let G be a connected $\{butterfly, hammer, (K_1 \cup K_2) + K_p\}$-free graph with $\omega(G) \geq 2p$, $p \geq 2$. Since $\omega(G) \geq 2p$, by (i) of Proposition 1, we see that $\langle V_1 \rangle$ is a complete multipartite graph. Next, let us show that V_2 is an independent set. Clearly $\omega(G) \neq 2$ and hence by Claim 1 of Theorem 5, if ab is an edge in G, then $a \in C_{m_1,n_1}$ and $b \in C_{m_2,n_2}$ where $(m_1, n_1) \neq (m_2, n_2)$. It is easy to see that by (iii) of Proposition 1, $|N_A(\{a, b\})| \leq 2p - 2$ and hence there exist at least two vertices $v_r, v_s \in A$ such that $[\{a, b\}, \{v_r, v_s\}] = \emptyset$. Without loss of generality, let us assume that $(m_1, n_1) <_L (m_2, n_2)$. If $m_1 < m_2$, then $bv_{m_1} \in E(G)$ and $\langle \{a, b, v_{m_1}, v_r, v_s\} \rangle \cong hammer$, a contradiction. If $m_1 = m_2$ and $n_1 < n_2$, then $bv_{n_1} \in E(G)$ and $\langle \{a, b, v_{n_1}, v_r, v_s\} \rangle \cong hammer$, again a contradiction. Hence V_2 is independent and thereby G is a multisplit graph.

One can notice that when $p = 0$ and 1, $(K_1 \cup K_2) + K_p \cong K_1 \cup K_2$ and $(K_1 \cup K_2) + K_p \cong paw$ respectively. As $(K_1 \cup K_2) \sqsubseteq P_4$, $(K_1 \cup K_2)$-free graphs are perfect. Also C. Brause et al. in [3] has shown that every multisplit graph is perfect. Hence as a consequence of Theorem 8 and Theorem 9 we get Corollary 8, Corollary 9 and Corollary 10.

Corollary 8. *Let p be a non-negative integer and G be a connected $\{butterfly, hammer, (K_1 \cup K_2) + K_p\}$-free graph with $\omega(G) \geq 2p$. If $\omega(G) \neq 2$, then G is perfect.*

Corollary 9. *[3] If G is a connected $\{2K_2, (K_1 \cup K_2) + K_p\}$-free graph with $\omega(G) \geq 2p$, $p \geq 2$, then G is a multisplit graph.*

Corollary 10. *[3] Let $p \geq 0$ be an integer and G be a $\{2K_2, (K_1 \cup K_2) + K_p\}$-free graph with $\omega(G) \geq 2p$. If $p \neq 1$ or $\omega(G) \neq 2$, then G is perfect.*

By using Theorem 6 and Theorem 9, one can find a linear χ-binding function for the class of $\{butterfly, hammer, (K_1 \cup K_2) + K_p\}$-free graphs.

Theorem 10. *Let p be a non-negative integer and G be a connected $\{butterfly, hammer, (K_1 \cup K_2) + K_p\}$-free graph with $\omega(G) \neq 2$. Then*

$$\chi(G) \leq \begin{cases} \omega(G) + \frac{p(p+1)}{2} & \text{for } 1 \leq \omega(G) \leq p+1 \\ \omega(G) + \frac{p(p-1)}{2} & \text{for } \omega(G) \geq p+2. \end{cases}$$

As a consequence of Corollary 1 and Theorem 10, we get Corollary 11 and as a result we see that we have a better χ-binding function than $\omega(G) + (2p-1)(p-1)$, the one given by C. Brause et al. in [3] for the class of $\{2K_2, (K_1 \cup K_2) + K_p\}$-free graphs, when $p \geq 2$.

Corollary 11. *Let p be a non-negative integer. If G is a $\{2K_2, (K_1 \cup K_2) + K_p\}$- free graph, then* $\chi(G) \leq \begin{cases} \omega(G) + \frac{p(p+1)}{2} \text{ for } 1 \leq \omega(G) \leq p+1 \\ \omega(G) + \frac{p(p-1)}{2} \text{ for } \omega(G) \geq p+2. \end{cases}$

3.4 $\{butterfly, hammer, 2K_1 + K_p\}$-free graphs

For connected $\{butterfly, hammer, 2K_1 + K_p\}$-free graphs with $\omega(G) \geq 2p$, by using similar techniques as in Theorem 9, we can show that they are split graphs.

Theorem 11. *Let p be a non-negative integer and G be a connected $\{butterfly, hammer, 2K_1 + K_p\}$-free graph. If $\omega(G) \geq 2p$, then G is a split graph.*

One can see that Corollary 12 follows from Theorem 11.

Corollary 12. *[3] If G is a connected $\{2K_2, 2K_1 + K_p\}$-free graph with $\omega(G) \geq 2p$ for some positive integer p, then G is a split graph.*

By using Theorem 11, without much difficulty one can find a linear χ-binding function for the class of connected $\{butterfly, hammer, 2K_1 + K_p\}$-free graphs with $\omega(G) \neq 2$.

Theorem 12. *Let p be a non-negative integer. If G is a connected $\{butterfly, hammer, 2K_1 + K_p\}$-free graph with $\omega(G) \neq 2$, then $\chi(G) \leq \omega(G) + \frac{p(p-1)}{2}$.*

As a simple consequence of Corollary 1 and Theorem 12, we get a χ-binding function better than $\omega + (2p - 1)(p - 1)$, which was given by C. Brause et al. in [3] for the class of $\{2K_2, 2K_1 + K_p\}$-free graphs. This can be seen in Corollary 13.

Corollary 13. *Let p be a non-negative integer and G be a connected $\{2K_2, 2K_1 + K_p\}$-free graph. Then $\chi(G) \leq \omega(G) + \frac{p(p-1)}{2}$.*

For the particular case when $\omega = 2p - 1$ and $\omega = 2p - 2$, with involved arguments one can observe that $\chi(G) = \omega(G)$.

Theorem 13. *Let p be a non-negative integer. If G is a connected $\{butterfly, hammer, 2K_1 + K_p\}$-free graph with $\omega(G) = 2p - 1$, then $\chi(G) = \omega(G)$.*

Theorem 14. *If p is a non-negative integer not equal to 2 or 3 and G is a connected $\{butterfly, hammer, 2K_1 + K_p\}$-free graph with $\omega(G) = 2p - 2$, then $\chi(G) = \omega(G)$.*

Finally, one can easily observe that Corollary 14, Corollary 15, Corollary 16 follows from Corollary 1, Corollary 12, Theorem 13 and Theorem 14.

Corollary 14. *If G is a $\{2K_2, 2K_1 + K_p\}$-free graph and p is a non-negative integer with $\omega(G) = 2p - 1$, then $\chi(G) = \omega(G)$.*

Corollary 15. *If G is a $\{2K_2, 2K_1 + K_p\}$-free graph with $\omega(G) = 2p - 2$ and p is a non-negative integer such that $p \neq \{2, 3\}$, then $\chi(G) = \omega(G)$.*

Corollary 16. *[1] If a graph G is $(2K_2, diamond)$-free, then $\chi(G) \leq 3$ for $\omega = 2, 3$ and G is perfect if $\omega \geq 4$.*

4   Conclusion

We conclude this paper with the following two open problems.

Problem 1: Let f^* and g^* denote the smallest χ-binding functions of $\{2K_2, H\}$-free graphs and connected $\{butterfly, hammer, H\}$-free graphs G with $\omega(G) \neq 2$. For what choices of H is $g^* = f^* + 3$?

Problem 2: Let $\mathcal{H} = \{H_1, H_2, \ldots, H_l\}$, where $\omega(H_i) \geq 3$ and $q = \min\{\omega(H_i)| \ i \in \{1, 2, \ldots, l\}\}$. If $\mathcal{K}$ denotes the class of all graphs with $\omega < q$ and $\mathcal{G}'(\mathcal{H})$ denotes the class of all connected $\mathcal{H}$-free graphs, then under what condition is $(\mathcal{G}'(\mathcal{H}) \backslash \mathcal{K})$ χ-bounded?

References

1. Bharathi, A.P., Choudum, S.A.: Colouring of $(P_3 \cup P_2)$-free graphs. Graphs Combin. **34**(1), 97–107 (2018)
2. Blázsik, Z., Hujter, M., Pluhár, A., Tuza, Z.: Graphs with no induced C_4 and $2K_2$. Discrete Math **115**(1–3), 51–55 (1993)
3. Brause, C., Randerath, B., Schiermeyer, I., Vumar, E.: On the chromatic number of $2K_2$-free graphs. Discrete Appl. Math **253**, 14–24 (2019)
4. Chung, F.R., Gyárfás, A., Tuza, Z., Trotter, W.T.: The maximum number of edges in $2K_2$-free graphs of bounded degree. Discrete Math **81**(2), 129–135 (1990)
5. Dhanalakshmi, S., Sadagopan, N., Manogna, V.: On $2K_2$-free graphs-structural and combinatorial view, pp. 1–15 . arXiv preprint arXiv:1602.03802 (2016)
6. Dong, W., Xu, B., Xu, Y.: On the chromatic number of some P_5-free graphs, pp. 1–18. arXiv preprint arXiv:2202.13177 (2022)
7. El-Zahar, M., Erdős, P.: On the existence of two non-neighboring subgraphs in a graph. Combinatorica **5**(4), 295–300 (1985)
8. Erdős, P.: Problems and results on chromatic numbers in finite and infinite graphs. In: Graph Theory with Applications to Algorithms and Computer Science (Kalamazoo, MI, 1984), pp. 201–213. Wiley-Intersci. Publ (1985)
9. Erdős, P.: Graph theory and probability. Can. J. Math. **11**, 34–38 (1959)
10. Gaspers, S., Huang, S.: $(2P_2, K_4)$-free graphs are 4-colorable. SIAM J. Discret. Math. **33**(2), 1095–1120 (2019)
11. Gyárfás, A.: Problems from the world surrounding perfect graphs. Zastosowania Matematyki Applicationes Mathematicae **19**(3–4), 413–441 (1987)
12. Karthick, T., Mishra, S.: Chromatic bounds for some classes of $2K_2$-free graphs. Discrete Math **341**(11), 3079–3088 (2018)

13. Mycielski, J.: Sur le coloriage des graphs. Colloquium Mathematicae **3**(2), 161–162 (1955)
14. Olariu, S.: Paw-free graphs. Inf. Process. Lett. **28**(1), 53–54 (1988)
15. Prashant, A., Francis, P., Raj, S.F.: Chromatic bounds for some subclasses of $(P_3 \cup P_2)$-free graphs. In: Balachandran, N., Inkulu, R. (eds.) CALDAM 2022. LNCS, vol. 13179, pp. 15–21. Springer, Cham (2022). https://doi.org/10.1007/978-3-030-95018-7_2
16. Prashant, A., Francis Raj, S., Gokulnath, M.: Chromatic bounds for the subclasses of pK_2-free graphs, pp. 1–16. arXiv preprint arXiv:2102.13458 (2021)
17. Randerath, B., Schiermeyer, I.: Vertex colouring and forbidden subgraphs-A survey. Graphs Combin. **20**(1), 1–40 (2004)
18. Schiermeyer, I.: On the chromatic number of $(P_5$, windmill)-free graphs. Opuscula Mathematica **37**(4), 609–615 (2017)
19. Schiermeyer, I., Randerath, B.: Polynomial χ-binding functions and forbidden induced subgraphs: A survey. Graphs Combin. **35**(1), 1–31 (2019)
20. Wagon, S.: A bound on the chromatic number of graphs without certain induced subgraphs. J. Combin. Theory B **29**(3), 345–346 (1980)
21. West, D.B.: Introduction to graph theory. Prentice-Hall of India Private Limited (2005)

The Weak $(2, 2)$-Labelling Problem for Graphs with Forbidden Induced Structures

Julien Bensmail[1]([✉]), Hervé Hocquard[2], and Pierre-Marie Marcille[2]

[1] Université Côte d'Azur, CNRS, Inria, I3S, Sophia-Antipolis, France
julien.bensmail.phd@gmail.com
[2] Univ. Bordeaux, CNRS, Bordeaux INP, LaBRI, 33400 Talence, France

Abstract. The Weak $(2,2)$-Conjecture is a graph labelling problem asking whether all connected graphs of at least three vertices can have their edges assigned red labels 1 and 2 and blue labels 1 and 2 so that any two adjacent vertices are distinguished either by their sums of incident red labels, or by their sums of incident blue labels. This problem emerged in a recent work aiming at proposing a general framework encapsulating several distinguishing labelling problems and notions, such as the well-known 1-2-3 Conjecture and so-called locally irregular decompositions.

In this work, we prove that the Weak $(2,2)$-Conjecture holds for two classes of graphs defined in terms of forbidden induced structures, namely claw-free graphs and graphs with no pair of independent edges. One main point of interest for focusing on such classes of graphs is that the 1-2-3 Conjecture is not known to hold for them. Also, these two classes of graphs have unbounded chromatic number, while the 1-2-3 Conjecture is mostly understood for classes with bounded and low chromatic number.

Keywords: Distinguishing labelling · 1-2-3 Conjecture · Sum distinction

1 Introduction

This work deals with several **distinguishing labelling problems**, taking part to a wide and vast area of research, as reported in several dedicated surveys on the topic, such as e.g. [7,10]. More particularly, we focus on a subset of these problems revolving around the so-called **1-2-3 Conjecture**, which can all be defined through the following unified terminology, introduced recently in [3].

Let G be a graph, and $\alpha, \beta \geq 1$ be two positive integers. An (α, β)-*labelling* of G is an assignment ℓ of labels from $\{1, \ldots, \alpha\} \times \{1, \ldots, \beta\}$ to the edges of G, where each edge e gets assigned a *label* $\ell(e) = (x, y)$ with *colour* $x \in \{1, \ldots, \alpha\}$ and *value* $y \in \{1, \ldots, \beta\}$. Now, for every vertex v of G and any $i \in \{1, \ldots, \alpha\}$, we denote by $\sigma_i(v)$ the *sum* of the values of the labels with colour i assigned to the edges incident to v, which we call the i-*sum* of v. We say that ℓ is *distinguishing* if for every two adjacent vertices u and v of G, there is an $i \in \{1, \ldots, \alpha\}$ such that the i-sums of u and v differ, that is, if $\sigma_i(u) \neq \sigma_i(v)$.

A. Bagchi and R. Muthu (Eds.): CALDAM 2023, LNCS 13947, pp. 204–215, 2023.
https://doi.org/10.1007/978-3-031-25211-2_16

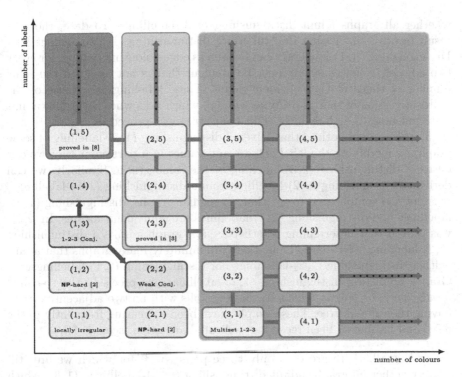

Fig. 1. The current knowledge we have on whether all graphs admit distinguishing (α, β)-labellings, for fixed $\alpha, \beta \geq 1$. For a pair (α, β), the associated box is green if all graphs were proved to admit the corresponding labellings, the box is red if it is known that not all graphs admit the corresponding labellings, while the box is blue if the status is unknown. Arrows indicate existential implications. (Color figure online)

Regarding these notions, it can be noted that if G is K_2, the complete graph of order 2, then there are no $\alpha, \beta \geq 1$ such that G admits distinguishing (α, β)-labellings. Apart from this peculiar case, it is not too complicated to prove that, for any fixed $\alpha \geq 1$, there is a $\beta \geq 1$ such that distinguishing (α, β)-labellings of any graph G exist. For these reasons, in the context of distinguishing labellings, we generally focus on *nice graphs*, which are those graphs with no (connected) component isomorphic to K_2. Therefore, throughout this work, every graph we consider is thus implicitly assumed nice.

A natural question, now, is whether, for some fixed $\alpha, \beta \geq 1$, every graph admits distinguishing (α, β)-labellings. It turns out, as mentioned earlier, that the literature actually provides answers for several values of α and β (see Fig. 1).

- Note that if α, β and α', β' are such that $\alpha' \geq \alpha$, $\beta' \geq \beta$, and $(\alpha, \beta) \neq (\alpha', \beta')$, then any distinguishing (α, β)-labelling is a distinguishing (α', β')-labelling.
- Distinguishing $(1, \beta)$-labellings are labellings where all labels are of the same colour, and all adjacent vertices should be distinguished according to their sums of incident labels. Such labellings are exactly those behind the so-called 1-2-3 Conjecture [9] of Karoński, Łuczak, and Thomason, which asks

whether all graphs admit distinguishing $(1,3)$-labellings. To date, the best result towards this is that they all admit distinguishing $(1,5)$-labellings [8].

- Distinguishing $(\alpha, 1)$-labellings can be seen as edge-colourings where, for every two adjacent vertices, there must be a colour that is not assigned the same number of times to their incident edges. These labellings are those of the multiset version of the 1-2-3 Conjecture [1], which asks whether graphs admit distinguishing $(3,1)$-labellings. This conjecture was proved in [11].
- In [3], the authors noticed that, given a distinguishing $(1,5)$-labelling of some graph, by modifying the label colours and values in a particular way, we can derive a distinguishing $(2,3)$-labelling of the same graph. Similarly, we can derive a distinguishing $(3,2)$-labelling from a distinguishing $(1,5)$-labelling.
- It is not too complicated to see that, in regular graphs, distinguishing $(1,2)$-labellings and distinguishing $(2,1)$-labellings are equivalent notions. In [2], it was proved that determining whether a cubic graph admits a distinguishing $(1,2)$-labelling is NP-hard. Thus, there are infinitely many graphs that admit neither distinguishing $(1,2)$-labellings nor distinguishing $(2,1)$-labellings.
- Graphs admitting distinguishing $(1,1)$-labellings are precisely the so-called *locally irregular graphs*, which are those graphs with no two adjacent vertices having the same degree. These graphs have been appearing frequently in the field, and have even been receiving dedicated attention, see e.g. [4].

From this all, there are thus only three pairs (α, β) for which we are still not sure whether all graphs admit distinguishing (α, β)-labellings: $(1,3)$, which corresponds to the original 1-2-3 Conjecture; $(1,4)$, which is weaker than the 1-2-3 Conjecture since more label values are available (while, similarly, all labels are of the same colour); and $(2,2)$, which is the only pair for which we have two label colours to deal with. The latter pair leads to the following conjecture [3].

Weak $(2,2)$-Conjecture. *Every graph admits a distinguishing $(2,2)$-labelling.*

At first glance, the 1-2-3 Conjecture and the Weak $(2,2)$-Conjecture might seem a bit distant. It is worth emphasising, however, that the former conjecture, if true, would imply the latter [5]. For this reason, the Weak $(2,2)$-Conjecture can be perceived as a weaker version of the 1-2-3 Conjecture. Also, to get progress towards these conjectures, one can thus investigate the Weak $(2,2)$-Conjecture for classes of graphs for which the 1-2-3 Conjecture is not known to hold. To date, the 1-2-3 Conjecture was mainly proved for 3-colourable graphs [10]. The weaker conjecture was mainly proved for 4-colourable graphs [5].

Theorem 1 ([5]). *The Weak $(2,2)$-Conjecture holds for 4-colourable graphs.*

Both conjectures were also proved for other classes of graphs, but not as significant. One reason why the chromatic number parameter appears naturally in this context is that having a proper vertex-colouring ϕ in hand can be helpful to design a distinguishing labelling, since ϕ informs on sets of vertices that are not required to be distinguished. One downside, however, is that making a labelling match ϕ somehow, might require lots of labels if ϕ itself contains lots of parts.

Here, we prove the Weak $(2, 2)$-Conjecture for two classes of graphs for which the 1-2-3 Conjecture (and thus the Weak $(2, 2)$-Conjecture, recall [5]) has not been proved. Besides, these classes have unbounded chromatic number, which, recall, is significant. Precisely, we prove the Weak $(2, 2)$-Conjecture for $K_{1,3}$-free graphs (with no induced claw) and $2K_2$-free graphs (with no pair of independent edges). Each result is proved by first dealing with the 5-colourable graphs of the class, before focusing on those with chromatic number at least 6.

Due to space limitation, in what follows we only present our result for $2K_2$-free graphs, its proof being a lighter, less technical version of that for $K_{1,3}$-free graphs, which require more involved arguments. We start in Sect. 2 with some preliminaries, before proving the Weak $(2, 2)$-Conjecture for $2K_2$-free graphs in Sect. 3. In concluding Sect. 4, we explain how to go from $2K_2$-free graphs to $K_{1,3}$-free graphs, summarising the arguments from our full-length paper [6].

2   Preliminaries

Let G be a graph, and ℓ be an (α, β)-labelling of G. If $\alpha=1$, then we will sometimes call ℓ a β-labelling for simplicity. Also, in such cases, instead of denoting the 1-sum of a vertex v by $\sigma_1(v)$, we will simply denote it as $\sigma(v)$, or as $\sigma_\ell(v)$ in case we want to emphasise that we refer to the labels assigned by ℓ. Now, when considering the Weak $(2, 2)$-Conjecture and, thus, $(\alpha, \beta) = (2, 2)$, it will be more convenient to see the labels with colour 1 as *red labels*, and similarly those with colour 2 as *blue labels*. We will thus refer, for any vertex v, to the *red sum* $\sigma_r(v)$ of v (being $\sigma_1(v)$), and to the *blue sum* $\sigma_b(v)$ of v (being $\sigma_2(v)$).

We now point out a situation where, assuming a partial labelling of a graph is given, we can extend it to some edges so that some properties are preserved.

Lemma 1. *Let G be a graph, H be a connected bipartite subgraph of G, and ℓ be a partial 2-labelling of G such that only the edges of H are not labelled. For any vertex w of H, there is a 2-labelling ℓ' of H such that, for every two adjacent vertices u and v of H with $w \notin \{u, v\}$, we have*

$$\sigma_\ell(u) + \sigma_{\ell'}(u) \neq \sigma_\ell(v) + \sigma_{\ell'}(v).$$

Proof. Let (U, V) denote the bipartition of H. We produce a 2-labelling ℓ' such that, for every vertex $u \neq w$ of H, we have $\sigma_\ell(u) + \sigma_{\ell'}(u) \equiv 0 \bmod 2$ if $u \in U$, and $\sigma_\ell(u) + \sigma_{\ell'}(u) \equiv 1 \bmod 2$ otherwise. This implies what we want to prove.

Start from all edges of H being assigned label 2 by ℓ'. Now, consider any vertex u of H for which $\sigma_\ell(u) + \sigma_{\ell'}(u)$ does not satisfy the required condition above. Since H is connected, there is a path P from u to w that uses edges of H only. Now turn all 1's assigned by ℓ' to the edges of P into 2's, and conversely turn all 2's into 1's. As a result, note that $\sigma_\ell(v) + \sigma_{\ell'}(v)$ is not altered for every vertex v of H with $v \notin \{u, w\}$, while both $\sigma_\ell(u) + \sigma_{\ell'}(u)$ and $\sigma_\ell(w) + \sigma_{\ell'}(w)$ had their parity altered. So $\sigma_\ell(u) + \sigma_{\ell'}(u)$ now verifies the desired condition.

Repeating those arguments until all vertices $u \neq w$ of H have $\sigma_\ell(u) + \sigma_{\ell'}(u)$ verifying the desired condition, we end up with ℓ' being as desired.   □

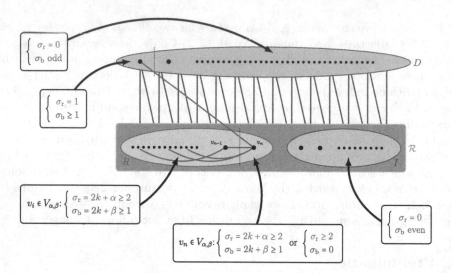

Fig. 2. Terminology used in the proof of Theorem 3, and the red sums and blue sums we aim at getting for the vertices by the designed $(2,2)$-labelling. In the depicted situation, it is assumed that an upward edge of R is assigned red label 1.

We also recall a nice tool that proved to be very useful towards proving the multiset version of the 1-2-3 Conjecture from [1]. Let G be a graph. A *balanced tripartition* of G is a partition V_0, V_1, V_2 of $V(G)$ fulfilling, for every vertex $v \in V_i$ with $i \in \{0,1,2\}$, that $d_{V_{i+1}}(v) \geq \max\{1, d_{V_i}(v)\}$ (all operations over the subscripts are modulo 3). That is, v has at least one neighbour in the next part V_{i+1}, and it actually has more neighbours in V_{i+1} than in V_i. It turns out that graphs with large chromatic number admit such balanced tripartitions.

Theorem 2 ([1]). *Every graph G with $\chi(G) > 3$ admits a balanced tripartition.*

3   Graphs with No Induced Pair of Independent Edges

As mentioned earlier, we prove the Weak $(2,2)$-Conjecture for $2K_2$-free graphs by treating the 5-chromatic ones first, and then those with large chromatic number.

Theorem 3. *Every $2K_2$-free graph with chromatic number 5 admits a distinguishing $(2,2)$-labelling.*

Proof. Let G be a $2K_2$-free graph with chromatic number 5. We construct a distinguishing $(2,2)$-labelling of G assigning red labels 1 and 2 and blue labels 1 and 2. We can assume G is connected, since its 5-chromatic components can be handled through what follows, while Theorem 1 applies for its 4-colourable ones.

Let D be a maximal independent set of G, and set $\mathcal{R} = G - D$. Note that every vertex v in $\mathcal{R}$ is incident to at least one *upward edge* vu, i.e., going to D (so, $u \in D$). We say that a component of $\mathcal{R}$ is *empty* if it contains no edges, while it is *non-empty* otherwise. Since G is $2K_2$-free, note that $\mathcal{R}$ contains at most one non-empty component. Actually, $\mathcal{R}$ must contain exactly one non-empty component

R as otherwise G would be bipartite, contradicting that its chromatic number is 5. Let now I denote the vertices from the empty components of $\mathcal{R}$, and let $\mathcal{H}$ be the subgraph of G induced by the edges incident to the vertices of I. Then $\mathcal{H}$ is bipartite, and because G is $2K_2$-free, $\mathcal{H}$ consists of only one component.

Since G is 5-chromatic, note that R is 4-chromatic; let thus $V_{0,0}, V_{0,1}, V_{1,0}, V_{1,1}$ be parts forming a proper 4-vertex-colouring ϕ of R. We modify ϕ, if needed, so that if v is a vertex of R with $d_R(v) = 1$, then v belongs to $V_{0,0}$ or $V_{0,1}$ (note that this is clearly possible, since v has exactly one neighbour in R). Now order the vertices $v_1, \ldots, v_n$ in any way satisfying that, for every $i \in \{1, \ldots, n-1\}$, vertex v_i is incident to at least one *forward edge* $v_i v_j$ (*i.e.*, with $j > i$, which is a *backward edge* from v_j's point of view). Such an ordering can be obtained e.g. by reversing the ordering in which vertices are encountered while performing a breadth-first search algorithm from any vertex (standing as the last vertex v_n).

We are now ready to start labelling the edges of G. We begin with all edges incident to the vertices of R. We consider the v_i's one by one, following the ordering above, and for every vertex v_i considered in that course, we assign a label to all upward edges (assigning them blue labels, except in one peculiar case) and forward edges (assigning them red labels only) incident to v_i so that some desired red sum and blue sum are realised at v_i. When proceeding that way, note that, whenever considering a new vertex as v_i, only its backward edges can be assumed to be labelled, with red labels. The procedure goes as follows:

- If $i \neq n$, then v_i is incident to forward edges. We start by assigning blue label 2 to all upward edges incident to v_i, and red label 2 to all forward edges incident to v_i. Assume $v_i \in V_{\alpha,\beta}$. If $\sigma_b(v_i) \not\equiv \beta \bmod 2$, then we change to blue label 1 the label assigned to any one upward edge incident to v_i. Likewise, if $\sigma_r(v_i) \not\equiv \alpha \bmod 2$, then we change to red label 1 the label assigned to any one forward edge incident to v_i. This way, we get $\sigma_r(v_i) \equiv \alpha \bmod 2$ and $\sigma_b(v_i) \equiv \beta \bmod 2$. In particular, by how we modified ϕ earlier, note that we must have $\sigma_r(v_i) \geq 2$ (either $d_R(v_i) \geq 2$ in which case this condition clearly holds; or $d_R(v_i) = 1$, in which case $\alpha = 0$ and thus the only inner edge incident to v_i is assigned red label 2, implying the condition).
- If $i = n$, then the only edges incident to v_n that remain to be labelled are upward edges. Recall, in particular, that all backward edges incident to v_n are assigned red labels. We consider two cases, assuming $v_n \in V_{\alpha,\beta}$:
  - If $\sigma_r(v_n) \equiv \alpha \bmod 2$, then we assign blue labels to all upward edges incident to v_n, their values being chosen so that $\sigma_b(v_n) \equiv \beta \bmod 2$. In that case, we thus have $\sigma_r(v_n) \equiv \alpha \bmod 2$ and $\sigma_b(v_n) \equiv \beta \bmod 2$. Again, by how ϕ was modified earlier, we must have $\sigma_r(v_n) \geq 2$.
  - If $\sigma_r(v_n) \not\equiv \alpha \bmod 2$, then we assign red label 1 to any one upward edge incident to v_n, while we assign blue labels to the other upward edges (if any) so that $\sigma_b(v_n) \equiv \beta \bmod 2$. Here, either $\sigma_b(v_n) \neq 0$ in which case $\sigma_r(v_n) \equiv \alpha \bmod 2$ and $\sigma_b(v_n) \equiv \beta \bmod 2$; or $\sigma_b(v_n) = 0$ in which case all edges incident to v_n are assigned red labels (implying that $\sigma_r(v_n) \geq 2$).

Note that, in all cases above, for all vertices $v_i \in V_{\alpha,\beta}$, we guarantee $2 \leq \sigma_r(v_i) \equiv \alpha \bmod 2$. Also, except maybe for v_n, we also guarantee $0 < \sigma_b(v_i) \equiv \beta \bmod 2$.

Regarding v_n, either $\sigma_{\mathrm{b}}(v_n) = 0$, in which case v_n is distinguished from all its neighbours in R through its blue sum, or $0 < \sigma_{\mathrm{b}}(v_n) \equiv \beta \bmod 2$, in which case v_n is distinguished from its neighbours in R through its red sum and/or blue sum. Regarding the vertices of D, only one of them can be incident to an edge being assigned a red label, 1. So, for every $u \in D$, we have $\sigma_{\mathrm{r}}(u) \leq 1$, while $\sigma_{\mathrm{r}}(v) \geq 2$ for every $v \in R$. Thus, currently, vertices of R are distinguished from their neighbours in D. If $\mathcal{H}$ has no edges, then all edges of G are labelled, and we have a distinguishing $(2,2)$-labelling. So, below, we can assume $\mathcal{H}$ has edges.

We are now left with labelling the edges of $\mathcal{H}$, which, recall, consists of exactly one component. We consider two main cases:

- Assume there is some vertex $w \in \mathcal{H}$ with $\sigma_{\mathrm{r}}(w) = 1$ (Fig. 2). Recall that there can be only one such vertex, which belongs to D and must be a neighbour of v_n. Recall also that the vertices of $D \cap V(\mathcal{H})$ can be incident to edges being currently assigned blue labels (being upward edges incident to vertices of R). Taking these labels into account, by Lemma 1 we can assign blue labels 1 and 2 to the edges of $\mathcal{H}$ so that any two of its adjacent vertices u and v with $w \notin \{u, v\}$ are distinguished by their blue sums.

  Since we did not modify labels assigned to edges incident to the vertices in R, and the edges of $\mathcal{H}$ are assigned blue labels only, the vertices of R remain distinguished from their neighbours due to arguments above. Regarding adjacent vertices of $\mathcal{H}$, they are either distinguished by their blue sums (if w is not involved), or because one of them has red sum 1 (if w is involved). So, here as well, we do not have conflicts.

- Assume no vertex of $\mathcal{H}$ currently has red sum at least 1. In this case, let w be any vertex of I. By Lemma 1, we can assign blue labels 1 and 2 to the edges of $\mathcal{H}$ so that, taking into account the other edges of G that are currently already assigned blue labels, and omitting w, any two adjacent vertices of $\mathcal{H}$ are distinguished by their blue sums. In case w has $d \geq 2$ neighbours $x_1, \ldots, x_d$ (which lie in D), then we further modify the labelling by changing to red label 1 the label assigned to $wx_1, \ldots, wx_d$.

  Again, we did not modify the red sums and blue sums of the vertices in R. Also, the only vertex of $D \cup I$ that might have red sum at least 2 is w (note that the x_i's, if they exist, have red sum 1), which lies in I, the set of isolated vertices of $\mathcal{R}$ and thus cannot be adjacent to the vertices of R. Since the vertices of R have red sum at least 2, they thus cannot be involved in conflicts. Now, if $d_G(w) = 1$, then, because G is not just an edge, the unique neighbour of w must have degree at least 2, meaning that w is necessarily distinguished from its unique neighbour. Otherwise, i.e., w has $d \geq 2$ neighbours $x_1, \ldots, x_d \in D$, then $\sigma_{\mathrm{r}}(w) = d \geq 2$ while the x_i's have red sum 1, and thus w cannot be involved in conflicts. Regarding the x_i's, they have red sum 1, so they cannot be in conflict with their neighbours of $\mathcal{H}$ different from w, since they have red sum 0. Finally, for every vertex of $\mathcal{H}$ not in $\{w, x_1, \ldots, x_d\}$, note that we did not modify its blue sum when introducing red labels. So we still have that any two such adjacent vertices are distinguished by their blue sums, by how we applied Lemma 1. So, no conflicts exist in G.

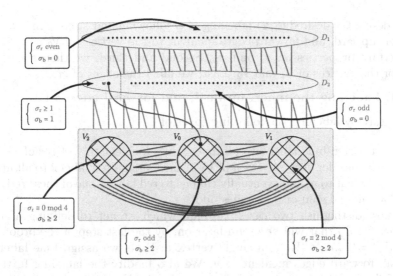

Fig. 3. Terminology used in the proof of Theorem 4, and the red sums and blue sums we aim at getting for the vertices by the designed $(2,2)$-labelling.

The resulting $(2,2)$-labelling of G is thus distinguishing, as desired.   □

Theorem 4. *Every $2K_2$-free graph with chromatic number at least 6 admits a distinguishing $(2,2)$-labelling.*

Proof. Let G be a $2K_2$-free graph with chromatic number at least 6. We construct a distinguishing labelling of G assigning red labels 1 and 2 and blue labels 1 and 2. Again, we may assume that G is connected.

Let D_1 be a maximal independent set of G. Note that every vertex of $G - D_1$ has at least one neighbour in D_1. Now let D_2 be a maximal independent set of $G - D_1$. Similarly, every vertex of $G - D_1 - D_2$ has at least one neighbour in D_2. Since $\chi(G) \geq 6$, note that $\chi(G - D_1 - D_2) \geq 4$. According to Theorem 2, there is thus a balanced tripartition V_0, V_1, V_2 of $G - D_1 - D_2$ (see Fig. 3). Note that D_1, D_2, V_0, V_1, and V_2 form a partition of $V(G)$. An *upward edge* of G is an edge with one end in $V_0 \cup V_1 \cup V_2$ and the other in $D_1 \cup V_2$. An *inner edge* of G is an edge with both ends in some V_i. If $u \in V_i$ and $u' \in V_{i+1}$ (where the operations over the subscripts of the V_i's are modulo 3) are adjacent for some $i \in \{0, 1, 2\}$, then uu' is a *forward edge* from u's perspective, and a *backward edge* from that of u'. Because G is $2K_2$-free, note that all three of $G[V_0]$, $G[V_1]$, and $G[V_2]$ contain at most one component with edges each.

We denote by $\mathcal{H}$ the set of the components of $G[D_1 \cup D_2]$. Since every vertex of D_2 has neighbours in D_1, note that $\mathcal{H}$ has edges. Since G is $2K_2$-free, there is exactly one component H of $\mathcal{H}$ that is *non-empty*, *i.e.*, contains edges. $\mathcal{H}$ can also contain *empty* components, which consist in a single vertex of D_1.

We design the desired $(2,2)$-labelling of G following four steps. First, we label all inner, upward, and forward edges incident to the vertices of V_0 so that they fulfil certain properties on σ_r and σ_b. Second and third, we then achieve the same for the vertices of V_1 and V_2. Last, we label the edges of $\mathcal{H}$.

Step 1: Labelling the inner, upward, and forward edges of V_0.

We start by labelling the following edges of G:

1. We first assign blue label 2 to all inner edges incident to vertices of V_0.
2. We then consider every vertex u of V_0 in turn, assign red label 2 to all upward edges incident to u, and eventually change to red label 1 one of these red labels so that the red sum of u becomes odd.
3. We now distinguish two cases, through which we get to defining a special vertex $w \in D_2$ that will be useful later on, by the last step of the proof.
   - $|V_0| = 1$, $i.e.$, $G[V_0]$ is a single vertex u. Here, we assign blue label 2 to all forward edges incident to u. We also modify the labelling further as follows. Set w as any neighbour of u in D_2. Note that, by swapping the red labels assigned to uw and another upward edge incident to u, we can, if necessary, assume uw is assigned red label 2. We then change the label assigned to uw to blue label 1.
   - Otherwise, $i.e.$ $|V_0| \geq 2$. Here, let $u_1, \ldots, u_n$ be an arbitrary ordering over the vertices of V_0, and consider the u_i's one by one in order. Since extra modifications must be made around u_1, let us consider that vertex specifically before describing the general case. Just as in the previous case, let w be any neighbour of u_1 in D_2. Again, we can swap labels assigned to upward edges, if necessary, so that u_1w is assigned red label 2. Then we change the label assigned to u_1w to blue label 1, before assigning blue label 2 to all forward edges incident to u_1. Now, for every subsequent u_i with $i \geq 2$, denote by $u_{i_1}, \ldots, u_{i_d}$ the $d \geq 0$ neighbours of u_i in V_0 that precede u_i in the ordering. If $d = 0$, then assign blue label 2 to all forward edges incident to u_i. Now, if $d \geq 1$, then recall that u_i is incident to $d_{V_1}(u_i) \geq d$ forward edges. By assigning red label 2 to none, one, two, etc., or all of these edges, and blue label 2 to all others, we can increase the red sum of u_i by any amount in $\{0, 2, \ldots, 2d_{V_1}(u_i)\}$, which set contains $d_{V_1}(u_i) + 1 \geq d + 1$ elements. There is thus a way to assign red label 2 to at most d forward edges incident to u_i, and blue label 2 to the rest, so that the red sum of u_i is different from the red sums of $u_{i_1}, \ldots, u_{i_d}$.

Once the steps above have been performed fully, note that all inner, upward, and forward edges incident to the vertices of V_0 are assigned a label. Also, for every vertex $u \in V_0$, we currently have $\sigma_r(u) \equiv 1 \bmod 2$, and it can be checked that also $\sigma_b(u) \geq 2$. Furthermore, every two adjacent vertices of V_0 currently have their red sums being different. Remark last that all upward edges incident to the vertices of V_0 are assigned red labels, except for exactly one edge incident to w, which is assigned blue label 1.

Step 2: Labelling the inner, upward, and forward edges of V_1.

Due to the previous step, note also that all backward edges incident to the vertices in V_1 are labelled with red label 2 and blue label 2. So, one should keep in mind that, currently, $\sigma_r(u)$ is even for every $u \in V_1$.

We now label more edges as follows:

1. First, we assign blue label 2 to all inner edges incident to vertices of V_1.
2. Second, we consider every vertex u of V_1 in turn. Recall that u is incident to at least two upward edges. We assign red label 2 to all these edges. If necessary, we change the label assigned to two of these edges to red label 1, so that $\sigma_r(u) \equiv 2 \bmod 4$.
3. Third, let $u_1, \ldots, u_n$ be an arbitrary ordering over the vertices of V_1, and consider the u_i's one by one in turn. For every u_i considered that way, denote by $u_{i_1}, \ldots, u_{i_d}$ the $d \geq 0$ neighbours of u_i in V_1 that precede u_i in the ordering. If $d = 0$, then assign blue label 2 to all forward edges incident to u_i. Now, if $d \geq 1$, then recall that u_i is incident to $d_{V_2}(u_i) \geq d$ forward edges. Thus, through assigning blue labels to these edges, we can make the blue sum of u_i vary by any amount in the set $\{d_{V_2}(u_i), \ldots, 2d_{V_2}(u_i)\}$, which contains $d_{V_2}(u_i) + 1 \geq d + 1$ elements. Thus, it is possible to assign blue labels to the forward edges incident to u_i so that its resulting blue sum is different from that of $u_{i_1}, \ldots, u_{i_d}$.

After completing the previous steps, all edges incident to the vertices in V_1 are labelled. For every vertex $u \in V_1$, we get $\sigma_r(u) \equiv 2 \bmod 4$, and also $\sigma_b(u) \geq 2$, because either $d_{V_1}(u) = 0$ and at least one forward edge incident to u is assigned blue label 2, or $d_{V_1}(u) > 0$ and at least one inner edge incident to u is assigned blue label 2. Also, adjacent vertices of V_1 are distinguished by their blue sums, and all upward edges incident to the vertices of V_1 are assigned red labels.

Step 3: Labelling the inner, upward, and forward edges of V_2.

Note that after performing the previous step, all backward edges incident to the vertices of V_2 are assigned blue labels; so, their red sum is currently 0.

We now perform the following:

1. We assign blue label 2 to all inner edges incident to vertices in V_2.
2. We then consider every vertex u of V_2 in turn, which, recall, is incident to at least two upward edges. We assign red label 2 to all these edges before, if necessary, changing the label assigned to two of these edges to red label 1, so that $\sigma_r(u) \equiv 0 \bmod 4$.
3. We finish off this step similarly as the previous one. Let $u_1, \ldots, u_n$ be any ordering over the vertices of V_2, and consider the u_i's one after the other. For every u_i, let $u_{i_1}, \ldots, u_{i_d}$ be the $d \geq 0$ neighbours of u_i in V_2 that precede u_i in the ordering. If $d = 0$, then assign blue label 2 to all forward edges incident to u_i. Otherwise, if $d \geq 1$, then recall that u_i is incident to $d_{V_0}(u_i) \geq d$ forward edges. Via assigning blue labels to these edges, we can thus make the blue sum of u_i increase by any value in $\{d_{V_0}(u_i), \ldots, 2d_{V_0}(u_i)\}$, which set contains

$d_{V_0}(u_i) + 1 \geq d + 1$ elements. Thus, we can assign blue labels to the forward edges incident to u_i so that its blue sum is different from that of $u_{i_1}, \ldots, u_{i_d}$.

Once this step achieves, all edges incident to vertices in $V_0 \cup V_1 \cup V_2$ are labelled. For every vertex $u \in V_2$, we have $\sigma_r(u) \equiv 0 \bmod 4$ and $\sigma_b(u) \geq 2$. Every two adjacent vertices of V_2 are distinguished by their blue sums, while all upward edges incident to the vertices in V_2 are assigned red labels. It is important to emphasise also that assigning blue labels to the edges joining vertices of V_2 and V_0 altered the blue sums of the vertices in V_0, which is not an issue since the adjacent vertices of V_0 are distinguished by their red sums, which were not altered. So, any two adjacent vertices in V_0 remain distinguished, and similarly for any two adjacent vertices in V_1. Finally, note that any two adjacent vertices in distinct V_i's are distinguished by their red sums having different values modulo 4.

Step 4: Labelling the edges of $\mathcal{H}$.

Recall that, at this point, we have $\sigma_b(v) = 0$ for every vertex $v \in D_1 \cup D_2 \setminus \{w\}$ and $\sigma_b(w) = 1$, while $\sigma_b(u) \geq 2$ for every vertex $u \in V_0 \cup V_1 \cup V_2$. In particular, if $v \in D_1$ belongs to an empty component of $\mathcal{H}$, then all edges incident to v are already labelled, and v is distinguished from its neighbours due to its blue sum.

Recall that H denotes the unique non-empty component of $\mathcal{H}$, and that H actually contains all edges of G that remain to be labelled. Recall also that H contains w, a special vertex we defined in the first labelling step, which is the only vertex of H having non-zero blue sum. According to Lemma 1, we can assign red labels 1 and 2 to the edges of H so that, even when taking into account the red labels assigned to the upward edges incident to the vertices in $V_0 \cup V_1 \cup V_2$, any two adjacent vertices of H different from w are distinguished by their red sums. Since $\sigma_b(w) = 1$ while $\sigma_b(v) = 0$ for every $v \in V(\mathcal{H}) \setminus \{w\}$, vertex w is also distinguished from its neighbours in $\mathcal{H}$. These conditions guarantee we have not introduced any conflicts involving vertices of $D_1 \cup D_2$ and vertices of $V_0 \cup V_1 \cup V_2$.

Thus, the resulting $(2,2)$-labelling of G is distinguishing. □

4   From $2K_2$-free Graphs to $K_{1,3}$-free Graphs, and Beyond

As mentioned earlier, we also proved the Weak $(2,2)$-Conjecture for claw-free graphs in the full paper [6], in a way that is quite reminiscent of how we proved Theorems 3 and 4. The structure of $2K_2$-free graphs being much more constrained than that of $K_{1,3}$-free graphs, this required involved refinements over our arguments; in particular:

– The main issue in order to adapt Theorem 3 to $K_{1,3}$-free graphs, is that $\mathcal{R}$ can now have several non-empty components, and similarly for $\mathcal{H}$. Then, by leading the proof the exact same way for every non-empty component of $\mathcal{R}$, multiple upward edges can now be assigned red label 1. Fortunately, the fact that G is claw-free implies that every $u \in D$ neighbours vertices from at most two components of $\mathcal{R}$, meaning that $\sigma_r(u) \leq 2$. So there can be conflicts between vertices of $\mathcal{R}$ and D, but we can get rid of those by altering the labelling around very local structures of G.

– Regarding adapting Theorem 4 to $K_{1,3}$-free graphs, the main issue is that $\mathcal{H}$ can now have several non-empty components, the most troublesome of which can consist of a single edge v_1v_2 with $v_1 \in D_1$ and $v_2 \in D_2$. The problem is that distinguishing v_1 and v_2 has nothing to do with the label assigned to v_1v_2; in particular, v_1 and v_2 might get in conflict because of how we labelled the upward edges, when dealing with the vertices of $V_0 \cup V_1 \cup V_2$. We deal with this issue through being extra cautious when labelling upward edges, to make sure such situations do not occur. This requires us to also modify our sum rules by a bit. For instance, we allow vertices of V_0 to have a red sum that is not odd, provided their neighbourhood satisfies some conditions.

In both cases, the final step of labelling the edges of $\mathcal{H}$ is also slightly trickier, due to the fact that we have less control over the labels assigned to upward edges. Fortunately, when G is claw-free, $\mathcal{H}$ is actually a bipartite graph with maximum degree at most 2, a structure which is quite favourable, and which we manage to deal with using algebraic tools (such as the Combinatorial Nullstellensatz).

To go farther on this topic, it could be interesting to investigate the Weak $(2, 2)$-Conjecture for more classes of graphs defined in terms of forbidden structures, such as triangle-free graphs, or graphs with large girth in general. One could wonder also about graphs in which many short cycles are present, such as chordal graphs. Another class could be that of P_4-free graphs (cographs).

References

1. Addario-Berry, L., Aldred, R.E.L., Dalal, K., Reed, B.A.: Vertex colouring edge partitions. J. Comb. Theory. Ser. B **94**(2), 237–244 (2005)
2. Ahadi, A., Dehghan, A., Sadeghi, M.-R.: Algorithmic complexity of proper labeling problems. Theor. Comput. Sci. **495**, 25–36 (2013)
3. Baudon, O., et al.: A general decomposition theory for the 1-2-3 Conjecture and locally irregular decompositions. Discret. Math. Theor. Comput. Sci. **21**(1), 2 (2019)
4. Baudon, O., Bensmail, J., Przybyło, J., Woźniak, M.: On decomposing regular graphs into locally irregular subgraphs. Eur. J. Comb. **49**, 90–104 (2015)
5. Bensmail, J.: On a graph labelling conjecture involving coloured labels. Discuss. Math. Graph Theory (in press)
6. Bensmail, J., Hocquard, H., Marcille, P.-M.: The weak (2, 2)-Labelling Problem for graphs with forbidden induced structures. Preprint (2022). https://hal.archives-ouvertes.fr/hal-03784687
7. Gallian, J.A.: A dynamic survey of graph labeling. Electron. J. Comb. 1, DS6 (2021)
8. Kalkowski, M., Karoński, M., Pfender, F.: Vertex-coloring edge-weightings: towards the 1-2-3 conjecture. J. Comb. Theory. Ser. B **100**, 347–349 (2010)
9. Karoński, M., Łuczak, T., Thomason, A.: Edge weights and vertex colours. J. Comb. Theory. Ser. B **91**, 151–157 (2004)
10. B. Seamone. The 1–2-3 Conjecture and related problems: a survey. Preprint (2012). http://arxiv.org/abs/1211.5122
11. Vučković, B.: Multi-set neighbor distinguishing 3-edge coloring. Discret. Math. **341**, 820–824 (2018)

Graph Connectivity

Short Cycles Dictate Dichotomy Status of the Steiner Tree Problem on Bisplit Graphs

A. Mohanapriya[1]([✉]), P. Renjith[2], and N. Sadagopan[1]

[1] Indian Institute of Information Technology, Design and Manufacturing,
Kancheepuram, Chennai, India
{coe19d003,sadagopan}@iiitdm.ac.in
[2] National Institute of Technology, Calicut, India
renjith@nitc.ac.in

Abstract. A graph G is said to be a bisplit graph if its vertex set can be partitioned into a stable set and a complete bipartite graph. The minimum Steiner tree problem (STREE) is defined as follows: given a connected graph G and a subset of vertices $R \subseteq V(G)$, the objective is to find a minimum cardinality set $S \subset V(G)$ such that the set $R \cup S$ induces a connected subgraph. In this paper, we present an interesting dichotomy result for STREE on bisplit graphs, we show that STREE is polynomial-time solvable for chordal bipartite bisplit graphs, and NP-complete otherwise. Further, we show that for chordal bisplit graphs, the problem is polynomial-time solvable. A revisit of our NP-complete reduction instances reveals that the instances are diameter at most 5 bipartite graphs. We also obtain one more dichotomy result for STREE on bisplit graphs which says that for diameter 5 the problem is NP-complete and polynomial-time solvable for diameter at most 4. On the parameterized complexity front, we show that the parameterized version of Steiner tree problem on bisplit graphs is fixed-parameter tractable when the parameter is the biclique size and is W[2]-hard on bisplit graphs if the parameter is the solution size. We conclude this paper by presenting structural results of bisplit graphs which will be of use to solve other combinatorial problems.

Keywords: Bisplit graphs · The Steiner tree problem · W-hardness · Dichotomy results

1 Introduction

The class of bisplit graphs was introduced by Brandstädt et al. in 2005 [1]. An undirected graph G is a bisplit graph if its vertex set can be partitioned into a stable set and a complete bipartite graph. Bisplit graphs have attracted researchers from both structural and algorithmic perspectives. Bisplit graphs can be recognized in $O(n^2)$ time [2].

© The Author(s), under exclusive license to Springer Nature Switzerland AG 2023
A. Bagchi and R. Muthu (Eds.): CALDAM 2023, LNCS 13947, pp. 219–230, 2023.
https://doi.org/10.1007/978-3-031-25211-2_17

Bisplit graphs resemble split graphs which are graphs whose vertex set can be partitioned into a clique and a stable set. Having seen this resemblance it is natural to ask for the computational complexity of classical problems on bisplit graphs which are NP-complete on split graphs [3]. In this paper, we look at one such problem namely the Steiner tree problem. We also highlight other combinatorial problems such as the Hamiltonian cycle problem [4], the outer-connected domination problem [5] are NP-complete on split graphs. Further, the complexity of these problems is analyzed in the subclass of split graphs. For example, STREE is NP-complete on $K_{1,5}$-free split graphs, and polynomial-time solvable on $K_{1,4}$-free split graphs, and a similar study was also made for Hamiltonian cycle [6] and Hamiltonian path [7]. As per our knowledge, there are very few combinatorial problems whose complexity status is known on bisplit graphs [8–10].

This paper aims to study the computational complexity of the Steiner tree problem on bisplit graphs and its subclasses. To the best of our knowledge, the computational complexity of the Steiner tree problem considered in this paper is not explored on bisplit graphs. The decision version of combinatorial problem which we consider in this paper is as defined below;

The Steiner tree problem (STREE)(G, R, k)

Instance: A graph G, a terminal set $R \subseteq V(G)$, and a positive integer k.

Question: Is there a set $S \subseteq V(G) \setminus R$ such that $|S| \leq k$, and $G[S \cup R]$ is connected ?

The Steiner tree problem is NP-complete on split graphs [3]. We investigate the complexity of STREE on bisplit graphs and show that STREE remain NP-complete for bisplit graphs as well. It is known that bisplit graphs are a subclass of comparability graphs, and the status of well-known combinatorial problems such as the dominating set problem [11], the Steiner tree problem [12], etc., are NP-complete in comparability graphs. Thus this paper reinforces the result of [12]. Further, a well-known subclass of bisplit graphs is bipartite chain graphs where the status of these problems are polynomial-time solvable [13].

Our research is driven by the following questions on bisplit graphs for STREE:

1. What is the complexity of STREE on bisplit graphs when the cycle length is bounded? For example, chordal bisplit, chordal bipartite bisplit, and other special bisplit graphs.
2. What is the complexity of STREE on bisplit graphs when the bisplit graphs have bounded diameter? For example, diameter 5 bisplit graphs.
3. Can cycle length or diameter dictate the dichotomy status of STREE on bisplit graphs?
4. What is the parameterized complexity of STREE on bisplit graphs when the parameters are the biclique size and the solution size?

Our Results:

1. We prove that the complexity of STREE on bisplit graphs is NP-complete. By using the reduction instances, we identify a subclass of bisplit graphs for which

STREE is polynomial-time solvable, which we shall discuss in Sect. 2. We also analyze the complexity of STREE on bisplit graphs with the parameter being the diameter, which shall be discussed in Sect. 3.

2. Our final result is from the parameterized complexity theory, to be presented in Sect. 4. We prove that the parameterized version of Steiner tree problem on bisplit graphs with the parameter being the solution size is W[2]-hard, and is fixed-parameter tractable when the parameter is the biclique size.

Graph Preliminaries: In this paper, we consider connected, undirected, unweighted, and simple graphs. For a graph G, $V(G)$ denotes the vertex set, and $E(G)$ represents the edge set. For a set $S \subseteq V(G)$, $G[S]$ denotes the subgraph of G induced on the vertex set S. The open neighborhood of a vertex v is $N_G(v) = \{u \mid \{u,v\} \in E(G)\}$ and the closed neighborhood of v is $N_G[v] = \{v\} \cup N_G(v)$. The degree of vertex v is $d_G(v) = |N_G(v)|$.

A graph G is a bipartite graph if $V(G)$ can be partitioned into $X \cup Y$ such that X, Y are disjoint independent sets. A bipartite graph G is called star-convex bipartite, if there is an associated star T on vertices of X, such that each vertex $y \in Y$, its neighborhood $N_G(y)$ induces a substar of T [14]. For such star-convex bipartite graphs, we say the convexity is on X. A bipartite graph is chordal bipartite if every cycle of length strictly greater than four has a chord. For any integer $k \geq 3$, a graph is k-chordal if it has no induced cycle of length greater than k [15]. For a bisplit graph, let $A \cup B$ represents complete bipartite graph, and I represents an independent set. Let $A = \{a_1, a_2, \ldots, a_n\}$, $B = \{b_1, b_2, \ldots, b_m\}$, and $I = \{x_1, x_2, \ldots, x_l\}$. For a connected graph, the distance $d(u, v)$ between u and v is the number of vertices in the shortest (u, v) path. For a connected graph, the diameter of G, $diam(G)$, is defined by $diam(G) = max\{d(u, v) : u, v \in V(G)\}$.

2   STREE in Bisplit Graphs

In this section, we show that STREE on bisplit graphs is NP-complete. We establish a classical hardness of STREE on bisplit graphs by presenting a polynomial-time reduction from STREE on split graphs. The decision version of Steiner tree problem is defined below:

$STREE\ (G, R, k)$
Instance: A graph G, a terminal set $R \subseteq V(G)$, and a positive integer k.
Question: Is there a set $S \subseteq V(G) \setminus R$ such that $|S| \leq k$, and $G[S \cup R]$ is connected ?

Theorem 1. *For bisplit graphs, STREE is NP-complete.*

Proof. **STREE is in NP** Given a bisplit graph G and a certificate $S \subseteq V(G)$, we show that there exists a deterministic polynomial-time algorithm for verifying the validity of S. Note that the standard Breadth First Search (BFS) algorithm can be used to check whether $G[S \cup R]$ is connected. It is easy to check whether

$|S| \leq k$. The certificate verification can be done in $O(|V(G)| + |E(G)|)$. Thus, we conclude that STREE is in NP.

STREE is NP-Hard It is known that STREE on split graphs is NP-hard. STREE on split graphs can be reduced in polynomial time to STREE on bisplit graphs using the following reduction. We map an instance of $(G, R = I, k)$ where $|K| = n$, $|I| = m$ to the corresponding instance of $(G', R', k' = 2k)$ as follows: $V(G') = V_1 \cup V_2 \cup V_3 \cup V_4$, $V_1 = \{v_i \mid v_i \in K, 1 \leq i \leq n\}$, $V_2 = \{u_i \mid v_i \in K, 1 \leq i \leq n\}$, $V_3 = \{w_i \mid y_i \in I, 1 \leq i \leq m\}$, $V_4 = \{x_i \mid y_i \in I, 1 \leq i \leq m\}$, and $E(G') = E' \cup \{\{v_i, w_i\}, \{u_i, x_i\} \mid \{v_i, y_i\} \in E(G), v_i \in K, y_i \in I\}$, $E' = \{\{v_i, u_j\} \mid v_i \in V_1, u_j \in V_2, 1 \leq i \leq j \leq n\}$. Let $R' = V_3 \cup V_4$, $k' = 2k$. Note that G' is a bisplit graph with $V_1 \cup V_2$ induces a biclique and $V_3 \cup V_4$ induces an independent set.

Claim 1. *(G, R, k) is an yes-instance of STREE in G if and only if $(G', R', k' = 2k)$ is an yes-instance of STREE in G'.*

Proof. Necessary: Suppose that there exists a Steiner tree in G for $R = I$. Hence the Steiner set $S \subseteq K$. The set S' of G' is $S' = \{v_i, u_i \mid v_i \in S, v_i \in V_2, u_i \in V_3\}$ is the Steiner set with $R' = V_3 \cup V_4$. Also, note that $|S'| = 2k$.

Sufficiency: Assume that there exists a Steiner tree of G' for R. We know that $S' \subseteq V_1 \cup V_2$. By our construction we know that for any $v_i \in V_1$, v_i is not adjacent to w_j, $w_j \in V_3$. Similarly $u_i \in V_2$, u_i is not adjacent to x_j, $x_j \in V_4$. Hence for $R \cap V_3$, we need k vertices from V_1 and for $R \cap V_4$, we need k vertices from V_2. Since $|S'| = 2k$, we now construct $S = \{v_i \mid v_i \in V_1, v_i \in S'\}$. Therefore, S is the corresponding solution of STREE in G. Since $k' = 2k$, observe that $|S' \cap V_1| \leq k$ or $|S' \cap V_2| \leq k$. Without loss of generality, let $|S' \cap V_1| \leq k$. Note that $\{v_i \in K \mid v_i \in S' \cap V_1\}$ forms a Steiner set of size at most k in G.   □

Therefore, STREE is NP-complete on bisplit graphs.   □

Corollary 1. *For bipartite graphs of diameter at most 5, STREE is NP-complete.*

Proof. Since bisplit graphs instance generated from Theorem 1 are bipartite graphs having diameter at most 5, STREE is NP-complete on bipartite graphs of diameter at most 5.   □

Theorem 2. *[16] For $K_{1,5}$-free split graphs, STREE is NP-complete.*

Theorem 3. *[16] For $K_{1,4}$-free split graphs, STREE is polynomial-time solvable.*

We extend the results of Theorem 2 and Theorem 3, for diameter at most 5 bipartite bisplit graphs. We have the following partitions in diameter at most 5 bipartite bisplit graphs; Let G be a diameter at most 5 bipartite bisplit graph. Let the biclique be $A \cup B$, and let the independent set be $D_1 \cup D_2$. Note that D_1 represents vertices in I that are adjacent to B, and D_2 represents vertices in I that are adjacent to A.

Corollary 2. *If the graph induced on $A \cup D_2$ is a $K_{1,5}$-free graph and the graph induced on $B \cup D_1$ is a $K_{1,5}$-free graph, then STREE is NP-complete.*

Proof. Since from Theorem 2, it is known that for $K_{1,5}$-free split graphs, STREE is NP-complete. When we take the input instance of the reduction to be $K_{1,5}$-free split graph, then the generated instances will have the property that the graph induced on $A \cup D_2$ is a $K_{1,5}$-free graph and the graph induced on $B \cup D_1$ is a $K_{1,5}$-free graph. Thus Corollary 2 is true because of Theorem 1 and Theorem 2.

Theorem 4. *If the graph induced on $A \cup D_2$ is a $K_{1,4}$-free graph and the graph induced on $B \cup D_1$ is a $K_{1,4}$-free graph, then STREE is polynomial-time solvable.*

Proof. Let G be a bipartite bisplit graph such that the graph induced on $A \cup D_2$ is a $K_{1,4}$-free graph and the graph induced on $B \cup D_1$ is a $K_{1,4}$-free graph. We solve for $R = I$ instance in G. For all other instances of R, the solution can be obtained by using $R = I$ algorithm. By considering A as clique and by using $K_{1,4}$-free split graph Steiner tree algorithm [16] and Theorem 3 as black box, we obtain a Steiner solution S_1 for D_2. Similarly for D_1, we obtain Steiner solution S_2 for D_1. Thus Steiner solution for G is $S_1 \cup S_2$. □

From Corollary 2 and Theorem 4, it is clear that we obtain a dichotomy on diameter at most 5 bipartite bisplit graph.

   Highlights:

1. Note that the reduction instances generated from Theorem 1 are odd cycle-tree bisplit graphs. Observe that the reduction instances can have cycles of any even length.
2. A natural question that can arise is that
   "What happens to the complexity of STREE on bisplit graphs if they are even-cycle free (chordal bisplit graphs)?"
   "What happens to the complexity of STREE on bisplit graphs if all even cycles are length four (chordal bipartite bisplit graphs)?"
   We answer these questions in the following sections.

2.1   Chordal Bisplit Graphs

We show that STREE on chordal bisplit graphs is polynomial-time solvable.

Theorem 5. *A graph G is a chordal bisplit graph, if and only if the following properties are satisfied.*

1. *The biclique in G is $K_{1,l}$, for some $l \geq 0$.*
2. *Each vertex $x \in I$, if $d_G(x) \geq 2$, then x is adjacent to a_1.*
3. *The graph induced on $B \cup I$ is a forest.*

Proof. (i) On the contrary, if G has $K_{2,l}$, for some $l \geq 0$. Then G has C_4 as an induced subgraph, which contradicts the fact that G is a chordal graph. Thus biclique of size $K_{i,l}$, $i \geq 2$, for some l is not possible. Without loss of generality we shall assume that $|A| = 1$, $A = \{a_1\}$.

(ii) Suppose that there exists a vertex, say $x \in I$ such that $d_G(x) \geq 2$ and x is not adjacent to a_1. Let the neighbour of x be v_1, v_2. Observe that $v_1, v_2 \in B$ and we know that $\{v_1, v_2\} \notin E(G)$. Then a_1, v_1, x, v_2 is a cycle of length 4, which contradicts the fact that G is a chordal graph.

(iii) Since G is a chordal graph, even cycle is not possible in $G[B \cup I]$. Further, since G is a bisplit graph, an odd cycle of length greater than or equal to five is not possible by Theorem 14. The only possible cycle length on $G[B \cup I]$ is three. Suppose that $G[B \cup I]$ has cycle of length three, it must the case that either $\{b_i, b_j\} \in E(G)$, $b_i, b_j \in B$, $1 \leq i < j \leq m$ or $\{x_u, x_v\} \in E(G)$ $x_u, x_v \in I$, $1 \leq u < v \leq t$. It contradicts the fact that G is a bisplit graph.   □

Let the vertices in biclique be $\{a_1, b_1, \ldots, b_l\}$ such that $A = \{a_1\}$, $B = \{b_1, \ldots, b_l\}$. In Theorem 6, we show that for chordal bisplit graphs, finding a minimum Steiner tree is polynomial-time solvable. For the polynomial-time solvable cases of STREE, we consider the case $R = I$, for other cases such as $R \subseteq A \cup B$, $R \subseteq A$, $R \subseteq B$, can be found by using the case $R = I$.

Algorithm 1. *STREE for chordal bisplit graphs*

1: **Input**: A connected chordal bisplit graph G with $R = I$.
2: **Output**: A Steiner solution S for G
3: Let the vertices in biclique be $a_1, b_1, \ldots, b_l$
4: **if** there exist a pendent vertex v in R **then**
5:     $S = S \cup N(v)$
6:     $R = R \setminus N(v)$
7: **end if**
8: **if** $R = \emptyset$ and $G[R \cup S]$ is connected **then**
9:     Stop the algorithm
10: **else**
11:     $S = S \cup \{a_1\}$
12: **end if**
13: Return S

Theorem 6. *Finding a minimum Steiner tree on chordal bisplit graphs is polynomial-time solvable.*

Proof. We consider the case $R \subseteq I$. Suppose if there exist a pendent vertex v in R, then by Step 5 of Algorithm 1 in order to connect v with rest of the vertices in R, $N(v)$ is included in S. If any vertex in R whose neighbor is not in S or the vertices included so far do not induce a connected graph, then by Step 11 of the algorithm, including a_1 in S as by Theorem 5, a_1 is adjacent to vertices of degree at least 2, a_1 is adjacent to vertices in B. Therefore, $G[R \cup S]$ is connected, and $|S|$ is the minimum.   □

Since we check for the presence of pendent vertices in G, the check can be done in linear time. Thus the time complexity of Algorithm 1 is $O(n)$, where n is the number of vertices in G.

2.2   Chordal Bipartite Bisplit Graphs

Since finding a minimum Steiner set for a given R on chordal bisplit graphs turns out to be polynomial-time solvable. We extend this line of study further, by restricting cycle length. Further, it is known [17] that on chordal bipartite graphs, STREE is NP-complete. Hence we are interested to study the complexity of STREE on chordal bipartite bisplit graphs.

We prove that finding a minimum Steiner tree on chordal bipartite bisplit graphs is polynomial-time solvable. We use the algorithm of finding a minimum Steiner tree on strongly chordal split graphs as the black box.

Theorem 7. *[3] Finding a minimum Steiner tree on strongly chordal graphs is polynomial-time solvable.*

Corollary 3. *Finding a minimum Steiner tree on strongly chordal split graphs is polynomial-time solvable.*

Since all other cases such as $R \subseteq (A \cup B)$ or $R \subseteq (A \cup B \cup I)$ can be reduced to the case $R = I$, it is enough to consider the case when $R = I$.

Theorem 8. *A minimum Steiner tree on chordal bipartite bisplit graphs can be found in polynomial time.*

Proof. Let G be chordal bipartite bisplit graphs. To find the solution set S for R, we transform the graph G to G' such that $V(G') = V(G)$, $E(G') = E(G) \cup \{\{a_i, a_j\} \mid a_i, a_j \in A, 1 \le i < j \le n\} \cup \{\{b_i, b_j\} \mid b_i, b_j \in B, 1 \le i < j \le m\}$, $R' = I$. Observe that the resultant graph is a strongly chordal split graph [4]. It is known that finding a minimum Steiner tree on strongly chordal split graphs is polynomial-time solvable. Thus a minimum Steiner set for G' can be obtained in polynomial-time. Let S' be a minimum Steiner set for G'. Moreover, the transformation of G to G' is a solution preserving reduction. Using R' of G', we obtain a minimum Steiner set to R of G. Let S' be a minimum Steiner set for R' of G. Suppose if $A \cap S' \ne \emptyset$ and $B \cap S' \ne \emptyset$, then S' is also a minimum Steiner set for G. Suppose if $G[R \cup S']$ is disconnected, then one of $A \cap S' = \emptyset$ and $B \cap S' = \emptyset$. In that case, we include arbitrary vertex $a \in A$ in S', if $A \cap S' = \emptyset$ or an arbitrary vertex from B is included in S'. Therefore, S' is a minimum Steiner set of G for $R = I$.   □

Remarks:

1. Thus we obtain a dichotomy with cycle length as the parameter of interest.
2. It is known [18] that finding a minimum Steiner tree on strictly chordality k graphs, $k \ge 5$ is polynomial-time solvable.

3   Complexity of STREE on Bisplit Graphs with Diameter as the Parameter

On bipartite bisplit graphs with diameter at most 5, it known from Theorem 1 that STREE is NP-complete. A natural question is to analyze "the complexity of STREE on diameter 4 bipartite bisplit graphs?" We answer this question in Theorem 10. An edge $\{u, v\}$ is said to be a dominating edge, if $(N_G(u) \cup N_G(v)) = V(G)$.

Theorem 9. *If bipartite bisplit graphs have diameter 4, then it contains a dominating edge.*

Proof. Let G be a diameter 4 bipartite bisplit graph. It is clear that the distance between any two vertices must be at most 4. Hence there does not exist two vertices in A, say u, v such that $N_G(v) \neq N_G(u)$. Suppose that $N_G(v) \neq N_G(u)$. Let $y \in N_G(v), z \in N_G(u)$. Then the path between y, z must be of length 5. This contradicts the fact that G is a diameter 4 bipartite bisplit graph. Similarly, the argument is true with respect to B as well.   □

Theorem 10. *Let G be a bipartite bisplit graph with diameter 4. Then the minimum Steiner set can be found in linear time.*

Proof. If $|I| = 1$, then clearly $S = \emptyset$. If $|I| \geq 2$, by the structure of G and by Theorem 9, clearly one of $\{u\}$ or $\{v\}$ or $\{u, v\}$ is a desired Steiner set.   □

4 Parameterized Complexity Results

In this section, we study the complexity of the parameterized version of Steiner tree problem on split graphs with the parameter being the solution size and with respect to another parameter being the size of the biclique.

4.1 Parameterized Intractability

We consider the following parameterized version of the Steiner tree problem.

The parameterized version of Steiner tree problem (PSTREE)
PSTREE(G, R, k)
Instance: A graph G, a terminal set R, a positive integer k
Parameter: k
Question: Does there exist a set $S \subseteq V(G) \setminus R$ such that $|S| \leq k$, and $G[R \cup S]$ is connected?

We now prove that our reduction establishes a stronger result: Theorem 1 is indeed a parameter preserving reduction which we establish in Theorem 11.

Theorem 11. *For bisplit graphs, PSTREE is W[2]-hard with parameter being the solution size.*

Proof. It is known that the parameterized version of Steiner tree problem on split graphs with the parameter being solution size is W[2]-hard. Note that the reduction presented in Theorem 1 maps (G, R, k) to $(G', R', k' = 2k)$. From Claim 1 of Theorem 1, we can observe that the reduction is a solution preserving reduction. Hence the reduction is a deterministic polynomial-time parameterized reduction. Therefore, PSTREE on split graphs is W[2]-hard.   □

4.2   FPT Algorithm for Bisplit Graph

We consider the following parameterized version of the Steiner tree problem.

The parameterized version of Steiner tree problem (PSTREE1)
PSTREE(G, R, k)
Instance: A bisplit graph G with biclique size l, a terminal set R, a positive integer k
Parameter: k, l
Question: Does there exist a set $S \subseteq V(G) \setminus R$ such that $|S| \leq k$, and $G[R \cup S]$ is connected??

We now prove that with respect to the parameter being the size of biclique, PSTREE1 is in FPT.

Let the vertices in biclique L be $x_1, \ldots, x_n, y_1, \ldots, y_m$, and let $R = I$. For each vertex in L, we decide whether to include that vertex in the solution or not. The basic idea is that for every vertex in L, we consider two cases: for every vertex $l \in L$, include l in the solution and reduce the graph $G = G - N_G^I[l]$ or $G = G - l$. If we can find the Steiner set of size $k - 1$ in either $G = G - N_G^I[l]$ or $G = G - l$, we can construct the Steiner solution for G.

Algorithm 2. *STREE for bisplit graphs*

1: **Input**: A connected bisplit graph G with $R = C$.
2: **Output**: A Steiner solution S for G
3: Let the vertices in biclique L be $x_1, \ldots, x_n, y_1, \ldots, y_m$
4: **for** each vertex l in L **do**
5:     Include $l \in S$, continue with $G = G - N_G^C[l]$
6:     Remove l, continue with $G = G - l$
7: **end for**
8: Pick a branch for which $|S| \leq k$

The key Observation can be stated as follows:

Observation 1. *Let G be a bisplit graph and $l \in L$ be a vertex in the biclique. The graph G has a Steiner solution of size k if and only if either $G = G - N_G^C[l]$ has a Steiner solution of size k or $G = G - l$ has a Steiner solution of size $k - 1$.*

Theorem 12. *PSTREE1 on bisplit graphs can be computed in time $O(2^l.n^{\Theta(1)})$.*

Proof. Since $R = I$, $S \subseteq L$. Let $|I| = t$. By our approach, we choose an arbitrary vertex in $v \in L$ such that $d_G^I(v) \neq 0$, and we branch by having $v \in S$ and another branch with $v \notin S$. We can observe that length of the tree is $|L|$ and the number of leaves is at most $2^{|L|}$. Since all possible paths from the root to leaves are explored, one of the paths from the root to a leaf yields a minimum Steiner set.

Observe that Algorithm 2 lists all feasible solutions. The running time of the algorithm is bounded by the number of nodes $(2^{|L|})$ and the time taken at each node is $n^{\Theta(1)}$. □

5   Structural Results on Bisplit Graphs

In this section, we shall present some structural results on bisplit graphs, which will be of independent interest, however can be used in analyzing computational complexity status of other combinatorial problems on bisplit graphs.

Theorem 13. *Bisplit graphs have diameter at most 5.*

Proof. Suppose that there exists a bisplit graph G having diameter 6. It must be the case that there exist at least two vertices having the shortest distance between them being 6. For any two vertices from biclique, their distance is at most three. For a vertex from I and a vertex from $A \cup B$, their distance is at most four. If there are two vertices from I, say y, z, and four vertices are from biclique, say u, v, w, x. It is clear that two vertices from A and from B. Without loss of generality, $u \in N_G(y)$, and $v \in N_G(z)$, then the shortest path between y, z is y, u, w, v, z, a path of length five, which contradicts the fact that there exist at least two vertices having shortest distance between them is 6. Therefore, Bisplit graphs have diameter at most 5.   □

Theorem 14. *If a graph G is a bisplit graph, then it does not contain an odd cycle of length greater than or equal to 5 as an induced subgraph.*

Proof. On the contrary, suppose that G contains C_5 as an induced subgraph, then the maximum stable set in C_5 is 2. Let the cycle C_5 in G be c_1, c_2, c_3, c_4, c_5. Without loss of generality, c_3, c_5 are in I, then c_1, c_2, c_4 are in biclique $K_{m,n}$. Let the vertices in biclique be partitioned as $A \cup B$ such that the graph induced on A is a stable set and the graph induced on Y is a stable set. From the structure of C_5, it is known that at least one vertex has to be in each of the partitions, then the biclique edges in G form a chord in the cycle. Therefore, bisplit graphs do not contain odd cycles of length greater than or equal to 5.   □

Theorem 15. *Every minimum vertex separator S in a bisplit graph G is a subset of $A \cup B$, where $A \cup B$ forms a biclique in G.*

Proof. From the structure of bisplit graphs, it is known that any minimum vertex separator cannot be from stable set I and $I \cap S = \emptyset$. Suppose that $S \subset I$, then the vertices in $V(G) \setminus S$ form a connected subgraph. Thus, $S \subset I$ is a minimum vertex separator is not possible. Suppose that S is a minimum vertex separator $I \cap S \neq \emptyset$, then $S' = S \setminus I$ is also a vertex separator and $|S'| < |S|$, which is a contradiction that S is a minimum vertex separator. Thus every minimum vertex separator S in a bisplit graph G is a subset of vertices in a biclique.   □

Observation 2. *Every induced subgraph of a bisplit graph is a bisplit graph.*

Observation 3. *Bisplit graphs are a subclass of wheel-free graphs(W_n-free graphs, $n \geq 3$).*

Observation 4. *Bisplit graphs are K_4-free perfect graphs.*

Theorem 16. *If G is a bisplit graph, then $\overline{G}$ is a P_7-free chordality 4 graph.*

Proof. By the definition of the bisplit graph, we know that A, B, C are stable sets. Thus, in $\overline{G}$ we obtain three cliques, for a path of maximum length each clique can contribute at most two vertices. The maximum length possible in $\overline{G}$ is P_6. Hence $\overline{G}$ is a P_7-free graph.

Next we show that $\overline{G}$ is a chordality 4 graph. Suppose that there exists a cycle of length greater than or equal to 5 as an induced subgraph in $\overline{G}$, then at least three vertices are from C, which forms a chord in the cycle. Thus $\overline{G}$ has a cycle of length greater than or equal to five as an induced subgraph, which is a contradiction.

Therefore, $\overline{G}$ is a P_7-free chordality 4 graph.   □

Observation 5. *If G is a bisplit graph, then the diameter of $\overline{G}$ is at most 6.*

Definition 1. *A graph G is a strict star-convex bipartite graph if it is a star-convex bipartite graph and each vertex in Y is adjacent to the root of the imaginary star.*

Lemma 1. *Strict star-convex bipartite graphs are a subclass of bisplit graphs.*

Proof. By the structure of strict star-convex bipartite graphs, there exists a vertex say x which is adjacent to all the vertices of degree at least two in the partition of $V(G)$ not containing x. Let G be a strict star-convex bipartite graph with two partitions X, Y. Without loss of generality assume that x is in the partition X of G and all the vertices in Y are adjacent to x.

Now we show that any strict star-convex bipartite graph is a bisplit graph G', by showing three partitions A, B, C such that $A = \{u\}$, where u is adjacent all vertices in Y in G, $B = Y$, $C = X \setminus \{x\}$ and $E(G') = E(G)$.   □

Conclusion and Directions for Further Research:

Algorithmic and complexity aspects of STREE, were investigated on bisplit graphs.

- We have investigated the classical complexity of STREE on bisplit graphs, chordal bipartite bisplit graphs, and chordal bisplit graph.
- We have analyzed parameterized complexity of the Steiner tree problem on bisplit graphs with respect to the solution size as the parameter. Further, we also analyzed the parameterized complexity of the Steiner tree problem on bisplit graphs with respect to biclique as the parameter.
- One can investigate the complexity of other combinatorial problems on bisplit graphs.

References

1. Brandstädt, A., Hammer, P.L., Lozin, V.V., et al.: Bisplit graphs. Discrete Math. **299**(1–3), 11–32 (2005)

2. Abueida, A., Sritharan, R.: A note on the recognition of bisplit graphs. Discret. Math. **306**(17), 2108–2110 (2006)
3. White, K., Farber, M., Pulleyblank, W.: Steiner trees, connected domination and strongly chordal graphs. Networks **15**(1), 109–124 (1985)
4. Müller, H.: Hamiltonian circuits in chordal bipartite graphs. Discret. Math. **156**(1), 291–298 (1996)
5. Mark Keil, J., Pradhan, D.: Computing a minimum outer-connected dominating set for the class of chordal graphs. Inf. Process. Lett. **113**(14), 552–561 (2013)
6. Renjith, P., Sadagopan, N.: Hamiltonicity in split graphs - a dichotomy. In: Gaur, D., Narayanaswamy, N.S. (eds.) CALDAM 2017. LNCS, vol. 10156, pp. 320–331. Springer, Cham (2017). https://doi.org/10.1007/978-3-319-53007-9_28
7. Renjith, P., Sadagopan, N.: Hamiltonian path in split graphs-a dichotomy. arXiv preprint arXiv:1711.09262 (2017)
8. Padamutham, C., Palagiri, V.S.R.: Complexity of Roman 2-domination and the double Roman domination in graphs. AKCE Int. J. Graphs Comb. **17**(3), 1081–1086 (2020)
9. Guo, L., Lin, B.L.S.: Vulnerability of super connected split graphs and bisplit graphs. Discret. Contin. Dynam. Syst.-S **12**(4&5), 1179 (2019)
10. Rosenfeld, M., Kahn, G., Beaudou, L.: Bisplit graphs satisfy the Chen-Chvátal conjecture. Discret. Math. Theor. Comput. Sci. **21** (2019)
11. Corneil, D.G., Perl, Y.: Clustering and domination in perfect graphs. Discret. Appl. Math. **9**(1), 27–39 (1984)
12. D'Atri, A., Moscarini, M.: Distance-hereditary graphs, steiner trees, and connected domination. SIAM J. Comput. **17**(3), 521–538 (1988)
13. Aadhavan, S., Mahendra Kumar, R., Renjith, P., Sadagopan, N.: Hamiltonicity: Variants and Generalization in P_5-free Chordal Bipartite graphs. arXiv preprint arXiv:2107.04798 (2021)
14. Chen, H., Lei, Z., Liu, T., Tang, Z., Wang, C., Ke, X.: Complexity of domination, hamiltonicity and treewidth for tree convex bipartite graphs. J. Comb. Optim. **32**(1), 95–110 (2016)
15. Krithika, R., Mathew, R., Narayanaswamy, N.S., Sadagopan, N.: A Dirac-type characterization of k-chordal graphs. Discre. Math. **313**(24), 2865–2867 (2013)
16. Renjith, P., Sadagopan, N.: The Steiner tree in $K_{1,r}$-free split graphs-A Dichotomy. Discret. Appl. Math. **280**, 246–255 (2020)
17. Müller, H., Brandstädt, A.: The NP-completeness of Steiner tree and dominating set for chordal bipartite graphs. Theoret. Comput. Sci. **53**(2–3), 257–265 (1987)
18. Dhanalakshmi, S., Sadagopan, N.: On strictly chordality-k graphs. Discret. Appl. Math. **280**, 93–112 (2020)

Some Insights on Dynamic Maintenance of Gomory-Hu Tree in Cactus Graphs and General Graphs

Vineet Malik[1]($\boxtimes$) and Sushanta Karmakar[2]

[1] Department of Computer Science, Purdue University, West Lafayette, USA
malik83@purdue.edu
[2] Department of Computer Science and Engineering, Indian Institute of Technology, Guwahati, India
sushantak@iitg.ac.in

Abstract. For any flow network, min(s, t)-cut query is a fundamental graph query that asks for a minimum weight cut that separates vertices s and t. Gomory and Hu [13] proposed a data structure which is an undirected weighted tree that compactly stores min(s, t)-cut for all (s, t) pairs of an undirected weighted graph. Although there have been some research towards the problem of dynamically maintaining Gomory-Hu tree of a graph [4, 18], an efficient dynamic (incremental or decremental) algorithm for general graphs remains elusive. Also efficient dynamic algorithms for maintenance of Gomory-Hu tree for special graphs has not been investigated sufficiently. In this paper we propose algorithms for Gomory-Hu tree for a special class of graphs, known as *cactus graphs*. First we show that Gomory-Hu tree for a cactus graph can be constructed in linear time. Then we provide both incremental and decremental algorithms for maintaining a Gomory-Hu tree of a cactus graph. The algorithms use relations between blocks of a graph and its Gomory-Hu tree. For the incremental algorithm the amortized update time is $O(\log n)$ and for the decremental algorithm the worst-case update time is $O(\log n)$. For general graphs with integral weights, we present a data structure requiring $O(mn^2)$ space that helps us create a new Gomory-Hu tree if the weights of some edges of a given graph are changed by some integral amounts. Specifically, if the weights of k edges are changed by $w_1, w_2, ..., w_k$ units respectively, then a new Gomory-Hu tree of the modified graph can be reconstructed in $O((\sum_{i=1}^{k} w_i)mn)$ time.

Keywords: Min-cut · Gomory-Hu tree · Cactus graph · Dynamic graph

1 Introduction

In a flow network, computing the *min-cut* is a fundamental graph problem. In a weighted graph, the minimum weight cut that separates two vertices s and t

---

V. Malik—Partially supported by Samsung Research Fellowship.

A. Bagchi and R. Muthu (Eds.): CALDAM 2023, LNCS 13947, pp. 231–244, 2023.
https://doi.org/10.1007/978-3-031-25211-2_18

is denoted as min(s,t)-cut. The Max-flow Min-cut Theorem [10] states that for all (s,t) pair of vertices, the weight of a min(s,t)-cut is equal to the maximum flow from s to t in the flow network. This gave rise to many *min-cut* algorithms based on finding a *max-flow* in the flow network.

About half a century ago, Gomory and Hu [13] proposed a tree representation for all-pair *min-cut* of an undirected weighted graph. The tree, popularly known as the *Gomory-Hu tree*, is constructed over the set of the vertices of the given graph and has a strong property that for all (s,t) pair of vertices, any min(s,t)-cut in the tree, is also a *min(s,t)-cut* in the original graph. Hence the problem of finding a min(s,t)-cut in a undirected weighted graph can be reduced to finding a min(s,t)-cut in a undirected weighted tree. We know that any two vertices on a tree are connected by a unique path. Hence, an edge of minimum weight on the unique s to t path corresponds to the min(s,t)-cut in the tree and the weight of that edge is equal min(s,t)-cut value in the original graph. Note that removing the min(s,t)-cut edge in the Gomory-Hu tree will disconnect the tree into two components. The vertex sets of the two components is the cut bipartition. The cutset of this min(s,t)-cut set would be the set of edges that join vertices of the two vertex sets. This allows construction of a data structure that can pre-process over a Gomory-Hu tree and can answer min(s,t)-cut query for any (s,t) pair in amortized logarithmic time [2].

Gomory and Hu [13] also gave an algorithm to compute a Gomory-Hu tree of a graph by making $n-1$ different min(s,t)-cut computations on auxiliary graphs. Gusfield [15] modified this algorithm to make those $n-1$ different min(s,t)-cut computations on the original graph. The time complexity of this algorithm is $O((n-1) \times$ (time for a single min(s,t)-cut computation)) $+ O(n^2)$. Amir Abboud et al. [2] recently showed that there exists a data structure with near-linear pre-processing time that can answer min(s,t)-cut queries in amortized polylogarithmic time if and only if Gomory-Hu tree of a graph can be constructed in near-linear time. This essentially showed that Gomory-Hu trees are optimal for answering min-cut queries. Any solution to the min(s,t)-cut problem for all (s,t) pairs requires at least $n-1$ different min(s,t)-cut computations. The remarkable thing about Gomory-Hu tree is that it takes exactly $n-1$ machine words to encode at least one min(s,t)-cut, for each (s,t) pairs. Hence Gomory-Hu trees are also optimal in space.

1.1   Related Work and Motivation

The max-flow min-cut theorem [10] gives the fundamental relation between *flows* and *cuts* in a flow network. The best algorithms for *min(s,t)-cut* in general graphs follow from algorithms for *max-flow*. The Ford-Fulkerson algorithm [11], which is one of the earliest algorithm for $max - flow$, has a time complexity of $O(mC)$ (here C is the upper bound of the max-flow). The *Push-relabel* algorithm by Andrew V. Goldberg and Robert Tarjan [12] has a time complexity of $O(mn^2)$. This is better than the complexity of Edmonds-Karp algorithm [9] which has a time complexity of $O(nm^2)$. The current best algorithm for the max flow problem [7] is almost linear, computing max-flow in $m^{1+o(1)}$ time.

The current best known time complexity of computing the Gomory-Hu tree for a general weighted graph is $\tilde{O}(n^2)$ shown recently in [1]. Significant progress has also been made on special cases of the problem. When the largest edge weight U is small, offline algorithms [21,22] that run in time $\tilde{O}(min(m^{\frac{10}{7}}U^{\frac{1}{7}}, m^{\frac{11}{8}}U^{\frac{1}{4}}))$ are closer to the optimal. For planar graphs, Gomory-Hu tree can be constructed in near-linear time [6].

There has been some work on dynamically maintaining a global min-cut of a graph. A cut in a graph is called a global min-cut of the graph iff no other cut in the graph has value less than the value of that cut. When the global min-cut size is small, Thorup [24] demonstrated that the global min-cut of a graph can be exactly maintained in $\tilde{O}(\sqrt{n})$ worst-case update time and $(1 + \epsilon)$-approximately otherwise. A different analysis of the similar problem was done later [14]. Karger [19] proposed a fully dynamic randomised algorithm for maintaining minimum cut in $\tilde{O}(n^{\frac{1}{2}+\epsilon})$ expected amortized time per edge insertion with an approximation factor of $(1 + \frac{2}{\epsilon})^{\frac{1}{2}}$. In planar graphs with arbitrary edge weights, Lacki and Sankowski [20] proposed algorithms having sub-linear worst case update and query time for maintaining the exact minimum cut. Dinitz and Westbrook [8] gave an incremental algorithm for maintaining the classes of k-edge-connectivity (for arbitrary k) in a $(k - 1)$-edge-connected graph.

Relating to the problem of dynamic maintenance of all-pair $min(s,t)$-cut, Hartmann et al. [18] gave a fully dynamic algorithm for Gomory-Hu tree. Though the algorithm performs well on some real world data, in the worst case the algorithm is as bad as computing the Gomory-Hu tree from scratch. Barth et al. [4] showed how to compute the Gomory-Hu tree using two pre-computed Gomory-Hu trees in $O(n^2)$ time when weight of exactly one edge varies. If weights of k edges vary then their method requires 2^k pre-computed Gomory-Hu trees. Whether the problem can be solved using polynomial number of trees was left as an open question. Recently Baswana et al. [5] presented a compact data structure that can report those (s,t)-pairs of vertices for which the $min(s,t)$-cut changes due to insertion of an edge in optimal query time. The data structure works for a static graph. However if the graph changes, due to insertion of an edge, then the data structure is not dynamically updated making it incapable of answering subsequent queries. In recent years significant advances have been made in the design and analysis of dynamic algorithms for various graph problems. Dynamic algorithms have been proposed for shortest path, connectivity, reachability, matching, clustering and various other graph theoretic problems [17]. However, almost no dynamic algorithm is known for the Gomory-Hu tree construction problem. This paper is an attempt to fill that gap.

Our Contribution. In this work we first show that Gomory-Hu tree for a cactus graph can be constructed optimally in linear time. Then we provide both incremental and decremental algorithms for Gomory-Hu tree for cactus graphs. The algorithms use relations between blocks of a graph and its Gomory-Hu tree. For the incremental algorithm the amortized update time is $O(\log n)$ and for the decremental algorithm the worst-case update time is $O(\log n)$. For general graphs with integral weights, we present a data structure requiring $O(mn^2)$ space that helps us create a new Gomory-Hu tree if the weights of some edges of the given graph are changed by some integral amounts. Specifically, if the weights of k edges are changed by $w_1, w_2, ..., w_k$ units respectively, then a new Gomory-Hu tree of the modified graph can be constructed in $O((\sum_{i=1}^{k} w_i)mn)$ time.

Paper Organization. In Sect. 3, we introduce relation between block decomposition of a graph and its Gomory-Hu tree. Section-4 contains dynamic algorithms (both incremental and decremental) for maintaining Gomory-Hu tree of a cactus graph. Sect. 5 presents a sensitivity data structure for Gomory-Hu tree for general graphs with integral weights. Finally we conclude in Sect. 6.

2   Preliminary

All graphs considered in this work are edge-weighted undirected simple graphs. The weights associated with edges are positive real numbers. Throughout the paper, for a graph G, $V(G)$ denotes the set of vertices, $E(G)$ denotes the set of edges and $\forall e \in E(G), W_G(e)$ denotes the weight of edge e. When the graph G is clear from the context, we use V, E and $W(e)$ for the respective purposes.

Cut: A cut in a graph is a bi-partition of the set of vertices. Let S be a non-empty proper subset of V. The cut corresponding to the bi-partition of V into S and $V \backslash S$ is denoted by (S)-cut. Cutset of (S)-cut is the set of edges with one endpoint in S and the other endpoint in $V \backslash S$. Value of (S)-cut is defined as the sum of the weights of the edges in its cutset. Let $s, t \in V$, $s \neq t$. $\forall S \subset V$, (S)-cut is called a (s, t)-cut iff $(s \in S$ and $t \in V \backslash S)$ or $(t \in S$ and $s \in V \backslash S)$. A (s, t)-cut is called a $\min(s, t)$-cut iff the value of that (s, t)-cut is not greater than the value of any other (s, t)-cut.

Definition 1 (Gomory-Hu tree). *Let G be an undirected weighted graph. Then an undirected weighted tree T over the same set of vertices as G is called a Gomory-Hu tree of G iff: $\forall$ distinct pair of vertices $s, t \in V$ and $\forall$ subsets S of V, if (S)-cut is a $\min(s, t)$-cut in T, then (S)-cut is a $\min(s, t)$-cut in G and value of (S)-cut in T is equal to value of (S)-cut in G.*

Gomory-Hu tree of a connected graph may not be unique and in case of disconnected graphs, we have a forest of Gomory-Hu trees of its connected components.

3   Gomory-Hu Trees of Blocks of a Graph

In this section, we derive the relationship between Gomory-Hu tree of a graph and Gomory-Hu trees of blocks of the graph. We first present a result about Gomory-Hu tree of the union of two graphs having single common vertex, which is used to derive the main result of this section in Theorem-1.

Lemma 1. *Let G_1 and G_2 be two connected graphs having exactly one common vertex. Then the following hold:*

1. *If T_1, T_2 are Gomory-Hu trees of G_1, G_2 respectively, then $T_1 \cup T_2$ is a Gomory-Hu tree of $G_1 \cup G_2$.*
2. *If T is a Gomory-Hu tree of $G_1 \cup G_2$, then $\exists$ trees T_1, T_2 such that T_1, T_2 are Gomory-Hu trees of G_1, G_2 respectively and $T_1 \cup T_2 = T$.*

A decomposition of a connected graph is a set of subgraphs such that every edge of the graph belongs to exactly one subgraph in the set. The decomposition in which every subgraph in the collection is a block of the graph is called the block decomposition of the graph. We have followed standard definition for blocks, as provided in [25].

Theorem 1. *Let G be a connected graph and $H_1, H_2, ..., H_k$ be all the blocks of G. Then the following hold:*

1. *If $T_1, T_2, ..., T_k$ are Gomory-Hu trees of $H_1, H_2, ..., H_k$ respectively, then $\cup_{i=1}^k T_i$ is a Gomory-Hu tree of G.*
2. *If T is a Gomory-Hu tree of G, then $\exists$ trees $T_1, T_2, ..., T_k$ such that $\forall 1 \leq i \leq k, T_i$ is a Gomory-Hu tree of H_i and $\cup_{i=1}^k T_i = T$.*

The first part of Theorem-1 shows a way to compute Gomory-Hu tree of a graph using Gomory-Hu trees of blocks of the graph. The second part of the Theorem-1 gives a decomposition of Gomory-Hu tree of a graph into Gomory-Hu trees of blocks of the graph.

4   Gomory-Hu Tree of a Cactus Graph

In this section we combine results of the previous section with the properties of cactus graphs to obtain efficient algorithms for Gomory-Hu trees of cactus graphs. A connected graph G is a cactus graph iff every block of G is either an edge or a simple cycle. A disconnected graph is a cactus graph iff every connected component of it is a cactus graph. Equivalently, it can be defined as a graph in which no two cycles share an edge. Cactus graphs are sparse graphs, i.e., $|E(G)| = O(n)$, where n is the number of vertices in the graph.

Lemma 2. *Let G be a simple cycle. Let e_{min} be an edge of minimum weight in G. Let G' be a graph such that $V(G') = V(G)$ and $E(G') = E(G) - \{e_{min}\}$ with weight function as $\forall e \in E(G'), W_{G'}(e) = W_G(e) + W_G(e_{min})$. Then G' is a Gomory-Hu tree of G.*

The converse of Lemma-2 is not true in general, i.e., not every Gomory-Hu tree of a simple cycle is of the form shown in the lemma. However, we show that if a simple cycle has a unique edge of minimum weight, then it has a unique Gomory-Hu tree in Lemma-3.

Lemma 3. *Let G be a connected cactus graph. G has a unique Gomory-Hu tree iff every cycle of G has a unique edge of minimum weight.*

Lemma-3 gives a necessary and sufficient condition for the uniqueness of Gomory-Hu tree for cactus graphs. Many recent papers ([3,6,23]) use a sufficient condition for uniqueness of Gomory-Hu tree, however to the best of our knowledge no necessary and sufficient condition is known for general graphs. The following theorem is stated without proof due to space limitations:

Theorem 2. *Let G be a cactus graph. A Gomory-Hu tree of G can be constructed in linear time.*

4.1   Dynamically Maintaining a Gomory-Hu Tree

Gomory-Hu Tree Representation: The dynamic algorithms presented here maintain the *adjacency list array* of a Gomory-Hu tree of a cactus graph. In this *adjacency list array*, for all vertices, their adjacency list are stored using *height-balanced* BST (Binary Search tree), which allows insertion, look-up, and deletion of edges in $O(\log n)$ time. So the entry indexed to each vertex in the *adjacency list array* stores pointer to the root node of the BST storing its adjacency list.

The Gomory-Hu tree of the cactus graph maintained by our dynamic algorithm is of the form described in Lemma-2, i.e., the Gomory-Hu tree is a subgraph of the cactus graph. The weight of any edge in this Gomory-Hu tree corresponding to a cycle of the cactus graph is equal to the sum of the weight of that edge in the cactus graph and the weight of the minimum weight edge of the cycle.

Finding the weight of an edge of the Gomory-Hu tree: For any graph structure, one of the fundamental queries is to find the weight of any edge. The naive way is to store the weight of each edge as an additional data in the nodes of the BSTs in the *adjacency list array*, allowing finding the weight of any edge in $O(log(n))$ time. We use a different approach for storing the weights, requiring only $O(n)$ more space. This approach allows us to increase or decrease the weights of all the edges of the Gomory-Hu tree corresponding to the same cycle in the cactus graph in total $O(log(n))$ time.

Our Approach for Storing the Weights: We store two values in every node of the BST, one being the weight of the corresponding edge in the cactus graph and second being the cycle number the edge corresponds to, i.e., we assign each cycle in the graph a distinct number from 1 to k where the total number of cycles in the graph is k. The edges that do not correspond to any cycle can be assigned a cycle number of 0. We also maintain an additional array referred as the *cycle array* and the value stored at index i of this array is the weight of minimum weight edge of cycle i.

Space Complexity: Since no two cycles of a cactus graph share an edge, it implies that any two cycles share at most 1 vertex. Hence the maximum number of cycles in a cactus graph is bounded by n. Hence the size of the *cycle array* is $O(n)$. Notice that finding all the cycles of a cactus graph and assigning each edge their cycle number can be done in $O(n)$ time. To get the weight of an edge in the Gomory-hu tree, we simply add the weight of that edge in the graph which is stored in the node and the value corresponding to the index in the *cycle array*.

The query time of both incremental and decremental algorithm is $O(1)$ which is taken as the time to return the pointer to these array structures. In the rest of the analysis we discuss the update time, i.e. time taken to update these arrays.

4.2   Incremental Algorithm for Gomory-Hu Tree

Incremental algorithms deal with updates which increase the size of the graph by insertion of new edges. This can be divided into two cases:

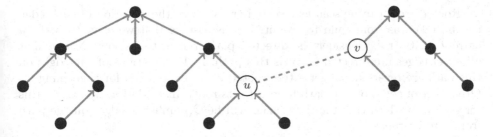

Fig. 1. Neither of u or v is a root vertex in their tree.

- **Case-1**: Inserted edge connects vertices of *different* connected components.
- **Case-2**: Inserted edge connects vertices of the *same* connected component.

Maximal connected subgraphs of a graph are called its connected componenets. Identifying which of the above two case occurs can be done using *Union-Find* data structure. It takes amortized $O(\alpha(n))$ time, where $\alpha(n)$ is the extremely slow-growing inverse Ackermann function. For any practical purposes, it can be considered as amortized constant time.

Our incremental algorithm maintains a rooted tree structure of the Gomory-hu tree, using an array *Par* of size n where $Par[i]$ = parent of vertex i for all i except the root vertex and $Par[j] = 0$ where j is a root vertex. In case of a forest of Gomory-Hu trees, there will be multiple root vertices, one for each tree. The incremental algorithm uses this representation to compute the unique path between any two vertices of the same connected component in $O(l)$ time, where l denotes the length of the path between those two vertices. This rooted tree is only used in our incremental algorithm and is not maintained in the decremental algorithm (Sect.-4.3).

Case-1: Inserted edge connects vertices of *different* connected components.

Since the newly inserted edge connects two connected components of the graph, it will form a block containing this single edge of the modified graph. Hence, the Gomory-Hu forest of the modified graph can be obtained by joining the Gomory-Hu trees of the two connected components by this edge. Inserting an edge in the adjacency array takes $O(log(n))$ time. However, since we maintain a rooted tree representation for internal computations, we also require to update it.

Let (u, v) be the inserted edge. The following sub-cases are possible:

- If u is the root in the Gomory-Hu tree of its component, then changing $Par[u]$ from 0 to v gives the new Gomory-Hu tree.
- If v is the root in the Gomory-Hu tree of its component, then changing $Par[v]$ from 0 to u gives the new Gomory-Hu tree.
- When neither of u and v are roots in their trees, as shown in Fig.-1, the above changes will not lead to a rooted tree.

We handle this by first making one of u or v as the roots of their Gomory-Hu tree, then it can be handled easily as stated below.

Rooted trees can be seen as directed trees where the direction of each edge can be taken as from child to parent. Let us assume that we want to make u as root of its tree. Consider the directed path from u to its root. Notice that all other edges are directed towards this path. If the directions of the edges on this path are reversed, we get a tree rooted at u. Time taken for this change is O(number of edges on the path from u to its root). Instead of u, we can do this for v also, in which case the time taken will be O(number of edges on the path from v to its root).

We traverse the directed paths leaving from u and v, simultaneously. The path on which we first reach at a root vertex first is the path with lesser number of edges and we will reverse the direction of that path.

Once the path is chosen, reversing the direction of edges on that path can be done in O(path length). Hence the time complexity is O(min(path length from u to its root, path length from v to its root). This is of the form: $T(a + b) = T(a) + T(b) + c \times min(a, b)$, where a and b are the number of vertices in the two trees and c is some constant. Then it can be shown that $T(r)$ is $O(r\log(r))$. Hence, the amortized time complexity to handle Case-(i) is $O(\log(n))$.

Case-2: Inserted edge connects vertices of the same connected component.

Let G be the graph and T its Gomory-Hu tree and (u, v) be the edge to be inserted. Let the modified graph be G'. Since u and v belong to the same connected component, there is a path from u to v in G. Hence this newly inserted edge forms a cycle in the connected component. The path from u to v in G must be unique, otherwise insertion of edge (u, v) will violate the property of cactus graph that no two cycles share an edge. Hence all the edges on the path from u to v in G form blocks of single edge. Hence the Gomory-Hu tree T contains those edges. Hence T contains that path.

Finding path from u to v in T: Since the tree T is stored as a rooted tree, to obtain the path between u, v we can simply traverse in parallel from u and v along the parent edges towards the root. The first common vertex on the path from u to root and the path from v to root is their LCA. The path between u, v is given by the joining the path of u to LCA and v to LCA. Let l denote the length of this u, v path. So time complexity of this process is $O(l)$.

Let e be an edge of minimum weight on the path from u to v. We can find e by traversing over the path. This is shown in Fig.-2. We compare the weight of edge e with the weight of the inserted edge (u, v). The following cases are possible:

Subcase-2.1: If $w_G(e) \geq w_{G'}((u, v))$, then (u, v) is an edge of minimum weight in its cycle in G'. Hence, the path from u to v in G with the weights of the edges on the path increased by $w_{G'}((u, v))$, is a Gomory-Hu tree of the newly formed cycle. Hence increasing the weights of the edges of the path from u to v in T by $w_{G'}((u, v))$ gives a Gomory-Hu tree of G'. This can be done in $O(l)$.

Subcase-2.2: Here $w_G(e) < w_{G'}((u, v))$. Let C be the cycle containing edge (u, v) in G'. The path obtained by removing edge e from C and the weight of every edge on the path increased by $w_G(e)$ is a Gomory-Hu tree of C. To get this, we need to remove edge e from T and insert edge (u, v), with the weight of the edges modified as required. Edge e lies either on the path u to LCA or on the

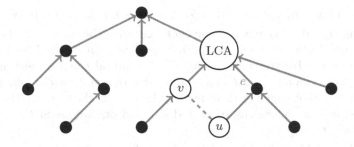

Fig. 2. e is the edge of minimum weight on the path from u to v.

path from v to LCA in T. We traverse both of them and check on which of the two it lies on. If it lies on the path from u to LCA then we remove e by assigning $Par[\text{child in e}] = 0$. This creates a new rooted tree containing u. Note that We represent the Gomory-Hu tree as a rooted tree. Therefore we make the new rooted tree containing u, rooted at u before inserting the edge (u, v). This is done by reversing the direction of the edges from the breaking point to u. This will take $O(l)$. Now, we set $Par[u] = v$ to insert the edge (u, v). Finally, we increase weights of the edges by $w_G(e)$. This gives the modified Gomory-Hu tree in $O(l)$.

Time Complexity:

- Case-1: Update time is amortized $O(\log n)$.
- Case-2: Let (u, v) be the edge to be inserted. Let ℓ be the number of edges on the path from u to v in G. Then, update time is $O(\ell \, log(n))$. The ℓ edges on the path from u to v in G, were not part of any cycle. Hence, insertion of those ℓ edges came under Case-1. As cycles of a cactus graph are edge disjoint, the amortized analysis gives the update time as amortized $O(log(n))$.

4.3 Decremental Algorithm for Gomory-Hu Tree

Decremental algorithm deal with updates that decrease the size of the graph by deletion of an edge from the graph. This can be divided into two cases:

1. Deleted edge belongs to some cycle.
2. Deleted edge doesn't belong to any cycle.

Pre-processing: We do a pre-processing to obtain the list of all the cycles of the graph. Corresponding to each edge e we store a number $CycleNumber[e]$ which is the index of the cycle where the edge e belongs to. We maintain an array indexed with these cycle numbers. The value stored at index i of this array is the weight of the minimum weight edge of the i th cycle in the graph. Also, while computing the initial Gomory-Hu tree (can be done in linear time), we obtain an array called $MinEdge$ such that $MinEdge[i] = (r, p)$ means that edge (r, p) was removed while obtaining a Gomory-Hu tree of the ith cycle.

The steps taken by the algorithm for the cases are described below:

Case-1: Deleted edge belongs to some cycle. Let (u, v) be the edge to be deleted and $i = CycleNumber[(u, v)]$ and $MinEdge[i] = (r, p)$. Let G be the graph and T its Gomory-Hu tree and the modified graph after edge deletion be G'. Hence the path from r to p in T is a Gomory-Hu tree of ith cycle. The modified Gomory-Hu tree is obtained by deleting edge (u, v) from the Gomory-Hu tree and inserting edge (r, p). These insertion and deletion operations in the adjacency array will take $O(log(n))$.

Updating the weights of the edges: We only need to set the value indexed at i in the cycle weight array to 0. As the edges corresponding to cycle i use this value when we query their weight, setting it to 0 reduces the weights of all those edges simultaneously in $O(1)$ time.

Case-2: Deleted edge does not belong to any cycle. Let (u, v) be the edge to be deleted. As (u, v) does not belong to any cycle, deleting this edge disconnects the graph. Hence the modified Gomory-Hu tree can be obtained by deleting this edge from the Gomory-Hu tree. Hence, it can be done in $O(log(n))$ time.

Hence, the time complexity of the update operation in both cases is $O(log(n))$.

5   Gomory-Hu Tree Sensitivity Data Structure

In this section we provide a sensitivity data structure for Gomory-Hu tree construction for graphs with integer weights. Theorem-3 presents this key result.

Theorem 3. *For any integer weighted undirected graph G, there exists a data structure of size $O(n^2 m_G)$ that can report a Gomory-Hu tree of any other integer weighted undirected graph G' having the same set of vertices as G, in time $O(n \times max(m_G, m_{G'}) \times WeightedHammingDistance(G, G'))$.*

Here, $m_G = |E(G)|$ and $m_{G'} = |E(G')|$. Weighted hamming distance between G, G' is defined as the L_1 distance of their weighted adjacency matrices. Since G' is any graph over the same set of vertices as G, G' can be obtained from G by some sequence of the following 4 elementary graph edit operations: (i) insert a new edge e of weight 1 (ii) delete an existing edge e of weight 1 (iii) increase the weight of an existing edge e by 1 unit (iv) decrease the weight of an existing edge e of weight w by 1 unit (here $w > 1$).

The minimum number of the above elementary graph edit operations required to modify G to G' is same as the $WeightedHammingDistance(G, G')$. If we consider only the operations of increasing/decreasing the weights of existing edges, then we get Corollary-1. The proof of these results come from the analysis of the below proposed data structure which we have omitted here due to space limitations.

Corollary 1. *For any integer weighted undirected graph G, there exists a data structure of size $O(n^2 m)$ that can report a new Gomory-Hu tree if weights of some k number of edges of G is changed by $w_1, w_2, ..., w_k$ integral amounts respectively, in time $O((\sum_{i=1}^{k} w_i)mn)$.*

Algorithm 1: Computing a Gomory-Hu tree of G'

Input: G' (an integer weighted undirected graph with $V(G') = V(G)$)

Output: T (a Gomory-Hu tree of G')

1 Compute a minimum length sequence of *elementary graph edit operations* $(r_1, r_2, ..., r_k)$ that when applied on G modifies it to G'.

2 Run Gusfield's Algorithm [15] to compute a Gomory-Hu tree T of G', with Algorithm-2 being used as calls to min-cut oracle.

Proposed Data Structure: For all pairs of distinct vertices (s, t) in the graph G, the data structure stores an integer-max(s, t)-flow. Hence, the space complexity of the data structure is $O(n^2 \times m)$ and construction time is $O(n^2 \times$ (Time to compute integer-max(s, t)-flow)).

Query to the data structure: The query comes with an integer weighted graph G' and the output is a Gomory-Hu tree of G', which is computed as shown in Algorithm-1 in time $O(n \times max(m_G, m_{G'}) \times WeightedHammingDistance (G, G'))$.

The Algorithm-1 uses Gusfield's Algorithm [15] for computing Gomory-Hu tree. The Gusfield's algorithm shows that a Gomory-Hu tree of a graph can be computed with $n - 1$ calls to an oracle for the graph, that when given 2 distinct vertices $s, t \in V$ returns a min(s, t)-cut of the graph. Along with the $n - 1$ calls to a min-cut oracle, the Gusfield's algorithm takes additional computation time of $O(n^2)$, which leads to the total time complexity of computing a Gomory-Hu tree being $O((n - 1) \times$ (Time to compute min(s, t)-cut)) $+ O(n^2)$.

We use our data structure as the min-cut oracle. Algorithm-2 shows the steps to compute a min(s, t)-cut of G' using our data structure. Hence the time complexity to return the output for any query, i.e. the time complexity of Algorithm-1 is $O((n - 1) \times$ (Time taken by Algorithm-2) $+ O(n^2))$.

The best-known static algorithm (until 2022) for computing a Gomory-Hu tree of a general weighted graph was the Gusfield's algorithm where instead of calls to some oracle, it used the current best known static algorithm for computing a min(s, t)-cut of a graph [7], that has a time complexity of $O(m^{1+o(1)})$. Hence the time complexity for computing a Gomory-Hu tree becomes $O(nm^{1+o(1)})$. However the recent result by Abboud et al. [1] provides a Monte Carlo randomized algorithm for computing a Gomory-Hu tree having a time complexity of $\tilde{O}(n^2)$. Though this result outperforms our scheme for dense graphs, however for sparse graphs our scheme performs better when the number of edge modifications is bounded by a constant. Moreover our data structure is deterministic and has a much simpler implementation.

Algorithm 2: Computing a min(s,t)-cut of G'

Input: two distinct vertices $s, t \in V$ and a sequence of elementary graph edit
operations $(r_1, r_2, ..., r_k)$ that modify G to G'
Output: min(s,t)-cut of G'

1 Let $H \leftarrow G$.
2 Let $f \leftarrow$ max(s,t)-flow of G, stored in our data structure.
3 **for** *graph edit operation r_i, i from 1 to k* **do**
4   Let $H' \leftarrow$ graph obtained if H is modified by applying r_i.
5   Compute f', a max(s,t)-flow of H'.
6     $H \leftarrow H'$ and $f \leftarrow f'$. // Update H and f for next iteration.
7 **end**
8 Compute a min(s,t)-cut of G' using f.

---

6   Conclusion and Future Works

In this work we presented both incremental and decremental algorithms for
Gomory-Hu trees of cactus graphs were also given. For general graphs with
integral weights, we present a data structure requiring $O(mn^2)$ space that helps
us create a new Gomory-Hu tree if the weights of some edges of the given graph
are changed by some integral amounts. Specifically, if the weights of k edges are
changed by $w_1, w_2, ..., w_k$ units respectively, then a new Gomory-Hu tree of the
modified graph can be reconstructed in $O((\sum_{i=1}^{k} w_i)mn)$ time.

There are two key directions in which our work can be extended:

- Cactus graphs are special cases of outerplanar graphs. In a recent paper [23],
it was shown that an outerplanar graph with no cut vertices has a Gomory-Hu
tree as its subgraph. Efficient algorithms may be obtained for such families
of graphs using similar ideas.
- Block of a graph is a maximal subgraph such that if a vertex is removed
from it, it still remains connected. As a graph can be broken into blocks, a
block can be decomposed into triconnected components [16]. A triconnected
component is a maximal subgraph such that if two vertices are removed from
it, it still remains connected. This hints the possibility for some generalization
of Theorem-1 for such smaller decompositions.

References

1. Abboud, A., Krauthgamer, R., Li, J., Panigrahi, D., Saranurak, T., Trabelsi,
O.: Breaking the cubic barrier for all-pairs max-flow: Gomory-hu tree in nearly
quadratic time. In: FOCS (2022)
2. Abboud, A., Krauthgamer, R., Trabelsi, O.: Cut-equivalent trees are optimal for
min-cut queries. In: 61st IEEE Annual Symposium on Foundations of Computer
Science, FOCS 2020, Durham, NC, USA, 16–19 Nov 2020, pp. 105–118. IEEE
(2020)

3. Abboud, A., Krauthgamer, R., Trabelsi, O.: New algorithms and lower bounds for all-pairs max-flow in undirected graphs. In: Proceedings of the Fourteenth Annual ACM-SIAM Symposium on Discrete Algorithms, pp. 48–61. SIAM (2020)
4. Barth, D., Berthomé, P., Diallo, M., Ferreira, A.: Revisiting parametric multi-terminal problems: Maximum flows, minimum cuts and cut-tree computations. Discret. Optim. **3**(3), 195–205 (2006)
5. Baswana, S., Gupta, S., Knollmann, T.: Mincut sensitivity data structures for the insertion of an edge. In: 28th Annual European Symposium on Algorithms (ESA 2020). Schloss Dagstuhl-Leibniz-Zentrum für Informatik (2020)
6. Borradaile, G., Sankowski, P., Wulff-Nilsen, C.: Min st-cut oracle for planar graphs with near-linear preprocessing time. ACM Trans. Algor. (TALG) **11**(3), 1–29 (2015)
7. Chen, L., Kyng, R., Liu, Y.P., Peng, R., Gutenberg, M.P., Sachdeva, S.: Maximum flow and minimum-cost flow in almost-linear time. arXiv preprint arXiv:2203.00671 (2022)
8. Dinitz, Y., Westbrook, J.: Maintaining the classes of 4-edge-connectivity in a graph on-line. Algorithmica **20**(3), 242–276 (1998)
9. Edmonds, J., Karp, R.M.: Theoretical improvements in algorithmic efficiency for network flow problems. J. ACM (JACM) **19**(2), 248–264 (1972)
10. Elias, P., Feinstein, A., Shannon, C.: A note on the maximum flow through a network. IRE Trans. Inf. Theory **2**(4), 117–119 (1956)
11. Ford, L.R., Fulkerson, D.R.: Maximal flow through a network. Can. J. Math. **8**, 399–404 (1956)
12. Goldberg, A.V., Tarjan, R.E.: A new approach to the maximum-flow problem. J. ACM (JACM) **35**(4), 921–940 (1988)
13. Gomory, R.E., Hu, T.C.: Multi-terminal network flows. J. Soc. Ind. Appl. Math. **9**(4), 551–570 (1961)
14. Goranci, G., Henzinger, M., Thorup, M.: Incremental exact min-cut in polyloga-rithmic amortized update time. ACM Trans. Algorithms **14**(2), 17:1–17:21 (2018)
15. Gusfield, D.: Very simple methods for all pairs network flow analysis. SIAM J. Comput. **19**(1), 143–155 (1990)
16. Gutwenger, Carsten, Mutzel, Petra: A linear time implementation of SPQR-trees. In: Marks, Joe (ed.) GD 2000. LNCS, vol. 1984, pp. 77–90. Springer, Heidelberg (2001). https://doi.org/10.1007/3-540-44541-2_8
17. Hanauer, K., Henzinger, M., Schulz, C.: Recent advances in fully dynamic graph algorithms (invited talk). In: 1st Symposium on Algorithmic Foundations of Dynamic Networks, SAND 2022, 28–30 March 2022, pp. 1–47 (2022)
18. Hartmann, Tanja, Wagner, Dorothea: Fast and simple fully-dynamic cut tree con-struction. In: Chao, Kun-Mao., Hsu, Tsan-sheng, Lee, Der-Tsai. (eds.) ISAAC 2012. LNCS, vol. 7676, pp. 95–105. Springer, Heidelberg (2012). https://doi.org/10.1007/978-3-642-35261-4_13
19. Karger, D.R.: Using randomized sparsification to approximate minimum cuts. In: SODA, vol. 94, pp. 424–432 (1994)
20. Łącki, Jakub, Sankowski, Piotr: Min-cuts and shortest cycles in planar graphs in $O(n \log\log n)$ time. In: Demetrescu, Camil, Halldórsson, Magnús M.. (eds.) ESA 2011. LNCS, vol. 6942, pp. 155–166. Springer, Heidelberg (2011). https://doi.org/10.1007/978-3-642-23719-5_14
21. Liu, Y.P., Sidford, A.: Faster energy maximization for faster maximum flow. In: Proceedings of the 52nd Annual ACM SIGACT Symposium on Theory of Com-puting, pp. 803–814 (2020)

22. Madry, A.: Computing maximum flow with augmenting electrical flows. In: 2016 IEEE 57th Annual Symposium on Foundations of Computer Science (FOCS), pp. 593–602. IEEE (2016)
23. Naves, G., Shepherd, B.: When do gomory-hu subtrees exist? SIAM J. Discret. Math. **36**(3), 1567–1585 (2022)
24. Thorup, M.: Fully-dynamic min-cut. Combinatorica **27**(1), 91–127 (2007)
25. West, D.B., et al.: Introduction to graph theory, vol. 2. Prentice hall Upper Saddle River (2001)

Monitoring Edge-Geodetic Sets in Graphs

Florent Foucaud[1]([✉]) [iD], Krishna Narayanan[2][iD],
and Lekshmi Ramasubramony Sulochana[2][iD]

[1] Université Clermont-Auvergne, CNRS, Mines de Saint-Étienne,
Clermont-Auvergne-INP, LIMOS, 63000 Clermont-Ferrand, France
florent.foucaud@uca.fr
[2] PSG College of Technology, Coimbatore, India
rs.amcs@psgtech.ac.in

Abstract. We introduce a new graph-theoretic concept in the area of network monitoring. In this area, one wishes to monitor the vertices and/or the edges of a network (viewed as a graph) in order to detect and prevent failures. Inspired by two notions studied in the literature (edge-geodetic sets and distance-edge-monitoring sets), we define the notion of a monitoring edge-geodetic set (MEG-set for short) of a graph G as an edge-geodetic set $S \subseteq V(G)$ of G (that is, every edge of G lies on some shortest path between two vertices of S) with the additional property that for every edge e of G, there is a vertex pair x, y of S such that e lies on *all* shortest paths between x and y. The motivation is that, if some edge e is removed from the network (for example if it ceases to function), the monitoring probes x and y will detect the failure since the distance between them will increase.

We explore the notion of MEG-sets by deriving the minimum size of a MEG-set for some basic graph classes (trees, cycles, unicyclic graphs, complete graphs, grids, hypercubes, ...) and we prove an upper bound using the feedback edge set of the graph.

1 Introduction

We introduce a new graph-theoretic concept, that is motivated by the problem of network monitoring, called *monitoring edge-geodetic sets*. In the area of network monitoring, one wishes to detect or repair faults in a network; in many applications, the monitoring process involves distance probes [1–3,8]. Our networks are modeled by finite, undirected simple connected graphs, whose vertices represent systems and whose edges represent the connections between them. We wish to monitor a network such that when a connection (an edge) fails, we can detect the said failure by means of certain probes. To do this, we select a small subset of vertices (representing the probes) of the network such that all connections are covered by the shortest paths between pairs of vertices in the network. Moreover, any two probes are able to detect the current distance that separates them. The

Research financed by the French government IDEX-ISITE initiative 16-IDEX-0001 (CAP 20-25) and by the ANR project GRALMECO (ANR-21-CE48-0004).

goal is that, when an edge of the network fails, some pair of probes detects a change in their distance value, and therefore the failure can be detected. Our inspiration comes from two areas: the concept of *geodetic sets* in graphs and its variants [9], and the concept of *distance edge-monitoring sets* [7,8].

We now proceed with some necessary definitions. A *geodesic* is a shortest path between two vertices u, v of a graph G [14]. The length of a geodesic between two vertices u, v in G is the *distance* $d_G(u, v)$ between them. For an edge e of G, we denote by $G - e$ the graph obtained by deleting e from G. An edge e in a graph G is a *bridge* if $G - e$ has more connected components than G. A vertex of a graph is said to be a *leaf* if its neighborhood contains exactly one vertex. The *open neighborhood* of a vertex $v \in V(G)$ is $N_G(v) = \{u \in V \mid uv \in E(G)\}$ and its *closed neighborhood* is the set $N_G[v] = N_G(v) \cup \{v\}$.

Monitoring Edge-Geodetic Sets. We now formally define of our main concept.

Definition 1. *Two vertices x, y monitor an edge e in graph G if e belongs to all shortest paths between x and y. A set S of vertices of G is called a* monitoring edge-geodetic set *of G (MEG-set for short) if, for every edge e of G, there is a pair x, y of vertices of S that monitors e.*

We denote by $\text{meg}(G)$ the size of a smallest MEG-set of G. We note that $V(G)$ is always an MEG-set of G, thus $\text{meg}(G)$ is always well-defined.

Related Notions. A set S of vertices of a graph G is a *geodetic set* if every vertex of G lies on some shortest path between two vertices of S [9]. An *edge-geodetic set* of G is a set $S \subseteq V(G)$ such that every edge of G is contained in a geodesic joining some pair of vertices in S [13]. A *strong edge-geodetic set* of G is a set S of vertices of G such that for each pair u, v of vertices of S, one can select a shortest $u - v$ path, in a way that the union of all these $\binom{|S|}{2}$ paths contains $E(G)$ [12]. It follows from these definitions that any strong edge-geodetic set is an edge-geodetic set, and any edge-geodetic set is a geodetic set (if the graph has no isolated vertices). In fact, every MEG-set is a strong edge-geodetic set. Indeed, given an MEG-set S, one can choose any shortest path between each pair of vertices of S, and the set of these paths covers $E(G)$. Indeed, every edge of G is contained in *all* shortest paths between some pair of S. Hence, MEG-sets can be seen as an especially strong form of strong edge-geodetic sets.

A set S of vertices of a graph G is a *distance-edge monitoring* set if, for every edge e, there is a vertex x of S and a vertex y of G such that e lies on all shortest paths between x and y [7,8]. Thus, it follows immediately that any MEG-set of a graph G is also a distance-edge monitoring set of G.

Our Results. We start by presenting some basic lemmas about the concept of MEG-sets in Sect. 2, that are helpful for understanding this concept. We then study in Sect. 3 the optimal value of $\text{meg}(G)$ when G is a tree, cycle, unicyclic graph, complete (multipartite) graph, hypercubes and grids. In Sect. 4, we show that $\text{meg}(G)$ is bounded above by a linear function of the *feedback edge set number* of G (the smallest number of edges of G needed to cover all cycles of G, also called *cyclomatic number*) and the number of leaves of G. This implies that meg(G) is bounded above by a function of the *max leaf number* of G (the maximum number of leaves in a spanning tree of G). These two parameters are popular in structural graph theory and in the design of algorithms. We refer to Fig. 1 for the relations between parameter meg and other graph parameters. Finally, we conclude in Sect. 5.

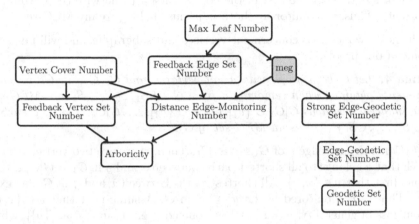

Fig. 1. Relations between the parameter meg and other structural parameters in graphs (with no isolated vertices). For the relationships of distance edge-monitoring sets, see [7, 8]. Arcs between parameters indicate that the value of the bottom parameter is upper-bounded by a function of the top parameter.

2   Preliminary Lemmas

We now give some useful lemmas about the basic properties of MEG-sets.

A vertex is *simplicial* if its neighborhood forms a clique. In particular, a leaf is simplicial.

Lemma 2. *In a graph G with at least one edge, any simplicial vertex belongs to any edge-geodetic set and thus, to any MEG-set of G.*

Proof. Let us consider by contradiction an MEG-set of G that does not contain said simplicial vertex v. Any shortest path passing through its neighbors will not pass through v, because all the neighbors are adjacent, hence leaving the edges incident to v uncovered, a contradiction.   □

Two distinct vertices u and v of a graph G are *open twins* if $N(u) = N(v)$ and *closed twins* if $N[u] = N[v]$. Further, u and v are *twins* in G if they are open twins or closed twins in G.

Lemma 3. *If two vertices are twins of degree at least 1 in a graph G, then they must belong to any MEG-set of G.*

Proof. For any pair u, v of open twins in G, for any shortest path passing through u, there is another one passing through v. Thus, if u, v were not part of the MEG-set, then the edges incident to u and v would remain unmonitored, a contradiction.

If u, v are closed twins, if some shortest path contains the edge uv, then it must be of length 1 and consist of the edge uv itself (otherwise there would be a shortcut). Thus, to monitor uv, both u, v must belong to any MEG-set.   □

The next two lemmas concern cut-vertices and subgraphs, and will be useful in some of our proofs.

Lemma 4. *Let G be a graph with a cut-vertex v and $C_1, C_2, \ldots, C_k$ be the k components obtained when removing v from G. If $S_1, S_2, \ldots, S_k$ are MEG-sets of the induced subgraphs $G[C_1 \cup \{v\}], G[C_2 \cup \{v\}], \ldots, G[C_k \cup \{v\}]$, then $S = (S_1 \cup S_2, \ldots, \cup S_k) \setminus \{v\}$ is an MEG-set of G.*

Proof. Consider any edge e of G, say in C_1. Then, there are two vertices x, y of S_1 such that e belongs to all shortest paths between x and y in $G_1 = G[C_1 \cup \{v\}]$. Assume first that $v \notin \{x, y\}$. All shortest paths between x and y in G also exist in G_1. Thus, e is monitored by $\{x, y\} \subseteq S$ in G. Assume next that $v \in \{x, y\}$: without loss of generality, $v = x$. At least one edge exists in $G[C_2 \cup \{v\}]$, which implies that $S_2 \setminus \{v\}$ is nonempty, say, it contains z. Then, e is monitored by y and z, since $z \in S$. Thus, S monitors all edges of G, as claimed.   □

3   Basic Graph Classes and Bounds

In this section, we study MEG-sets for some standard graph classes.

3.1   Trees

Theorem 5. *For any tree T with at least one edge, the only optimal MEG-set of T consists of the set of leaves of T.*

Proof. The fact that all leaves must be part of any MEG-set follows from Lemma 2, as they are simplicial. For the other side, let L be the set of leaves of T. Let $e = xy$ be an edge of T and consider two leaves of T, l_x and l_y, such that l_x is closer to x than to y and that l_y is closer to y than to x. We note that e belongs to the unique (shortest) path between l_x and l_y, thus e is monitored by L. Hence, L is an MEG-set of T.   □

Corollary 6. *For any path graph P_n, where $n \geqslant 2$, we have $\mathrm{meg}(P_n) = 2$.*

This provides a lower bound which is tight for path graphs, which have order n and exactly 2 leaves.

Corollary 7. *For any tree T of order $n \geqslant 3$, we have $2 \leqslant \mathrm{meg}(T) \leqslant n - 1$.*

The upper bound is tight for star graphs, which have order n and $n-1$ leaves.

3.2   Cycle Graphs

Theorem 8. *Given an n-cycle graph C_n, for $n = 3$ and $n \geqslant 5$, meg(C_n) = 3. Moreover, meg(C_4) = 4.*

Proof. Let us first prove that we need at least three vertices to monitor any cycle. By contradiction, let us assume that two vertices suffice. For any arbitrary vertex pair in the cycle graph, there are two paths joining them, but there is either one single shortest path or two equidistant shortest paths between them. Thus, the edges on at least one of the two paths between the pair will not be monitored by it. Hence, we need at least three vertices in any MEG-set of C_n ($n \geqslant 3$).

We now prove the upper bound. Let $n \geqslant 5$ or $n = 3$, with the vertices of C_n from v_0 to v_{n-1}. Consider the set $S = \{v_0, v_{\lfloor \frac{n}{3} \rfloor}, v_{\lfloor \frac{2n}{3} \rfloor}\}$. We show that S is an MEG-set of C_n.

Consider every edge of C_n between a vertex pair v_x and v_y in S, then we note that they lie on every (unique) shortest path between these vertices, which has a length at least one for $n \leqslant 5$ and at least 2 otherwise, and at most $\lceil \frac{n}{3} \rceil$. Thus, meg($C_n$) = 3 when $n \geqslant 5$ or $n = 3$.

In the case of C_4, the above construction does not work. Consider a set of three vertices, say v_0, v_1, v_2 without loss of generality due to the symmetries of C_4. Notice that the edge v_0v_3 is unmonitored by this set. Thus, we have mcg(C_4) = 4. □

3.3   Unicyclic Graphs

A *unicyclic graph* is a connected graph containing exactly one cycle [10]. We now determine the optimal size of an MEG-set of such graphs.

Theorem 9. *Let G be a unicyclic graph where the only cycle C has length k and whose set of leaves is $L(G)$, $|L(G)| = l$. Let V_c^+ be the set of vertices of C with degree at least 3. Let $p(G) = 1$ if $G[V(C) \backslash V_c^+]$ contains a path whose length is at least $\lfloor \frac{k}{2} \rfloor$, and $p(G) = 0$ otherwise.*

Then, if $k \in \{3, 4\}$,
$$\text{meg}(G) = l + k - |V_c^+|.$$

Otherwise ($k \geqslant 5$), then

$$\text{meg}(G) = \begin{cases} 3, & \text{if } |V_c^+| = 0 \\ l + 2, & \text{if } |V_c^+| = 1 \\ l + p(G), & \text{if } |V_c^+| > 1 \end{cases}$$

Proof. Let G be a unicyclic graph where the only cycle C has length k and whose set of leaves is $L(G)$. By Lemma 2, all leaves are part of any MEG-set of G. This implies that meg(G) is at least l. If $|V_C^+| = 0$ (i.e. $l = 0$), we are done by Theorem 8, so let us assume $|V_C^+| > 0$ and thus, $l > 0$.

Similarly as in the proof of Lemma 4, for every vertex v of V_C^+, we know that at least one leaf will exist in the tree component T_v formed if we remove the neighbors of v in C from G. Informally speaking, towards the rest of the graph, this leaf simulates the fact that v is in the solution set.

If $k \in \{3, 4\}$, we consider $S = L(G)$ and we add to S all vertices of C that are of degree 2 in G. One can easily check that this is an MEG-set. Moreover, one can see that adding these degree 2 vertices is necessary by using similar arguments as in the proof of Theorem 8 on cycles.

Next, we assume that $k \geqslant 5$. Let $v_0, \ldots, v_{k-1}$ be the vertices of C.

When $|V_c^+| = 1$, without loss of generality, consider the vertex in V_c^+ to be v_0. Then, the vertices $\{v_{\lfloor \frac{k}{3} \rfloor}, v_{\lfloor \frac{2k}{3} \rfloor}\}$ on the cycle are sufficient to monitor the graph, in the same way as in Theorem 8. Moreover, by the same arguments as in the proof of Theorem 8, one can see that if at most one vertex on C is chosen in the MEG-set, some edge will not be monitored.

If $|V_c^+| > 1$ and $p(G) = 0$, the l leaves are sufficient to monitor G. Indeed, consider an edge e. If e is not on C, let v be the vertex of V_C^+ closest to e, and let $w \neq v$ be the vertex of V_C^+ closest to v (it exists because $|V_c^+| > 1$). Consider a leaf f of G such that e lies on some path from v to f. Since $p(G) = 0$, the path from w to f is a unique shortest path, and thus, e is monitored by f and some leaf whose closest vertex on C is w.

If e is an edge of C, e lies on a path between two vertices v, w of V_C^+. Since $p(G) = 0$, this path is a shortest path, and e is monitored by two leaves, each of which has v and w as its closest vertex of C, respectively.

Finally, consider the case where $p(G) = 1$ and $|V_c^+| > 1$. Since $p(G) = 1$, $G[V(C) \setminus V_c^+]$ contains a path P whose length is at least $\lfloor \frac{k}{2} \rfloor$ and thus, the edges of P are not monitored by the set of leaves of G, which implies that $\operatorname{meg}(G) \geqslant l + 1$. To show that $\operatorname{meg}(G) \leqslant l + 1$, we select as an MEG-set, the set of leaves together with the middle vertex of P (if P has even length) or one of the middle vertices of P (if P has odd length). One can see that this is an MEG-set by similar arguments as in the previous case.   □

3.4 Complete Graphs

The following follows immediately from Lemma 2, since every vertex of a complete graph is simplicial.

Theorem 10. *For any $n \geqslant 2$, we have $\operatorname{meg}(K_n) = n$.*

3.5 Complete Multipartite Graphs

The complete k-partite graph $K_{p_1, p_2, \ldots, p_k}$ consists of k disjoint sets of vertices of sizes $p_1, p_2, \ldots, p_k$, with an edge between any two vertices from distinct sets.

Theorem 11. *We have $\operatorname{meg}(K_{p_1, p_2, \ldots, p_k}) = |V(K_{p_1, p_2, \ldots, p_k})|$, with the exceptional case of a bipartite graph $K_{1,p}$ with an independent set of size 1 (a star graph), for which $\operatorname{meg}(K_{1,p}) = p$.*

Proof. In a complete k-partite graph, all vertices in a given partite set are twins. Therefore, by Lemma 3, all vertices of ant partite set of size at least 2 need to be a part of any MEG-set.

If we have several partite sets of size 1, then the vertices from these sets are closed twins, and again by Lemma 3 they all belong to any MEG-set.

Thus, we are done, unless there is a unique partite set of size 1, whose vertex we call v. If there are at least three partite sets, then note that v is never part of a unique shortest path, and thus the edges incident with v cannot be monitored if v is not part of the MEG-set.

On the other hand, if the graph is bipartite, it is a star $K_{1,p}$. Here, we know by Theorem 5 that $meg(G) = p$, as claimed.   □

3.6   Hypercubes

The *hypercube of dimension* n, denoted by Q_n, is the undirected graph consisting of $k = 2^n$ vertices labeled from 0 to $2^n - 1$ and such that there is an edge between any two vertices if and only if the binary representations of their labels differ by exactly one bit [15]. The *Hamming distance* $H(A, B)$ between two vertices A, B of a hypercube is the number of bits where the two binary representations of its vertices differ.

We next show that not only C_4 has the whole vertex set as its only MEG-set (Theorem 8), but that this also holds for all hypercubes.

Theorem 12. *For a hypercube graph Q_n with $n \geqslant 2$, we have $meg(Q_n) = 2^n$.*

Proof. Assume by contradiction that there is an MEG-set M of size at most $2^n - 1$. Let $v \in V(G)$ be a vertex that is not in M. It is known that for every vertex pair $\{v_x, v\}$ with $H(v_x, v) \leqslant n$, there are $H(v_x, v)$ vertex-disjoint paths of length $H(v_x, v)$ between them [15]. Thus, there is no vertex pair in M with a unique shortest path going through the edges incident with v, and M is not an MEG-set, a contradiction.   □

3.7   Grid Graphs

The graph $G \square H$ is the Cartesian product of graphs G and H and with vertex set $V(G \square H) = V(G) \times V(H)$, and for which $\{(x, u), (y, v)\}$ is an edge if $x = y$ and $\{u, v\} \in E(H)$ or $\{x, y\} \in E(G)$ and $u = v$. The *grid graph* $G(m, n)$ is the Cartesian product $P_m \square P_n$ with vertex set $\{(i, j) \mid 1 \leqslant i \leqslant m, 1 \leqslant j \leqslant n\}$.

Theorem 13. *For any $m, n \geqslant 2$, we have $meg(G(m, n)) = 2(m + n - 2)$.*

Proof. We claim that the set $S = \{(i, j) \in V(G(m, n)) \; i \in \{1, m\}$ and $1 \leqslant j \leqslant n$ or $j \in \{1, n\}$ and $1 \leqslant i \leqslant m\}$ of $2(m + n - 2)$ vertices of $G(m, n)$ that form the boundary vertices of the grid, form the only optimal MEG-set.

For the necessity side, let us assume that some vertex $v = (i, j)$ of S is not part of the MEG-set. If v is a corner vertex (without loss of generality say $v = (1, 1)$, the two edges incident with v are not monitored, as for any shortest path going through them, there is another one going through vertex $(2, 2)$. If v is not a corner vertex (without loss of generality say $v = (1, j)$ with $2 \leqslant j \leqslant n-1$), then the edge e between $v = (1, j)$ and $(2, j)$ is not monitored, indeed for any shortest path containing e, there is another one avoiding it, either going through vertex $(2, j - 1)$ or through $(2, j + 1)$.

To see that S is an MEG-set, first see that each boundary edge is monitored by its endpoints. Next, consider an edge e that is not a boundary edge, without loss of generality, e is between (i, j) and $(i+1, j)$. Then, it is monitored by $(1, j)$ and (m, j), whose unique shortest path goes through e.   □

4   Relation to Feedback Edge Set Number

A feedback edge set of a graph G is a set of edges which when removed from G leaves a forest. The smallest size of such a feedback edge set of G is denoted by $\text{fes}(G)$ and is sometimes called the *cyclomatic number* of G.

We next introduce the following terminology from [6]. A vertex is a *core vertex* if it has degree at least 3. A path with all internal vertices of degree 2 and whose end-vertices are core vertices is called a *core path*. Do note that we allow the two end-vertices to be equal, but that every other vertex must be distinct. A core path that is a cycle (that is, both end-vertices are equal) is a *core cycle*. For the sake of distinction, a core path that is not a core cycle is called a *proper core path*. We say that a (non-empty) path from a core vertex u to a leaf v is a *leg* of u if all internal vertices of the path have degree 2 (u is not considered to be a part of the leg). The *base graph* of a graph G is the graph of minimum degree 2 obtained from G by iteratively removing vertices of degree 1. A *hanging tree* is a connected subtree of G which is the union of some legs removed from G during the process of creating the base graph G_b of G. Thus, G can be decomposed into its base graph and a set of maximal hanging trees. The root of such a maximal hanging tree T is the vertex common to T and G_b.

See Fig. 2 for a graph whose core vertices are in red. It has two hanging trees, four core cycles, three proper core paths of length 4, and six proper core paths of length 1.

Based on the aforementioned, we have the following lemma.

Lemma 14 ([6,11]). *Let G be a graph with $\text{fes}(G) = k \geqslant 2$. The base graph of G has at most $2k$ - 2 core vertices, that are joined by at most $3k$ - 3 edge-disjoint core paths. Equivalently, G can be obtained from a multigraph H of order at most $2k - 2$ and size at most $3k - 3$ by subdividing its edges an arbitrary number of times and iteratively adding degree 1 vertices.*

Lemma 15. *Let S be an MEG-set of the base graph G_b of G and $L(G)$ be the set of leaves in G. Then, $S \cup L(G)$ is an MEG-set of G.*

Proof. Let G_b be a base graph of G. Consider all vertices that are roots of maximal hanging trees on G_b. By Theorem 5, the optimal MEG-set of each tree

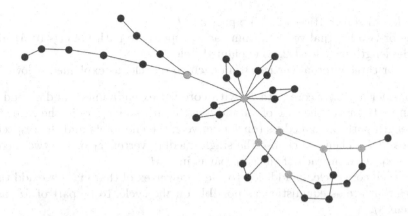

Fig. 2. Example of a graph G with its core vertices in red. (Color figure online)

consists of all leaves. We repeatedly apply Lemma 4 to G, where for each application of Lemma 4, the cut-vertex is the root of a hanging tree in consideration. □

Lemma 2, Theorem 5 and Lemma 15 together imply that if $\mathrm{fes}(G) = 0$, then $\mathrm{meg}(G) \leqslant \mathrm{fes}(G) + |L(G)|$. Moreover, if $\mathrm{fcs}(G) = 1$, then $\mathrm{meg}(G) \leqslant \mathrm{fes}(G) + |L(G)| + 3$, where $|L(G)|$ is the number of leaves of G. We next give a similar bound when $\mathrm{fes}(G) \geqslant 2$.

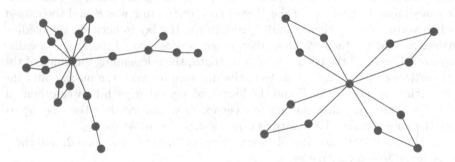

Fig. 3. Example of a graph G and its base graph G_b with four core cycles.

Theorem 16. *If* $\mathrm{fes}(G) \geqslant 2$, *then* $\mathrm{meg}(G) \leqslant 9\,\mathrm{fes}(G) + |L(G)| - 8$ *where* $|L(G)|$ *is the number of leaves of* G.

Proof. Let $k = \mathrm{fes}(G)$. We show how to construct a MEG-set M of G_b of order at most $9k - 8$ and, by applying Lemma 15 to G, of order $9k - 8 + |L(G)|$ for G. If an edge e is part of a maximal hanging tree, then by Lemma 2 and Lemma 4, it is monitored by the leaves of G on the maximal hanging tree. M is constructed as follows.

- We let all core vertices of G_b be part of M.
- One or two internal vertices from each proper core path belongs to M, only if the length is at least 2, as explained below.
- Two or three internal vertices from each core cycle, as explained below.

Consider a proper core path P, with core vertex endpoints c and c', and the median vertex x_1 in the case of an odd-length path and x_1, x_2 in the case of an even-length path, with d edges (on P) between the endpoints and the respective medians in P. Then, we choose the single median vertex x_1 or the two median vertices x_1, x_2 from each of the core paths into M.

For each core cycle, in addition to the core vertex of that cycle, we add three vertices that are as equidistant as possible on the cycle, to be part of M (as in Theorem 8).

Let e be any edge of G. We now show that our construction M monitors any such edge in G_b. If e lies on a core cycle, assume an origin core vertex of v_0. Then, based on Lemma 4 and Theorem 8, we deduce that in the worst case, four vertices together suffice to monitor the edges.

If the edge e lies on a proper core path P, then we have the following cases. Let c and c' be the core vertex endpoints of P, and the median vertex x_1 in the case of an odd-length path and x_1, x_2 in the case of an even-length path and d edges of P between the end points and the respective medians in P. Without loss of generality, let us say that e lies on the path P such that its closest core vertex is c and closest median x_1 in the event of an even-length path. Suppose first that d is even. Given that the distance between c and x_1 is d in P, the length of any other path between them must be at least $d + 2$. Therefore, c and x_1 monitor e. We can similarly argue that if the closest core vertex to e was c' and the closest median vertex was x_2, then c' and x_2 monitor e. If e lay in between the median vertices x_1 and x_2, then we know that those vertices would monitor e because they are adjacent. If the path was of odd length, then depending on which of the core vertices c and c' was closest to e, the distance between the median and the core vertices would be d in P and the length of any other path between them at least $d + 1$, ensuring that the median vertex x_1 would monitor the edge apart from the core vertices. This justifies our construction of M for G_b.

By Lemma 14, the number of core vertices of G_b is at most $2k - 2$, and there are at most $3k - 3$ core paths.

If we have core cycles in our graph, then we must note that there can be at most k such cycles in the graph. Indeed, if there were $k + 1$ core cycles in the graph, since they are all edge-disjoint, we need at least $k + 1$ edges to be removed from G to obtain a forest, a contradiction to the fact that $\mathrm{fes}(G) = k$.

Let n_c be the number of core cycles and n_p be the number of proper core paths. We have $|M| = 3n_c + 2n_p + 2k - 2$. Since $n_c \leqslant k$ and $n_c + n_p \leqslant 3k - 3$ by Lemma 14, we get $|M| \leqslant 3k + 2(2k - 3) + 2k - 2 = 9k - 8$. $\square$

Recall that the *max leaf number* of G, denoted $\mathrm{mln}(G)$, is the maximum number of leaves in a spanning tree of G. It can be seen as a refinement of the feedback edge set number of G [4,5]. We get the following corollary.

Corollary 17. *For any graph G, we have* $\mathrm{meg}(G) \leqslant 10\,\mathrm{mln}(G)$, *where* $\mathrm{mln}(G)$ *is the max leaf number of G.*

Proof. It is known that $\mathrm{fes}(G) \leqslant \mathrm{mln}(G)$ [4], and clearly, $|L(G)| \leqslant \mathrm{mln}(G)$, thus the bound follows from Theorem 16. □

Proposition 18. *For any integer $k \geqslant 2$, there exists a graph G with* $\mathrm{fes}(G) = k$ *and* $\mathrm{meg}(G) = 3k + |L(G)|$.

Proof. Consider G and its base graph G_b in Fig. 3. We know that the leaves must be part of any MEG-set by Lemma 2. The MEG-set for G_b consists of all the vertices in each of the core cycles (each a C_4) in G_b, except the common core vertex. It is easy to check that no smaller set can work. The size of the optimal MEG-set in this example is $3k + |L(G)|$ and therefore, this is an instance where this proposition holds. □

5 Conclusion

Inspired by a network monitoring application, we have defined the new concept of MEG-sets of a graph, which is a common refinement of the popular concept of a geodetic set and its variants, and of the previously studied distance edge-monitoring sets.

We have studied the concept on basic graph classes. It is interesting to note that there are many graph classes which require the entire vertex set in any MEG-set: complete graphs, complete multipartite graphs, and hypercubes. It could thus be a difficult, but interesting, question, to characterize all such graphs.

Our upper bound using the feedback edge set number is probably not tight. What is a tight bound on this regard?

Finally, it remains to investigate computational aspects of the problem.

Acknowledgements. Florent Foucaud thanks Ralf Klasing and Tomasz Radzik for initial discussions which inspired the present study.

References

1. Bampas, E., Biló, D., Drovandi, G., Gualá, L., Klasing, R., Proietti, G.: Network verification via routing table queries. J. Comput. Syst. Sci. **81**(1), 234–248 (2015)
2. Beerliova, Z., et al.: Network discovery and verification. IEEE J. Sel. Areas Commun. **24**(12), 2168–2181 (2006)
3. Bejerano, Y., Rastogi, R.: Robust monitoring of link delays and faults in IP networks. IEEE/ACM Trans. Networking **14**(5), 1092–1103 (2006)
4. Eppstein, D.: Metric dimension parameterized by max leaf number. J. Graph Algorithms Appl. **19**(1), 313–323 (2015)
5. Fellows, M.R., Lokshtanov, D., Misra, N., Mnich, M., Rosamond, F., Saurabh, S.: The complexity ecology of parameters: an illustration using bounded max leaf number. Theory Comput. Syst. **45**(4), 822–848 (2009)

6. Epstein, L., Levin, A., Woeginger, G.J.: The (weighted) metric dimension of graphs: hard and easy cases. Algorithmica **72**(4), 1130–1171 (2015)
7. Foucaud, F., Klasing, R., Miller, M., Ryan, J.: Monitoring the edges of a graph using distances. In: Changat, M., Das, S. (eds.) CALDAM 2020. LNCS, vol. 12016, pp. 28–40. Springer, Cham (2020). https://doi.org/10.1007/978-3-030-39219-2_3
8. Foucaud, F., Kao, S., Klasing, R., Miller, M., Ryan, J.: Monitoring the edges of a graph using distances. Discret. Appl. Math. **319**, 424–438 (2022)
9. Harary, F., Loukakis, E., Tsouros, C.: The geodetic number of a graph. Math. Comput. Model. **17**, 89–95 (1993)
10. Harary, F.: Graph Theory. Addison-Wesley, Reading (1994)
11. Kellerhals, L., Koana, T.: Parameterized complexity of geodetic set. In: Proceedings of the 15th International Symposium on Parameterized and Exact Computation (IPEC 2020). Leibniz International Proceedings in Informatics (LIPIcs), vol. 180, pp. 20:1–20:14 (2020)
12. Manuel, P., Klavžar, S., Xavier, A., Arokiaraj, A., Thomas, E.: Strong edge geodetic problem in networks. Open Math. **15**(1), 1225–1235 (2017)
13. Santhakumaran, A.P., John, J.: Edge geodetic number of a graph. J. Discret. Math. Sci. Cryptogr. **10**, 415–432 (2007)
14. Skiena, S.: Implementing Discrete Mathematics: Combinatorics and Graph Theory with Mathematica. Addison-Wesley, Reading (1990)
15. Saad, Y., Schultz, M.: Topological properties of hypercubes. IEEE Trans. Comput. **37**(7), 867–872 (1988)

Cyclability, Connectivity and Circumference

Niranjan Balachandran[1] and Anish Hebbar[2]([✉])

[1] Indian Institute of Technology Bombay, Mumbai, India
niranj@iitb.ac.in
[2] Indian Institute of Science, Bangalore, India
anishhebbar@iisc.ac.in

Abstract. In a graph G, a subset of vertices $S \subseteq V(G)$ is said to be cyclable if there is a cycle containing the vertices in some order. G is said to be k-cyclable if any subset of $k \geq 2$ vertices is cyclable. If any k *ordered* vertices are present in a common cycle in that order, then the graph is said to be k-ordered. We show that when $k \leq \sqrt{n+3}$, k-cyclable graphs also have circumference $c(G) \geq 2k$, and that this is best possible. Furthermore when $k \leq \frac{3n}{4} - 1$, $c(G) \geq k+2$, and for k-ordered graphs we show $c(G) \geq \min\{n, 2k\}$. We also generalize a result by Byer et al. [4] on the maximum number of edges in nonhamiltonian k-connected graphs, and show that if G is a k-connected graph of order $n \geq 2(k^2 + k)$ with $|E(G)| > \binom{n-k}{2} + k^2$, then the graph is hamiltonian, and moreover the extremal graphs are unique.

Keywords: Cyclability · Connectivity · Circumference · Hamiltonicity

1 Introduction

We consider only finite, undirected, simple graphs throughout this paper. The vertex and edge sets of G will be denoted by $V(G)$ and $E(G)$ respectively, the graph complement by $\overline{G}$. The length of the longest cycle in the graph G, also known as the circumference, will be denoted by $c(G)$. The minimum degree, independence number and connectivity of a graph will denoted by $\delta(G), \alpha(G)$ and $\kappa(G)$ respectively. We will also use $d_H(v)$ for the degree of v in H. The set of neighbours of a vertex $v \in V(G)$ will denoted by $N(v)$, and the closed neighbourhood of v, viz. $N(v) \cup \{v\}$ will be denoted by $N[v]$. The join of two graphs G_1, G_2, denoted $G_1 \vee G_2$ is simply a copy of G_1 and G_2, with all edges between $V(G_1)$ and $V(G_2)$ also being present.

A subset $S \subseteq V(G)$ of vertices in a graph G is said to be cyclable if G has a cycle containing the vertices of S in some order, possibly including other vertices. A graph G is said to be k-cyclable if any $k \geq 2$ vertices of G lie on a common cycle. Note that the problem of determining the hamiltonicity of a graph is a special case of cyclability, namely when $k = n$. Cyclability and connectivity are interlinked, as was shown by Dirac [8] who proved for every

A. Bagchi and R. Muthu (Eds.): CALDAM 2023, LNCS 13947, pp. 257–268, 2023.
https://doi.org/10.1007/978-3-031-25211-2_20

$k \geq 2$, k-connected graphs are also k-cyclable. In fact, for $k = 2$ connectivity and cyclability are equivalent, but in general for $k \geq 3$ it is not necessarily true that every k-cyclable graph is also k-connected, as can be seen by considering the graph $K_2 \vee 2K_k$ which has connectivity exactly 2 and is also k-cyclable. For a brief survey of results involving conditions for cycles to contain a particular set, refer to [12].

There is a rich literature on conditions guaranteeing the presence of long cycles in graphs, the most classical one being that of Dirac [7] who showed that in 2-connected graphs, the circumference is at least$c(G) \geq \min\{n, 2\delta(G)\}$. Moreover, k-connected graphs have a circumference of at least $\min\{n, 2k\}$ from an easy consequence of Menger's theorem, and this is tight. A famous result by Chvátal and Erdős [5] relates the connectivity and independence number of a graph to hamiltonicity, and says that if the connectivity of a graph G is at least its independence number, then the graph is hamiltonian. However, not much is known when the requirement of connectivity is weakened to cyclability. Bauer et al. [1] obtained lower bounds for the length of the longest cycle in 3-cyclable graphs in terms of the minimum degree and independence number, but not much else is known for k-cyclable graphs for arbitrary k.

Cyclability has also received interest from an algorithmic and complexity theoretic point of view as it is a 'hard' parameter that can be thought of as a more quantitative measure of hamiltonicity. Since the classical HAMILTO-NIAN CYCLE problem is NP-complete, the problem of determining whether a graph is k-cyclable (CYCLABILITY) is NP-complete as well. The problem of determining whether a given subset S of vertices is cyclable (TERMINAL CYCLABILITY) has been studied in the Parameterized Complexity framework (FPT) (parameterized by $|S|$) and the best known algorithm has running time $O(2^{|S|}n^{O(1)})$ [2]. For some special classes of graphs such as interval graphs and bipartite permutation graphs, Crespelle and Golovach [6] showed that both these problems can be solved in polynomial time. For $|S| = O((\log \log n)^{1/10})$, Kawarabayashi [14] obtained a polynomial time algorithm for TERMINAL CYCLABILITY.

Note that k-connectivity guarantees $c(G) \geq \min\{n, 2k\}$ and also ensures k-cyclability. Thus, a natural question to ask is whether the same bound on the circumference can be obtained when the connectivity criteria is weakened to cyclability. When $k = n - 1$, we would require any set of $n - 1$ vertices of G to lie on a common cycle. It turns out that in this case, it is not necessary that the graph is hamiltonian. Indeed, the existence of hypohamiltonian graphs [9] of order n is known for all $n \geq 18$. Our first result in this paper gives a similar circumference bound for a wide range of k:

Theorem 1. *Let G be a k-cyclable graph, where $2 \leq k \leq n$. Then,*

$$c(G) \geq \begin{cases} 2k & \text{if } k \leq \sqrt{n+3} \\ k+2 & \text{if } k \leq \frac{3n}{4} - 1 \end{cases}$$

Moreover, for $2 \leq k \leq \sqrt{n+3}$, this bound on the circumference is best possible.

Note that for $k \geq \frac{n}{2}$ it is still possible that one can have a bound of the form $c(G) \geq (1 + \gamma)k$ for some fixed positive constant $\gamma < 1$ as long as $k \neq n - o(n)$.

A related notion is the orderedness of a graph, a strong hamiltonian property that was first introduced by Ng and Schultz [16]. A graph G is said to be k-ordered if any sequence of distinct vertices $T = \{v_1, \ldots, v_k\}$ are present in some common cycle in that order, possibly including other vertices. Note that k-ordered graphs are naturally also k-cyclable, and it is also easy to see that they are $(k-1)$-connected. For a comprehensive survey of results on k-ordered graphs, see [11]. We show that for k-orderedness, the same circumference bound as k-connectivity holds for all $2 \leq k \leq n$.

Theorem 2. *Let G be a k-ordered graph, $2 \leq k \leq n$. Then, $c(G) \geq \min\{n, 2k\}$.*

Our second pursuit in this paper is to obtain Turán-type results for the circumference of k-connected graphs, specifically the maximum number of edges in nonhamiltonian k-connected graphs. A classical result states that if G is a graph of order n with $|E(G)| > \binom{n-1}{2} + 1$, then G is hamiltonian. This was generalized by [4] for $k \leq 3$, where they showed that if G is k-connected and satisfies $|E(G)| > \binom{n-k}{2} + k^2$ with n sufficiently large, then the graph is hamiltonian and the extremal graphs are unique. We further generalize their result and extend it to any k satisfying $n \geq 2(k^2 + k)$.

Theorem 3. *Let G be a k-connected graph of order $n \geq 2(k^2 + k)$. If $|E(G)| > \binom{n-k}{2} + k^2$, then G is hamiltonian. Moreover, the extremal graphs are unique.*

The rest of the paper is organized as follows. We lay out some preliminaries in the next section, and give the proofs of Theorems 1, 2, and 3 in the following section. We conclude with some remarks and open questions.

2   Preliminaries

When the underlying graph is clear, we will use δ, κ, α instead of $\delta(G), \kappa(G), \alpha(G)$ for brevity, and also omit the subscript in $d_H(v)$. We also use the following well-known lemma attributed to Dirac repeatedly throughout the paper, and provide an outline of the proof for completeness.

Lemma 4 ([8]). *Any k-connected graph G is k-cyclable. Moreover, it satisfies $c(G) \geq \min\{n, 2k\}$*

Proof Sketch. Suppose some subset S of vertices with $|S| = k$ was not fully contained in any cycle. Then, take a cycle C containing as many of the vertices of S as possible, and pick some $v \in S$ that is not in C. By Menger's theorem, we can choose k vertex-disjoint paths from v to the cycle C, and these endpoints divide C into k segments. Since there are strictly less than k vertices of S in C, one of the segments does not contain any vertex from S, and thus we can extend this segment with the 2 disjoint paths from v at the ends of the segment to obtain a cycle containing more vertices of S, contradiction.

Now consider the longest cycle C in G and suppose its length is strictly less than $\min\{n, 2k\}$. Pick some $v \in V(G)$ not in C, and by Menger's theorem there are k vertex disjoint paths from v to C. By the pigeonhole principle, some two endpoints of these k paths must be adjacent on the cycle C, giving a contradiction as we can replace the edge between these endpoints with the 2 paths to obtain a longer cycle.   □

A famous result by Chvátal and Erdős states the following

Theorem 5 ([5]). *If in a graph G, $\alpha(G) \leq \kappa(G)$, then G is hamiltonian.*

A natural generalization of the above is to flip the condition $\alpha(G) \leq \kappa(G)$, and instead ask for lower bounds on the circumference of a graph G where $\alpha(G) \geq \kappa(G)$. Foquet and Jolivet [13] conjectured the following, which was later proven by Suil O, Douglas B. West and Hehui Wu.

Theorem 6 ([17]). *If G is a k-connected n-vertex graph with independence number α and $\alpha \geq k$, then G has a cycle of length at least $\frac{k(n+k-\alpha)}{\alpha}$.*

The following result by Dirac is well-known and was a precursor to a number of results involving the length of the longest cycle in a graph.

Theorem 7 ([7]). *If G is 2-connected and has minimum degree δ, $c(G) \geq \min\{2\delta, n\}$.*

Note that 2-connectivity is equivalent to 2-cyclability. Bauer et al. obtained a bound on the circumference of 3-cyclable graphs in terms of the minimum degree and independence number.

Theorem 8 ([1]). *If G is 3 cyclable, then*

$$c(G) \geq \min\{n, 3\delta - 3, n + \delta - \alpha\}.$$

Ng and Schultz studied a related hamiltonian property termed k-orderedness, and showed the following connectivity result. Once again, we include the proof for completeness.

Lemma 9 ([16]). *Let G be a k-ordered graph. Then, G is $(k-1)$-connected.*

Proof. If not, there exists a set S of $k-2$ vertices whose removal disconnects G, breaking it into at least 2 components. Take 2 vertices u, v in different components, then any path from u to v must go through some vertex of S. Thus, let T consist of u, v and then the vertices of S, in that order. These vertices must appear in some cycle in that order, giving a contradiction.   □

We will also need the concept of graph closure introduced by Bondy and Chvátal. Define the closure of G, denoted $cl(G)$, to be the graph obtained by repeatedly joining any two nonadjacent vertices x, y that satisfy $d(x) + d(y) \geq n$ in G. They showed that $cl(G)$ is well-defined (independent of the order in which nonadjacent vertex pairs are considered), and that G is hamiltonian if and only if $cl(G)$ is also hamiltonian.

Lemma 10 ([3]). *Suppose $cl(G) = G$ for a nonhamiltonian graph G of order n. Then $d(x) + d(y) \leq n - 1$ for any pair $\{x, y\}$ of nonadjacent vertices.*

This was later generalized to obtain results for higher order connectivity, the bounds now also involving the independence number. We define

$$\sigma_k(G) = \min\{\sum_{i=1}^{k} d(x_i), \{x_1, \ldots x_k\} \text{ an independent set of size k in G}\}$$

Note that $\sigma_1(G)$ simply corresponds to the minimum degree δ, and Ore's theorem [18] states that if $\sigma_2(G) \geq n$, then the graph is hamiltonian.

Theorem 11 ([15]). *Let G be a k-connected graph of order n and independence number α. If $\sigma_{k+1}(G) \geq n + (k-1)\alpha - (k-1)$, then G is hamiltonian.*

3   Proofs of the Results

Proof of Theorem 1.
We will first prove the bound for the regime $2 \leq k \leq \sqrt{n+3}$.

Consider any k-cyclable graph with $\alpha(G) \geq k$. Then, let S be a set of k independent vertices, and consider the cycle containing it. This gives us a cycle of length at least $2k$, as any 2 independent vertices are not adjacent to each other. Thus, we can assume $\alpha(G) \leq k - 1$. Let the connectivity of the graph be κ. Using Theorem 6, it suffices to show

$$\frac{\kappa(n + \kappa - \alpha)}{\alpha} \geq 2k \iff n \geq 2k(\frac{\alpha}{\kappa}) + (\alpha - \kappa)$$

As k-cyclable graphs are also 2-cyclable, and thus 2-connected, we must have $\kappa \geq 2$. Hence, it is sufficient to show the stronger inequality

$$n \geq 2k(\frac{k-1}{\kappa}) + k - 3$$

which is always true when

$$n \geq k^2 - 3 \iff k \leq \sqrt{n+3}$$

Note that if we only ask for an improvement of the form $c(G) \geq (1+\gamma)k$ for some positive constant $\gamma < 1$, we can improve the range of k for which the result holds. Once again, let S be any set of at least $\frac{(1+\gamma)k}{2}$ many independent vertices, and consider the cycle containing S. This corresponds to a cycle containing at least $(1+\gamma)k$ many vertices since any two independent vertices are not adjacent, and thus we get $\alpha < \frac{(1+\gamma)k}{2}$. Similar to the previous argument, if the connectivity of the graph is κ, by Theorem 6 it suffices to show

$$\frac{\kappa(n + \kappa - \alpha)}{\alpha} \geq (1+\gamma)k \iff n \geq (1+\gamma)k(\frac{\alpha}{\kappa}) + (\alpha - \kappa)$$

Using $\kappa \leq 2$ and $\alpha < \frac{(1+\gamma)k}{2}$, we are done as long as

$$n \geq \frac{(1+\gamma)^2 k^2}{4} + \frac{(1+\gamma)k}{2} - 2 \iff \frac{\sqrt{4n+9}}{1+\gamma} \geq k$$

So the above argument only yields a linear improvement in $c(G)$ for k up to around $2\sqrt{n}$.

Now, suppose $2 \leq k \leq \frac{3n}{4} - 1$, and assume to the contrary that $c(G) < k + 2$. We must have $k \geq 3$ as 2-cyclable graphs are 2-connected and hence have circumference at least 4 for $n \geq 4$. By Theorem 7, we must have $\delta \leq \frac{k+1}{2}$. Moreover, $\alpha \leq \frac{k+1}{2}$ as otherwise we could simply take a cycle containing α many independent vertices. Consider a vertex v with minimum degree δ, with neighbourhood $N(v)$ satisfying $|N(v)| = \delta$. Now, choose v and any $k-1$ vertices from $V \backslash N[v]$, which is possible as long as $k - 1 \leq n - 1 - \delta$. Then, any cycle containing these vertices must also contain some 2 neighbours of v, giving $c(G) \geq k + 2$, and we are done.

Thus, we must have $k + \delta > n$. Note that when $2 \leq k \leq \frac{3n}{4} - 1$, $n \geq k+2$ if $n \geq 4$. So, we must either have $3\delta - 3 \leq k + 1$ or $n + \delta - \alpha \leq k + 1$, otherwise we are done by Theorem 8.

The former inequality gives $\delta \leq \frac{k+4}{3}$, which gives

$$n < k + \delta \leq \frac{4k+4}{3} \implies \frac{3n-4}{4} < k$$

a contradiction. Hence, we must have $\delta \geq \frac{k+5}{3}$, $\alpha \leq \frac{k+1}{2}$ giving

$$k + 1 \geq n + \delta - \alpha \geq n + \frac{k+5}{3} - \frac{k+1}{2} = n + \frac{7-k}{6}$$

or equivalently, $\frac{3n}{4} - 1 \geq k \geq \frac{6n+1}{7}$, which is again a contradiction. □
We now prove an analogous bound for the circumference of k-ordered graphs.

Proof of Theorem 2.
We know that k-ordered graphs are also $k - 1$ connected from Theorem 9, thus $\kappa \geq k - 1$. We also must have $\alpha \leq k - 1$, as otherwise we can simply take k independent vertices in any order to obtain a cycle of size at least $2k$, in which case we are done. Hence,

$$\kappa \geq k - 1 \geq \alpha$$

so by Theorem 5, we have that G is hamiltonian, and thus we are done in this case as well. □

In fact, it is not hard to see that the $\min\{n, 2k\}$ bound on the circumference is achieved for all $2 \leq k \leq n$. If $k > n/2$, simply consider the complete graph K_n which is clearly k-connected, k-ordered, k-cyclable and has circumference n. If $k \leq n/2$, consider the complete bipartite graph $G = K_{k,n-k} = (A, B, E)$, which is k-ordered, and hence k-cyclable. Indeed, take any sequence of k distinct vertices $T = (v_1, v_2, \ldots, v_k)$. We construct a cycle containing T in that order as follows.

Let T_A be the set of vertices in T and A, with T_B being defined similarly. Then, for any $v \in T_A$, if the next vertex in the sequence T is in T_B, then simply follow the edge joining them. Otherwise, first follow an edge to a vertex in $B \backslash T_B$, and then back to the next vertex which must have been in T_A. Follow the same procedure for vertices in T_B. At the end, follow the edge joining the first and last vertex. We cannot run out of vertices as the number of extra vertices outside T_A in A that are needed is at most $|T_B|$, and $|A| = k = |T_A| + |T_B|$. Similarly, $|B| = n - k \geq k = |T_A| + |T_B|$.

We now generalize a result by [4] on the maximal number of edges in a k-connected nonhamiltonian graph, for $k = 2, 3$. We will need the following short lemma which appears in [4].

Lemma 12 ([4]). *Let G be a nonhamiltonian, k-connected graph of order n. Then $k \leq \frac{n-1}{2}$ and $|E(\overline{G})| \geq \binom{k+1}{2} + (k-1)(n-k-1) - \sigma_{k+1}(G)$*

Proof. By Theorem 5, k-connected nonhamiltonian graphs must contain an independent set $I = \{x_1, \ldots, x_{k+1}\}$ of $k+1$ vertices. The graph is disconnected on removal of the $n - (k+1)$ vertices of $G - I$, thus we must have $n - (k+1) > k-1$, or $k \leq \frac{n-1}{2}$.

Now consider the independent set I satisfying $\sum_{i=1}^{k+1} d(x_i) = \sigma_{k+1}(G)$. Let the edges in $\overline{G}$ incident on at least one vertex of I be denoted X_I. Then X_I contains $\binom{k+1}{2}$ edges with both endpoints in I and $\sum_{i=1}^{k+1}(n - 1 - k - d_G(x_i))$ edges with exactly one endpoint in I. Thus, we obtain

$$|E(\overline{G})| \geq |X_I| = \binom{k+1}{2} + (k-1)(n-k-1) - \sigma_{k+1}(G)$$

□

Using a slight variation of the above result and Lemma 10, [4] also show the following result.

Lemma 13 ([4]). *Suppose $G = cl(G)$ for a nonhamiltonian graph G of order n, and $m \leq \alpha(G)$. Then*

$$|E(\overline{G})| \geq \begin{cases} \frac{m}{2}(n - m) & \text{for } n \text{ odd} \\ \frac{m}{2}(n - m) + \frac{m}{2} - 1 & \text{for } n \text{ even} \end{cases}$$

With the above results, we are ready to proceed to the proof of Theorem 3. The idea is that if n is not that much bigger than α, then we can get a sufficient lower bound on $|E(\overline{G})|$ using Lemma 13. Otherwise, n is much bigger than α, and we can use Theorem 11 and Lemma 12. To show the uniqueness of the extremal graphs, we will make use of the fact that these graphs must satisfy Lemma 10 *maximally*, i.e., addition of any further edge causes a violation of the condition.

Proof of Theorem 3.

First of all, assume $k \geq 2$ as we already know that when $|E(G)| > \binom{n-1}{2} + 1$, then G is hamiltonian and consequently connected as well. Assume G is non-hamiltonian. We may assume $G = cl(G)$, in which case $d(x) + d(y) \leq n - 1$ for any two nonadjacent vertices x, y, from Lemma 10. It suffices to prove that

$$|E(\overline{G})| \geq \binom{n}{2} - \left(\binom{n-k}{2} + k^2\right) = k \cdot n - \frac{3k^2 + k}{2}$$

Note first that if $\sigma_{k+1}(G) \leq n + k^2 - k - 1$, by Lemma 12

$$|E(\overline{G})| \geq \binom{k+1}{2} + (k+1)(n-k-1) - (n+k^2-k-1) = k \cdot n - \frac{3k^2+k}{2}$$

as desired. We now assume $\sigma_{k+1}(G) \geq n + k^2 - k$ and show that in this case, $|E(\overline{G})|$ is *strictly* greater than $k \cdot n - \frac{3k^2+k}{2}$. We will divide the problem into two cases, depending on the size of n compared to α.

Case 1: Assume $n > \frac{(k^2-1)\cdot\alpha+y}{k}$, where $y = \frac{-k^3+4k^2+3k+2}{2}$.

Let $I = \{x_1, x_2, \ldots, x_{k+1}\}$ be a set of $k + 1$ independent vertices satisfying $\sum_{i=1}^{k+1} d(x_i) = \sigma_{k+1}(G)$, and assume without loss of generality that

$$d(x_1) \geq \frac{\sigma_{k+1}(G)}{k+1} \geq \frac{n+k^2-k}{k+1}$$

Subcase 1a: Suppose $d(x_1) \geq n - 2k$. Note that $V(G) - I - N(x_1)$ is non-empty, as otherwise we would have $d(x_1) = n - k - 1$, giving $d(x_i) \leq k$ for $2 \leq i \leq k+1$ as $d(x_1) + d(x_i) \leq n-1$ for $2 \leq i \leq k+1$. This contradicts $\sigma_{k+1}(G) \geq n+k^2-k$. Thus, pick some $v \in V(G) - I - N(x_1)$, giving $d_{\overline{G}}(v) = n - 1 - d_G(v) \geq d_G(x_1) \geq n - 2k$. Therefore, $\overline{G}$ contains at least $n - 2k - |I| = n - 3k - 1$ edges with both endpoints not in I. Using the same bound we got in Lemma 12 but also including the extra edges in $\overline{G}$ incident with v (that have no endpoint in I) and using Theorem 11, we obtain

$$|E(\overline{G})| \geq \binom{k+1}{2} + (k+1)(n-k-1) + (n-3k-1) - \sigma_{k+1}(G)$$

$$\geq (k+2) \cdot n - \frac{k^2+9k+4}{2} - (n+(k-1)\alpha-k)$$

$$> k \cdot n - \frac{3k^2+k}{2} + \frac{3k^2+k}{2} - \frac{k^2+9k+4}{2} + k + \frac{(k^2-1)\cdot\alpha+y}{k} - (k-1)\alpha$$

$$= (k \cdot n - \frac{3k^2+k}{2}) + \frac{(k-1)\cdot\alpha+y+k(k^2-3k-2)}{k} > (k \cdot n - \frac{3k^2+k}{2})$$

as desired, where the last inequality follows from $y = \frac{-k^3+4k^2+3k+2}{2}$.

Subcase 1b: Suppose next that $d(x_1) \leq n - 2k - 1$. Then there exist distinct vertices $v_1, v_2 \ldots, v_k \in V(G) - I - N(x_1)$, and $\overline{G}$ contains at least

$$(d_{\overline{G}}(v_1) - k - 1) + (d_{\overline{G}}(v_2) - k - 2) + \cdots + (d_{\overline{G}}(v_k) - 2k) = \sum_{i=1}^{k} d_{\overline{G}}(v_i) - \frac{3k^2+k}{2}$$

edges with neither endpoint in I. Using $d(v_i) + d(x_1) \le n - 1$ as $G = cl(G)$, we get $d_{\overline{G}}(v_i) \ge d_G(x_1) \ge \frac{n + k^2 - k}{k + 1}$ for all $1 \le i \le k$. Consequently, we obtain at least

$$\frac{k(n + k^2 - k)}{k + 1} - \frac{3k^2 + k}{2}$$

edges in $\overline{G}$ with neither endpoint in I. Using Theorem 11 and Lemma 12 again, we get

$$
\begin{aligned}
|E(\overline{G})| &\ge \binom{k + 1}{2} + (k + 1)(n - k - 1) + \frac{k(n + k^2 - k)}{k + 1} - \frac{3k^2 + k}{2} - (n + (k - 1)\alpha - k) \\
&= (kn - \frac{3k^2 + k}{2}) + \frac{k}{k + 1}n - (k - 1)\alpha + \binom{k + 1}{2} - (k + 1)^2 + \frac{k(k^2 - k)}{k + 1} + k \\
&> (kn - \frac{3k^2 + k}{2}) + \frac{k}{k + 1} \frac{(k^2 - 1)\alpha + y}{k} - (k - 1)\alpha + \frac{-k^2 - k - 2}{2} + \frac{k(k^2 - k)}{k + 1} \\
&= (kn - \frac{3k^2 + k}{2}) + \frac{1}{k + 1}\left(\frac{-k^3 + 4k^2 + 3k + 2}{2} + \frac{(-k^2 - k - 2)(k + 1)}{2} + k^3 - k^2\right) \\
&= kn - \frac{3k^2 + k}{2}
\end{aligned}
$$

Case 2: Assume $n \le \frac{(k^2 - 1)\alpha + y}{k}$.

In this case, $\alpha \ge \frac{nk - y}{k^2 - 1}$. By Lemma 13, $|E(\overline{G})| \ge \frac{1}{2}\alpha(n - \alpha)$. This is a upward facing parabola for fixed n, so for $\frac{nk - y}{k^2 - 1} \le \alpha \le n - \frac{nk - y}{k^2 - 1}$, this function is minimized at $\alpha = \frac{nk - y}{k^2 - 1}$. Therefore, in this range

$$
\begin{aligned}
|E(\overline{G})| &\ge \frac{\alpha}{2}(n - \alpha) \ge \frac{1}{2}\left(\frac{nk - y}{k^2 - 1}\right)\left(\frac{n(k^2 - k - 1) + y}{k^2 - 1}\right) \\
&= \frac{n^2 k(k^2 - k - 1) + n(2k + 1 - k^2)y - y^2}{2(k^2 - 1)^2}
\end{aligned}
$$

If we want the above to be strictly greater than $kn - \frac{3k^2 + k}{2}$,

$$\frac{n^2 k(k^2 - k - 1)}{2(k^2 - 1)^2} \ge kn \iff n \ge \frac{2(k^2 - 1)^2}{k^2 - k - 1} = 2(k^2 + k + \frac{1 - k}{k^2 - k - 1})$$

suffices. This is because for $k \ge 5$, $y = \frac{-k^3 + 4k^2 3k + 2}{2} < 0$ and $2k + 1 - k^2 < 0$, giving $(2k + 1 - k^2)(y) > 0$. Similarly, $-y^2 = \frac{(-k^3 + 4k^2 + 3k + 2)^2}{4} > -(3k^2 + k)(k^2 - 1)^2$ for $k \ge 5$, so we only have to check the cases of $k = 2, 3, 4$ manually which is a routine check.

Now, it remains to consider the possibility that $\alpha > n - \frac{nk - y}{k^2 - 1} = \frac{n(k^2 - k - 1) + y}{k^2 - 1}$. In this case however, α is quite large compared to n, so the $\binom{\alpha}{2}$ edges in $\overline{G}$ between the vertices of an independent set of size α is strictly greater than $k \cdot n - \frac{3k^2 + k}{2}$ for all n. Indeed, we manually verify for $k \le 3$, and for $k \ge 4$ simply note that $\frac{nk}{2} + y \ge 0$, and hence when $n \ge 2(k^2 + k)$ we have

$$\alpha > \frac{n(k^2 - \frac{3k}{2} - 1)}{k^2 - 1} \ge \frac{9n}{15}, \quad \binom{9n/15}{2} > \frac{9n}{30} \cdot \frac{8n}{15} > kn$$

We now prove that the extremal nonhamiltonian k-connected graphs are unique for $n \geq 2(k^2 + k)$, by making use of Lemma 10. Recall that we may assume $G = cl(G)$ is a nonhamiltonian, k-connected graph of order $n \geq 2k^2 + 2k$ with $\sigma_{k+1}(G) = n + k^2 - k - 1$ as equality only holds if all the inequalities in the above proof are tight.

Thus, all the edges in $\overline{G}$ have atleast one endpoint in I. Let $I = \{x_1, x_2, \ldots, x_{k+1}\}$ be a set of independent vertices such that $k \leq d(x_1) \leq \ldots \leq d(x_{k+1})$. Note that k-connected graphs have minimum degree at least k as otherwise, the graph could be disconnected by removing at most $k-1$ vertices. As mentioned in the previous section, we may further assume that all edges in $\overline{G}$ have at least one endpoint in I, that is, if $x, y \in V(G) - I$, then $\{x, y\} \in E(G)$. We will now use the properties of graph closure repeatedly. First, note that we must have a clique on the remaining $n - k - 1$ vertices, each of which has degree at least $n - k - 2$.

- Say $d(x_k) \geq k + 1$ Consider the neighbours of x_k in the clique. These neighbours have degree at least $n - k - 1$, and hence since $G = cl(G)$, must be adjacent to x_{k+1} as well as $d(x_{k+1}) \geq k + 1$, But then, these neighbours have degree at least $n - k$, and hence must be adjacent to all of $x_1, \ldots, x_{k+1}$ by the same argument. Thus, I and $N(I)$ together form a complete bipartite graph with $|N(I)| \geq k + 1 = |I|$. If $d(x_{k+1}) > k + 1$, then it is easy to see that the graph is hamiltonian, and otherwise $k + 1 = d(x_i) \forall i \in [k+1]$, giving

$$\sigma_{k+1} = n + k^2 - k - 1 = (k+1)^2 \iff n = 3k + 2$$

which is false as we assumed $n \geq 2k^2 + 2k$.
- Otherwise $d(x_k) = k$, , and hence $d(x_{k+1}) = \sigma_{k+1} - k^2 = n - k - 1$, so we have a clique on the $n - k$ vertices in $G \backslash \{x_1, \ldots, x_k\}$. The neighbours of any $x_i, i \in [k]$ must have degree at least $n - k$, and hence are joined to all the x_i. Thus, we obtain the desired extremal graph with exactly $\binom{n-k}{2} + k^2$ many edges, namely a clique on $n - k$ vertices and k other independent vertices forming a complete bipartite graph with some k vertices from the clique. $\square$

4   Concluding Remarks

A simpler proof of Theorem 1 with a weaker constant can be obtained using Turán's theorem and a theorem of Erdős and Gallai [10] on the length of the longest cycle in a graph. Consider any k-cyclable graph with $\alpha(G) \geq k$. Then, let S be a set of k independent vertices, and consider the cycle containing it. This gives us a cycle of length atleast $2k$, as any two independent vertices are not adjacent to each other. Thus, we must have $\alpha(G) < k$. By a variant of Turán's theorem, we also have $\alpha > \frac{n}{\bar{d}+1}$, where $\bar{d}$ is the average degree. Thus, we obtain

$$\frac{2|E(G)|}{n} + 1 = \tilde{d} + 1 > \frac{n}{\alpha} \geq \frac{n}{k-1} \implies |E(G)| \geq \frac{1}{2} n \left(\frac{n}{k-1} - 1 \right)$$

which is larger than $\frac{1}{2}(2k-1)(n-1)$ if $n \geq 2k^2$. giving $c(G) \geq 2k$ when $k \leq \sqrt{n/2}$.

It is also interesting to understand what happens to the circumference of k-cyclable graphs for large values of k. As mentioned earlier in the introduction, it is not necessarily the case that $c(G) = n$ when $k = n - 1$ due to the existence of hypohamiltonian graphs. Thus, we have the following extremal problem.

Conjecture 1. *For a given n, let $f(n)$ be the largest value of k such that any k-cyclable graph satisfies $c(G) > k$. From the above, we have $f(n) < n - 1$ and from Theorem 1, we know $f(n) = \Omega(n)$. Is it the case that $f(n) = n - 2$?*

We can also ask for what regime of k as a function of n do results of the type in Theorem 1 hold.

Conjecture 2. *For a given n, let $g(n)$ be the largest value of k such that any k-cyclable graph satisfies $c(G) \geq 2k$. From Theorem 1 we know $g(n) = \Omega(\sqrt{n})$. Is it the case that $g(n) = O(\sqrt{n})$?*

Moreover, our results only give an improvement of the form $c(G) \geq (1+\gamma)k$, $0 < \gamma < 1$, for k up to around $2\sqrt{n}$, and it is natural to ask if such a linear bound on the circumference can be obtained for much larger regimes of k. Finally, note that the results of Theorem 3 only hold for $n \geq 2(k^2 + k)$. For fixed values of $k \leq 3$, [4] give a tight bound for the minimum value of n for this to hold. They also note that this bound cannot hold for $k = \Omega(n)$, in particular if $p = \lfloor \frac{n-1}{2} \rfloor$, the graph obtained by joining $n-p$ independent vertices to each vertex of K_p is k-connected and nonhamiltonian, with total number of edges more than $\binom{n-k}{2} + k^2$ when $\frac{n+1}{6} < k < \lfloor \frac{n-1}{2} \rfloor$. This still leaves a significant gap in the possible range of k for which k-connectivity and $|E(G)| > \binom{n-k}{2} + k^2$ implies hamiltonicity, as our result only applies for $k = O(\sqrt{n})$.

References

1. Bauer, D., McGuire, L., Trommel, H., Veldman, H.J.: Long cycles in 3-cyclable graphs. Discret. Math. **218**(1–3), 1–8 (2000). https://doi.org/10.1016/S0012-365X(99)00331-3
2. Björklund, A., Husfeldt, T., Taslaman, N.: Shortest cycle through specified elements. In: Proceedings of the Annual ACM-SIAM Symposium on Discrete Algorithms, pp. 1747–1753 (2012). https://doi.org/10.1137/1.9781611973099.139
3. Bondy, J.A., Chvatal, V.: A method in graph theory. Discret. Math. **15**(2), 111–135 (1976). https://doi.org/10.1016/0012-365X(76)90078-9
4. Byer, O.D., Smeltzer, D.L.: Edge bounds in nonhamiltonian k-connected graphs. Discret. Math. **307**(13), 1572–1579 (2007). https://doi.org/10.1016/j.disc.2006.09.008
5. Chvátal, V., Erdös, P.: A note on Hamiltonian circuits. Discret. Math. **2**(2), 111–113 (1972). https://doi.org/10.1016/0012-365X(72)90079-9
6. Crespelle, C., Golovach, P.A.: Cyclability in graph classes. Discret. Appl. Math. **313**, 147–178 (2022). https://doi.org/10.1016/j.dam.2022.01.021
7. Dirac, G.A.: Some theorems on abstract graphs. Proc. London Math. Soc. **s3-2**(1), 69–81 (1952). https://doi.org/10.1112/plms/s3-2.1.69

8. Dirac, G.A.: In abstrakten Graphen vorhandene vollständige 4-Graphen und ihre Unterteilungen. Math. Nachr. **22**(1–2), 61–85 (1960). https://doi.org/10.1002/mana.19600220107
9. Doyen, J., Van Diest, V.: New families of hypohamiltonian graphs. Discret. Math. **13**(3), 225–236 (1975). https://doi.org/10.1016/0012-365X(75)90020-5
10. Erdős, P., Gallai, T.: On maximal paths and circuits of graphs. Acta Mathematica Academiae Scientiarum Hungaricae **10**(3–4), 337–356 (1959). https://doi.org/10.1007/BF02024498
11. Faudree, R.J.: Survey of results on k-ordered graphs. Discret. Math. **229**(1–3), 73–87 (2001). https://doi.org/10.1016/S0012-365X(00)00202-8
12. Gould, R.J.: A look at cycles containing specified elements of a graph. Discret. Math. **309**(21), 6299–6311 (2009). https://doi.org/10.1016/j.disc.2008.04.017
13. J.L. Fouquet, J.J.: Probléme 438. Problémes combinatoires et théorie des graphes, Univ. Orsay, Orsay (1976)
14. Kawarabayashi, K.: An improved algorithm for finding cycles through elements. In: Lodi, A., Panconesi, A., Rinaldi, G. (eds.) IPCO 2008. LNCS, vol. 5035, pp. 374–384. Springer, Heidelberg (2008). https://doi.org/10.1007/978-3-540-68891-4_26
15. Li, H.: Generalizations of Dirac's theorem in Hamiltonian graph theory-a survey. Discret. Math. **313**(19), 2034–2053 (2013). https://doi.org/10.1016/j.disc.2012.11.025
16. Ng, L., Schultz, M.: k-ordered Hamiltonian graphs. J. Graph Theory **24**(1), 45–57 (1997). https://doi.org/10.1002/(SICI)1097-0118(199701)24:1⟨45::AID-JGT6⟩3.0.CO;2-J
17. Suil, O., West, D.B., Wu, H.: Longest cycles in k-connected graphs with given independence number. J. Comb. Theory Ser. B **101**(6), 480–485 (2011). https://doi.org/10.1016/j.jctb.2011.02.005
18. Ore, O.: Note on Hamilton circuits. Am. Math. Mon. **67**(1), 55 (1960). https://doi.org/10.2307/2308928

Graph Domination

On Three Domination-Based
Identification Problems in Block Graphs

Dipayan Chakraborty[1] , Florent Foucaud[1][(✉)] , Aline Parreau[2],
and Annegret K. Wagler[1]

[1] Université Clermont-Auvergne, CNRS, Mines de Saint-Étienne,
Clermont-Auvergne-INP, LIMOS, 63000 Clermont-Ferrand, France
{dipayan.chakraborty,florent.foucaud,annegret.wagler}@uca.fr
[2] Univ Lyon, CNRS, INSA Lyon, UCBL, Centrale Lyon, Univ Lyon 2, LIRIS,
UMR5205, F-69622 Villeurbanne, France
aline.parreau@univ-lyon1.fr

Abstract. The problems of determining the minimum-sized *identifying*, *locating-dominating* and *open locating-dominating codes* of an input graph are special search problems that are challenging from both theoretical and computational viewpoints. In these problems, one selects a dominating set C of a graph G such that the vertices of a chosen subset of $V(G)$ (i.e. either $V(G) \setminus C$ or $V(G)$ itself) are uniquely determined by their neighborhoods in C. A typical line of attack for these problems is to determine tight bounds for the minimum codes in various graph classes. In this work, we present tight lower and upper bounds for all three types of codes for *block graphs* (i.e. diamond-free chordal graphs). Our bounds are in terms of the number of maximal cliques (or *blocks*) of a block graph and the order of the graph. Two of our upper bounds verify conjectures from the literature - with one of them being now proven for block graphs in this article. As for the lower bounds, we prove them to be linear in terms of both the number of blocks and the order of the block graph. We provide examples of families of block graphs whose minimum codes attain these bounds, thus showing each bound to be tight.

Keywords: identifying code · locating-dominating · domination number · block graph · maximal clique · order of a graph · articulation

1 Introduction

For a graph (or network) G that models a facility or a multiprocessor network, detection devices can be placed at its vertices to locate an intruder (like a faulty processor, a fire or a thief). Depending on the features of the detection devices,

This work was sponsored by a public grant overseen by the French National Research Agency as part of the "Investissements d'Avenir" through the IMobS3 Laboratory of Excellence (ANR-10-LABX-0016) and the IDEX-ISITE initiative CAP 20-25 (ANR-16-IDEX-0001). We also acknowledge support of the ANR project GRALMECO (ANR-21-CE48-0004).

A. Bagchi and R. Muthu (Eds.): CALDAM 2023, LNCS 13947, pp. 271–283, 2023.
https://doi.org/10.1007/978-3-031-25211-2_21

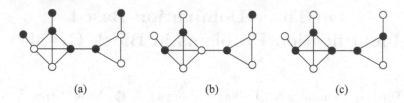

Fig. 1. Examples of (a) ID-code, (b) LD-code and (c) OLD-code. The set of black vertices in each of the three graphs constitute the respective code of the graph.

different types of dominating sets can be used to determine the optimum distributions of these devices across the vertices of G. In this article, we study three problems arising in this context, namely three types of dominating sets - the *identifying codes*, *locating-dominating codes* and *open locating-dominating codes* - of a given graph. Each of these problems has been extensively studied during the last decades. These three types of codes are among the most prominent notions within the larger research area of identification problems in discrete structures pioneered by Rényi [24], with numerous applications, for example in fault-diagnosis [23], biological testing [21] or machine learning [8].

Let $G = (V(G), E(G))$ be a graph, where $V(G)$ and $E(G)$ denote the set of vertices (also called the *vertex set*) and the set of edges (also called the *edge set*), respectively, of G. The *(open) neighborhood* of a vertex $u \in V(G)$ is the set $N_G(u)$ of all vertices of G adjacent to u; and the set $N_G[u] = \{u\} \cup N_G(u)$ is called the *closed neighborhood* of u.

A vertex subset $C \subseteq V(G)$ is called an *identifying code* [20] (or an *ID-code* for short) of G if

(1) $N_G[u] \cap C \neq \emptyset$ for each vertex u (the property of *domination*); and
(2) $N_G[u] \cap C \neq N_G[v] \cap C$ for all distinct vertices $u, v \in V(G)$ (the property of *closed-separation* in G).

See Fig. 1(a) for an example of an ID-code. A graph G admits an ID-code if and only if G has no *closed-twins* (i.e. a pair of distinct vertices $u, v \in V(G)$ with $N_G[u] = N_G[v]$).

A subset $C \subseteq V(G)$ is called a *locating-dominating code* [26] (or an *LD-code* for short) of G if

(1) $N_G[u] \cap C \neq \emptyset$ for each vertex u (the property of domination); and
(2) $N_G(u) \cap C \neq N_G(v) \cap C$ for all distinct vertices $u, v \in V(G) \setminus C$ (the property of *location* in G).

See Fig. 1(b) for an example of an LD-code. Every graph has an LD-code.

Finally, a subset $C \subseteq V(G)$ is called an *open locating-dominating code* [25] (or an *OLD-code* for short) of G if

(1) $N_G(u) \cap C \neq \emptyset$ for each vertex u (the property of *open-domination*); and
(2) $N_G(u) \cap C \neq N_G(v) \cap C$ for all distinct vertices $u, v \in V(G)$ (the property of *open-separation* in G).

See Fig. 1(c) for an example of an OLD-code. A graph G admits an OLD-code if and only if G has neither isolated vertices nor *open-twins* (i.e. a pair of distinct vertices $u, v \in V(G)$ with $N_G(u) = N_G(v)$).

A graph with no open twins, no closed twins or neither open- nor closed-twins is also called *open-twin-free*, *closed-twin-free* and *twin-free*, respectively. For the rest of this article, we often simply use the word *code* to mean any of the above three ID-, LD- or OLD-codes without distinction. Given a graph G, the *identifying code number* $\gamma^{ID}(G)$ (or *ID-number* for short), the *locating-dominating number* $\gamma^{LD}(G)$ (or *LD-number* for short) and the *open locating-dominating number* $\gamma^{OLD}(G)$ (or *OLD-number* for short) of G are the minimum cardinalities among all ID-codes, LD-codes and OLD-codes, respectively, of G. In other words, for simplicity, for any symbol $X \in \{ID, LD, OLD\}$, we have the X-number: $\gamma^X(G) = \min\{|C| : C \text{ is an X-code of } G\}$. In the case that all three codes are addressed together as one unit anywhere in the text, i.e. any specific symbol for $X \in \{ID, LD, OLD\}$ is irrelevant to the context, we then simply refer to the X-numbers as the *code numbers* of G.

For any two sets A and B, let $A \triangle B = (A \setminus B) \cup (B \setminus A)$ denote the *symmetric difference* between A and B. Then, for a vertex subset $C \subset V(G)$ and distinct vertices $u, v \in V(G)$, if there exists a vertex $w \in (N_G(u) \cap C) \triangle (N_G(v) \cap C)$ (resp. $(N_G[u] \cap C) \triangle (N_G[v] \cap C)$), then both C and the vertex w are said to *open-separate* (resp. *closed-separate*) the vertices u and v (in C).

Known Results. Given a graph G, determining $\gamma^{ID}(G)$, $\gamma^{LD}(G)$ or $\gamma^{OLD}(G)$ is, in general, NP-hard [7,25] and remains so for several graph classes like bipartite graphs [7], split and interval graphs [15] where other hard problems become easy to solve. The problems are also hard to approximate within a factor of $\log |V(G)|$ [10]. As these problems are computationally hard, a typical line of attack is to determine bounds on the code numbers for specific graph classes. Lower bounds for all three code numbers for several graph classes like interval graphs, permutation graphs, cographs [14] and lower bounds for ID-numbers for trees [5], line graphs [12], planar graphs [22] and many others of bounded VC-dimension [6] have been determined. Upper bounds for ID-codes (See e.g. [4,9]), LD-codes (see e.g. [4,13,16]) and OLD-codes (see [18]) for certain graph classes have also been obtained.

Our Work. In this paper, we consider the family of block graphs, a subclass of chordal graphs defined by Harary in [17] (see also [19] for equivalent characterizations). A *block graph* is a graph in which every maximal 2-connected subgraph (or *block*) is complete. Linear-time algorithms to compute all three code numbers in block graphs have been presented in [2]. In this paper, we complement these results by determining tight lower and upper bounds on all three code numbers for block graphs. We give bounds using (i) the number of vertices, i.e. the order of a graph, as has been done for several other classes of graphs; and (ii) the number of blocks of a block graph, a quantity equally relevant to block graphs. In doing so, we also prove the following conjectures.

Conjecture 1 ([1], Conjecture 1). The ID-number of a closed-twin-free block graph is bounded above by the number of blocks in the graph.

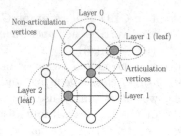

Fig. 2. Example of different layer numbers, articulation vertices (grey) and non-articulation vertices (white) of a connected block graph.

Conjecture 2 ([13,16], Conjecture 2). Every twin-free graph G with no isolated vertices satisfies $\gamma^{LD}(G) \leq \frac{|V(G)|}{2}$.

Terminologies. For a block graph G, let $\mathcal{K}(G)$ denote the set of all blocks of G, i.e. the set of all complete subgraphs of G of maximal order. Note that the vertex sets of any two distinct blocks of G can intersect at a single vertex at most; and any such vertex at the intersection of the vertex sets of two distinct blocks is called an *articulation vertex* of both the blocks. Any vertex that is *not* an articulation vertex, is called a *non-articulation vertex* of G. For a connected block graph G, we fix a *root block* K_0 of G and define a system of assigning numbers to every block of G depending on "how far" the latter is from K_0. So, define a *layer function* $f : \mathcal{K}(G) \to \mathbb{Z}$ on G by: $f(K_0) = 0$ and, for any other (*non-root*) block K, define inductively $f(K) = i$ if $V(K) \cap V(K') \neq \emptyset$ for some block K' other than K such that $f(K') = i - 1$. Any block K with $f(K) = i$ is said to be in *Layer i*. See Fig. 2 for a demonstration of the layers.

For a pair of distinct blocks K, K' of G such that their vertex sets intersect and that $f(K) = f(K') + 1$, we call the (only) vertex in the intersection $V(K) \cap V(K')$ *the negative articulation vertex* of K and *a positive articulation vertex* of K'. Note that the root block does not have any negative articulation vertex and every other block has exactly *one* negative articulation vertex. Finally, any block of G that has exactly one articulation vertex is called a *leaf block*, and whereas any block that is *not* a leaf block is called a *non-leaf block* of G.

Structure of the Paper. Sections 2 and 3 of this paper are dedicated to our results on the upper bounds and lower bounds, respectively, on the code numbers of block graphs. We conclude the paper in Sect. 4. In this extended abstract, Theorems 4 and 6 are presented with their proof sketches only, whereas Theorem 9 and all lemmas are presented with their statements only. Theorems 3 and 8, however, are presented with their proofs in full. For the purposes of this abstract, all results marked with ($\star$) are either presented with only their statements or with only sketches of their proofs.

2   Upper Bounds

In this section, we establish upper bounds on the ID-, LD- and OLD-numbers for block graphs. Two of these upper bounds are proving Conjectures 1 and 2.

2.1 Identifying Codes

The number of blocks is as relevant a quantity for block graphs as is the number of vertices for trees. Next, we prove Conjecture 1 to provide an upper bound on $\gamma^{ID}(G)$ for a block graph G in terms of its number of blocks.

Theorem 3. *Let G be a closed-twin-free block graph and let $\mathcal{K}(G)$ be the set of all blocks of G. Then $\gamma^{ID}(G) \leq |\mathcal{K}(G)|$.*

Proof. As the ID-number of a graph is the sum of the ID-numbers of all its components, it is enough to assume that the block graph G is connected. Now, assume by contradiction that there is a closed-twin-free block graph G of minimum order such that $\gamma^{ID}(G) > |\mathcal{K}(G)|$. We also assume that G has at least four vertices since it can be easily checked that the theorem is true for block graphs with at most three vertices. Suppose that $K \in \mathcal{K}(G)$ is a leaf-block of G. Due to the closed-twin-free property of G, one can assume that $V(K) = \{x, y\}$. Without loss of generality, suppose that x is the non-articulation and y the negative articulation vertex of K. Let $G' = G - x$ be the graph obtained by deleting the vertex $x \in V(G)$ (and the edge incident on x) from G. Then G' is a block graph with $|\mathcal{K}(G')| = |\mathcal{K}(G)| - 1$. We now consider the following two cases.

Case 1 (G' is closed-twin-free). By the minimality of the order of G, there is an ID-code C' of G' such that $|C'| \leq |\mathcal{K}(G')| = |\mathcal{K}(G)| - 1$. First, assume that $y \notin C'$. Then by the property of domination of C', there exists a vertex $z \in V(G')$ such that $z \in N_{G'}(y) \cap C'$. We claim that $C = C' \cup \{x\}$ is an ID-code of G. First of all, that C is a dominating set of G is clear from the fact that C' is a dominating set of G'. To prove that C is a closed-separating set of G, we see that x is closed-separated in C from all vertices in $V(G') \setminus \{y\}$ by itself and is closed-separated in C from y by the vertex $z \in C'$. Moreover, all other pairs of distinct vertices closed-separated by C' and are also closed-separated by C. Thus, C, indeed, is an ID-code of G. This implies that $\gamma^{ID}(G) \leq |C| \leq |\mathcal{K}(G)|$, contrary to our assumption.

We therefore assume that $y \in C'$. If again, there exists a vertex $z \in N_{G'}(y) \cap C'$, then by the same reasoning as above, $C = C' \cup \{x\}$ is an ID-code of G. Otherwise, we have $N_{G'}[y] \cap C' = \{y\}$. Now, since G is connected, we have $deg_G(y) > 1$ and therefore, there exists a vertex $w \in N_G(y) \setminus \{x\}$. Then $C = C' \cup \{w\}$ is an ID-code of G. To prove so, we only need to check that C closed-separates x from every vertex in $V(G')$. Now, y closed-separates x from every vertex in $V(G') \setminus \{y, w\}$ in C; w closed-separates x from y in C; and w closed-separates itself from x in C. Moreover, C is clearly also a dominating set of G. Hence, this leads to the same contradiction as before.

Case 2 (G' has closed-twins). Assume that vertices $u, v \in V(G')$ are a pair of closed-twins of G'. Since u and v were not closed-twins in G, it means that x is adjacent to u, say, without loss of generality. This implies that $u = y$. Note that v is then unique with respect to being a closed-twin with y in G'. This is because, if y and some vertex $v'(\neq v) \in V(G')$ were also closed-twins in G', then it would mean that v and v' were closed-twins in G, contrary to our assumption. Now, let $G'' = G' - v$. We claim the following.

Claim 2A. G" is closed-twin-free.

Proof of Claim 2A. Toward a contradiction, if vertices $z, w \in V(G'')$ were a pair of closed-twins in G'', it would mean that $z \in N_{G'}(v)$, without loss of generality, and $w \notin N_{G'}(v)$. This would, in turn, imply that $z \in N_{G'}(y)$ (since y and v are closed-twins in G'). Or, in other words, $y \in N_{G''}(z)$. Now, since z and w are closed-twins in G'', we have $w \in N_{G'}(y)$. Again, by virtue of y and v being closed-twins in G', we have $w \in N_{G'}(v)$, contrary to our assumption. ■

We also note here that the vertices y and v must be from the same block for them to be closed-twins in G'. Thus, G'' is a connected closed-twin-free block graph. Therefore, by the minimality of the order of G, there is an ID-code C'' of G'' such that $|C''| \leq |\mathcal{K}(G'')| < |\mathcal{K}(G)|$. If $y \notin C''$, then we claim that $C = C'' \cup \{x\}$ is an identifying code of G. This is true because, firstly, C is a dominating set of G (note that, by the property of domination of C'' in G'', there exists a vertex $z \in N_{G''}(y) \cap C''$; and since y and v are closed-twins in G', we have $z \in N_G(v) \cap C$). Moreover, x is closed-separated in C from every other vertex in $V(G) \setminus \{y\}$ by x itself; and x and y are closed-separated in C by some vertex in $N_{G''}(y) \cap C''$ that dominates y. The vertices y and v are closed-separated in C by x; y is closed-separated in C'' from all vertices in $V(G'') \setminus \{y\}$ and so is v, since y and v have the same closed neighborhood in G'. Finally, every pair of distinct vertices closed-separated by C'' still remain so by C. Thus, C, indeed, is an ID-code of G. This implies that $\gamma^{ID}(G) \leq |C| \leq |\mathcal{K}(G)|$; again a contradiction.

Let us, therefore, assume that $y \in C''$. This time, we claim that $C = (C' \setminus \{y\}) \cup \{x, v\}$ is an ID-code of G. That C is a dominating set of G is clear. As for the closed-separating property of C, as before, x is closed-separated in C from every vertex in $V(G) \setminus \{y\}$ by x itself; and x and y are closed-separated in C by v. Vertices y and v are closed-separated in C by x; and v and x are closed-separated in C by v. Since y and v have the same closed neighbourhood in G' and since y is closed-separated in C'' from every other vertex in $V(G'')$, both v and y are each closed-separated in C from every vertex in $V(G'') \setminus \{v, y\}$. Finally, every pair of distinct vertices of G'' closed-separated by C'' still remain so by C. This proves that C is an ID-code of G and hence, again, we are led to the contradiction that $\gamma^{ID}(G) \leq |C| \leq |\mathcal{K}(G)|$. This proves the theorem. □

Besides for stars, the upper bound in Theorem 3 is attained by the ID-numbers of thin headless spiders [3]. These graphs, therefore, serve as examples to show that the bound in Theorem 3 is tight.

2.2 Locating-Dominating Codes

In our next result, we prove Conjecture 2 for block graphs.

Theorem 4 (⋆). *Let G be a twin-free block graph with no isolated vertices. Then we have $\gamma^{LD}(G) \leq \frac{|V(G)|}{2}$.*

Proof (sketch). It is enough to prove the theorem for a connected twin-free block graph G. The proof follows from partitioning the vertex set of G into two *parts*

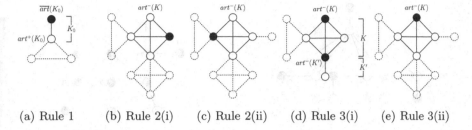

(a) Rule 1   (b) Rule 2(i)   (c) Rule 2(ii)   (d) Rule 3(i)   (e) Rule 3(ii)

Fig. 3. The rules in the proof of Theorem 4. The symbols $art^+(K)$, $art^-(K)$ and $\overline{art}(K)$ represent the set of all positive articulation, negative articulation and non-articulation vertices, respectively of K. The black and white vertices represent those picked in the sets C^* and D^*, respectively. The blocks with dashed edges represent those that are yet to be analysed for their choices of vertices in C^* and D^*.

C^* and D^* and showing that both the parts are LD-codes of G. So, assign a leaf block of G to be the root block and define a layer function f on G with the root block in Layer 0. Then construct the sets C^* and D^* by the following rules.

(1) The root block is of size 2, as G is twin-free. So, pick the positive articulation vertex of the root block in D^* and the other vertex in C^*. See Fig. 3(a). Next, assume that K is non-root block of G.

(2) Let the negative articulation vertex of K be in D^*. (i) If K has one non-articulation vertex, pick it in C^*. Moreover, pick all positive articulation vertices of K in D^*. (ii) If K has no non-articulation vertices, pick *one* of its positive articulation vertices in C^*, and the rest in D^*. See Figs. 3(b) and 3(c).

(3) Let the negative articulation vertex of K be in C^*. (i) If K has one non-articulation vertex, pick it in D^*. Pick *one* positive articulation vertex (if available) of K in C^*, and the rest in D^*. (ii) If K has no non-articulation vertices, pick all its positive articulation vertices in D^*. See Figs. 3(d) and 3(e).

Clearly, the sets C^* and D^* are complements of each other in $V(G)$; and every block of G has at least one vertex in each of them. Thus, both are dominating sets of G. Next, we show that both C^* and D^* are locating sets of G each. We start with C^* and show that any two distinct vertices $u, v \in D^*$ are open-separated in C^*. As G is twin-free, there exist distinct blocks $K, K' \in \mathcal{K}(G)$ such that $u \in V(K)$ and $v \in V(K')$. Then, it is enough to show the following claim.

Claim: Either u or v is an articulation vertex of K or K', respectively.

Proof of Claim. Toward a contradiction, let us assume that both u and v are non-articulation vertices of K and K', respectively. Since both $V(K)$ and $V(K')$ have non-empty intersection with C^*, the only non-trivial case to investigate is

$$V(K) \cap C^* = V(K') \cap C^* = V(K) \cap V(K'). \quad (1)$$

Case 1 ($f(K') = f(K) + 1$): Here, K must be a non-root block (by Rule 1) and has its negative articulation vertex in D^*. Since u is a non-articulation vertex of K, by Rule 2(i), u must belong to C^*, a contradiction to our assumption $u \in D^*$.

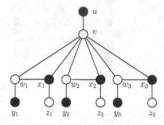   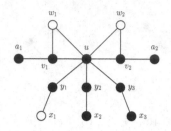

(a) Block graph H_3 whose LD-number attains the upper bound in Theorem 4.   (b) Block graph $G_{2,3}$ whose OLD-number attains the upper bound in Theorem 6.

Fig. 4. The black vertices constitute a minimum respective code of each graph.

Case 2 ($f(K) = f(K')$). Here, the negative articulation vertices of both K and K' are the same and is in C^*. Assume K to be a non-leaf block (as one of K, K' must be, for G to be twin-free). Since v is a non-articulation vertex of K', by Rule 3(i), K has a positive articulation vertex in C^*, which contradicts (1). ∎

This proves the above claim and that C^* is a locating set of G. The proof for D^* being a locating set of G is carried out in a very similar manner.   □

The trees whose LD-codes attain the bound in Theorem 4 were characterized in [13]. There are also arbitrarily large connected twin-free block graphs - that are not trees - and whose LD-numbers attain the bound in Theorem 4. Examples of such graphs are, for instance, those of the type in Fig. 4(a). We therefore have the following proposition.

Proposition 5 (⋆).   *There exist arbitrarily large connected twin-free block graphs whose LD-numbers are equal to half the number of vertices.*

2.3   Open Locating-Dominating Codes

We now focus our attention on upper bounds on OLD-numbers of block graphs.

Theorem 6 (⋆).   *Let G be a connected open-twin-free block graph such that G is neither a copy of P_2 nor of P_4. Let $m_Q(G)$ be the number of non-leaf blocks of G with at least one non-articulation vertex. Then $\gamma^{OLD}(G) \leq |V(G)| - m_Q(G) - 1$.*

Proof (sketch). It is easy to check that the result holds when G is iomorphic to a bull graph (a K_3 with two leaves each adjacent to a distinct vertex of the K_3; see Fig. 5(a)); So, we assume that G is not a bull graph. We define a particular type of "join" of two graphs: Assume G' to be any graph and X to be either a 4-path or a bull graph. For a fixed vertex $q \in V(G')$, we define a new graph $G' \rhd_q X$ to be the graph obtained by identifying a vertex $q \in V(G')$ with an articulation vertex of X. Next, we choose a root block of G according to whether $G \cong G' \rhd_q X$ or $G \not\cong G' \rhd_q X$, for some block graph G'. Thereafter, we construct a particular vertex subset $C \subset V(G)$ and, through various case analyses, show that C indeed is an OLD-code of G and is of size $|V(G)| - m_Q(G) - 1$.   □

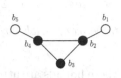

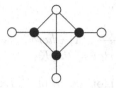

(a) The Bull graph.

(b) Z: Graph of largest size whose minimum OLD-code is a 3-clique.

Fig. 5. The black vertices constitute a minimum respective code of each graph.

Foucaud et al. [11] have shown that, for any open-twin-free graph G with no isolated vertices, $\gamma^{OLD}(G) \leq |V(G)| - 1$ unless G is a special kind of bipartite graph called a *half-graph* (a *half-graph* is a bipartite graph with both parts of the same size, where each part can be ordered so that the open neighbourhoods of consecutive vertices differ by exactly one vertex). Since P_2 and P_4 are the only block graphs that are half-graphs, Theorem 6 is a refinement of their result for block graphs.

We now show that the upper bound on the OLD-numbers for block graphs in Theorem 6 is tight and is attained by arbitrarily large connected block graphs of the type in Fig. 4(b).

Proposition 7 ($\star$). *There exist arbitrarily large connected open-twin free block graphs whose OLD-numbers equal the upper bound in Theorem 6.*

3 Lower Bounds

The general lower bound for the size of an identifying code using the number of vertices is $\gamma^{ID}(G) \geq \lceil \log_2(|V(G)| + 1) \rceil$ [20]. However, to reach this bound, a graph needs to have a large VC-dimension [6] (the *VC-dimension* of a graph G is the size of a largest *shattered set*, that is, a set S of vertices such that for every subset S' of S, some closed neighbourood in G intersects S exactly at S'). Indeed, if a graph has VC-dimension c, then any identifying code has size at least $O(|V(G)|^{1/c})$ [6]. The value $1/c$ is not always tight, see for example the case of line graphs which have VC-dimension at most 4 but for which the tight order for the lower bound is $\Omega(|V(G)|^{1/2})$ [12]. Similar results hold for LD- and OLD-codes (using the same techniques as in [6]). Block graphs have VC-dimension at most 2 (one can check that a shattered set of size 3 would imply the existence of an induced 4-cycle or diamond), and thus, using results from [6], their ID-number is lower bounded by $\Omega(|V(G)|^{1/2})$. In this section, we improve this lower bound to a linear one which is also tight. Our first result of this section is the following.

Theorem 8. *Let G be a connected block graph. Then we have*

$\gamma^{ID}(G) \geq \frac{|V(G)|}{3} + 1$, $\gamma^{LD}(G) \geq \frac{|V(G)|+1}{3}$ and, for G not isomorphic to Z,

$\gamma^{OLD}(G) \geq \frac{|V(G)|}{3} + 1$; where Z is the graph K_4 with three leaves each adjacent to a distinct vertex of the K_4.

See Fig. 5(b) for the graph Z. Extremal cases where these bounds are attained can be constructed as follows (see Fig. 6). Consider the graph with one path on vertices $u_1, ..., u_k$ (the vertices in the code) and attach further vertices as follows.

(1) for an ID-code C: attach a single vertex to each u_i and vertices to the pairs u_i, u_{i+1} for $1 < i < k - 1$,
(2) for an OLD-code C: attach a single vertex to u_1, u_k and each u_i for $2 < i < k - 1$ and vertices to all the pairs u_i, u_{i+1},
(3) for an LD-code C: attach a single vertex to each u_i and vertices to all the pairs u_i, u_{i+1}.

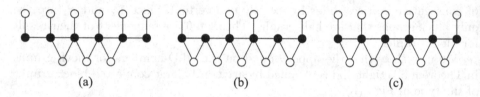

(a)   (b)   (c)

Fig. 6. Extremal cases where the lower bounds are attained, black vertices form a minimum (a) ID-code, (b) OLD-code, (c) LD-code.

Note that the graphs presented here are all the possible extremal cases for ID-codes, whereas further extremal graphs for OLD-codes and for LD-codes exist. If we now consider the parameter $|\mathcal{K}(G)|$, we can use the relation $|V(G)| \geq |\mathcal{K}(G)| + 1$ to obtain a similar lower bound. However, this lower bound can be improved as our next theorem shows.

Theorem 9 ($\star$). *Let G be a connected block graph and $\mathcal{K}(G)$ be the set of all blocks of G. Then we have*

$$\gamma^{ID}(G) \geq \frac{3(|\mathcal{K}(G)|+2)}{7}, \quad \gamma^{LD}(G) \geq \frac{|\mathcal{K}(G)|+2}{3} \text{ and } \gamma^{OLD}(G) \geq \frac{|\mathcal{K}(G)|+3}{2}.$$

To prove Theorems 8 and 9, we introduce the following notations and terminologies. By $n_i(G)$ we shall mean the number of vertices of degree i in a graph G. For a given code C of a connected block graph G, let the subgraph $G[C]$ of G have k components and that $C_1, C_2, ..., C_k$ are all of its components. Note that each C_i is a block graph and so is $G[C]$, therefore. Then, $V(G)$ is partitioned into the four following parts. Starting with $V_1 = C$, we define the other parts.

(1) $V_2 = \{v \in V(G) \setminus V_1 : |N(v) \cap C| = 1\}$,
(2) $V_3 = \{v \in V(G) \setminus V_1 :$ there exist distinct $i, j \leq k$ such that $N(v) \cap C_i \neq \emptyset$ and $N(v) \cap C_j \neq \emptyset\}$, and
(3) $V_4 = V(G) \setminus (V_1 \cup V_2 \cup V_3)$. Note that, for all $v \in V_4$, $N(v) \cap C \subset V(C_i)$ for some i and that $|N(v) \cap V(C_i)| \geq 2$.

Our next lemmas establish upper bounds on the sizes of V_1, V_2, V_3 and V_4.

Lemma 10 ($\star$). *Let G be a connected block graph and C be a code of G. Then following are upper bounds on the size of the vertex subset V_2 of G.*

(1) $|V_2| \leq |C| - n_0(G[C])$ *if C is an ID-code.*
(2) $|V_2| \leq |C|$ *if C is an LD-code.*
(3) $|V_2| \leq |C| - n_1(G[C])$ *if C is an OLD-code.*

Lemma 11 ($\star$). *Let G be a connected block graph and C be a code of G such that $G[C]$ has k components. Then, we have $|V_3| \leq k - 1$.*

Lemma 12 ($\star$). *Let G be a connected block graph and C be a code of G such that $G[C]$ has k components. Then, we have $|V_4| \leq |C| - k$. In particular,*

(1) $|V_4| \leq |C| - 3k + 2n_0(G[C])$ *if C is an ID-code;*
(2) $|V_4| \leq |\mathcal{K}(G[C])| \leq |C| - 2k_1 - 3k_2 + n_1(G[C])$ *if C is an OLD-code; where $k_1 = |\{C_i : C_i$ is a component of $G[C]$ and $C_i \cong K_3\}|$ and $k_2 = k - k_1$.*

Proof of Theorem 8. Let C be a code of G and that $G[C]$ have k components. We prove the theorem using the relation $|V(G)| = |C| + |V_2| + |V_3| + |V_4|$ and the upper bounds for $|V_2|$ $|V_3|$ and $|V_4|$ in Lemmas 10, 11 and 12, respectively.

If C is an ID-code, then we have

$$
\begin{aligned}
|V(G)| &= C| + |V_2| + |V_3| + |V_4| \\
&\leq |C| + |C| - n_0(G[C]) + k - 1 + |C| - 3k + 2n_0(G[C]) \\
&= 3C| - 2k - 1 + n_0(G[C]).
\end{aligned}
$$

Now, there must be at least as many components of $G[C]$ as there are isolated vertices in $G[C]$, i.e. we have $k \geq n_0(G[C])$. This implies that $|V(G)| \leq 3|C| - k - 1$. Thus, for $k \geq 2$, the result holds. Moreover, when $k = 1$, we must have $n_0(G[C]) = 0$ and so, again, the result holds.

If C is an LD-code, then the result holds because we have

$$|V(G)| = |C| + |V_2| + |V_3| + |V_4| \leq |C| + |C| + k - 1 + |C| - k = 3|C| - 1.$$

Finally, if C is an OLD-code, then we have

$$
\begin{aligned}
|V(G)| &= |C| + |V_2| + |V_3| + |V_4| \\
&\leq |C| + |C| - n_1(G[C]) + k_1 + k_2 - 1 + |C| - 2k_1 - 3k_2 + n_1(G[C]) \\
&= 3C| - k_1 - 2k_2 - 1.
\end{aligned}
$$

This implies that the result holds when either $k_1 \geq 2$ or when $k_2 \geq 1$.

If however, $k_1 = 1$ and $k_2 = 0$, then $G[C]$ is isomorphic to K_3. If $n \leq 6$, the result holds since $|V(G)| \leq 3|C| - 3$. Thus, let $|V(G)| = 7$. Since no vertex $v \in V(G) \setminus C$ can be adjacent to exactly two vertices of C (or else, the last vertex of C would not be open-separated from v), each vertex in $V(G) \setminus C$ must be adjacent to either exactly one or all three vertices of C. Therefore, $G \cong Z$ in Fig. 5(b). Hence, the result holds for all connected block graphs $G \not\cong Z$. $\square$

The proof of Theorem 9 is by using similar bounding techniques as in the proof of Theorem 8, but on $|\mathcal{K}(G)|$ instead of $|V(G)|$. Using $|\mathcal{K}(G)| = |E(G)| = |V(G)| - 1$ for any tree G, the bounds in Theorem 9 are equivalent to the known lower bounds for trees in terms of number of vertices (see [5] for ID-codes, [26] for LD-codes and [25] for OLD-codes). In fact, the code numbers of infinite families of trees attain the three bounds in Theorem 9.

4 Conclusion

Block graphs form a subclass of chordal graphs for which all three considered identification problems can be solved in linear time [2]. In this paper, we complemented this result by presenting tight lower and upper bounds for the optimum sizes of all the three types of codes. We gave bounds in terms of both the number of vertices - as it has been done for several other classes of graphs - and also the number of blocks of G - a parameter more fitting for block graphs. In particular, we verified Conjecture 1 on an upper bound on the ID-number for block graphs from [1] and Conjecture 2 on the LD-numbers from [16] for the special case of block graphs. Moreover, we addressed the questions to find block graphs where the provided lower and upper bounds are attained.

The structural properties of block graphs have enabled us to prove interesting bounds for the three considered problems. It would be further interesting to study other structured classes in a similar way. It would also be interesting to prove Conjecture 2 for a larger class of graphs, like chordal graphs, for example.

References

1. Argiroffo, G.R., Bianchi, S.M., Lucarini, Y., Wagler, A.K.: On the identifying code number of block graphs. In: Proceedings of ICGT 2018, Lyon, France (2018)
2. Argiroffo, G.R., Bianchi, S.M., Lucarini, Y., Wagler, A.K.: Linear-time algorithms for three domination-based separation problems in block graphs. Discret. Appl. Math. **281**, 6–41 (2020)
3. Argiroffo, G., Bianchi, S., Wagler, A.: Study of identifying code polyhedra for some families of split graphs. In: Fouilhoux, P., Gouveia, L.E.N., Mahjoub, A.R., Paschos, V.T. (eds.) ISCO 2014. LNCS, vol. 8596, pp. 13–25. Springer, Cham (2014). https://doi.org/10.1007/978-3-319-09174-7_2
4. Balbuena, C., Foucaud, F., Hansberg, A.: Locating-dominating sets and identifying codes in graphs of girth at least 5. Electron. J. Comb. **22**, P2.15 (2015)
5. Bertrand, N., Charon, I., Hudry, O., Lobstein, A.: 1-identifying codes on trees. Australas. J Comb. **31**, 21–36 (2005)
6. Bousquet, N., Lagoutte, A., Li, Z., Parreau, A., Thomassé, S.: Identifying codes in hereditary classes of graphs and VC-dimension. SIAM J. Discret. Math. **29**(4), 2047–2064 (2015)
7. Charon, I., Hudry, O., Lobstein, A.: Minimizing the size of an identifying or locating-dominating code in a graph is NP-hard. Theoret. Comput. Sci. **290**(3), 2109–2120 (2003)

8. Chlebus, B.S., Nguyen, S.H.: On finding optimal discretizations for two attributes. In: Polkowski, L., Skowron, A. (eds.) RSCTC 1998. LNCS (LNAI), vol. 1424, pp. 537–544. Springer, Heidelberg (1998). https://doi.org/10.1007/3-540-69115-4_74

9. Foucaud, F., Lehtilä, T.: Revisiting and improving upper bounds for identifying codes. SIAM J. Discret. Math. **36**(4), 2619–2634 (2022)

10. Foucaud, F.: Decision and approximation complexity for identifying codes and locating-dominating sets in restricted graph classes. J. Discret. Algorithms **31**, 48–68 (2015). https://doi.org/10.1016/j.jda.2014.08.004

11. Foucaud, F., Ghareghani, N., Roshany-Tabrizi, A., Sharifani, P.: Characterizing extremal graphs for open neighbourhood location-domination. Discret. Appl. Math. **302**, 76–79 (2021)

12. Foucaud, F., Gravier, S., Naserasr, R., Parreau, A., Valicov, P.: Identifying codes in line graphs. J. Graph Theory **73**(4), 425–448 (2013)

13. Foucaud, F., Henning, M.A.: Location-domination and matching in cubic graphs. Discret. Math. **339**(4), 1221–1231 (2016)

14. Foucaud, F., Mertzios, G.B., Naserasr, R., Parreau, A., Valicov, P.: Identification, location-domination and metric dimension on interval and permutation graphs. I. bounds. Theor. Comput. Sci. **668**, 43–58 (2017). https://doi.org/10.1016/j.tcs.2017.01.006

15. Foucaud, F., Mertzios, G.B., Naserasr, R., Parreau, A., Valicov, P.: Identification, location-domination and metric dimension on interval and permutation graphs. II. Algorithms and complexity. Algorithmica **78**(3), 914–944 (2017)

16. Garijo, D., González, A., Márquez, A.: The difference between the metric dimension and the determining number of a graph. Appl. Math. Comput. **249**, 487–501 (2014)

17. Harary, F.: A characterization of block-graphs. Can. Math. Bull. **6**(1), 1–6 (1963)

18. Henning, M.A., Yeo, A.: Distinguishing-transversal in hypergraphs and identifying open codes in cubic graphs. Graphs Comb. **30**, 909–932 (2014)

19. Howorka, E.: On metric properties of certain clique graphs. J. Comb. Theory Ser. B **27**(1), 67–74 (1979)

20. Karpovsky, M.G., Chakrabarty, K., Levitin, L.B.: On a new class of codes for identifying vertices in graphs. IEEE Trans. Inf. Theory **44**(2), 599–611 (1998)

21. Moret, B.M.E., Shapiro, H.D.: On minimizing a set of tests. SIAM J. Sci. Stat. Comput. **6**(4), 983–1003 (1985)

22. Rall, D.F., Slater, P.J.: On location-domination numbers for certain classes of graphs. Congr. Numer. **45**, 97–106 (1984)

23. Rao, N.: Computational complexity issues in operative diagnosis of graph-based systems. IEEE Trans. Comput. **42**(4), 447–457 (1993)

24. Rényi, A.: On random generating elements of a finite boolean algebra. Acta Scientiarum Mathematicarum Szeged **22**, 75–81 (1961)

25. Seo, S.J., Slater, P.J.: Open neighborhood locating dominating sets. Australas. J. Comb. **46**, 109–120 (2010)

26. Slater, P.J.: Domination and location in acyclic graphs. Networks **17**(1), 55–64 (1987)

Computational Aspects of Double Dominating Sequences in Graphs

Gopika Sharma[✉] and Arti Pandey

Department of Mathematics, Indian Institute of Technology Ropar,
Rupnagar, Punjab, India
{2017maz0007,arti}@iitrpr.ac.in

Abstract. In a graph $G = (V, E)$, a vertex $u \in V$ dominates a vertex $v \in V$ if $v \in N_G[u]$. A sequence $S = (v_1, v_2, \ldots, v_k)$ of vertices of G is called a double dominating sequence of G if (i) for each i, the vertex v_i dominates at least one vertex $u \in V$ which is dominated at most once by the previous vertices of S and, (ii) all vertices of G have been dominated at least twice by the vertices of S. GRUNDY DOUBLE DOMINATION problem asks to find a double dominating sequence of maximum length for a given graph G. In this paper, we prove that the decision version of the problem is NP-complete for bipartite and co-bipartite graphs. We look for the complexity status of the problem in the class of chain graphs which is a subclass of bipartite graphs. We use dynamic programming approach to solve this problem in chain graphs and propose an algorithm which outputs a Grundy double dominating sequence of a chain graph G in linear-time.

Keywords: Double Dominating Sequences · Bipartite Graphs · Chain Graphs · NP-completeness

1 Introduction

For a graph $G = (V, E)$, a set $D \subseteq V$ is called a *dominating set* of G, if for each vertex $x \in V$, $N_G[x] \cap D \neq \emptyset$. The MINIMUM DOMINATION problem is to find a dominating set of a graph G having minimum cardinality. One of the fundamental problems in graph theory is the MINIMUM DOMINATION problem and there is a huge amount of literature on this topic, see [7–10]. Further, Fink and Jacobson introduced the concept of double domination [4,5]. For a graph G with no isolated vertices, a set $D \subseteq V$ is called a *double dominating set* of G, if for every vertex $x \in D$, $|N_G[x] \cap D| \geq 2$.

In 2014, Brešar et al. introduced the concept of dominating sequences. A motivation for introducing dominating sequences came from the well known domination game in which we get a vertex sequence as an outcome of a two-player game, played on a graph. For detailed description, one may refer [2].

Formally, a *dominating sequence* of G is a sequence S of vertices of G such that (i) each vertex of S dominates at least one vertex of G which was not dominated by any of the previous vertices of S, and (ii) every vertex of G is dominated by at least one vertex of S. The GRUNDY DOMINATION problem is to find a longest dominating sequence of a given graph G. The GRUNDY DOMINATION DECISION (GDD) problem is the decision version of the GRUNDY DOMINATION problem.

Recently Haynes et al. proposed various kinds of vertex sequences, each of which is specified in terms of some conditions that must be satisfied by every subsequent vertex in the sequence [6]. Predictably, double domination in the sequence context is one of these variations. Before formally presenting the definitions related to this variant, we mention that for a sequence S, consisting of distinct vertices of a graph G, the corresponding set of vertices is denoted by $\widehat{S}$.

A sequence S is called a *double neighborhood sequence* of G if for each i, the vertex v_i dominates at least one vertex u of G which is dominated at most once by the vertices $v_1, v_2, \ldots, v_{i-1}$. If $\widehat{S}$ is a double dominating set of G, then we call S a *double dominating sequence* of G. A double dominating sequence of G with maximum length is called a *Grundy double dominating sequence* of G. The length of a Grundy double dominating sequence is the *Grundy double domination number* of G and is denoted by $\gamma_{gr}^{\times 2}(G)$. Given a graph G with no isolated vertices, the GRUNDY DOUBLE DOMINATION (GD2) problem asks to find a Grundy double dominating sequence of G. The decision version of the GRUNDY DOUBLE DOMINATION problem is as follows.

Decision Version: GRUNDY DOUBLE DOMINATION DECISION (GD2D) Problem
Input: A graph $G = (V, E)$ with no isolated vertices and $k \subset \mathbb{Z}^+$.
Question: Is there a double dominating sequence of G of length at least k?

This concept was introduced in a slightly different manner by Haynes et al. in [6]. In their version, S_i denotes the subsequence $(v_1, v_2, \ldots, v_i)$ which consists of the first i vertices of S. If for each i, the vertex $v_i \in \widehat{S}$ dominates at least one vertex $x \in V(G) \setminus \widehat{S_{i-1}}$ which is dominated at most once by the vertices in $\widehat{S_{i-1}}$ and S is of maximal length, then S is called a *double dominating sequence* of G. This definition does not obey the property that $\widehat{S}$ is a double dominating set of G. Brešar et al. introduced the former definition of double dominating sequences and argued that two invariants are equal in all graphs [3]. So, in this paper, we only consider the former version of double dominating sequences.

The Grundy double domination number of a tree T is exactly the number of vertices of T [6]. Recently, Brešar et al. proved that the GD2D problem is NP-complete for split graphs and can be solved efficiently for threshold graphs [3]. Here, we extend the literature of this variant by studying it for bipartite graphs.

The structure of the paper is as follows. In Sect. 2, we give some basic definitions and notations used throughout the paper. In Sect. 3, we prove that the GD2D problem is NP-complete even when restricted to bipartite and co-bipartite graphs. On the positive note, we present a linear-time algorithm for determining the Grundy double domination number of chain graphs in Sect. 4. Finally, we conclude the paper in Sect. 5.

2   Preliminaries

All graphs considered in this paper are simple, undirected and connected. Let $[n] = \{1, \ldots, n\}$ for any positive integer n. Given a graph G, the *open neighborhood* of a vertex x is $N_G(x) = \{y \in V(G) : xy \in E(G)\}$, while the *closed neighborhood* of x is $N_G[x] = N_G(x) \cup \{x\}$. For a graph $G = (V, E)$, the subgraph induced on a set $U \subseteq V$, denoted by $G[U]$, is the subgraph of G whose vertex set is U and whose edge set consists of all edges in G that have both endpoints in U.

A *complete graph* is a graph in which every two vertices are adjacent. A complete graph on n vertices is denoted by K_n. An *independent set* of G is a set of vertices $A \subseteq V(G)$ such that no two vertices of A are adjacent in G. A *bipartite graph* is a graph whose vertex set can be partitioned into two independent sets. The *complement* of G, denoted by $\overline{G}$, is the graph obtained by removing the edges of G and adding the edges that are not in G. A *co-bipartite graph* is a graph which is the complement of a bipartite graph. A bipartite graph $G = (X, Y, E)$ is a *chain graph* if there exists an ordering $\alpha = (x_1, x_2, \ldots, x_{n_1}, y_1, y_2, \ldots, y_{n_2})$ of vertices of G such that $N(x_1) \subseteq N(x_2) \subseteq \cdots \subseteq N(x_{n_1})$ and $N(y_1) \supseteq N(y_2) \supseteq \cdots \supseteq N(y_{n_2})$, where $X = \{x_1, x_2, \ldots, x_{n_1}\}$ and $Y = \{y_1, y_2, \ldots, y_{n_2}\}$. The ordering α is called a *chain ordering* of G and it can be computed in linear-time [11].

Recall that a relation on a set A is a subset of $A \times A$. We define a relation R on the vertex set of a chain graph $G = (X, Y, E)$ such that two vertices u and v of G are related if and only if $N_G(u) = N_G(v)$. It is easy to see that R is an equivalence relation so it provides a partition P of $V(G)$. Let $\{X_1, X_2, \ldots, X_k\}$ and $\{Y_1, Y_2, \ldots, Y_k\}$ be the parts obtained from the relation R for the X and Y side respectively. We write the partition P as $\{X_1, X_2, \ldots, X_k, Y_1, Y_2, \ldots, Y_k\}$. We keep the order of the sets in P so that it is satisfied that $N(X_1) \subset N(X_2) \subset \cdots \subset N(X_k)$ and $N(Y_1) \supset N(Y_2) \supset \cdots \supset N(Y_k)$. For each $i, j \in [k]$, it is easy to see that $N(X_i) = \cup_{r=1}^{i} Y_r$ and $N(Y_j) = \cup_{r=j}^{k} X_r$.

For two vertex sequences $S_1 = (v_1, \ldots, v_n)$ and $S_2 = (u_1, \ldots, u_m)$, in G, the *concatenation* of these two sequences is defined by the sequence $S_1 \oplus S_2 = (v_1, \ldots, v_n, u_1, \ldots, u_m)$. For an ordered set $A = \{u_1, u_2, \ldots, u_k\}$ of vertices, (A) denotes the sequence of vertices $(u_1, u_2, \ldots, u_k)$.

Proofs of the results marked with $\star$ are omitted due to space constraints.

3   NP-Completeness

3.1   Bipartite Graphs

Recall that the GD2D problem is NP-complete for general graphs [3]. In this subsection, we prove that the problem remains NP-complete for bipartite graphs.

Let $\mathcal{H} = (\mathcal{X}, \mathcal{E})$ be a hypergraph with no isolated vertices. An *edge cover* of $\mathcal{H}$ is a set of hyperedges from $\mathcal{E}$ that covers all vertices of $\mathcal{X}$. A *legal hyperedge sequence* of $\mathcal{H}$ is a sequence of hyperedges $\mathcal{C} = (C_1, \ldots, C_r)$ of $\mathcal{H}$ such that, for each i, $i \in [r]$, C_i covers a vertex not covered by C_j, for each $j < i$. In addition, if the set $\widehat{C}$ is an edge cover of $\mathcal{H}$, then $\mathcal{C}$ is called an *edge covering sequence* of $\mathcal{H}$. The maximum length of an edge covering sequence of $\mathcal{H}$ is denoted by $\rho_{gr}(\mathcal{H})$. The GRUNDY COVERING problem asks to find an edge covering sequence of $\mathcal{H}$ having size $\rho_{gr}(\mathcal{H})$. The GRUNDY COVERING DECISION (GCD) problem is the decision version of the GRUNDY COVERING problem.

It is known that GCD problem is NP-hard in general graphs [1]. For $k \leq 2$, we can find an edge covering sequence of the hypergraph $\mathcal{H}$ of length at least k in polynomial time. So, the GCD problem is NP-complete for $k \geq 3$.

Theorem 1. *The GD2D problem is NP-complete for bipartite graphs.*

Proof. It is clear that the GD2D problem is in class NP. To show the NP-hardness, we give a polynomial reduction from the GCD problem in hypergraphs which is known to be NP-hard [1]. Given a hypergraph $\mathcal{H} = (\mathcal{X}, \mathcal{E})$ with $|\mathcal{X}| = n$ and $\mathcal{E} = \{\mathcal{E}_1, \mathcal{E}_2, \ldots, \mathcal{E}_m\}$, $(n, m \geq 2)$, we construct an instance $G = (X^*, Y^*, E^*)$ of the GD2D problem, where G is a bipartite graph, as follows. $X^* = I \cup X'$ and $Y^* = \mathcal{E}'$, where $I = \{v_1, v_2, \ldots, v_m\}$, $X' = \{x_1, x_2, \ldots, x_n\}$ and $\mathcal{E}' = \{\alpha, e_1, e_2, \ldots, e_m\}$. A vertex of X' corresponds to a vertex of $\mathcal{X}$ in the hypergraph $\mathcal{H}$ and the vertex e_i of $\mathcal{E}'$ corresponds to the hyperedge $\mathcal{E}_i$ of $\mathcal{H}$. Now, a vertex x of X' is adjacent to a vertex of $e_i \in \mathcal{E}'$ in G if and only if $x \in \mathcal{E}_i$ in $\mathcal{H}$. Each vertex of I is adjacent to each vertex of $\mathcal{E}'$ in G. Clearly, G is a bipartite graph. Figure 1 illustrates the construction of G when $\mathcal{H}$ is the hypergraph given by $(X = \{x_1, x_2, x_3, x_4\}, \mathcal{E} = \{\mathcal{E}_1, \mathcal{E}_2, \mathcal{E}_3, \mathcal{E}_4\})$, where $\mathcal{E}_1 = \{x_1, x_2, x_4\}$, $\mathcal{E}_2 = \{x_2, x_3\}$, $\mathcal{E}_3 = \{x_1, x_2\}$ and $\mathcal{E}_4 = \{x_2, x_3, x_4\}$.

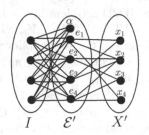

Fig. 1. Construction of bipartite graph G from the hypergraph H.

Now, we show that $\rho_{gr}(\mathcal{H}) \geq k$ if and only if $\gamma_{gr}^{\times 2}(G) \geq n + m + k + 1$, for $k \geq 3$. First, let $(\mathcal{E}_{i_1}, \mathcal{E}_{i_2}, \ldots, \mathcal{E}_{i_{k'}})$ be an edge covering sequence of size at least k in $\mathcal{H}$. Then the sequence $(x_1, x_2, \ldots, x_n, v_1, v_2, \ldots, v_m, \alpha, e_{i_1}, e_{i_2}, \ldots, e_{i_{k'}})$ is a double dominating sequence of size at least $n + m + k + 1$ in G. So, we have $\gamma_{gr}^{\times 2}(G) \geq n + m + k + 1$.

For the converse part, we give a claim first.

Claim 1* *There exists a double dominating sequence of G of size at least $n + m + k + 1$ in which the first vertex from $\mathcal{E}'$ is the vertex α.*

Let S be a double dominating sequence of size at least $n + m + k + 1$ in G satisfying Claim 1. Note that $|\hat{S} \cap \mathcal{E}'| \geq k + 1$.

As $k \geq 3$, let e be the second vertex coming from $\mathcal{E}'$ in S. Now, let A denotes the set of vertices appearing before the vertex α in S, B denotes the set of vertices appearing after the vertex α and before the vertex e. Finally, C denotes the set of vertices appearing after the vertex e in S.

Claim 2* $|I \cap (A \cup B)| \geq 2$.

Claim 3* There exists a double dominating sequence S_0 of G of size at least $n + m + k + 1$ satisfying Claim 1 such that $\widehat{S_0} \cap X' = X'$ and all vertices of X' appear before the vertex e in the sequence S_0.

Claim 3 ensures that we can assume that $\hat{S} \cap X' = X'$ and all vertices of X' appear before the vertex e in the sequence S. Combining all claims, we get that $|\hat{S} \cap (\mathcal{E}' \setminus \{\alpha\})| \geq k$ and these vertices of $(\mathcal{E}' \setminus \{\alpha\})$ are appearing only to dominate vertices of X' second time. So, these vertices of $\hat{S} \cap (\mathcal{E}' \setminus \{\alpha\})$ correspond to a legal hyperedge sequence of size at least k in the hypergraph $\mathcal{H}$. So, $\rho_{gr}(\mathcal{H}) \geq k$.

Therefore, the GD2D problem is NP-complete for bipartite graphs.   □

3.2 Co-bipartite Graphs

In this subsection, we prove that the problem also remains NP-complete for co-bipartite graphs. For this, we give a polynomial reduction from the GDD problem in general graphs when $k \geq 4$, which is already known to be NP-complete [1]. Given a graph $G = (V, E)$ with $V = \{v_1, v_2, \ldots, v_n\}$ $(n \geq 2)$, we construct an instance $G' = (V', E')$ of the GD2D problem in the following way.

Define the vertex set V' as $V' = V_1 \cup V_2 \cup V_3$, where $V_r = \{v_i^r : i \in [n]\}$ for each r, $1 \leq r \leq 3$. Add the edges in G' in the following way. (i) Add the edges so that $G'[V_1]$ and $G'[V_2 \cup V_3]$ are complete subgraphs of G'. (ii) If $v_j \in N_G[v_i]$, then add an edge between v_i^1 and v_j^2. (iii) For each $i \in [n]$, add the edge $v_i^1 v_i^3$ in G'. Formally, define $E' = \{v_i^1 v_j^1, v_i^2 v_j^2, v_i^3 v_j^3 : 1 \leq i < j \leq n\} \cup \{v_i^2 v_j^3 : 1 \leq i \leq j \leq n\} \cup \{v_i^1 v_j^2 : v_j \in N_G[v_i]\} \cup \{v_i^1 v_i^3 : i \in [n]\}$. Clearly, G' is a co-bipartite graph. Figure 2 illustrates the construction of G' from a graph G.

To prove the NP-hardness of the GD2D problem in co-bipartite graphs, it is enough to prove the following theorem.

Theorem 2.* Let G' be the co-bipartite graph constructed from a graph $G = (V, E)$ with $V = \{v_1, v_2, \ldots, v_n\}$ $(n \geq 2)$ as explained above. Then, $\gamma_{gr}(G) \geq k$ if and only if $\gamma_{gr}^{\times 2}(G') \geq n + k$, for $k \geq 4$.

4   Algorithm for Chain Graphs

In this section, we present a linear-time algorithm to solve the GD2 problem in chain graphs. Let $G = (X, Y, E)$ denotes a chain graph and P is the partition of $V(G)$ obtained by the relation R. Recall that, $P =$

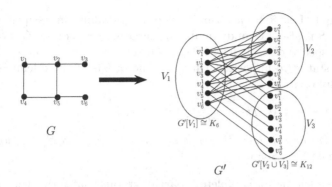

Fig. 2. Construction of co-bipartite graph G' from the graph G.

$(X_1, X_2, \ldots, X_k, Y_1, Y_2, \ldots, Y_k)$. Let $|X| = n_1$ and $|Y| = n_2$. For $i \in [k]$, x^i denotes the vertex of X_i having minimum index in the chain ordering of G. Similarly, y^i denotes the vertex of Y_i having maximum index in the chain ordering of G. Below, we give a result which gives the Grundy double domination number of a complete bipartite graph.

Proposition 1.* *Let $G = (X, Y, E)$ be a complete bipartite graph. Then $\gamma_{gr}^{\times 2}(G) = \max\{|X|, |Y|\} + 1$.*

For technical reasons, we actually consider a slightly more generalized problem in chain graphs. Let $G = (X, Y, E)$ be a chain graph and $M \subseteq V(G)$. Vertices of M are called *marked vertices* of G. All remaining vertices of G are called *unmarked vertices*. We denote the set of unmarked vertices of G by V_0 and the subgraph of G induced on the set V_0 by G_0. The set of marked vertices satisfy all the conditions written in Eq. 1.

$$M \subseteq (X_k \cup Y_1), \ |M \cap X_k| \leq 1, \ |M \cap Y_1| \leq 1, \ |X_k \setminus M| \geq 1, \ |Y_1 \setminus M| \geq 1 \ (1)$$

A sequence $S = (v_1, v_2, \ldots, v_k)$, where $v_i \in V_0$ for each $i \in [k]$, is called an *M-double neighborhood sequence* of (G, M) if for each i, the vertex v_i dominates at least one vertex u of G which is dominated at most once by its preceding vertices in the sequence S. In addition, if $\hat{S}$ is a double dominating set of G_0, then we call S an *M-double dominating sequence* of (G, M). Note that $\hat{S}$ may not be a double dominating set of G. An M-double dominating sequence with maximum length is called a *Grundy M-double dominating sequence* of (G, M). The length of a Grundy M-double dominating sequence of (G, M) is called the *Grundy M-double domination number* of (G, M) and is denoted by $\gamma_{grm}^{\times 2}(G, M)$. Given a chain graph G and $M \subseteq V(G)$ satisfying Eq. 1, the GRUNDY M-DOUBLE DOMINATION (GMD2) problem asks to compute a Grundy M-double dominating sequence of (G, M).

Throughout this section, $\mathcal{G} = (G, M)$ denotes an instance of the GMD2 problem, where $G = (X, Y, E)$ is a chain graph and M is a subset of $V(G)$

satisfying Eq. 1. Let S be a Grundy M-double dominating sequence of $\mathcal{G}$. If $M = \emptyset$ then, S is also a Grundy double dominating sequence of G. So, the GD2 problem is a special case of the GMD2 problem.

Now, we state two important lemmas. The proofs of these lemmas are easy and, hence are omitted.

Lemma 1. *Let $M \neq \emptyset$. Then, $\gamma_{grm}^{\times 2}(\mathcal{G}) \leq \gamma_{gr}^{\times 2}(G)$.*

Lemma 2. *For any Grundy M-double dominating sequence S of $\mathcal{G}$, we have that $X_k \cap \hat{S} \neq \emptyset$ and $Y_1 \cap \hat{S} \neq \emptyset$.*

We prove a lemma for complete bipartite graphs that forms the basis of our algorithm.

Lemma 3. *$\gamma_{grm}^{\times 2}(\mathcal{G}) \in \{\max\{n_1, n_2\}, \max\{n_1, n_2\} + 1\}$, for a complete bipartite graph G.*

Proof. There are four cases to consider.

Case 1: $\underline{M = \emptyset}$:
In this case, $\gamma_{grm}^{\times 2}(\mathcal{G}) = \gamma_{gr}^{\times 2}(G)$. Using Proposition 1, we have $\gamma_{grm}^{\times 2}(\mathcal{G}) = \max\{n_1, n_2\} + 1$.

Case 2: $M \cap X_k = \{x_{n_1}\}$ and $M \cap Y_1 = \emptyset$:
Since $|M \cap X_k| = 1$, we have that $n_1 \geq 2$. We consider two subcases now.

Subcase 2.1: $n_1 = \max\{n_1, n_2\}$:
Here, we have that $\gamma_{gr}^{\times 2}(G) = n_1 + 1$. Now, if $n_2 = 1$, $\gamma_{grm}^{\times 2}(\mathcal{G}) \leq |X| - 1 + |Y| = n_1$. As the sequence $(x_1, x_2, \ldots, x_{n_1-1}, y_1)$ is an M-double dominating sequence of $\mathcal{G}$ of length n_1. So, $\gamma_{grm}^{\times 2}(\mathcal{G}) = n_1 = \max\{n_1, n_2\}$. Otherwise, if $n_2 > 1$, the sequence $(x_1, x_2, \ldots, x_{n_1-1}, y_1, y_2)$ is an M-double dominating sequence of $\mathcal{G}$ of length $n_1 + 1$. Thus, we have that $\gamma_{grm}^{\times 2}(\mathcal{G}) = n_1 + 1 = \max\{n_1, n_2\} + 1$ using Lemma 1.

Subcase 2.2: $n_2 = \max\{n_1, n_2\}$:
Here, we have that $\gamma_{gr}^{\times 2}(G) = n_2 + 1$. Since $n_1 \geq 2$, we have that $n_2 \geq 2$. The sequence $(y_1, y_2, \ldots, y_{n_2}, x_1)$ is an M-double dominating sequence of $\mathcal{G}$ of length $n_2 + 1$. Thus, we have $\gamma_{grm}^{\times 2}(\mathcal{G}) = n_2 + 1 = \max\{n_1, n_2\} + 1$ using Lemma 1.

Case 3: $M \cap Y_1 = \{y_1\}$ and $M \cap X_k = \emptyset$:
This case is similar to case 2.

Case 4: $M \cap X_k = \{x_{n_1}\}$ and $M \cap Y_1 = \{y_1\}$:
Clearly, $n_1 \geq 2$ and $n_2 \geq 2$. We again consider two subcases.

Subcase 4.1: $n_1 = \max\{n_1, n_2\}$:
Here, we have that $\gamma_{gr}^{\times 2}(G) = n_1 + 1$. If $n_2 \geq 3$, the sequence $(x_1, x_2, \ldots, x_{n_1-1}, y_2, y_3)$ is an M-double dominating sequence of $\mathcal{G}$ of length

$n_1 - 1 + 2 = n_1 + 1$. So, $\gamma_{grm}^{\times 2}(\mathcal{G}) = n_1 + 1 = \max\{n_1, n_2\} + 1$ using Lemma 1. But, if $n_2 = 2$, the sequence $(x_1, x_2, \ldots, x_{n_1-1}, y_2)$ is an M-double dominating sequence of $\mathcal{G}$ of length $n_1 - 1 + 1 = n_1$. So, $\gamma_{grm}^{\times 2}(\mathcal{G}) = n_1 = \max\{n_1, n_2\}$ using the fact that $\gamma_{grm}^{\times 2}(G) \leq |X| - 1 + |Y| - 1 = n_1 - 1 + 2 - 1 = n_1$.

Subcase 4.2: $n_2 = \max\{n_1, n_2\}$:
Similar to the subcase 4.1, we can prove that $\gamma_{grm}^{\times 2}(\mathcal{G})$ is either n_2 or $n_2 + 1$. ☐

Algorithm 1 computes a Grundy M-double dominating sequence of $\mathcal{G}$ based on the Lemma 3, when G is a complete bipartite graph. Next, we state some lemmas for $\mathcal{G}$, when G is not a complete bipartite graph, that is, $k \geq 2$.

Lemma 4.* *If there exists a Grundy M-double dominating sequence S^* of $\mathcal{G}$ such that $|X_k \cap \widehat{S^*}| \geq 3$, then exactly one of the following is true:*

(1) $\gamma_{grm}^{\times 2}(\mathcal{G}) = |X| + k$.
(2) $\gamma_{grm}^{\times 2}(\mathcal{G}) = |X| + k - 1$.

Similar to Lemma 4, we state another lemma for the Y side of G. Proof of Lemma 5 is simlar to the Lemma 4.

Lemma 5. *If there exists a Grundy M-double dominating sequence S^* of $\mathcal{G}$ such that $|Y_1 \cap \widehat{S^*}| \geq 3$, then exactly one of the following is truc:*

(1) $\gamma_{grm}^{\times 2}(\mathcal{G}) = |Y| + k$.
(2) $\gamma_{grm}^{\times 2}(\mathcal{G}) = |Y| + k - 1$.

Lemma 6.* *Let $\mathcal{G}$ be an instance of the GMD2 problem such that there is no Grundy M-double dominating sequence S^* of $\mathcal{G}$ satisfying $|X_k \cap \widehat{S^*}| \geq 3$ or $|Y_1 \cap \widehat{S^*}| \geq 3$. Assume that S is a Grundy M-double dominating sequence of $\mathcal{G}$ such that $|X_k \cap \widehat{S}| = 2$. Then either $|Y_1 \cap \widehat{S}| = 1$ or there exists another Grundy M-double dominating sequence S' of $\mathcal{G}$ satisfying one of the following:*
(1) $|X_k \cap \widehat{S'}| = 2$ and $|Y_1 \cap \widehat{S'}| = 1$. (2) $|X_k \cap \widehat{S'}| = 1$ and $|Y_1 \cap \widehat{S'}| = 2$.

The proof of the next lemma is easy and, hence is omitted.

Lemma 7. *Let $\mathcal{G}$ be an instance of the GMD2 problem such that there is no Grundy M-double dominating sequence S^* of $\mathcal{G}$ satisfying $|X_k \cap \widehat{S^*}| \geq 3$ or $|Y_1 \cap \widehat{S^*}| \geq 3$. Assume that S is a Grundy M-double dominating sequence of $\mathcal{G}$ such that $|X_k \cap \widehat{S}| = 1$. Then $Y_k \subseteq \widehat{S}$.*

Similar to Lemma 7, we state another lemma for G.

Lemma 8. *Let $\mathcal{G}$ be an instance of the GMD2 problem such that there is no Grundy M-double dominating sequence S^* of $\mathcal{G}$ satisfying $|X_k \cap \widehat{S^*}| \geq 3$ or $|Y_1 \cap \widehat{S^*}| \geq 3$. Assume that S is a Grundy M-double dominating sequence of $\mathcal{G}$ such that $|Y_1 \cap \widehat{S}| = 1$. Then $X_1 \subseteq \widehat{S}$.*

Using Lemmas 6, 7 and 8, we can directly state the following result.

Algorithm 1: $S = \text{GrundyM1}(G, M)$

Input: $\mathcal{G} = (G, M)$, where $G = (X, Y, E)$ is a complete bipartite graph and $M \subseteq V(G)$ satisfying equation 1,
$X = \{x_1, \ldots, x_{n_1}\}$ and $Y = \{y_1, \ldots, y_{n_2}\}$.

Output: A Grundy M-double dominating sequence S of $\mathcal{G}$.

if $M = \emptyset$ then
    if $n_1 \geq n_2$ then
     $S = (x_1, x_2, \ldots, x_{n_1}, y_1)$
    else
     $S = (y_1, y_2, \ldots, y_{n_2}, x_1)$

if $M \cap X_k = \{x_{n_1}\}$ and $M \cap Y_1 = \emptyset$ then
    if $n_1 \geq n_2$ then
     if $n_2 = 1$ then
      $S = (x_1, x_2, \ldots, x_{n_1-1}, y_1)$
     else
      $S = (x_1, x_2, \ldots, x_{n_1-1}, y_1, y_2)$
    else
     $S = (y_1, y_2, \ldots, y_{n_2}, x_1)$

if $M \cap Y_1 = \{y_1\}$ and $M \cap X_k = \emptyset$ then
    if $n_2 \geq n_1$ then
     if $n_1 = 1$ then
      $S = (y_2, y_3, \ldots, y_{n_2}, x_1)$
     else
      $S = (y_2, y_3, \ldots, y_{n_2}, x_1, x_2)$
    else
     $S = (x_1, x_2, \ldots, x_{n_2}, y_2)$

if $M \cap X_k = \{x_{n_1}\}$ and $M \cap Y_1 = \{y_1\}$ then
    if $n_1 \geq n_2$ then
     if $n_2 \geq 3$ then
      $S = (x_1, x_2, \ldots, x_{n_1-1}, y_2, y_3)$
     else
      $S = (x_1, x_2, \ldots, x_{n_1-1}, y_2)$
    else
     if $n_1 \geq 3$ then
      $S = (y_2, y_3, \ldots, y_{n_2}, x_1, x_2)$
     else
      $S = (y_2, y_3, \ldots, y_{n_2}, x_1)$

return S.

Lemma 9. *Let $\mathcal{G}$ be an instance of the GMD2 problem such that there is no Grundy M-double dominating sequence S^* of $\mathcal{G}$ satisfying $|X_k \cap \widehat{S^*}| \geq 3$ or $|Y_1 \cap \widehat{S^*}| \geq 3$. Then one of the following is true:*

(1) There exists a Grundy M-double dominating sequence S of $\mathcal{G}$ such that $|X_k \cap \widehat{S}| = 1$ and $Y_k \subseteq \widehat{S}$.
(2) There exists a Grundy M-double dominating sequence S of $\mathcal{G}$ such that $|Y_1 \cap \widehat{S}| = 1$ and $X_1 \subseteq \widehat{S}$.

Let $\mathcal{G}$ be an instance of the GMD2 problem such that there is no Grundy M-double dominating sequence S^* of $\mathcal{G}$ satisfying $|X_k \cap \widehat{S^*}| \geq 3$ or $|Y_1 \cap \widehat{S^*}| \geq 3$. We call a Grundy M-double dominating sequence S of $\mathcal{G}$ as a *type 1 optimal sequence* of $\mathcal{G}$ if it satisfies that $|X_k \cap \widehat{S}| = 1$ and $Y_k \subseteq \widehat{S}$. Similarly, We call a Grundy M-double dominating sequence S of $\mathcal{G}$, a *type 2 optimal sequence* of $\mathcal{G}$ if it satisfies that $|Y_1 \cap \widehat{S}| = 1$ and $X_1 \subseteq \widehat{S}$.

Lemma 10.* *Let $\mathcal{G}$ be an instance of the GMD2 problem. Then one of the following is true:*
(1) There exists a type 1 optimal sequence of $\mathcal{G}$.
(2) There exists a type 2 optimal sequence of $\mathcal{G}$.

Finally, we state the lemma which completely characterizes the structure of an optimal solution for an instance of the GMD2 problem. The proof is easy and hence, is omitted.

Lemma 11. *Let $\mathcal{G}$ be an instance of the GMD2 problem. Then one of the following is true:*
(1) There exists a type 1 optimal sequence S of $\mathcal{G}$ in which the vertex of $X_k \cap \widehat{S}$ appear in the last.
(2) There exists a type 2 optimal sequence S of $\mathcal{G}$ in which the vertex of $Y_1 \cap \widehat{S}$ appear in the last.

We use a dynamic programming approach to solve the GMD2 problem for an instance $\mathcal{G}$ in Algorihtm 2. Through Lemma 11, we characterized the structure of an optimal solution. Next, we define the optimal solution of the problem recursively in terms of the optimal solutions to subproblems. For GMD2 problem, we pick the subproblems as the problem of finding a Grundy M-double dominating sequence of $\mathcal{G}' = (G', M')$, where G' is a subgraph of G and $M' \subseteq V(G')$ satisfying Eq. 1.
Let S be a Grundy M-double dominating sequence of $\mathcal{G}$ which is a type 1 optimal sequence of $\mathcal{G}$ and the vertex of $X_k \cap \widehat{S}$ appear in the last. We also assume that all vertices of Y_k appear together just before the vertex of X_k. Let G_1 denotes the subgraph of G induced on the set of vertices $(X \setminus X_k) \cup \{x_{t+1}\} \cup (Y \setminus Y_k)$, where $t = |X| - |X_k|$. Let $M_1 = \{x_{t+1}\} \cup (M \cap Y_1)$. Then the subsequence of S obtained by removing the last $|Y_k| + 1$ vertices of S is a Grundy M-double dominating sequence of (G_1, M_1).
Similarly, if S is a type 2 optimal sequence of $\mathcal{G}$ having the vertex of $Y_1 \cap \widehat{S}$ in the last and G_2 denotes the subgraph of G induced on the set of vertices $(Y \setminus Y_1) \cup \{y_t\} \cup (X \setminus X_1)$, where $t = |M \cap Y_1| + 1$. Again, assume that all vertices of X_1 appear together just before the vertex of Y_1. Let $M_2 = \{y_t\} \cup (M \cap X_k)$.

Then the subsequence of S obtained by removing the last $|X_1| + 1$ vertices of S is a Grundy M-double dominating sequence of (G_2, M_2).

Now, we give the algorithm to compute a Grundy M-double dominating sequence of $\mathcal{G}$.

Algorithm 2: $S = \mathrm{GrundyM}(G, M)$

Input: $\mathcal{G} = (G, M)$, where $G = (X, Y, E)$ is a chain graph and $M \subseteq V(G)$ satisfying equation 1. $X = \{x_1, \ldots, x_{n_1}\}$ and $Y = \{y_1, \ldots, y_{n_2}\}$.

Output: A Grundy M-double dominating sequence S of $\mathcal{G}$.

if $k = 1$ then
$\quad$ $S = \mathrm{GrundyM1}(G, M)$;
$\quad$ return S;

else
$\quad$ $t = |X| - |X_k|$, $X'_{k-1} = X_{k-1} \cup \{x_{t+1}\}$;
$\quad$ if $k \geq 3$ then
$\quad\quad$ $X' = \cup_{i=1}^{k-2} X_i \cup X'_{k-1}$;
$\quad$ else
$\quad\quad$ $X' = X'_{k-1}$;
$\quad$ $G^1_{k-1} = G[X' \cup (Y \setminus Y_k)]$, $M \cap X_k = \{x_{t+1}\}$;
$\quad$ $S_1 = \mathrm{GrundyM}(G^1_{k-1}, M) \oplus (Y_k) \oplus x_{t+1}$;
$\quad$ $t = |M \cap Y_1| + 1$, $Y'_1 = Y_2 \cup \{y_t\}$;
$\quad$ if $k \geq 3$ then
$\quad\quad$ $Y' = \cup_{i=3}^{k} Y_i \cup Y'_1$;
$\quad$ else
$\quad\quad$ $Y' = Y'_1$;
$\quad$ $G^2_{k-1} = G[(X \setminus X_1) \cup Y']$, $M \cap Y_1 = \{y_t\}$;
$\quad$ $S_2 = \mathrm{GrundyM}(G^2_{k-1}, M) \oplus (X_1) \oplus y_t$;
$\quad$ if $|\widehat{S_1}| \geq |\widehat{S_2}|$ then
$\quad\quad$ return S_1;
$\quad$ else
$\quad\quad$ return S_2;

Algorithm 2 computes a Grundy M-double dominating sequence of $\mathcal{G} = (G, M)$ by recursively appending some vertices at the end of the Grundy M-double dominating sequence of (G', M'), where G' is a subgraph of G. Note that this task can be performed in linear-time.

Based on the above discussion, we directly state the following theorem.

Theorem 3. *Algorithm 2 outputs a Grundy M-double dominating sequence of $\mathcal{G} = (G, M)$ in linear-time, where G is a chain graph.*

To solve the GD2 problem in a chain graph G, we compute a Grundy M-double dominating sequence of $(G, \emptyset)$ using Algorithm 2. So, we can state the following theorem.

Theorem 4. *A Grundy double dominating sequence of a chain graph G can be computed in linear-time.*

5   Conclusion

We studied the GD2D problem in this paper. We proved that the problem is NP-complete for bipartite graphs and co-bipartite graphs. We also proved that the GD2D problem is efficiently solvable for chain graphs. We solved this problem in chain graphs using a dynamic programming approach. Since the class of chain graphs is a subclass of bipartite graphs, the gap between the efficient algorithms and NP-completeness in the subclasses of bipartite graphs has been narrowed a little. To find the status of the problem in the graph classes such as bipartite permutation graphs, convex bipartite graphs and chordal bipartite graphs can be the next research direction. These graph classes are subclasses of bipartite graphs and superclasses of chain graphs. Various types of vertex sequences were proposed for which computational complexities are still unknown in many graph classes [6]. These kind of vertex sequences are open for further research.

Acknowledgement. We would like to thank Prof. Boštjan Brešar for his suggestion to work on this problem. We are also grateful to him for providing many useful comments leading to the improvements in the paper.

References

1. Brešar, B., Gologranc, T., Milanič, M., Rall, D.F., Rizzi, R.: Dominating sequences in graphs. Discret. Math. **336**, 22–36 (2014)
2. Brešar, B., Klavžar, S., Rall, D.F.: Domination game and an imagination strategy. SIAM J. Discret. Math. **24**(3), 979–991 (2010)
3. Brešar, B., Pandey, A., Sharma, G.: Computational aspects of some vertex sequences of Grundy domination-type. Indian J. Disc. Math. **8**, 21–38 (2022)
4. Fink, J.F.: n-domination in graphs. In: Graph Theory with Applications to Algorithms and Computer Science, pp. 282–300. Wiley (1985)
5. Fink, J.F., Jacobson, M.S.: On n-domination, n-dependence and forbidden subgraphs. In: Graph Theory with Applications to Algorithms and Computer Science, pp. 301–311 (1985)
6. Haynes, T., Hedetniemi, S.: Vertex sequences in graphs. Discrete Math. Lett **6**, 19–31 (2021)
7. Haynes, T.W., Hedetniemi, S.T., Henning, M.A. (eds.): Topics in Domination in Graphs. DM, vol. 64. Springer, Cham (2020). https://doi.org/10.1007/978-3-030-51117-3
8. Haynes, T.W., Hedetniemi, S.T., Henning, M.A. (eds.): Structures of domination in graphs, Developments in Mathematics, vol. 66. Springer, Cham (2021). https://doi.org/10.1007/978-3-030-58892-2
9. Haynes, T.W., Hedetniemi, S.T., Slater, P.J. (eds.): Domination in graphs, Monographs and Textbooks in Pure and Applied Mathematics, vol. 209. Marcel Dekker Inc, New York (1998), advanced topics

10. Haynes, T.W., Hedetniemi, S.T., Slater, P.J.: Fundamentals of domination in graphs, Monographs and Textbooks in Pure and Applied Mathematics, vol. 208. Marcel Dekker Inc, New York (1998)
11. Heggernes, P., Kratsch, D.: Linear-time certifying recognition algorithms and forbidden induced subgraphs. Nord. J. Comput. **14**(1–2), 87–108 (2007)

Relation Between Broadcast Domination and Multipacking Numbers on Chordal Graphs

Sandip Das[1], Florent Foucaud[2][ORCID], Sk Samim Islam[1]([✉]),
and Joydeep Mukherjee[3]

[1] Indian Statistical Institute, Kolkata, India
samimislam08@gmail.com
[2] Université Clermont-Auvergne, CNRS, Mines de Saint-Étienne,
Clermont-Auvergne-INP, LIMOS, 63000 Clermont-Ferrand, France
florent.foucaud@uca.fr
[3] Ramakrishna Mission Vivekananda Educational and Research Institute,
Howrah, India

Abstract. For a graph $G = (V, E)$ with a vertex set V and an edge set E, a function $f : V \to \{0, 1, 2, ..., diam(G)\}$ is called a *broadcast* on G. For each vertex $u \in V$, if there exists a vertex v in G (possibly, $u = v$) such that $f(v) > 0$ and $d(u, v) \leq f(v)$, then f is called a dominating broadcast on G. The cost of the dominating broadcast f is the quantity $\sum_{v \in V} f(v)$. The minimum cost of a dominating broadcast is the broadcast domination number of G, denoted by $\gamma_b(G)$.

A multipacking is a set $S \subseteq V$ in a graph $G = (V, E)$ such that for every vertex $v \in V$ and for every integer $r \geq 1$, the ball of radius r around v contains at most r vertices of S, that is, there are at most r vertices in S at a distance at most r from v in G. The multipacking number of G is the maximum cardinality of a multipacking of G and is denoted by $\mathrm{mp}(G)$.

It is known that $\mathrm{mp}(G) \leq \gamma_b(G)$ and that $\gamma_b(G) \leq 2\,\mathrm{mp}(G)+3$ for any graph G, and it was shown that $\gamma_b(G) - \mathrm{mp}(G)$ can be arbitrarily large for connected graphs (as there exist infinitely many connected graphs G where $\gamma_b(G)/\mathrm{mp}(G) = 4/3$ with $\mathrm{mp}(G)$ arbitrarily large). For strongly chordal graphs, it is known that $\mathrm{mp}(G) = \gamma_b(G)$ always holds.

We show that, for any connected chordal graph G, $\gamma_b(G) \leq \left\lceil \frac{3}{2}\,\mathrm{mp}(G) \right\rceil$. We also show that $\gamma_b(G) - \mathrm{mp}(G)$ can be arbitrarily large for connected chordal graphs by constructing an infinite family of connected chordal graphs such that the ratio $\gamma_b(G)/\mathrm{mp}(G) = 10/9$, with $\mathrm{mp}(G)$ arbitrarily large. This result shows that, for chordal graphs, we cannot improve the bound $\gamma_b(G) \leq \left\lceil \frac{3}{2}\,\mathrm{mp}(G) \right\rceil$ to a bound in the form $\gamma_b(G) \leq c_1 \cdot \mathrm{mp}(G) + c_2$, for any constant $c_1 < 10/9$ and c_2.

Keywords: Chordal graph · Multipacking · Dominating broadcast

This research was financed by the IFCAM project "Applications of graph homomorphisms" (MA/IFCAM/18/39).
F. Foucaud—Research financed by the French government IDEX-ISITE initiative 16-IDEX-0001 (CAP 20-25) and by the ANR project GRALMECO (ANR-21-CE48-0004).

A. Bagchi and R. Muthu (Eds.): CALDAM 2023, LNCS 13947, pp. 297–308, 2023.
https://doi.org/10.1007/978-3-031-25211-2_23

1   Introduction

Covering and packing problems are fundamental in graph theory and algorithms [6]. In this paper, we study two dual covering and packing problems
called *broadcast domination* and *multipacking*. The broadcast domination problem has a natural motivation in telecommunication networks: imagine a network
with radio emission towers, where each tower can broadcast information at any
radius r for a cost of r. The goal is to cover the whole network by minimizing the total cost. The multipacking problem is its natural packing counterpart
and generalizes various other standard packing problems. Unlike many standard
packing and covering problems, these two problems involve arbitrary distances
in graphs, which makes them challenging. The goal of this paper is to study the
relation between these two parameters in the class of chordal graphs, which are
those graphs that do not contain any induced cycle of a length at least 4.

For a graph $G = (V, E)$ with a vertex set V, an edge set E and the diameter
$diam(G)$, a function $f : V \rightarrow \{0, 1, 2, ..., diam(G)\}$ is called a *broadcast* on G.
Suppose G be a graph with a broadcast f. Let $d(u, v) = $ the length of a shortest
path joining the vertices u and v in G. We say $v \in V$ is a *tower* of G if $f(v) > 0$.
Suppose $u, v \in V$ (possibly, $u = v$) such that $f(v) > 0$ and $d(u, v) \leq f(v)$, then
we say v *broadcasts* (or *dominates*) u and u *hears* the broadcast from v.

For each vertex $u \in V$, if there exists a vertex v in G (possibly, $u = v$)
such that $f(v) > 0$ and $d(u, v) \leq f(v)$, then f is called a *dominating broadcast*
on G. The *cost* of the broadcast f is the quantity $\sigma(f)$, which is the sum of
the weights of the broadcasts over all vertices in G. So, $\sigma(f) = \sum_{v \in V} f(v)$.
The minimum cost of a dominating broadcast in G (taken over all dominating
broadcasts) is the *broadcast domination number* of G, denoted by $\gamma_b(G)$. So,
$$\gamma_b(G) = \min_{f \in D(G)} \sigma(f) = \min_{f \in D(G)} \sum_{v \in V} f(v), \text{ where } D(G) = \text{set of all dominating}$$
broadcasts on G.

Suppose f is a dominating broadcast with $f(v) \in \{0, 1\} \; \forall v \in V(G)$, then
$\{v \in V(G) : f(v) = 1\}$ is a *dominating set* on G. The minimum cardinality of a
dominating set is the *domination number* which is denoted by $\gamma(G)$.

An *optimal broadcast* or *optimal dominating broadcast* on a graph G is a
dominating broadcast with a cost equal to $\gamma_b(G)$. A dominating broadcast is
efficient if no vertex hears a broadcast from two different vertices. So, no tower
can hear a broadcast from another tower in an efficient broadcast. There is
a theorem that says, for every graph there is an optimal efficient dominating
broadcast [7]. Define a ball of radius r around v by $N_r[v] = \{u \in V(G) : d(v, u) \leq
r\}$. Suppose $V(G) = \{v_1, v_2, v_3, \ldots, v_n\}$. Let c and x be the vectors indexed by
(i, k) where $v_i \in V(G)$ and $1 \leq k \leq diam(G)$, with the entries $c_{i,k} = k$ and
$x_{i,k} = 1$ when $f(v_i) = k$ and $x_{i,k} = 0$ when $f(v_i) \neq k$. Let $A = [a_{j,(i,k)}]$ be a
matrix with the entries

$$a_{j,(i,k)} = \begin{cases} 1 & \text{if } v_j \in N_k[v_i] \\ 0 & \text{otherwise.} \end{cases}$$

Hence, the broadcast domination number can be expressed as an integer linear program:

$$\gamma_b(G) = \min\{c.x : Ax \geq 1, x_{i,k} \in \{0,1\}\}.$$

The *maximum multipacking problem* is the dual integer program of the above problem. Moreover, multipacking is a generalization of packing problems. A *multipacking* is a set $M \subseteq V$ in a graph $G = (V, E)$ such that $|N_r[v] \cap M| \leq r$ for each vertex $v \in V(G)$ and for every integer $r \geq 1$. The *multipacking number* of G is the maximum cardinality of a multipacking of G and it is denoted by $\mathrm{mp}(G)$. A *maximum multipacking* is a multipacking M of a graph G such that $|M| = \mathrm{mp}(G)$. If M is a multipacking, we define a vector y with the entries $y_j = 1$ when $v_j \in M$ and $y_j = 0$ when $v_j \notin M$. So,

$$\mathrm{mp}(G) = \max\{y.1 : yA \leq c, y_j \in \{0,1\}\}.$$

Broadcast domination is a generalization of domination problems and multipacking is a generalization of packing problems. Erwin [8,9] introduced broadcast domination in his doctoral thesis in 2001. Multipacking was introduced in Teshima's Master's Thesis [15] in 2012 (also see [3,6,7,14]). For general graphs, an optimal dominating broadcast can be found in polynomial-time $O(n^6)$ [12]. The same problem can be solved in linear time for trees [4]. However, until now, there is no known polynomial-time algorithm to find a maximum multipacking of general graphs (the problem is also not known to be NP-hard). However, polynomial-time algorithms are known for trees and more generally, strongly chordal graphs [4] See [10] for other references concerning algorithmic results on the two problems.

It is known that $\mathrm{mp}(G) \leq \gamma_b(G)$, since broadcast domination and multipacking are dual problems [5]. It is known that $\gamma_b(G) \leq 2\,\mathrm{mp}(G) + 3$ [1] and it is a conjecture that $\gamma_b(G) \leq 2\,\mathrm{mp}(G)$ for every graph G [1]. Hartnell and Mynhardt [11] constructed a family of connected graphs such that the difference $\gamma_b(G) - \mathrm{mp}(G)$ can be arbitrarily large and in fact, for which the ratio $\gamma_b(G)/\mathrm{mp}(G) = 4/3$. Therefore, for general connected graphs,

$$\frac{4}{3} \leq \lim_{\mathrm{mp}(G) \to \infty} \sup \left\{ \frac{\gamma_b(G)}{\mathrm{mp}(G)} \right\} \leq 2.$$

A natural question comes to mind: What is the optimal bound on this ratio for other graph classes? It is known that $\gamma_b(G) = \mathrm{mp}(G)$ holds for strongly chordal graphs [4]. Thus, a natural class to study is the class of chordal graphs.

In this paper, we establish an improved relation between $\gamma_b(G)$ and $\mathrm{mp}(G)$ for connected chordal graphs by showing that $\gamma_b(G) \leq \lceil \frac{3}{2}\mathrm{mp}(G) \rceil$. We then construct a family of connected chordal graphs such that the difference $\gamma_b(G) - \mathrm{mp}(G)$ can be arbitrarily large and the ratio $\gamma_b(G)/\mathrm{mp}(G) = 10/9$ for every member G of that family. Thus, for chordal connected graphs G, we have:

$$\frac{10}{9} \leq \lim_{\mathrm{mp}(G) \to \infty} \sup \left\{ \frac{\gamma_b(G)}{\mathrm{mp}(G)} \right\} \leq \frac{3}{2}.$$

We also make a connection with the *fractional* versions of the two concepts, as introduced in [2].

In Sect. 2, we show that for any connected chordal graph G, $\gamma_b(G) \leq \lceil \frac{3}{2} \mathrm{mp}(G) \rceil$ and there is a polynomial-time algorithm to construct a multipacking of G of size at least $\lceil \frac{2\,\mathrm{mp}(G)-1}{3} \rceil$. In Sect. 3, we prove our main result which says that the difference $\gamma_b(G) - \mathrm{mp}(G)$ can be arbitrarily large for connected chordal graphs, and we conclude in Sect. 4.

2   An Inequality Linking Broadcast Domination and Multipacking Numbers of Chordal Graphs

In this section, we use results from the literature to show that the general bound connecting multipacking number and broadcast domination number can be improved for chordal graphs.

Theorem 1 ([11]). *If G is a connected graph of order at least 2 having diameter d and multipacking number $\mathrm{mp}(G)$, where $P = v_0, \ldots, v_d$ is a diametral path of G, then the set $M = \{v_i : i \equiv 0 \pmod 3, i = 0, 1, \ldots, d\}$ is a multipacking of G of size $\lceil \frac{d+1}{3} \rceil$ and $\lceil \frac{d+1}{3} \rceil \leq \mathrm{mp}(G)$.*

Theorem 2 ([9,15]). *If G is a connected graph of order at least 2 having radius r, diameter d, multipacking number $\mathrm{mp}(G)$, broadcast domination number $\gamma_b(G)$ and domination number $\gamma(G)$, then $\mathrm{mp}(G) \leq \gamma_b(G) \leq min\{\gamma(G), r\}$.*

Theorem 3 ([13]). *If G is a connected chordal graph with radius r and diameter d, then $2r \leq d + 2$.*

Proposition 1. *If G is a connected chordal graph, then $\gamma_b(G) \leq \lceil \frac{3}{2} \mathrm{mp}(G) \rceil$.*

Proof. From Theorem 1, $\lceil \frac{d+1}{3} \rceil \leq \mathrm{mp}(G)$ which implies that $d \leq 3\,\mathrm{mp}(G) - 1$. Moreover, from Theorem 2 and Theorem 3, $\gamma_b(G) \leq r \leq \lfloor \frac{d+2}{2} \rfloor \leq \lfloor \frac{(3\,\mathrm{mp}(G)-1)+2}{2} \rfloor = \lfloor \frac{3}{2}\,\mathrm{mp}(G) + \frac{1}{2} \rfloor$. Therefore, $\gamma_b(G) \leq \lfloor \frac{3}{2}\,\mathrm{mp}(G) + \frac{1}{2} \rfloor = \lceil \frac{3}{2}\,\mathrm{mp}(G) \rceil$. $\square$

The proof of Proposition 1 has the following algorithmic application.

Proposition 2. *If G is a connected chordal graph, there is a polynomial-time algorithm to construct a multipacking of G of size at least $\lceil \frac{2\,\mathrm{mp}(G)-1}{3} \rceil$.*

Proof. If $P = v_0, \ldots, v_d$ is a diametrical path of G, then the set $M = \{v_i : i \equiv 0 \pmod 3, i = 0, 1, \ldots, d\}$ is a multipacking of G of size $\lceil \frac{d+1}{3} \rceil$ by Theorem 1. We can construct M in polynomial-time since we can find a diametral path of a graph G in polynomial-time. Moreover, from Theorem 1, Theorem 2 and Theorem 3, $\lceil \frac{2\,\mathrm{mp}(G)-1}{3} \rceil \leq \lceil \frac{2r-1}{3} \rceil \leq \lceil \frac{d+1}{3} \rceil \leq \mathrm{mp}(G)$. $\square$

Example 1. *The connected chordal graph S_3 (Fig. 1) has $\mathrm{mp}(S_3) = 1$ and $\gamma_b(S_3) = 2$. So, here $\gamma_b(S_3) = \lceil \frac{3}{2} \mathrm{mp}(S_3) \rceil$.*

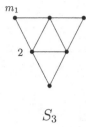

$$S_3$$

Fig. 1. S_3 is a connected chordal graph with $\gamma_b(S_3) = 2$ and $\mathrm{mp}(S_3) = 1$

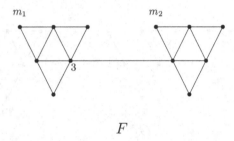

$$F$$

Fig. 2. F is a connected chordal graph with $\gamma_b(F) = 3$ and $\mathrm{mp}(F) = 2$

Example 2. *The connected chordal graph F (Fig. 2) has* $\mathrm{mp}(F) = 2$ *and* $\gamma_b(F) = 3$. *So, here* $\gamma_b(F) = \lceil \frac{3}{2} \mathrm{mp}(F) \rceil$.

Example 3. *The connected chordal graph H (Fig. 3) has* $\mathrm{mp}(H) = 4$ *and* $\gamma_b(H) = 6$. *So, here* $\gamma_b(H) = \lceil \frac{3}{2} \mathrm{mp}(H) \rceil$.

We could not find an example of connected chordal graph with $\mathrm{mp}(G) = 3$ and $\gamma_b(G) = \lceil \frac{3}{2} \mathrm{mp}(G) \rceil = 5$.

3   Unboundedness of the Gap Between Broadcast Domination and Multipacking Numbers of Chordal Graphs

Here we prove that the difference between broadcast domination number and multipacking number of connected chordal graphs can be arbitrarily large. We state the theorem formally below.

Theorem 4. *The difference $\gamma_b(G) - \mathrm{mp}(G)$ can be arbitrarily large for connected chordal graphs.*

Consider the graph G_1 as in Fig 4. Let B_1 and B_2 be two isomorphic copies of G_1. Join $b_{1,21}$ of B_1 and $b_{2,1}$ of B_2 by an edge (Fig. 5 and 6). We denote this new graph by G_2 (Fig. 5). In this way, we form G_k by joining k isomorphic copies of G_1: $B_1, B_2, \cdots, B_k$ (Fig. 6). Here B_i is joined with B_{i+1} by joining $b_{i,21}$ and $b_{i+1,1}$. We say that B_i is the i-th block of G_k. B_i is an induced subgraph of G_k as

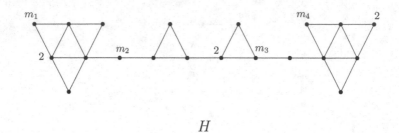

$$H$$

Fig. 3. H is a connected chordal graph with $\gamma_b(H) = 6$ and $\mathrm{mp}(H) = 4$

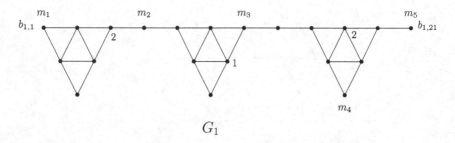

$$G_1$$

Fig. 4. G_1 is a connected chordal graph with $\gamma_b(G_1) = 5$ and $\mathrm{mp}(G_1) = 5$. $M_1 = \{m_i : 1 \le i \le 5\}$ is a multipacking of size 5.

given by $B_i = G_k[\{b_{i,j} : 1 \le j \le 21\}]$. Similarly, for $1 \le i \le 2k-1$, we define $B_i \cup B_{i+1}$, induced subgraph of G_{2k}, as $B_i \cup B_{i+1} = G_{2k}[\{b_{i,j}, b_{i+1,j} : 1 \le j \le 21\}]$. We prove Theorem 4 by establishing that $\gamma_b(G_{2k}) = 10k$ and $\mathrm{mp}(G_{2k}) = 9k$. Then we can say, for all natural numbers k, $\gamma_b(G_{2k}) - \mathrm{mp}(G_{2k}) = k$, so the difference can be arbitrarily large.

3.1   Proof of Theorem 4

Our proof of Theorem 4 is accomplished through a set of lemmas which are stated and proved below. We begin by observing a basic fact about multipacking in a graph. We formally state it in Lemma 1 for ease of future reference.

Lemma 1. *Suppose M is a multipacking in a graph G. If $u, v \in M$ and $u \neq v$, then $d(u, v) \ge 3$.*

Proof. If $d(u, v) = 1$, then $u, v \in N_1[v] \cap M$, then M cannot be a multipacking. So, $d(u, v) \neq 1$. If $d(u, v) = 2$, then there exists a common neighbour w of u and v. So, $u, v \in N_1[w] \cap M$, then M cannot be a multipacking. So, $d(u, v) \neq 2$. Therefore, $d(u, v) > 2$. □

Lemma 2. $\mathrm{mp}(G_{2k}) \ge 9k$, *for each positive integer k.*

Proof. Consider the set $M_{2k} = \{b_{2i-1,1}, b_{2i-1,7}, b_{2i-1,13}, b_{2i-1,18}, b_{2i-1,21}, b_{2i,4}, b_{2i,8}, b_{2i,14}, b_{2i,18} : 1 \le i \le k\}$ (Fig. 6) of size $9k$. We want to show that M_{2k} is a

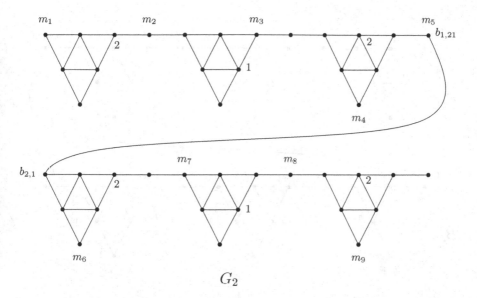

$$G_2$$

Fig. 5. Graph G_2 with $\gamma_b(G_2) = 10$ and $\mathrm{mp}(G_2) = 9$. $M = \{m_i : 1 \le i \le 9\}$ is a multipacking of size 9.

multipacking of G_{2k}. So, we have to prove that, $|N_r[v] \cap M_{2k}| \le r$ for each vertex $v \in V(G_{2k})$ and for every integer $r > 1$. We prove this statement using induction on r. It can be checked that $|N_r[v] \cap M_{2k}| \le r$ for each vertex $v \in V(G_{2k})$ and for each $r \in \{1, 2, 3, 4\}$. Now assume that the statement is true for $r = s$, we want to prove that, it is true for $r = s+4$. Observe that, $|(N_{s+4}[v] \setminus N_s[v]) \cap M_{2k}| \le 4$ for every vertex $v \in V(G_{2k})$. Therefore, $|N_{s+4}[v] \cap M_{2k}| \le |N_s[v] \cap M_{2k}| + 4 \le s + 4$. So, the statement is true. Therefore, M_{2k} is a multipacking of G_{2k}. So, $\mathrm{mp}(G_{2k}) \ge |M_{2k}| = 9k$. □

Lemma 3. $\mathrm{mp}(G_1) = 5$.

Proof. $V(G_1) = N_3[b_{1,7}] \cup N_2[b_{1,17}]$. Suppose M is a multipacking on G_1 such that $|M| = \mathrm{mp}(G_1)$. So, $|M \cap N_3[b_{1,7}]| \le 3$ and $|M \cap N_2[b_{1,17}]| \le 2$. Therefore, $|M \cap (N_3[b_{1,7}] \cup N_2[b_{1,17}])| \le 5$. So, $|M \cap V(G)| \le 5$, that implies $|M| \le 5$. Let $M_1 = \{b_{1,1}, b_{1,7}, b_{1,13}, b_{1,18}, b_{1,21}\}$. Since $|N_r[v] \cap M| \le r$ for each vertex $v \in V(G_1)$ and for every integer $r \ge 1$, so M_1 is a multipacking of size 5. Then $5 = |M_1| \le |M|$. So, $|M| = 5$. Therefore, $\mathrm{mp}(G_1) = 5$. □

So, now we have $\mathrm{mp}(G_1) = 5$. Using this fact we prove that $\mathrm{mp}(G_2) = 9$.

Lemma 4. $\mathrm{mp}(G_2) = 9$.

Proof. As mentioned before, $B_i = G_k[\{b_{i,j} : 1 \le j \le 21\}]$, $1 \le i \le 2$. So, B_1 and B_2 are two blocks in G_2 which are isomorphic to G_1. Let M be a multipacking of G_2 with size $\mathrm{mp}(G_2)$. So, $|M| \ge 9$ by Lemma 2. Since M is a multipacking of

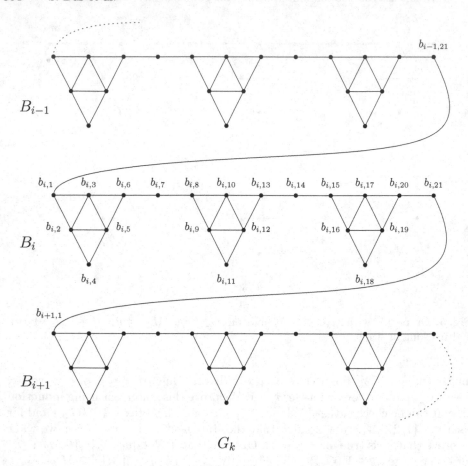

Fig. 6. Partial depiction of graph G_k.

G_2, so $M \cap V(B_1)$ and $M \cap V(B_2)$ are multipackings of B_1 and B_2, respectively. Let $M \cap V(B_1) = M_1$ and $M \cap V(B_2) = M_2$. Since $B_1 \cong G_1$ and $B_2 \cong G_1$, so $\mathrm{mp}(B_1) = 5$ and $\mathrm{mp}(B_2) = 5$ by Lemma 3. This implies $|M_1| \le 5$ and $|M_2| \le 5$. Since $V(B_1) \cup V(B_2) = V(G_2)$ and $V(B_1) \cap V(B_2) = \phi$, so $M_1 \cap M_2 = \phi$ and $|M| = |M_1| + |M_2|$. Therefore, $9 \le |M| = |M_1| + |M_2| \le 10$. So, $9 \le |M| \le 10$.

We establish this lemma by using contradiction on $|M|$. In the first step, we prove that if $|M_1| = 5$, then the particular vertex $b_{1,21} \in M_1$. Using this, we can show that $|M_2| \le 4$. In this way we show that $|M| \le 9$.

For the purpose of contradiction, we assume that $|M| = 10$. So, $|M_1| + |M_2| = 10$, and also $|M_1| \le 5$, $|M_2| \le 5$. Therefore, $|M_1| = |M_2| = 5$.

Claim 4.1. If $|M_1| = 5$, then $b_{1,21} \in M_1$.

Proof of Claim. Suppose $b_{1,21} \notin M$. Let $S = \{b_{1,7}, b_{1,14}\}$, $S_1 = \{b_{1,r} : 1 \le r \le 6\}$, $S_2 = \{b_{1,r} : 8 \le r \le 13\}$, $S_3 = \{b_{1,r} : 15 \le r \le 20\}$. If $u, v \in S_t$, then $d(u,v) \le 2$, this holds for each $t \in \{1,2,3\}$. So, by Lemma 1, u, v together cannot be in a multipacking. Therefore $|S_t \cap M_1| \leq 1$ for $t = 1, 2, 3$ and $|S \cap M_1| \leq |S| = 2$. Now, $5 = |M_1| = |M_1 \cap [V(G_1) \setminus \{b_{1,21}\}]| = |M_1 \cap (S \cup S_1 \cup S_2 \cup S_3)| = |(M_1 \cap S) \cup (M_1 \cap S_1) \cup (M_1 \cap S_2) \cup (M_1 \cap S_3)| \leq |M_1 \cap S| + |M_1 \cap S_1| + |M_1 \cap S_2| + |M_1 \cap S_3| \leq 2 + 1 + 1 + 1 = 5$. Therefore, $|S_t \cap M_1| = 1$ for $t = 1, 2, 3$ and $|S \cap M_1| = 2$, so $b_{1,7}, b_{1,14} \in M_1$. Since $|S_2 \cap M_1| = 1$, there exists $w \in S_2 \cap M_1$. Then $N_2[b_{1,10}]$ contains three vertices $b_{1,7}, b_{1,14}, w$ of M_1, which is not possible. So, this is a contradiction. Therefore, $b_{1,21} \in M_1$.   ◁

Claim 4.2. If $|M_1| = 5$, then $|M_2| \leq 4$.

Proof of Claim. Let $S' = \{b_{2,14}, b_{2,21}\}$, $S_4 = \{b_{2,r} : 1 \leq r \leq 6\}$, $S_5 = \{b_{2,r} : 8 \leq r \leq 13\}$, $S_6 = \{b_{2,r} : 15 \leq r \leq 20\}$. By Lemma 1, $|S_t \cap M_2| \leq 1$ for $t = 4, 5, 6$ and also $|S' \cap M_2| \leq |S'| = 2$.

Observe that, if $S_4 \cap M_2 \neq \phi$, then $b_{2,7} \notin M_2$ (i.e. if $b_{2,7} \in M_2$, then $S_4 \cap M_2 = \phi$). [Suppose not, then $S_4 \cap M_2 \neq \phi$ and $b_{2,7} \in M_2$, so, there exists $u \in S_4 \cap M_2$. Then $N_2[b_{2,3}]$ contains three vertices $b_{1,21}, b_{2,7}, u$ of M, which is not possible. This is a contradiction].

Suppose $S_4 \cap M_2 \neq \phi$, then $b_{2,7} \notin M_2$. Now, $5 = |M_2| = |M_2 \cap [V(B_2) \setminus \{b_{2,7}\}]| = |M_2 \cap (S' \cup S_4 \cup S_5 \cup S_6)| = |(M_2 \cap S') \cup (M_2 \cap S_4) \cup (M_2 \cap S_5) \cup (M_2 \cap S_6)| \leq |M_2 \cap S'| + |M_2 \cap S_4| + |M_2 \cap S_5| + |M_2 \cap S_6| \leq 2 + 1 + 1 + 1 = 5$. Therefore $|S_t \cap M_2| = 1$ for $t = 4, 5, 6$ and $|S' \cap M_2| = 2$. Since $|M_2 \cap S_6| = 1$, so there exists $u_1 \in M_2 \cap S_6$. Then $N_2[b_{2,17}]$ contains three vertices $b_{2,14}, b_{2,21}, u_1$ of M_2, which is not possible. So, this is a contradiction.

Suppose $S_4 \cap M_2 = \phi$, then either $b_{2,7} \in M_2$ or $b_{2,7} \notin M_2$. First consider $b_{2,7} \notin M_2$, then $5 = |M_2| = |M_2 \cap (S' \cup S_5 \cup S_6)| = |(M_2 \cap S') \cup (M_2 \cap S_5) \cup (M_2 \cap S_6)| \leq |M_2 \cap S'| + |M_2 \cap S_5| + |M_2 \cap S_6| \leq 2 + 1 + 1 = 4$. So, this is a contradiction. And if $b_{2,7} \in M_2$, then $5 = |M_2| = |M_2 \cap (S' \cup S_5 \cup S_6 \cup \{b_{2,7}\})| = |(M_2 \cap S') \cup (M_2 \cap S_5) \cup (M_2 \cap S_6) \cup (M_2 \cap \{b_{2,7}\})| \leq |M_2 \cap S'| + |M_2 \cap S_5| + |M_2 \cap S_6| + |M_2 \cap \{b_{2,7}\}| \leq 2 + 1 + 1 + 1 = 5$. Therefore $|S_t \cap M_2| = 1$ for $t = 5, 6$ and $|S' \cap M_2| = 2$. Since $|M_2 \cap S_6| = 1$, so there exists $u_2 \in M_2 \cap S_6$. Then $N_2[b_{2,17}]$ contains three vertices $b_{2,14}, b_{2,21}, u_2$ of M_2, which is not possible. So, this is a contradiction. So, $|M_1| = 5 \implies |M_2| \leq 4$.   ◁

Recall that for contradiction, we assume $|M| = 10$, which implies $|M_2| = 5$. In the proof of the above claim, we established $|M_2| \leq 4$, which in turn contradicts our assumption. So, $|M| \neq 10$. Therefore, $|M| = 9$.   □

Notice that graph G_{2k} has k copies of G_2. Moreover, we have $\mathrm{mp}(G_2) = 9$. If $\mathrm{mp}(G_{2k}) > 9k$, then we will use the Pigeonhole principle to show that $\mathrm{mp}(G_{2k}) = 9k$.

Lemma 5. $\mathrm{mp}(G_{2k}) = 9k$, *for each positive integer* k.

Proof. For $k = 1$ it is true by Lemma 4. Moreover, we know $\mathrm{mp}(G_{2k}) \geq 9k$ by Lemma 2. Suppose $k > 1$ and assume $\mathrm{mp}(G_{2k}) > 9k$. Let $\hat{M}$ be a multipacking of G_{2k} such that $|\hat{M}| > 9k$. Let $\hat{B}_j$ be a subgraph of G_{2k} defined as $\hat{B}_j = B_{2j-1} \cup B_{2j}$ where $1 \leq j \leq k$. So, $V(G_{2k}) = \bigcup_{j=1}^{k} V(\hat{B}_j)$ and $V(\hat{B}_p) \cap V(\hat{B}_q) = \phi$ for all $p \neq q$ and $p, q \in \{1, 2, 3, \ldots, k\}$. Since $|\hat{M}| > 9k$, so by the Pigeonhole principle there exists a number $j \in \{1, 2, 3, \ldots, k\}$ such that $|\hat{M} \cap \hat{B}_j| > 9$. Since $\hat{M} \cap \hat{B}_j$ is a multipacking of $\hat{B}_j$, so $\mathrm{mp}(\hat{B}_j) > 9$. But $\hat{B}_j \cong G_2$ and $\mathrm{mp}(G_2) = 9$ by Lemma 4, so $\mathrm{mp}(\hat{B}_j) = 9$, which is a contradiction. Therefore, $\mathrm{mp}(G_{2k}) = 9k$. □

R. C. Brewster and L. Duchesne [2] introduced fractional multipacking in 2013 (also see [16]). Suppose G is a graph with $V(G) = \{v_1, v_2, v_3, \ldots, v_n\}$ and $w : V(G) \to [0, \infty)$ is a function. So, $w(v)$ is a weight on a vertex $v \in V(G)$. Let $w(S) = \sum_{u \in S} w(u)$ where $S \subseteq V(G)$. We say w is a *fractional multipacking* of G, if $w(N_r[v]) \leq r$ for each vertex $v \in V(G)$ and for every integer $r \geq 1$. The *fractional multipacking number* of G is the value $\max_w w(V(G))$ where w is any fractional multipacking and it is denoted by $mp_f(G)$. A *maximum fractional multipacking* is a fractional multipacking w of a graph G such that $w(V(G)) = mp_f(G)$. If w is a fractional multipacking, we define a vector y with the entries $y_j = w(v_j)$. So,

$$mp_f(G) = \max\{y.\mathbf{1} : yA \leq c, y_j \geq 0\}.$$

So, this is a linear program which is the dual of the linear program $\min\{c.x : Ax \geq 1, x_{i,k} \geq 0\}$. Let,

$$\gamma_{b,f}(G) = \min\{c.x : Ax \geq 1, x_{i,k} \geq 0\}.$$

Using the strong duality theorem for linear programming, we can say that

$$\mathrm{mp}(G) \leq mp_f(G) = \gamma_{b,f}(G) \leq \gamma_b(G).$$

Lemma 6. *If k is a positive integer, then $mp_f(G_k) \geq 5k$.*

Proof. We define a function $w : V(G_k) \to [0, \infty)$ where $w(b_{i,1}) = w(b_{i,6}) = w(b_{i,7}) = w(b_{i,8}) = w(b_{i,13}) = w(b_{i,14}) = w(b_{i,15}) = w(b_{i,20}) = w(b_{i,21}) = \frac{1}{3}$ and $w(b_{i,4}) = w(b_{i,11}) = w(b_{i,18}) = \frac{2}{3}$ for each $i \in \{1, 2, 3, \ldots, k\}$ (Fig. 7). So, $w(G_k) = 5k$. We want to show that w is a fractional multipacking of G_k. So, we have to prove that $w(N_r[v]) \leq r$ for each vertex $v \in V(G_k)$ and for every integer $r \geq 1$. We prove this statement using induction on r. It can be checked that $w(N_r[v]) \leq r$ for each vertex $v \in V(G_k)$ and for each $r \in \{1, 2, 3, 4\}$. Now assume that the statement is true for $r = s$, we want to prove that it is true for $r = s + 4$. Observe that, $w(N_{s+4}[v] \setminus N_s[v]) \leq 4, \forall v \in V(G_k)$. Therefore, $w(N_{s+4}[v]) \leq w(N_s[v]) + 4 \leq s + 4$. So, the statement is true. So, w is a fractional multipacking of G_k. Therefore, $mp_f(G_k) \geq 5k$. □

Lemma 7. *If k is a positive integer, then $mp_f(G_k) = \gamma_b(G_k) = 5k$.*

Proof. Define a broadcast f on G_k as $f(b_{i,j}) = \begin{cases} 2 & \text{if } 1 \leq i \leq k \text{ and } j = 6, 17 \\ 1 & \text{if } 1 \leq i \leq k \text{ and } j = 12 \\ 0 & \text{otherwise} \end{cases}$.

Here f is an efficient dominating broadcast and $\sum_{v \in V(G_k)} f(v) = 5k$. So, $\gamma_b(G_k) \leq 5k, \forall k \in \mathbb{N}$. So, by the strong duality theorem and Lemma 6, $5k \leq mp_f(G_k) = \gamma_{b,f}(G_k) \leq \gamma_b(G_k) \leq 5k$. Therefore, $mp_f(G_k) = \gamma_b(G_k) = 5k$. □

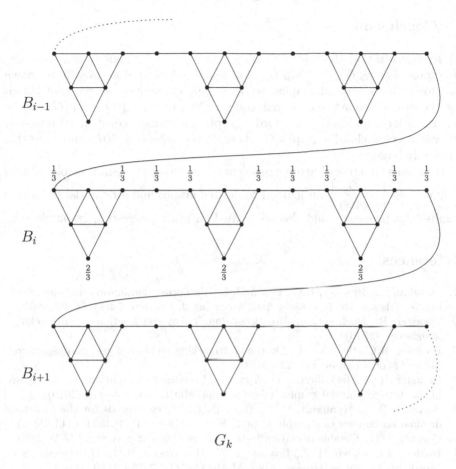

$$B_{i-1}$$

$$B_i$$

$$B_{i+1}$$

$$G_k$$

Fig. 7. Fractional multipacking of G_k.

So, $\gamma_b(G_{2k}) = 10k$ by Lemma 7 and $\mathrm{mp}(G_{2k}) = 9k$ by Lemma 5. So, we can say that for all positive integers k, $\gamma_b(G_{2k}) - \mathrm{mp}(G_{2k}) = k$. Therefore, this proves Theorem 4. So, the difference $\gamma_b(G) - \mathrm{mp}(G)$ can be arbitrarily large for connected chordal graphs.

Corollary 1. *The difference $mp_f(G) - \mathrm{mp}(G)$ can be arbitrarily large for connected chordal graphs.*

Proof. We get $mp_f(G_{2k}) = 10k$ by Lemma 7 and $\mathrm{mp}(G_{2k}) = 9k$ by Lemma 5. Therefore, $mp_f(G_{2k}) - \mathrm{mp}(G_{2k}) = k$ for all positive integers k. □

Corollary 2. *For every integer $k \geq 1$, there is a connected chordal graph G_{2k} with $\mathrm{mp}(G_{2k}) = 9k$, $mp_f(G_{2k})/\mathrm{mp}(G_{2k}) = 10/9$ and $\gamma_b(G_{2k})/\mathrm{mp}(G_{2k}) = 10/9$.*

Corollary 3. *For connected chordal graphs G,*

$$\frac{10}{9} \leq \lim_{\mathrm{mp}(G)\to\infty} \sup\left\{ \frac{\gamma_b(G)}{\mathrm{mp}(G)} \right\} \leq \frac{3}{2}.$$

4   Conclusion

We have shown that the bound $\gamma_b(G) \leq 2\,\mathrm{mp}(G) + 3$ for general graphs G can be improved to $\gamma_b(G) \leq \lceil \frac{3}{2}\mathrm{mp}(G) \rceil$ for connected chordal graphs. It is known that for strongly chordal graphs, $\gamma_b(G) = \mathrm{mp}(G)$, we have shown that this is not the case for connected chordal graphs. Even more, $\gamma_b(G) - \mathrm{mp}(G)$ can be arbitrarily large for connected chordal graphs, as we have constructed infinitely many connected chordal graphs G where $\gamma_b(G)/\mathrm{mp}(G) = 10/9$ and $\mathrm{mp}(G)$ is arbitrarily large.

It remains an interesting open problem to determine the best possible value of

$$\lim_{\mathrm{mp}(G)\to\infty} \sup\left\{ \frac{\gamma_b(G)}{\mathrm{mp}(G)} \right\}$$ for general connected graphs and for chordal connected graphs. This problem could also be studied for other interesting graph classes.

References

1. Beaudou, L., Brewster, R.C., Foucaud, F.: Broadcast domination and multipacking: bounds and the integrality gap. Australas. J. Combin. **74**(1), 86–97 (2019)
2. Brewster, R., Duchesne, L.: Broadcast domination and fractional multipackings. Manuscript (2013)
3. Brewster, R.C., Beaudou, L.: On the multipacking number of grid graphs. Discret. Math. Theor. Comput. Sci. **21** (2019)
4. Brewster, R.C., MacGillivray, G., Yang, F.: Broadcast domination and multipacking in strongly chordal graphs. Discret. Appl. Math. **261**, 108–118 (2019)
5. Brewster, R.C., Mynhardt, C.M., Teshima, L.E.: New bounds for the broadcast domination number of a graph. Central Eur. J. Math. **11**(7), 1334–1343 (2013)
6. Cornuéjols, G.: Combinatorial optimization: packing and covering. SIAM (2001)
7. Dunbar, J.E., Erwin, D.J., Haynes, T.W., Hedetniemi, S.M., Hedetniemi, S.T.: Broadcasts in graphs. Discret. Appl. Math. **154**(1), 59–75 (2006)
8. Erwin, D.J.: Dominating broadcasts in graphs. Bull. Inst. Combin. Appl. **42**(89), 105 (2004)
9. Erwin, D.J.: Cost domination in graphs. Ph.D. thesis, Western Michigan University (2001)
10. Foucaud, F., Gras, B., Perez, A., Sikora, F.: On the complexity of broadcast domination and multipacking in digraphs. Algorithmica **83**(9), 2651–2677 (2021)
11. Hartnell, B.L., Mynhardt, C.M.: On the difference between broadcast and multipacking numbers of graphs. Utilitas Math. **94**, 19–29 (2014)
12. Heggernes, P., Lokshtanov, D.: Optimal broadcast domination in polynomial time. Discret. Math. **306**(24), 3267–3280 (2006)
13. Laskar, R., Shier, D.: On powers and centers of chordal graphs. Discret. Appl. Math. **6**(2), 139–147 (1983)
14. Meir, A., Moon, J.W.: Relations between packing and covering numbers of a tree. Pac. J. Math. **61**(1), 225–233 (1975)
15. Teshima, L.E.: Broadcasts and multipackings in graphs. Ph.D. thesis (2012)
16. Teshima, L.E.: Multipackings in graphs. arXiv preprint arXiv:1409.8057 (2014)

Cops and Robber on Oriented Graphs with Respect to Push Operation

Sandip Das[1], Harmender Gahlawat[2(✉)], Ashwin Ramgopal[3],
Uma Kant Sahoo[1], and Sagnik Sen[4]

[1] Indian Statistical Institute, Kolkata, India
[2] Ben-Gurion University of the Negev, Beer-sheba, Israel
harmendergahlawat@gmail.com
[3] The University of Chicago, Chicago, USA
[4] Indian Institute of Technology Dharwad, Dharwad, India

Abstract. *Graph Searching games* are extensively studied in the literature for their vast number of applications in artificial intelligence, robot motion planning, game planning, distributed computing, and graph theory. In particular, COPS AND ROBBER is one of the most well-studied graph searching game, where a set of cops try to *capture* the position of a single robber. The *cop number* of a graph is the minimum number of cops required to capture the robber on the graph.

In an oriented graph $\overrightarrow{G}$, the *push* operation on a vertex v reverses the orientation of all arcs incident on v. We define and study a variant of the game of COPS AND ROBBER on oriented graphs, where the players also have the ability to push the vertices of the graph.

1 Introduction

The first formulation of *graph searching games* is due to Parsons [31,32], who formulated pursuit-evasion in graphs to model the search for a person trapped in a complicated system of dark caves. Since then, graph searching and pursuit-evasion have been studied extensively, having applications in artificial intelligence [20,24], constrained satisfaction problems and database theory [14,15], distributed computing [3,6] and network decontamination [30], and significant implications in graph theory and algorithms [1,36].

COPS AND ROBBER is one of the most intensively studied graph searching game played on graphs, where a set of cops pursue a single robber. In this article, we study the game of COPS AND ROBBER on oriented graphs. Classically, the game in the oriented setting has the following rules. The game starts with the cops placing themselves on the vertices of an oriented graph $\overrightarrow{G}$. More than one cop may simultaneously occupy the same vertex of the graph. Then the robber chooses a vertex to start. Now the cops and the robber make alternating moves beginning with the cops. In a cop move, each cop can either stay on the same vertex or move to a vertex in its out-neighborhood. In the robber move, the robber can either stay on the same vertex or move to a vertex in its out-neighborhood. If at some point in the game, one of the cops occupies the same vertex as the robber, we call it the *capture*. If the cops can ensure the capture, then we say that the cops win, and if the robber can evade the capture forever, then the robber wins.

Next, we define a few necessary parameters. The *cop number* $c(\overrightarrow{G})$ of an oriented graph $\overrightarrow{G}$ is the minimum number of cops needed by the Cop Player to have a winning strategy. We say that an oriented graph $\overrightarrow{G}$ is *k-copwin* if k cops have a winning strategy in $\overrightarrow{G}$. For brevity, we say that $\overrightarrow{G}$ is *cop-win* if $\overrightarrow{G}$ is 1-copwin. Most research in oriented (or directed) graphs considers the model defined above. However, there is some research concerning variations of the game in oriented graphs [8].

Let $\overrightarrow{uv}$ be an arc of an oriented[1] graph $\overrightarrow{G}$. We say that u is an *in-neighbor* of v and v is an *out-neighbor* of u. Let $N^-(u)$ and $N^+(u)$ denote the set of in-neighbors and out-neighbors of u, respectively. Moreover, let $N^+[v] = N^+(v) \cup \{v\}$ and $N^-[v] = N^-(v) \cup \{v\}$. A vertex without any in-neighbor is a *source* and a vertex without any out-neighbor is a *sink*. A vertex v is said to be dominating if $N^+[v] = V(\overrightarrow{G})$. For a vertex v, the *push operation on v*, denoted by $push(v)$, reverses the orientation of each arc incident on v. Notice that the push operation is a well-studied modification operation on directed or oriented graphs [10,25,26,28,29,33–35]. In this work, for convenience and for the sake of better readability, we retain the name of an oriented graph even after some vertices have been pushed, allowing a slight abuse of notation. However, there is no scope of confusion to the best of our knowledge.

In this paper, we consider the game of COPS AND ROBBER on oriented graphs with respect to the push operation. We define some variations of the game where the agents might have the ability to push the vertices of the graph. For that purpose, we define two kinds of push ability.

1. *Weak push*: Let A be an agent (cop/robber) having the weak push ability, and let A be on a vertex v. Then in its turn, A can either move to a vertex $u \in N^+[v]$ or can push the vertex v.
2. *Strong push*: Let A be an agent (cop/robber) having the strong push ability, and let A be on a vertex v. Then in its turn, A can either move to a vertex $u \in N^+[v]$ or can push any vertex of the graph.

We have the following immediate observation.

Observation 1. *Let $\overrightarrow{G}$ be an oriented graph without a dominating vertex. Then one cop, even with the strong push ability, cannot win if the robber has the weak push ability.*

Proof. We give a strategy for $\mathcal{R}$ to evade the capture forever. Let the cop, say, $\mathcal{C}$, starts at a vertex v. Since $\overrightarrow{G}$ does not have a dominating vertex, there is a

---

[1] An oriented graph is a directed graph without self-loops and 2-cycles.

vertex $u \in V(\rightarrow)$ such that $u \notin N^+[v]$. Then, $\mathcal{R}$ enters at a u and stays on u forever. Now, whenever $\mathcal{R}$ is attacked by $\mathcal{C}$ (i.e., $\mathcal{C}$ is on a vertex w such that $u \in N^+(w)$), $\mathcal{R}$ pushes the vertex u (to make $u \in N^-(w)$). Now, $\mathcal{C}$ cannot capture $\mathcal{R}$ immediately, and either has to move to some other vertex from where it is attacking $\mathcal{R}$ or push its current vertex. In any case, $\mathcal{R}$ pushes u whenever it is attacked, and this goes on forever. Hence one cop can never capture $\mathcal{R}$. □

Hence, in this work, we restrict our attention to the variations where the robber does not have the push ability, but the cops either have the strong push ability or the weak push ability. We would also like to note here that if neither cops nor the robber has the push ability, then this game is equivalent to the classical COPS AND ROBBER game on oriented graphs.

Let $c_{sp}(\overrightarrow{G})$ be the cop number of $\overrightarrow{G}$ when the cops have the strong push ability and let $c_{wp}(\overrightarrow{G})$ be the cop number of $\overrightarrow{G}$ when the cops have the weak push ability. The following observation is obvious.

Observation 2. *Let* $\overrightarrow{G}$ *be an oriented graph. Then* $c_{sp}(\overrightarrow{G}) \leq c_{wp}(\overrightarrow{G})$.

Our Contribution. In this paper, we consider the game of cops and robber on oriented graphs where cops either have the weak push or the strong push ability.

In Sect. 3, we consider the game where the cop player has the strong push ability. We consider multiple graph classes that are cop-win in this game variant but have higher cop number in classical COPS AND ROBBER. We begin by showing, in Theorem 1, that if $\overrightarrow{G}$ is an orientation of a complete multipartite graph, then $c_{sp}(\overrightarrow{G}) = 1$. Second, we show that for a graph $\overrightarrow{G}$ such that its underlying graph G is a subcubic graph, $c_{sp}(\overrightarrow{G}) = 1$ in Theorem 2. Finally, we show, in Theorem 3, that for a graph $\overrightarrow{G}$ such that its underlying graph G is an interval graph, $c_{sp}(\overrightarrow{G}) = 1$.

Related Work. The COPS AND ROBBER game is well studied on both directed and undirected graphs. Hamidoune [17] considered the game on Cayley digraphs. Frieze et al. [11] studied the game on digraphs and gave an upper bound of $\mathcal{O}\left(\frac{n(\log \log n)^2}{\log n}\right)$ for cop number in digraphs. Loh and Oh [27] considered the game on strongly connected planar digraphs and proved that every n-vertex strongly connected planar digraph has cop number $\mathcal{O}(\sqrt{n})$. Moreover, they constructively proved the existence of a strongly connected planar digraph with cop number greater than three, which is in contrast to the case of undirected graphs where the cop number of a planar graph is at most three [2]. The computational complexity of determining the cop number of a digraph (and undirected graphs also) is a challenging question in itself. Goldstein and Reingold [13] proved that deciding whether k cops can capture a robber is EXPTIME-complete for a variant of COPS AND ROBBER and conjectured that the same holds for classical COPS AND ROBBER as well. Later, Kinnersley [22] proved that conjecture and established that determining the cop number of a graph or digraph is EXPTIME-complete. Kinnersley [23] also showed that n-vertex strongly connected cop-win digraphs can have capture time $\Omega(n^2)$, whereas for undirected cop-win graphs the capture time is at most $n - 4$ moves [12].

Hahn and MacGillivray [16] gave an algorithmic characterization of the cop-win finite reflexive digraphs and showed that any k-cop game can be reduced to 1-cop game, resulting in an algorithmic characterization for k-copwin finite reflexive digraphs. However, these results do not give a structural characterization of such graphs. Darlington et al. [7] tried to structurally characterize cop-win oriented graphs and gave a conjecture that was later disproved by Khatri et al. [21], who also studied the game in oriented outerplanar graphs and line digraphs.

Recently, the cop number of planar Eulerian digraphs and related families was studied in several articles [9,18,19]. In particular, Hosseini and Mohar [19] considered the orientations of integer grid that are vertex-transitive and showed that at most four cops can capture the robber on arbitrary finite quotients of these directed grids. De la Maza et al. [9] considered the *straight-ahead* orientations of 4-regular quadrangulations of the torus and the Klein bottle and proved that their cop number is bounded by a constant. They also showed that the cop number of every k-regularly oriented toroidal grid is at most 13.

Bradshaw et al. [5] proved that the cop number of directed and undirected Cayley graphs on abelian groups has an upper bound of the form of $\mathcal{O}(\sqrt{n})$. Modifying this construction, they obtained families of graphs and digraphs with cop number $\Theta(\sqrt{n})$. The family of digraphs thus obtained has the largest cop number in terms of n of any known digraph construction.

2   Preliminaries

For a natural number ℓ, let $[\ell]$ denote the set $\{1, \ldots, \ell\}$.

Graph Theory. In this paper, we consider the game on oriented graphs whose underlying graph is simple, finite, and connected. Let $\overrightarrow{G}$ be an oriented graph with G as the underlying undirected graph of $\overrightarrow{G}$. We say that $\overrightarrow{G}$ is an *orientation* of G. We consider the push operation on the vertices of $\overrightarrow{G}$, and hence the orientations of arcs in $\overrightarrow{G}$ might change. So, what we refer to $\overrightarrow{G}$ is the graph with the current orientations. Note that although the orientations of the arcs in $\overrightarrow{G}$ might change, the underlying graph G remains the same. Moreover, it is worth noting that given $\overrightarrow{G}$ and $\overrightarrow{H}$ such that $\overrightarrow{G}$ and $\overrightarrow{H}$ have the same underlying graph, it might be possible that there is no sequence of pushing vertices in $\overrightarrow{G}$ that yields $\overrightarrow{H}$.

A graph G is *subcubic* if each vertex $v \in V(G)$ has degree at most three. An *interval representation* of a graph G is a set $\mathcal{I} = \{[x_u^-, x_u^+] : u \in V(G)\}$ of intervals where each interval in $\mathcal{I}$ corresponds to a vertex, and two intervals intersect if and only if the corresponding vertices share an edge. A graph is an *interval graph* if it has an interval representation. A k-*partite graph* is a graph whose vertex set can be partitioned into k independent sets. A *complete k-partite graph* is a k-partite graph such that there is an edge between every pair of vertices from different independent sets. A *complete multipartite graph* is a graph that is complete k-partite for some $k > 1$.

Let v be vertex of $\vec{G}$ and S is a subset of vertices of $\vec{G}$ (i.e., $S \subseteq V(\vec{G})$). Then, we say that v is a *source in* S if $S \subseteq N^+[v]$. Moreover, we say that $|N^+(v)|$ is the *out-degree* of v, $|N^-(v)|$ is the *in-degree* of v, and $|N^+(v)| + |N^-(v)|$ is the *degree* of v.

The Game. We say that a vertex v is *safe* if no vertex in $N^-[v]$ is occupied by a cop. A vertex v is said to be attacked if there is some cop in $N^-(v)$. When we have a single cop, we denote the cop by $\mathcal{C}$. We denote the robber by $\mathcal{R}$ throughout the paper. Let $V' \subseteq V(G)$ be a set of vertices. We say that $\mathcal{R}$ is *restricted* to V' if $\mathcal{R}$ cannot move to a vertex $u \in V(G) \setminus V'$ without getting captured in the next cop move.

2.1 Preliminary Results

First, we have the following easy (but useful) observation.

Observation 3. *Let $\vec{G}$ be an oriented graph. Then a cop with strong push ability can make any vertex v a source vertex.*

Proof. $\mathcal{C}$ can achieve this by pushing every vertex in $N^-(v)$. $\square$

A corollary of Observation 3 is that the tournaments are cop-win in the strong cop model. This is in contrast to the fact that the normal cop number of even strongly connected tournaments is unbounded [37]. Next, we have the following definition.

Definition 1 (Trapping $\mathcal{R}$). *If $\mathcal{R}$ is at a vertex v such that $|N^+(v)| = 0$, then we say that $\mathcal{R}$ is trapped at v.*

In the following lemma, we show that if $\mathcal{R}$ can be trapped at a vertex v, then $\mathcal{R}$ will be captured by $\mathcal{C}$ (having either the strong push or the weak push ability) in a finite number of rounds.

Lemma 1. *Let $\vec{G}$ be an oriented graph and $\mathcal{R}$ is trapped at a vertex $v \in V(\vec{G})$. Then $\mathcal{R}$ will be captured by $\mathcal{C}$, having push ability, in a finite number of rounds.*

Proof. Let $\mathcal{C}$ be at a vertex u when $\mathcal{R}$ gets trapped at v. If $u \in N^-(v)$, then $\mathcal{C}$ can capture $\mathcal{R}$ in the next move of $\mathcal{C}$. So, we suppose that $u \notin N^-(v)$. Since G (underlying graph of $\vec{G}$) is a connected graph, there is a shortest u,v-path, say, P, in G. Moreover, observe that there is a vertex $w \in N^-(v)$ such that w lies on this path. Now, $\mathcal{C}$ will move along this path to the vertex w in a manner so that it neither pushes v or a vertex in $N^-(v)$ (hence, ensuring that $\mathcal{R}$ stays trapped at v).

Let $u = u_1, \dots, u_\ell = w$ be an ordering of vertices of path P, along the path P from u to w. If $\overrightarrow{u_i u_{i+1}}$ (for $i \in [\ell - 1]$) is an arc, then $\mathcal{C}$ moves along this arc from u_i to u_{i+1}. Else, $\overrightarrow{u_{i+1} u_i}$ is an arc and $\mathcal{C}$ pushes the vertex u_i to reverse the orientation of the arc to get the arc $\overrightarrow{u_i u_{i+1}}$, and move along it to u_{i+1} in the next cop move. Using this strategy, note that $\mathcal{C}$ will reach w in a finite number of (at most $2n$) cop moves. Moreover, since P is a shortest path, observe that P contains at most one vertex from $N^-(v)$. Hence, the way $\mathcal{C}$ pushes vertices, no vertex in $N^-[v]$ is pushed, and therefore, $\mathcal{R}$ stays trapped at v. Thus, $\mathcal{C}$ will reach w in a finite number of rounds and capture $\mathcal{R}$ in the next cop move.   □

3   Cop-Win Classes Under Strong Push

In this section, we consider the game where the cops have the ability of strong push. In this regard, we consider various graph classes that have high cop number in classical Cops and Robber game on oriented graphs and show that these classes are cop-win if the cop has the ability of strong push. For brevity, in this section, if a graph $\overrightarrow{G}$ is cop-win, given that $\mathcal{C}$ has the ability of strong push, we simply call $\overrightarrow{G}$ cop-win. We mention here that we do not have any construction of an oriented graph that is not cop-win under this model. We have the following positive results.

3.1   Complete Multipartite Graphs

In this section, we consider the orientations of complete multipartite graphs and show that they are cop-win in the strong push model. In particular, we have the following theorem.

Theorem 1. *Let $\overrightarrow{G}$ be an oriented graph such that the underlying graph G of $\overrightarrow{G}$ is a complete multipartite graph with partitions $A_1, \ldots, A_k$. Then, $c_{sp}(\overrightarrow{G}) = 1$.*

Proof. We give a strategy for $\mathcal{C}$. $\mathcal{C}$ chooses a vertex $v \in A_1$ and makes it a source vertex (by Observation 3). Once v is a source vertex, observe that the only safe place for $\mathcal{R}$ is to be on vertices of A_1 and $\mathcal{R}$ cannot move out of A_1 as long as v is a source vertex and $\mathcal{C}$ is on v.

Let $\mathcal{R}$ be on a vertex $u \in A_1$. For each vertex $w \in N^+(u)$, $\mathcal{C}$ pushes w. We remark that during this process, $\mathcal{C}$ might push a vertex $w \in N^+(u)$ such that $w \in N^+(v)$, and then v is no longer a source vertex, but this does not hurt the strategy of $\mathcal{C}$ because, at every instance, $N^+(u) \subseteq N^+(v)$. Note that once $\mathcal{C}$ finishes pushing all vertices in $N^+(u)$, $\mathcal{R}$ gets trapped at u since $N^+(u) = \emptyset$. Since $\mathcal{R}$ is trapped at a vertex $u \in V(\overrightarrow{G})$, due to Lemma 1, $\mathcal{C}$ can capture $\mathcal{R}$ in a finite number of cop moves.   □

3.2   Subcubic Graphs

In this section, we show that the orientations of subcubic graphs are cop-win if the cop has the strong push ability. Due to Lemma 1, to show that subcubic graphs are cop-win, it is sufficient to show that one cop can trap $\mathcal{R}$ in a graph $\overrightarrow{G}$, such that the underlying graph G is subcubic. First, we have the following lemma that we will use to prove Theorem 2.

Lemma 2. *Let v be a vertex of an oriented graph $\overrightarrow{G}$ such that $|N^+(v)| + |N^-(v)| \leq 3$. Moreover, let $\overrightarrow{H}$ be an induced subgraph of $\overrightarrow{G}$ we get after deleting v. If $\overrightarrow{H}$ is cop-win, then $\overrightarrow{G}$ is cop-win.*

Proof. Let $\mathcal{C}$ have a winning strategy in $\overrightarrow{H}$. We will use this strategy in $\overrightarrow{G}$ when $\mathcal{R}$ is on a vertex $u \neq v$. We show that if $\mathcal{R}$ is on/moves to vertex v, then (in the next move of $\mathcal{R}$) we can either trap $\mathcal{R}$ or force $\mathcal{R}$ to move out of v while keeping the orientation of each arc that does not have v as an endpoint same as it was before $\mathcal{R}$ moved to v. We have the following cases depending on the in-degree of v when $\mathcal{R}$ moves to v:

1. $|N^-(v)| = 3$: In this case, v is a sink vertex, and if $\mathcal{R}$ enters v, then $\mathcal{R}$ cannot move from v and is trapped at v. Hence, due to Lemma 1, $\mathcal{C}$ can move to capture $\mathcal{R}$ in a finite number of rounds.
2. $|N^-(v)| = 2$: In this case, observe that $|N^+(v)| \leq 1$. If $|N^+(v)| = 0$, then $\mathcal{R}$ is trapped at v and $\mathcal{C}$ can move to capture $\mathcal{R}$. If $|N^+(v)| = 1$, then let u be the unique out-neighbor of v. If $\mathcal{R}$ moves to v, then $\mathcal{C}$ simply pushes u. This traps $\mathcal{R}$ at the vertex v (since $|N^+(v)| = 0$ now). Hence, due to Lemma 1, $\mathcal{C}$ can move to capture $\mathcal{R}$ in a finite number of rounds.
3. $|N^-(v)| = 1$: In this case, first, we show that $\mathcal{C}$ can force $\mathcal{R}$ to move out of v in the next move of $\mathcal{R}$ while keeping the orientation of each arc not incident on v the same. Moreover, once $\mathcal{R}$ has moved out of v, in the next move of $\mathcal{R}$, $\mathcal{R}$ cannot move to v. Later, we show how $\mathcal{C}$ can use this step and the winning strategy for $\overrightarrow{H}$ to get a winning strategy for $\overrightarrow{G}$. We have the following two cases:
    (a) $\mathcal{R}$ begins the game at the vertex v: In this case, $\mathcal{C}$ pushes the vertex v. Now, observe that $|N^+(v)| = 1$. Let $N^+(v) = \{u\}$. Now, if $\mathcal{R}$ does not move to the vertex u in this robber turn, then $\mathcal{C}$ can push u to trap $\mathcal{R}$ at v. This leads to the capture of $\mathcal{R}$ (due to Lemma 1). So, assume $\mathcal{R}$ moves the vertex u. Note that for each arc $\overrightarrow{a}$ such that v is not an endpoint of $\overrightarrow{a}$, the orientation of $\overrightarrow{a}$ is not changed (reversed) in this step. Now, $\mathcal{C}$ will use the winning strategy of $\overrightarrow{H}$ as long as $\mathcal{R}$ does not move to v.
    (b) $\mathcal{R}$ moves to the vertex v from some vertex u: In this case also $\mathcal{C}$ pushes the vertex v. Note that this operation does not change the orientation of any arc that do not have v as an endpoint. Now, observe that $N^+(v) = \{u\}$. Similarly to the previous case (Case 3a), if $\mathcal{R}$ does not move to the vertex u in this robber turn, then $\mathcal{C}$ can push u to trap $\mathcal{R}$ at v. This leads to the capture of $\mathcal{R}$ (due to Lemma 1). So, assume $\mathcal{R}$ moves the vertex u. Now, $\mathcal{R}$ is back to the vertex u, and $\mathcal{C}$ again continues with its winning strategy for $\overrightarrow{H}$. Moreover, in the next move of $\mathcal{R}$, $\mathcal{R}$ cannot move to v (as $u \in N^+(v)$ now). So, the next move of $\mathcal{R}$ will be to a vertex (possibly staying at u) that is also present in $\overrightarrow{H}$.

Let $\mathcal{C}$ have a strategy to win in $\overrightarrow{H}$ using at most ℓ moves. In $\overrightarrow{G}$, each time $\mathcal{R}$ moves to the vertex v from some vertex u (u is also a vertex in $\overrightarrow{H}$), in the next robber move, it has to move back to u. Moreover, to force $\mathcal{C}$ out of v

(unless $\mathcal{C}$ traps or captures $\mathcal{R}$) $\mathcal{C}$ never pushes a vertex that is not v. Thus, the orientation of each arc that do not have v as an endpoint remains the same after these two moves ($\mathcal{R}$ moving to v from u and then coming back to u from v). Moreover, in the next robber move, $\mathcal{R}$ cannot move to v. Therefore, the next move of $\mathcal{R}$ in $\overrightarrow{G}$ can be translated to a valid move of $\mathcal{R}$ in $\overrightarrow{H}$ as well. Hence, by moving to v, the best $\mathcal{R}$ can do is to waste at most two moves of the cop. Therefore, $\mathcal{R}$ will be captured in $\overrightarrow{G}$ using at most 3ℓ cop moves.

4. $|N^-(v)| = 0$. In this case, notice that the vertex v is not accessible to $\mathcal{R}$.

Therefore, if $\overrightarrow{H}$ is cop-win, then $\overrightarrow{G}$ is cop-win as well.   □

Theorem 2. *Let $\overrightarrow{G}$ be an oriented graph such that its underlying graph G is a subcubic graph. Then $c_{sp}(\overrightarrow{G}) = 1$.*

Proof. We will prove this using contradiction arguments. Consider a minimal graph $\overrightarrow{G}$ such that its underlying graph G is subcubic and $\overrightarrow{G}$ is not cop-win (i.e., every induced subgraph of $\overrightarrow{G}$ we get after deleting at least one vertex is cop-win). Note that $\overrightarrow{G}$ contains at least two vertices as a single vertex graph is trivially cop-win. Let $\overrightarrow{H}$ be the induced subgraph of $\overrightarrow{G}$ we get after deleting a vertex $v \in V(\overrightarrow{G})$. By our assumption that $\overrightarrow{G}$ is a minimal graph having cop number at least two, we have that $c_{sp}(\overrightarrow{H}) = 1$. Then, due to Lemma 2, $c_{sp}(\overrightarrow{G}) = 1$, which contradicts our assumption that $\overrightarrow{G}$ is not cop-win.

Therefore, any oriented graph $\overrightarrow{G}$ such that its underlying graph G is subcubic is cop-win.   □

3.3   Interval Graphs

In this section, we show that if graph $\overrightarrow{G}$ is an orientation of an interval graph G, then $c_{sp}(\overrightarrow{G}) = 1$. To prove this result, we use *nice path decomposition*, a well-known tool for designing dynamic programming algorithms for graphs of bounded pathwidth. For an interval graph, the nice path decomposition can be computed in linear time [4].

Definition 2. *A nice path decomposition of an interval graph G is a path $T = (t_1, \ldots, t_k)$ where each node t_i is associated to a subset B_i of $V(G)$ called a bag, and each internal node t_i has exactly two neighbors t_{i-1} and t_{i+1}, with the following properties.*

1. *The nodes of T containing a given vertex of G form a nonempty connected subpath of T.*
2. *Any two adjacent vertices of G appear in the bag of a common node of T.*
3. *For each node t_i of T, B_i is a clique.*
4. *Each node of T belongs to one of the following types: introduce, forget, or leaf.*
5. *An introduce node t_i is such that $B_i \setminus \{v\} = B_{i-1}$, for some vertex $v \in B_i$.*

6. A forget node t_i is such that $B_i = B_{i-1} \setminus \{v\}$, for some vertex $v \in B_{i-1}$.
7. A leaf node t_i is a leaf of T with $B_i = \{v\}$ for some vertex v of G.
8. There are two leaf nodes in T: t_1 and t_k.
9. Each vertex $v \in V(G)$ is introduced exactly once (in some introduce node) and forgotten exactly once (in some forget node).

For a nice path decomposition and a node t_i of T, we define $V_{\leq t_i}$ as the union of all bags corresponding to nodes in the $(t_1, \ldots, t_{i-1})$ subpath of T. More formally, $V_{\leq t_i} = \bigcup_{j \in [i-1]} B_j$. We can similarly define $V_{<t_i} = V_{\leq t_i} \setminus B_i$, $V_{\geq t_i} = V(G) \setminus V_{<t_i}$, and $V_{>t_i} = V(G) \setminus V_{\leq t_i}$.

First, we present an overall idea of our winning strategy for $\mathcal{C}$. We define an *image of the cop* $\mathcal{C}$, denoted by $\mathcal{IC}$. We will think of $\mathcal{C}$ moving over the vertices of G and $\mathcal{IC}$ moving over the nodes of T. If $\mathcal{C}$ is on a vertex $u \in V(G)$, then we fix a node $t_i \in T$ as the position of $\mathcal{IC}$ such that $u \in B_i$. We note that a graph vertex v might be contained in multiple bags $\{B_{j+1}, \ldots, B_{j+\ell}\}$, but we always specify which node $t \in \{t_{j+1}, \ldots t_{j+\ell}\}$ is occupied by $\mathcal{IC}$. To begin with, $\mathcal{IC}$ will start at the node t_1 and $\mathcal{C}$ at the graph vertex in B_1. Now, after every finite number of rounds, we will move $\mathcal{IC}$ from a node t_i to the node t_{i+1} such that the following invariant is maintained: When $\mathcal{IC}$ is on a node s, $\mathcal{R}$ is restricted to $V_{\geq s}$. Finally, using this strategy, $\mathcal{IC}$ will reach the node t_k, and since B_k is a single vertex, say, u, $\mathcal{R}$ is restricted to u. Since $\mathcal{C}$ is also on u, $\mathcal{R}$ is finally captured. Now, we discuss these ideas more formally.

First, we have the following lemma.

Lemma 3. *Let $\overrightarrow{G}$ be an oriented graph such that its underlying graph G is an interval graph. Let T be a nice path decomposition of G. If $\mathcal{IC}$ is at a node t_i ($i < k$), $\mathcal{C}$ at a vertex $v \in B_i$ such that $N^+[v] = B_i$, and $\mathcal{R}$ is restricted to $V_{>t_i}$, then after a finite number of rounds, we can move $\mathcal{IC}$ to the node t_{i+1}, $\mathcal{C}$ to a vertex $u \in B_{i+1}$ such that $N^+[u] = B_{i+1}$, and $\mathcal{R}$ is restricted to $V_{>t_{i+1}}$. Moreover, during this whole procedure, $\mathcal{R}$ is restricted to $V_{\geq t_i}$.*

Proof. We have the following cases depending on the type of the node t_i.

1. Introduce node: Let the vertex introduced in the bag B_{i+1} be x (i.e., $B_{i+1} = B_i \cup \{x\}$). Since B_{i+1} is a clique, there is an edge between v and x in G. Therefore, either $\overrightarrow{vx}$ is an arc in $\overrightarrow{G}$, or $\overrightarrow{xv}$ is an arc in $\overrightarrow{G}$. If the arc is oriented as $\overrightarrow{vx}$, then $\mathcal{C}$ does nothing. Otherwise, if the arc is oriented $\overrightarrow{xv}$, then the cop pushes x to orient the arc $\overrightarrow{vx}$. Note that now v is a source in B_{i+1} as well. Observe that during these moves of the cop, $\mathcal{R}$ cannot enter a vertex of B_i since $\mathcal{C}$ is present at a vertex v such that $N^+[v] = B_i$. Finally, we move $\mathcal{IC}$ to the node t_{i+1} and set $u = v$. Note that at this point, since $\mathcal{C}$ is at a source vertex of B_{i+1}, $\mathcal{R}$ is restricted to $V_{>t_{i+1}}$. Hence, all the required conditions of the lemma are satisfied.
2. Forget node: Let the vertex forgotten in the bag B_{i+1} be x (i.e., $B_{i+1} = B_i \setminus \{x\}$). If $x \neq v$, then $\mathcal{C}$ does nothing. We simply move $\mathcal{IC}$ to node t_{i+1} and set $u = v$. Observe that all the conditions of the lemma are satisfied.

If $x = v$, then we do the following. Let u be a vertex in B_{i+1} such that if u was introduced in the node B_j, then any vertex $y \in B_{i+1} \setminus \{u\}$ was introduced in some node $B_{j'}$ where $j' > j$ (i.e., for each $\ell \le i$, if $y \in B_\ell$, then $u \in B_\ell$). Since each bag B_i is a clique, note that for any vertex $y \in B_{i+1} \setminus \{u\}$, $N(y) \cap V_{<t_{i+1}} \subseteq N(u) \cap V_{<t_{i+1}}$. Now, for each vertex $z \in N^-(u)$ such that z has been forgotten, push z. This makes u dominate all the vertices that have been forgotten before B_i and are adjacent to some vertex in $B_{i+1} \setminus \{u\}$. Now C moves to u. Observe that $\mathcal{R}$ cannot move directly to v from some vertex in $w \in V_{>t_i}$ because u and w cannot be adjacent.

Note that though in this round, $\mathcal{R}$ can move to a vertex in B_{i+1}, we will ensure that it is restricted to $V_{\ge t_i}$ (i.e., it cannot move to a vertex in $V_{<t_i}$). In the next move of C, C pushes v. Now, again $\mathcal{R}$ cannot move to v without getting captured (as C is on u and $\overrightarrow{uv}$ is an arc). Moreover, $\mathcal{R}$ cannot move to a vertex $z \in V_{<t_i}$ from a vertex $y \in B_{i+1}$ without getting captured in the next cop move (because we ensured that $N(y) \cap V_{<t_{i+1}} \subseteq N(u) \cap V_{<t_{i+1}}$). Hence, though $\mathcal{R}$ can move to a vertex of B_i, it cannot move to a vertex in $V_{<t_i}$. In next few cop moves, for each vertex $y \in B_{i+1} \setminus \{u\}$, if the arc between u and y is not oriented as $\overrightarrow{uy}$, then C pushes y. Thus, after a finite number of rounds, u is a source in B_{i+1} (and B_i as well). Moreover, at this point, if $\mathcal{R}$ is on a vertex in B_i or B_{i+1}, then $\mathcal{R}$ will be captured in the next cop move. Hence, $\mathcal{R}$ is restricted to $V_{>t_{i+1}}$. Finally, we move $\mathcal{IC}$ from t_i to t_{i+1}. Note that all the conditions of our lemma are satisfied.

This completes the proof of our lemma.   □

Next, we use Lemma 3 to get the following theorem.

Theorem 3. *Let $\overrightarrow{G}$ be an oriented graph such that its underlying graph G is an interval graph. Then $c_{sp}(\overrightarrow{G}) = 1$.*

Proof. Consider a nice path decomposition T of the interval graph. Let $T = (t_1, \ldots, t_k)$ and let bag B_i be associated with node t_i.

We have the following winning strategy for C. Let $B_1 = \{v\}$. Then, C starts at vertex v and $\mathcal{IC}$ starts at t_1. Observe that all the conditions for Lemma 3 are satisfied. Now, we use Lemma 3 to move $\mathcal{IC}$ from t_i to t_{i+1}. Finally, using this strategy, $\mathcal{IC}$ reaches t_k. Let $B_k = \{u\}$. Observe that at this point C is on u, and due to Lemma 3, $\mathcal{R}$ is restricted to u. Thus, $\mathcal{R}$ gets captured.

Hence, C can follow this strategy to capture $\mathcal{R}$.   □

4   Conclusion

We studied COPS AND ROBBER on oriented graphs with respect to push operation. In particular, we studied the game between a single cop C and a single robber $\mathcal{R}$ where $\mathcal{R}$ has the ability of strong push. We established that if the underlying graph G of an oriented graph $\overrightarrow{G}$ is either a complete multipartite graph, a subcubic graph, or an interval graph, then $c_{sp}(\overrightarrow{G}) = 1$. We do not have any construction of an oriented graph $\overrightarrow{G}$ that is not cop-win when the cop has strong push ability. This might be an interesting question to either construct an oriented graph that is not cop-win in the strong push model or to give a winning strategy for the cop with strong push ability for any oriented graph.

Other exciting research directions might be to extend these results for further, more general graph classes. In particular, does a cop $\mathcal{C}$ with strong push ability always have a winning strategy in the orientations of planar graphs? Moreover, if G is an undirected cop-win graph, does $\mathcal{C}$ with strong push ability always have a winning strategy in any orientation $\overrightarrow{G}$ of G? Another interesting research direction might be to generalize these results for the weak-push cop.

Acknowledgement. This research was supported by the IFCAM project "Applications of graph homomorphisms" (MA/IFCAM/18/39).

References

1. Abraham, I., Gavoille, C., Gupta, A., Neiman, O., Talwar, K.: Cops, robbers, and threatening skeletons: Padded decomposition for minor-free graphs. SIAM J. Comput. **48**(3), 1120–1145 (2019)
2. Aigner, M., Fromme, M.: A game of cops and robbers. Discret. Appl. Math. **8**(1), 1–12 (1984)
3. Angelo, D., Navarra, A., Nisse, N.: A unified approach for gathering and exclusive searching on rings under weak assumptions. Distrib. Comput. **30**, 17–48 (2017)
4. Belmonte, R., Golovach, P.A., Heggernes, P., van't Hof, P., Kamiński, M., Paulusma, D.: Detecting fixed patterns in chordal graphs in polynomial time. Algorithmica **69**(3), 501–521 (2014)
5. Bradshaw, P., Hosseini, S.A., Turcotte, J.: Cops and robbers on directed and undirected abelian Cayley graphs. Eur. J. Comb. **97**, 103383 (2021)
6. Czyzowicz, J., Gąsieniec, L., Gorry, T., Kranakis, E., Martin, R., Pajak, D.: Evacuating robots via unknown exit in a disk. In: Kuhn, F. (ed.) DISC 2014. LNCS, vol. 8784, pp. 122–136. Springer, Heidelberg (2014). https://doi.org/10.1007/978-3-662-45174-8_9
7. Darlington, E., Gibbons, C., Guy, K., Hauswald, J.: Cops and robbers on oriented graphs. Rose-Hulman Undergraduate Math. J. **17**(1), 201–209 (2016)
8. Das, S., Gahlawat, H., Sahoo, U.K., Sen, S.: Cops and robber on some families of oriented graphs. Theor. Comput. Sci. **888**, 31–40 (2021)
9. de la Maza, S.G.H., Hosseini, S.A., Knox, F., Mohar, B., Reed, B.: Cops and robbers on oriented toroidal grids. Theoret. Comput. Sci. **857**, 166–176 (2021)
10. Fisher, D.C., Ryan, J.: Tournament games and positive tournaments. J. Graph Theory **19**(2), 217–236 (1995)
11. Frieze, A., Krivelevich, M., Loh, P.: Variations on cops and robbers. J. Graph Theory **69**(4), 383–402 (2012)
12. Gavenčiak, T.: Cop-win graphs with maximum capture-time. Discret. Math. **310**(10–11), 1557–1563 (2010)
13. Goldstein, A.S., Reingold, E.M.: The complexity of pursuit on a graph. Theoret. Comput. Sci. **143**(1), 93–112 (1995)
14. Gottlob, G., Leone, N., Scarcello, F.: A comparison of structural CSP decomposition methods. Artif. Intell. **124**(2), 243–282 (2000)

15. Gottlob, G., Leone, N., Scarcello, F.: The complexity of acyclic conjunctive queries. J. ACM **48**(3), 431–498 (2001)
16. Hahn, G., MacGillivray, G.: A note on k-cop, l-robber games on graphs. Discret. Math. **306**(19–20), 2492–2497 (2006)
17. Hamidoune, Y.O.: On a pursuit game on Cayley digraphs. Eur. J. Comb. **8**(3), 289–295 (1987)
18. Hosseini, S.A.: Game of cops and robbers on Eulerian digraphs. Ph.D. thesis, Simon Fraser University (2018)
19. Hosseini, S.A., Mohar, B.: Game of cops and robbers in oriented quotients of the integer grid. Discret. Math. **341**(2), 439–450 (2018)
20. Isaza, A., Lu, J., Bulitko, V., Greiner, R.: A cover-based approach to multi-agent moving target pursuit. In: Proceedings of the Fourth Artificial Intelligence and Interactive Digital Entertainment Conference, pp. 54–59. AAAI Press (2008)
21. Khatri, D., et al.: A study of cops and robbers in oriented graphs. arXiv:1811.06155 (2019)
22. Kinnersley, W.B.: Cops and robbers is exptime-complete. J. Comb. Theory Ser. B **111**, 201–220 (2015)
23. Kinnersley, W.B.: Bounds on the length of a game of cops and robbers. Discret. Math. **341**(9), 2508–2518 (2018)
24. Klein, K., Suri, S.: Catch me if you can: Pursuit and capture in polygonal environments with obstacles. In: Proceedings of the AAAI Conference on Artificial Intelligence (AAAI 2012), vol. 26, pp. 2010–2016 (2012)
25. Klostermeyer, W.F.: Pushing vertices and orienting edges. Ars Combinatoria **51**, 65–76 (1999)
26. Klostermeyer, W.F., et al.: Hamiltonicity and reversing arcs in digraphs. J. Graph Theory **28**(1), 13–30 (1998)
27. Loh, P., Oh, S.: Cops and robbers on planar directed graphs. J. Graph Theory **86**(3), 329–340 (2017)
28. MacGillivray, G., Wood, K.L.B.: Re-orienting tournaments by pushing vertices. Ars Combinatoria **57**, 33–47 (2000)
29. Mosesian, K.M.: Strongly Basable graphs (Russian). Akad. Nauk. Armian. SSR Dokl. **54**, 134–138 (1972)
30. Nisse, N.: Network decontamination. In: Flocchini, P., Prencipe, G., Santoro, N. (eds.) Distributed Computing by Mobile Entities. LNCS, vol. 11340, pp. 516–548. Springer, Cham (2019). https://doi.org/10.1007/978-3-030-11072-7_19
31. Parsons, T.D.: Pursuit-evasion in a graph. In: Alavi, Y., Lick, D.R. (eds.) Theory and Applications of Graphs. Lecture Notes in Mathematics, vol. 642, pp. 426–441. Springer, Heidelberg (1978). https://doi.org/10.1007/BFb0070400
32. Parsons, T.D.: The search number of a connected graph. In: Proceedings of the Ninth Southeastern Conference on Combinatorics, Graph Theory, and Computing, vol. XXI, pp. 549–554. Utilitas Mathematica (1978)
33. Pretzel, O.: On graphs that can be oriented as diagrams of ordered sets. Order **2**, 25–40 (1985)
34. Pretzel, O.: On reordering graphs by pushing down maximal vertices. Order **3**, 135–153 (1986)
35. Pretzel, O.: Orientations and edge functions on graphs. In: Surveys in Combinatorics. London Mathematical Society Lecture Notes, vol. 66, pp. 161–185 (1991)
36. Seymour, P.D., Thomas, R.: Graph searching and a min-max theorem for tree-width. J. Comb. Theory Ser. B **58**(1), 22–33 (1993)
37. Slivova, V.: Cops and robber game on directed complete graphs. Bachelor's thesis, Charles University in Prague (2015)

Mind the Gap: Edge Facility Location Problems in Theory and Practice

Moritz Beck[1(✉)], Joachim Spoerhase[2], and Sabine Storandt[1]

[1] Universität Konstanz, Konstanz, Germany
{beck,storandt}@inf.uni-konstanz.de
[2] Max Planck Institute for Informatics, Saarbrücken, Germany
jspoerha@mpi-inf.mpg.de

Abstract. Motivated by applications in urban planning, network analysis, and data visualization, we introduce center selection problems in graphs where the centers are represented by edges. This is in contrast to classic center selection problems where centers are usually placed at the nodes of a graph. Given a weighted graph $G(V, E)$ and a budget $k \in \mathbb{N}$, the goal is to select k edges from E such that the maximum distance from any point of interest in the graph to its nearest center is minimized. We consider three different problem variants, based on defining the points of interest either as the edges of G, or the nodes, or all points on the edges. We provide a variety of hardness results and approximation algorithms. A key difficulty of edge center selection is that the underlying distance function may not satisfy the triangle inequality, which is crucially used in approximation algorithms for node center selection. In addition, we introduce efficient heuristics that produce solutions of good quality even in large graphs, as demonstrated in our experimental evaluation.

Keywords: Facility location · Edge facility · K-center

1 Introduction

Center selection problems are well studied in geometric contexts as well as in graphs. In the classic k-CENTER problem, one has to select k out of n given points in the plane such that the maximum Euclidean distance from any of the n points to the nearest selected point is minimized [8]. In graphs, the analogue problem is to select k nodes as centers (e.g. for warehouses or hospitals) with the goal of minimizing the maximum shortest path distance from any node in the graph to its closest center [18]. The geometric version can also be phrased as graph problem by creating a complete graph on the point set with Euclidean edge weights. The geometric and the graph problem are both NP-hard, but a simple greedy selection strategy yields a 2-approximaton algorithm [3,7]. Many extensions of the problem have been investigated, including restrictions on the center locations, node weights that model center ("facility") opening costs, or capacities that limit how many customers can be served by a single center [1,4,9].

In this paper, we consider a family of center selection problems where the centers are represented by edges instead of nodes. There are several possible application scenarios where edge centers are desirable:

© The Author(s), under exclusive license to Springer Nature Switzerland AG 2023
A. Bagchi and R. Muthu (Eds.): CALDAM 2023, LNCS 13947, pp. 321–334, 2023.
https://doi.org/10.1007/978-3-031-25211-2_25

- *Urban planning.* Facilities with substantial areal impact in urban areas, such as parks or street parking zones, are inadequately modelled by node centers because they are not only accessible from a single entry point. Edge centers represent this aspect more faithfully [11].
- *Network analysis and design.* Computing the set of the top-k edge centers is a means to identify important connections in a given network. This complements work on group centrality measures which are typically focused on identifying important node sets [12].
- *Graph data visualization.* Displaying large amounts of graph data in digital maps is often too time-consuming for interactive usage and hence data reduction is applied first. One widely used reduction concept is graph simplification, e.g., by replacing long paths with single edges. However, these additional edges may induce topological inconsistencies and increase the drawing complexity over all zoom levels. Therefore, it was suggested in [17] to use graph sampling instead by choosing a proper subset of the edges to draw.

In all these applications, the goal is to choose an edge set which is in some sense nicely spread over the graph and hence there are no graph elements that are too far from the nearest chosen edge. The goal of the paper is to investigate such edge center selection problems from a theoretical and practical perspective.

1.1 Contribution

In this paper, we introduce and analyze k-EDGE-CENTER-SELECTION (ECS) problems in graphs, where the goal is to select a set of k edges such that the maximum distance of any point of interest in the graph to its nearest selected edge is as small as possible. We consider edges, nodes, and points on edges as possible locations of interest, and provide the following results for the three resulting problem variants, abbreviated as EECS, NECS, and PECS:

- We prove that NECS and PECS are NP-hard to approximate to a constant factor. For EECS, we prove the stronger result that it is NP-hard to get an f-approximation for any computable function f. A fine-grained complexity analysis shows that the hardness depends on the edge budget k. We identify for EECS and NECS thresholds for k for which the respective optimization problem becomes solvable in polytime.
- We instrument parametric pruning to design a 3-approximation algorithm for NECS and a 4-approximation algorithm for PECS.
- Inspired by the geometric setting, we also present a simple greedy algorithm for all three variants where the quality depends on ψ, which is defined as the ratio of the longest and the shortest edge in the input graph. We show that under assumptions that are sensible for the application of center selection in road networks, the greedy algorithm is a 3-approximation.
- In the experimental evaluation, we assess the running time and the solution quality of the greedy algorithm as well as heuristic approaches in dependency of the input graph type and the edge budget k, and show that sensible ECS solutions can be computed even in large graphs.

1.2 Related Work

In the geometric version of the problem, line segments serving as facilities, instead of points, have been considered in [11], motivated by a transportation network design problem. However, they allow to choose any line segment in the plane as facility while we have to choose an edge subset of a given graph. In [13–15], point, line segment and even polygonal facilities were discussed. But here again, facility location is permissible anywhere in continuous space, and geometric distance measures are used to assess the quality of the solution.

For graph settings, continuous center selection problems on graphs were investigated in [5]. There, nodes are weighted and centers may be chosen from the nodes of the graph and interior points of edges. A set of centers has to be chosen such that the weighted sum of all nodes to their nearest centers is minimized. It is shown that in this case it is sufficient to consider node centers.

In [16], several so called activation edge-cover problems were discussed. There, each edge is equipped with an activation function which depends on the values assigned to the nodes. The goal is then to find an assignment of small total cost, such that the edges activated by that assignment fulfill a certain cover constraint. Relationships to facility location problems were shown there; but only considering node facilities.

The problem of choosing a representative subset of edges of a given graph was thoroughly discussed in [10]. However, the problem is not phrased as a formal optimization problem there. Instead, several criteria of a good graph sample are listed, and then sampling strategies (mostly relying on random selection) are empirically evaluated and compared. In [17], algorithms for gap minimization in polylines (or path graphs) were introduced. The generalization to arbitrary graphs was left there as an open problem.

2 Edge Center Selection Problems

In this section, we formally define edge center selection problems in graphs. As a prerequisite, we generalize the geometric k-CENTER problem from point locations to segment locations.

Definition 1 (k-Segment-Center). *Given a set S of line segments in $\mathbb{R}^2$ and $k \in \mathbb{N}$, find a set of k segments $F \subseteq S$ such that the maximum Euclidean distance of any segment to its nearest segment in S is minimized.*

The distance between two segments is determined by the Euclidean length of the shortest connection between them. Just like the k-CENTER problem, we can rephrase the k-SEGMENT-CENTER problem as a graph problem. Here, we create again a complete graph in which now every segment is represented by a node, and the edge weights depend on the pairwise segment distances. In contrast to the k-CENTER problem, the resulting graph for k-SEGMENT-CENTER might not be metric, though. In particular, the triangle inequality might be violated

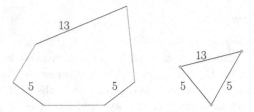

Fig. 1. k-SEGMENT-CENTER example with three segments (blue) and their pairwise Euclidean distances (black). The right image shows the graph representation, illustrating that the triangle inequality is not obeyed. (Color figure online)

as shown in Fig. 1. We will discuss how that affects the applicability of the 2-approximation algorithm for k-CENTER, where the quality guarantee relies on the triangle inequality, in Sect. 4.

However, for many application scenarios (as e.g. urban planning), the distances to facilities stem more realistically from (driving) distances in an underlying graph than from Euclidean distances in the plane. Therefore, we next define three problem variants that directly use a weighted graph as a basis. For that, we first need the notion of distance towards an edge in a given graph. Let $c(a, b)$ denote the shortest path cost from node a to b in G. Then the distance of a node a towards an edge $e = \{v, w\}$ is defined as $c(a, e) := \min\{c(a, v), c(a, w)\}$ and the distance of an edge $e = \{v, w\}$ towards another edge $e' = \{v', w'\}$ as $c(e, e') := \min\{c(v, e'), c(w, e')\}$.

Definition 2 (k-Edge/Node/Point-Edge-Center). *Given a connected, undirected, weighted graph $G(V, E, c)$ and a parameter $k \in \mathbb{N}$, find a set of k edges F of G such that the maximum shortest path distance from an edge/a node/any point on an edge to the nearest edge in F is minimized.*

So as the base definition, we consider distances between edges and edge centers (EECS). But if necessary for the application we can make the model more fine-grained by considering the distances between nodes and edge centers (NECS). And if one wants to take into account the driving distance from each individual address towards the next edge center without having to subsample long edges (and consequentially increase the graph size severely), the point-edge center selection model (PECS) can be used. The distance of a point to the end of an edge is thereby obtained by linear interpolation of the edge weight.

We note that in graphs with metric edge weights, we have $\text{OPT}_{\text{EECS}} \leq \text{OPT}_{\text{NECS}} \leq \text{OPT}_{\text{PECS}}$ by definition. Figure 2 shows that even on small example graphs the problem variants might produce different outcomes.

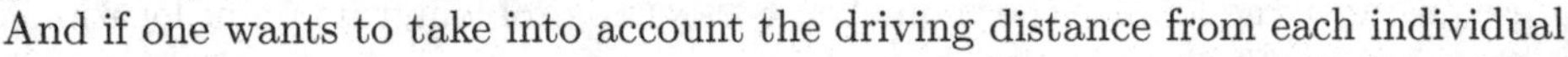

Fig. 2. Differing optimal solutions for k-NODE-EDGE-CENTER (left, OPT $= 1$) and k-POINT-EDGE-CENTER (right, OPT $= 2$) for $k = 2$.

3   Hardness and Tractability Results

In the geometric setting, the k-SEGMENT-CENTER problem naturally inherits the hardness of k-CENTER as it is a sound generalization thereof given that we can simply interpret points as segments of length zero.

For the k-EDGE-CENTER problems in graphs, we establish NP-hardness by reductions that also show APX-hardness.

Theorem 1. *It is NP-hard to approximate NECS (PECS) to a factor of $2 - \epsilon$ ($4/3 - \epsilon$) for any $\epsilon > 0$.*

Proof. Sketch. Per reduction from Set Cover. Let $((U, \mathcal{F}), k)$ be an instance of Set Cover. Construct a NECS/PECS instance $(G(V, E, w)), k')$ as follows. $V := \{r\} \cup \mathcal{F} \cup U$, $E := \{\{r, S\} \mid S \in \mathcal{F}\} \cup \{\{S, e\} \mid S \in \mathcal{F}, e \in S\}$, $w := 1$, $k' := k$.

If there is a set cover of size at most k, then there is a solution for the NECS (PECS) instance with maximum distance at most 1 (1.5). If there is no set cover of size at most k, then the optimum solution for the NECS/PECS instance has distance at least 2. Hence, the claim holds.   □

For EECS we reduce the MINIMUM EDGE DOMINATING SET (MEDS) problem.

Definition 3 (Minimum Edge Dominating Set). *In a graph $G(V, E)$, a set $D \subseteq E$ is called an edge dominating set if every edge not in D is adjacent to an edge in D. The optimization goal is to find a smallest set D with that property.*

As proven in [2], it is NP-hard to approximate MEDS to a factor $< 7/6$. A MEDS of size k is a EECS with distance $d = 0$. So it is NP-hard to decide for an EECS instance whether $d = 0$ is achievable and because any (mulitiplicative) approximation has to return 0 in that case (and necessarily a value larger than 0 otherwise), approximation is also NP-hard.

Theorem 2. *It is NP-hard to f-approximate EECS for any computable function f of n.*

Proof. Assume there is an f-approximation algorithm for k-EDGE-EDGE-CENTER for a computable function f. Then, given an instance of MEDS, we use binary search to find the smallest k such that the f-approximation computes a set of k edges with a maximum edge-edge distance of $d = 0$. This set of edges then forms a MEDS. Hence, we could determine the cardinality of an optimal MEDS in polytime which contradicts the NP-hardness of MEDS.   □

Next, we conduct a more fine-grained complexity analysis to show that despite the general hardness there are ranges for the edge budget k for which the respective ECS problems become tractable.

Lemma 1. *For $k \geq 2|D|$, where D is an optimal MEDS solution, EECS can be solved to optimality in polytime.*

Proof. If k is at least $2|D|$, then we can use the following algorithm to find the optimal solution: Every maximal matching in a given graph G is an edge dominating set by definition and a minimum maximal matching is a MEDS. We can thus simply compute any maximal matching (which then contains at most twice the number of edges as the optimum) to get an optimal solution for EECS with at most k centers and distance 0.   □

For $|D| \leq k < 2|D|$, the optimum distance for EECS is 0 as well. However, as proven in Theorem 2 (utilizing the hardness of MEDS) we cannot find an approximate solution for EECS for $k < 7/6|D|$ if P $\neq$ NP. For the range between $7/6|D|$ and $2|D|$, a finite approximation factor might be possible, though.

For the fine-grained analysis of NECS, we relate it to the MINIMUM EDGE COVER (MEC) problem.

Definition 4 (Minimum Edge Cover). *In a graph $G(V, E)$, an edge cover is a set of edges S such that every node in the graph is incident to at least one edge in S. The* MINIMUM EDGE COVER *problem demands to find the smallest set S.*

An optimal MEC can be found in polytime by first computing a maximum matching and then extending it greedily to cover remaining uncovered nodes [3].

Lemma 2. *For $k \geq |S|$, where S is an optimal MEC solution, NECS can be solved to optimality in polytime.*

Proof. If $k \geq |S|$, then the optimal solution distance is 0, as every node can have at least one incident center. As MEC can be solved to optimality in polytime, the same then is true for NECS.   □

The argument does not apply to PECS, as there even a MEC does not result in an objective function value of zero. But we will show in the next section, that PECS can be approximated within a constant factor for any edge budget k.

4 Approximation Algorithms

Motivated by the hardness results, we next investigate whether we can find approximate solutions for those edge center selection problems.

For k-CENTER, several algorithms exist which guarantee an approximation factor of 2, most prominently a *parametric pruning* and a *greedy* approach. As the problem cannot be approximated by a factor of $2 - \varepsilon$ for any $\varepsilon > 0$ unless P $=$ NP [6], those algorithms provide the best possible guarantee for k-CENTER. In the following, we discuss how these two approaches can be instrumented for ECS and analyze the resulting guarantees for each problem variant.

4.1 Parametric Pruning

The parametric pruning algorithm conducts a systematic search for a k-CENTER solution in a given complete, weighted graph $G(V, E)$ with edge costs $c : E \to \mathbb{R}^+$

as follows. The distance value OPT induced by the optimal selection of the k center nodes has to coincide with the cost value of some edge. Let R with $|R| \in \mathcal{O}(n^2)$ be the respective set of all edge costs. Then for any $r \in R$, a subgraph $G_r(V, E_r)$ of G is constructed where $E_r := \{e \in E \mid c(e) \leq r\}$. If a DOMINATING SET (DS) of size at most k exists in G_r, then it follows that OPT $\leq r$. The goal is hence to find the smallest r for which a sufficiently small DS exists. But as computing a DS of minimum size is an NP-hard problem itself, another step is needed to find an approximate solution in polytime. For that, the graph G_r^2 is considered, which contains an edge between two nodes if there is a path of hop length at most two in G_r. It can be proven that any MAXIMAL INDEPENDENT SET (MIS) in G_r^2 has a size of at most the size of a DS in G_r. At the same time for any node in G there is a node in the MIS at distance at most $2r$. Hence, returning the MIS for the smallest value $r \in R$ for which it has size at most k results in a 2-approximation for k-CENTER.

For ECS, we do not necessarily assume that the given graph is complete. Here, the set R is formed by the possible shortest path distances from the locations of interest towards the potential edge centers. Nevertheless, for NECS we can instrument parametric pruning to design a 3-approximation.

We compute R and for each $r \in R$ (in ascending order) we do the following: We construct G_r^2 and find a maximal independent set I in G_r^2. Then, we construct a graph $F = (I, E')$ with an initially empty edge set: For each edge $e \in E$, find $e' := \{z \in I \mid c(z, e) \leq r\}$ and add it to E' if $|e'| \geq 1$. (Every constructed edge e' has at most two incident nodes because the neighborhoods of its corresponding edge's endpoints contain at most one node of I each. So, e' can be a simple edge or a loop.) Compute a minimum edge cover S' for F and return $S := \{e \in E \mid e' \in S'\}$, i.e. the edges in the original graph corresponding to the edges in the edge cover, if it has at most k edges.

Theorem 3. *The parametric pruning algorithm for NECS guarantees an approximation factor of 3.*

Proof. The returned solution S contains at most k edges by design. Every node $v \in V$ has distance at most $3r$ from an edge $e \in S$: The distance of any node v to its nearest node $z \in I$ in the independent set is at most $2r$ by construction. The distance from any node $z \in I$ to its nearest chosen edge $e \in S$ is at most r because e' is incident to z in F. Hence, the total distance is at most $2r + r = 3r$.

$r \leq$ OPT: We show $|S'| \leq k$ if $r \geq$ OPT. Let S_* be an optimum solution to the NECS instance, and assume $r \geq$ OPT. For every node $z \in I$ there is an edge $e_z \in S_*$ at distance at most OPT. So the corresponding edge e'_z contains at least one node (namely z) and is contained in E'. Define the set $S'_* := \{e'_z \mid z \in I\}$. We have $|S'_*| \leq |S_*| = k$ and S'_* is a valid edge cover of F (because $r \geq$ OPT). Therefore, the minimum edge cover S' has at most k edges, too; so the algorithm returns when $r =$ OPT at the latest.

To sum up, the algorithm returns a set S of at most k edges such that every node has distance at most $3r$ to it and $r \leq$ OPT. $\qquad\square$

This parametric pruning algorithm can be adjusted to give a 4-approximation for PECS. See the appendix for details.

Theorem 4. *There is a 4-approximation algorithm for PECS.*

We remark that the parametric pruning technique is mostly interesting from a theoretical perspective unless the input graph and OPT are very small. Otherwise, already the construction of G_r^2 is quite time consuming despite the overall polynomial running time.

4.2   Greedy Selection

The arguably simplest algorithm for approximating k-CENTER is the greedy approach (GREEDY). It starts by selecting an arbitrary point as center in the first round. In every subsequent round, it always selects the point with maximum Euclidean distance to the already selected centers.

We can easily adapt the algorithm to deal with segments in the plane or edges in a graph, by always selecting the segment/edge with largest distance to the already selected ones. But now the question arises what quality guarantee can be shown. The proof of the approximation factor of 2 for k-CENTER relies on the triangle inequality. However, as shown in Fig. 1, distances between segments or edges do not have to obey the triangle inequality. To upper bound the violation that might occur, we will use the paramater $\psi := \frac{L}{\ell}$ in our analysis, where L is the length of the longest segment/edge in the input, and ℓ the length of the shortest segment/edge, respectively.

Theorem 5. *GREEDY is a $(\psi + 2)$-approximation algorithm for k-SEGMENT-CENTER.*

Proof. The proof follows the argumentation for the k-CENTER greedy algorithm. Suppose there is a segment with a distance larger than $(2 + \psi) \cdot$ OPT to the nearest selected segment. By construction, this means that the pairwise distance between selected segments is also larger than $(2 + \psi) \cdot$ OPT. So we have a set S of $k+1$ segments with pairwise distances larger than $(2 + \psi) \cdot$ OPT. At least two segments $a, b \in S$ are assigned to the same segment s in the optimum solution (pigeonhole principle). It holds:

$$c(a, b) \leq c(a, s) + L + c(s, b) \leq L + 2\,\text{OPT}$$
$$= \left(\frac{L}{\text{OPT}} + 2\right) \cdot \text{OPT} \leq \left(\frac{L}{\ell} + 2\right) \cdot \text{OPT} = (\psi + 2) \cdot \text{OPT}$$

This is a contradiction; hence the largest distance is at most $(\psi + 2) \cdot$ OPT.   □

We observe that the proof works whenever we can ensure that ℓ lower bounds OPT. Next, we investigate for the three ECS problems for which values of k this observation can be exploited.

Lemma 3. *For EECS, if $k < |D|$ where D is an optimal MEDS solution, then GREEDY guarantees a $\psi + 2$ approximation.*

Proof. Clearly, ℓ is a lower bound as long as k is smaller than the optimal MEDS as then there needs to be at least one edge which is not adjacent to any centre, and hence the distance of that edge to its nearest center is lower bounded by the length of the shortest edge in the graph.   □

Lemma 4. *NECS can be approximated within a factor of $\psi + 2$ for any k.*

Proof. Let S be an optimal MEC solution. For NECS, ℓ is a valid lower bound for the maximum distance of a node to its nearest center if $k < |S|$ as then there needs to exist a node which is not incident to a center. For $k \geq |S|$, a MEC constitutes an optimal ECS solution anyway, see Lemma 2.   □

Lemma 5. *PECS can be approximated within a factor of $\psi + 2$ for any k.*

Proof. If k is smaller than the optimal MEC solution S, then ℓ is a lower bound for the optimal PECS solution based on the same argument as given above for NECS and hence GREEDY provides the desired approximation guarantee. If $k \geq |S|$, then we can simply compute a MEC in polynomial time. This MEC yields a distance of at most $L/2$ while the optimum is at least $\ell/2$ as long as not all edges are allowed to be selected (then the problem would become trivial). Accordingly, also for large edge budgets, a PECS solution with distance at most $\psi + 2$ times the optimum can be computed in polytime.   □

Taking the length L of the longest (instead of shortest) edge as a lower bound on the optimum solution in the proof of Theorem 5 leads to the following:

Corollary 1. *For all ECS variants, if $\text{OPT} > L$, GREEDY guarantees a 3-approximation.*

The requirement that $\text{OPT} \geq L$ might appear to be rather restrictive at first glance and it would indeed be of no use if the input graph is complete. However, in our envisioned application scenario where we want to place facilities as parks on edges of a (typically sparse) road network, it seems to be reasonable to assume that the maximum driving distance to the nearest facility exceeds the maximum edge cost in the network.

5   Practical Computation

The greedy algorithm has a running time of $\mathcal{O}(k(n \log n + m))$ as we need to run Dijkstra's algorithm in every round to identify the edge furthest from the previously selected centers. This is viable in practice as long as the edge budget k is sufficiently small. Furthermore, we can accelerate the computation in practice by not always starting a completely new Dijkstra run in each round. Instead, we always keep the distance table but push the end nodes of the newly selected edge center with a distance of 0 in the priority queue and continue the search from there. This ensures that only nodes that are closer to the newly selected center than to the previous selected ones will be updated.

However, for large values of k, the algorithm might still take too long on large networks. Therefore, we next discuss two heuristics that are efficient for large edge budgets.

Random selection. As we strive for a set of centers that are nicely spread over the graph, random edge selection is actually a decent and fast approach; in particular for a large edge budget k. We will refer to this selection strategy as RAND and will use it as a baseline in the experiments.

MEC-based heuristic. For a more involved heuristic, we reconsider the concept of a MEC, see Definition 4. As an optimal MEC solution S can be computed in polytime, it is a good starting point for selecting a large set of centers. As already shown in the previous section, for NECS, this yields an optimal solution whenever $k \geq |S|$. For EECS, the same applies as a MEC always constitutes an edge dominating set and hence for $k \geq |S|$ we have distance of 0 from every edge to its nearest MEC center. For PECS, though, a budget $k > |S|$ might allow for a quality gain by adding centers to the MEC until the budget is exceeded. We select those additional edges again in a greedy fashion, but here simply by always choosing the longest edge that is not yet a center. This is the best possible extension strategy for a MEC, as the longest distance towards a center is always induced by the middle point of the currently longest non-center edge. We refer to this approach as MEC+.

6   Experiments

The algorithms proposed for practical application were implemented in Rust 1.54.0. Experiments were conducted on a single core of an AMD Ryzen Threadripper 3970X 32-Core Processor clocked at 3.7 GHz with 256 GB RAM. We use two types of graph data sets in our evaluation: Real-world *road networks* from the 9th DIMACS Implementation challenge[1] and *generated grids* with given row and column number, randomly chosen edge weights and some random deletion of edges to not only deal with full grids. The characteristics of some selected instances are provided in Table 1 and Fig. 3.

Table 1. Benchmark data samples.

Name	Type	n	m	ψ
NY	Road	264236	365050	36046
CAL	Road	1890815	2315222	215354
USA	Road	23947347	28854312	368855
100×80	Grid	8000	14238	7.5
200×150	Grid	30000	53685	1.25

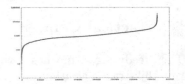

Fig. 3. Edge weight distribution in the NY instance.

---

[1] www.diag.uniroma1.it//challenge9/download.shtml.

6.1   Results for Small Edge Budgets

We first evaluated the RAND and the GREEDY algorithm for the ECS problems using edge budgets $k = 10, 50, 100, 500, 1000$. Compared to the RAND baseline, the greedy algorithm produces on average solutions that come with a distance that is only about 48% of the RAND distance on road networks and 51% on grids. In general, for larger k the performance gain of greedy in comparison to RAND was more pronounced.

The objective function values for EECS and NECS turned out to be very similar across all instances. However, we observe that the existence of long edges in the graphs tends to lead to a huge discrepancy between the solution distances of EECS/NECS and PECS due to the fact that long edges which do not become centers automatically lead to large driving distances for PECS but not necessarily for the other ones.

Regarding the running times of GREEDY, we observe that the algorithm actually scales better than linear in k. This is based on our Dijkstra implementation that does not reset the whole distance table but only relevant parts. The number of nodes that have to be settled in each round hence decreases significantly with growing k. For example, for the USA instance (the largest one considered), the computation took 106 s for $k = 10$, 263 s for $k = 100$ and 1338 s for $k = 1000$. While the latter is of course quite time-consuming, it's much better than the extrapolated time from $k = 10$, which would predict a running time around 10600 s for $k = 1000$.

6.2   Results for Large Edge Budgets

For sufficiently large edge budgets k, we know from our theoretical considerations that we can simply compute the optimal solution for EECS and NECS in polytime based on maximal matchings or MECs. For PECS, we use the MEC+ approach when the budget exceeds the size of the MEC. The MEC could be computed on all considered instances in at most one second; and much faster on the smaller instances. The MEC+ approach only demands to sort the remaining edges by weight and then performing a linear sweep. Hence MEC(+) is vastly faster than GREEDY. Furthermore, MEC+ also produces much better solutions on average for PECS. For example, for the NY instance, GREEDY produced a maximum distance of 15467 (in 7500 s), while using MEC+ resulted in a distance of 938.5 (in about 8 milliseconds), which is quality-wise better by a factor of over 15 (and in terms of running time better by about six orders of magnitude). For the CAL and USA instances, GREEDY has not produced a result after several hours, while MEC+ finished in 0.08 s and 0.96 s, respectively.

7   Conclusions and Future Work

We have studied several edge center selection problems in graphs and identified commonalities and differences. For NECS and PECS, we designed constant factor approximations, for EECS we ruled out the existence of one. Still, there is a gap between the lower and upper approximation bounds. We also designed a greedy algorithm suitable for practical purposes. In future work it would be interesting to consider extended problem settings, as e.g., capacitated edge facilities.

Appendix

Appendix A: Omitted Proof

Theorem 4. *There is a 4-approximation algorithm for PECS.*

Proof. We use an adjusted version of the parametric pruning algorithm described in Subsect. 4.1. We first need to argue that the set R of possible solution distance values is still polynomially bounded despite for PECS all points on edges are considered as locations of interest. To accomplish that we use the following observation: For any edge $e = \{v, w\} \in E$ the maximum distance of a point p on e towards the nearest center is determined either by $c(p, v)$ or $c(p, w)$ plus the cost of the respective end point towards its nearest center. Hence for any pair of edges $e_v, e_w \in E$ (including $e_v = e_w$) where e_v is assumed to be the nearest center for v and e_w the nearest center for w, the point on e which has maximum distance to e_v and e_w can be easily computed and the respective distance is added to R. Accordingly, we get $|R| \in \mathcal{O}(n^3)$.

Next, we make the crucial observation that if OPT $= r$, then all edges $e \in E$ with $c(e) > 2r$ have to be included in the PECS solution, as otherwise the midpoint of the edge would already have a too large distance to the end points. Hence only if the number of these heavy edges does not already exceed k, a PECS solution with the requested size can exist. Based on this observation, we propose the following modification: We identify the set of edges $H := \{e \in E \mid c(e) > 2r\}$. After computing F we find the smallest edge cover that contains all edges $H' := \{e' \mid e \in H\}$ by first removing all nodes incident to H' from F, computing a smallest edge cover on the resulting graph, adding H' and adding any incident edge for uncovered nodes (nodes are still not covered iff they became isolated by removing H' from F). As before, any node $v \in V$ has distance at most $3r$ from the set S containing the corresponding edges, and any point on an edge has distance at most r to its nearest node. Hence, this is a 4-approximation algorithm. □

Appendix B: Omitted Experimental Results

First, a remark about the benchmark instances: The ψ-values for the shown road networks are huge. It is important to note, though, that a small percentage of the edges is very short (length of 1) or very long ($L = \psi$) in these graphs, see Fig. 3 for an illustration of the edge weight distribution in NY. Considering the applications mentioned in the introduction, as placing parks in an urban area, such overly short or long edges would be non-sensical facilities. Hence when modelling suitable instances for such applications, merging short edges on degree-2 chains or subdividing long edges would be meaningful, and accordingly the ψ value would decrease. The grid instances hence better reflect the kind of input we would expect for facility placement in a city road network.

In the main text we observe that GREEDY works quite fast even on large graphs given that the edge budget is sufficiently small. And for graphs with small ψ value (e.g. grid instances), it also comes with a proper quality guarantee. For example, for the 200×150 grid instance, the approximation factor is at most 3.25.

But we also observe a disadvantage of using GREEDY for road networks. As the algorithm always selects the edge with the furthest distance to the previously chosen ones, it often chooses centers in dead-ends, as illustrated in Fig. 4. (The same would happen with node centers in the classical setting.) Obviously, if the path to the dead-end is not too long, then it would make more sense to select the center at the beginning of that path instead of the end, as this would decrease the driving distance from all non-path edges towards the center. One could hence improve the solution with a postprocessing step or integrate the observation directly into the algorithm.

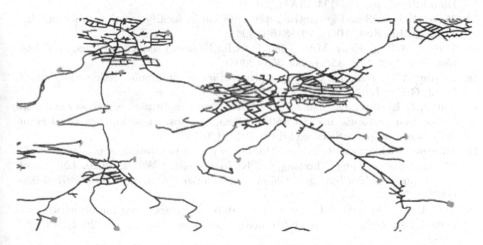

Fig. 4. Cutout of a road network with edge centers (blue) chosen by the greedy approximation algorithm. Note that many chosen edges in this cutout are at the end of dead-end streets. (Color figure online)

References

1. Batta, R., Lejeune, M., Prasad, S.: Public facility location using dispersion, population, and equity criteria. Eur. J. Oper. Res. **234**(3), 819–829 (2014)
2. Chlebík, M., Chlebíková, J.: Approximation hardness of edge dominating set problems. J. Comb. Optim. **11**(3), 279–290 (2006). https://doi.org/10.1007/s10878-006-7908-0
3. Garey, M.R., Johnson, D.S.: Computers and intractability, vol. 174. Freeman, San Francisco (1979)

4. Hajiaghayi, M.T., Mahdian, M., Mirrokni, V.S.: The facility location problem with general cost functions. Netw. Int. J. **42**(1), 42–47 (2003)
5. Hakimi, S.L.: Optimum distribution of switching centers in a communication network and some related graph theoretic problems. Oper. Res. **13**(3), 462–475 (1965)
6. Hochbaum, D.S.: Approximation algorithms for NP-hard problems. ACM SIGACT News **28**(2), 40–52 (1997)
7. Hochbaum, D.S., Shmoys, D.B.: A best possible heuristic for the k-center problem. Math. Oper. Res. **10**(2), 180–184 (1985)
8. Hsu, W.L., Nemhauser, G.L.: Easy and hard bottleneck location problems. Discret. Appl. Math. **1**(3), 209–215 (1979)
9. Khuller, S., Sussmann, Y.J.: The capacitated K-center problem. SIAM J. Discret. Math. **13**(3), 403–418 (2000)
10. Leskovec, J., Faloutsos, C.: Sampling from large graphs. In: Proceedings of the 12th ACM SIGKDD International Conference on Knowledge Discovery and Data Mining, pp. 631–636 (2006)
11. MacKinnon, R.D., Barber, G.M.: New approach to network generation and map representation: the linear case of the location-allocation problem (1972)
12. Medya, S., Silva, A., Singh, A., Basu, P., Swami, A.: Group centrality maximization via network design. In: Proceedings of the 2018 SIAM International Conference on Data Mining, pp. 126–134. SIAM (2018)
13. Miller, H.J.: GIS and geometric representation in facility location problems. Int. J. Geogr. Inf. Syst. **10**(7), 791–816 (1996)
14. Murray, A.T., O'Kelly, M.E., Church, R.L.: Regional service coverage modeling. Comput. Oper. Res. **35**(2), 339–355 (2008)
15. Murray, A.T., Tong, D.: Coverage optimization in continuous space facility siting. Int. J. Geogr. Inf. Sci. **21**(7), 757–776 (2007)
16. Nutov, Z., Kortsarz, G., Shalom, E.: Approximating activation edge-cover and facility location problems. In: 44th International Symposium on Mathematical Foundations of Computer Science (MFCS 2019) (2019)
17. Stankov, Toni, Storandt, Sabine: Maximum gap minimization in polylines. In: Di Martino, Sergio, Fang, Zhixiang, Li, Ki-Joune. (eds.) W2GIS 2020. LNCS, vol. 12473, pp. 181–196. Springer, Cham (2020). https://doi.org/10.1007/978-3-030-60952-8_19
18. Tansel, B.C., Francis, R.L., Lowe, T.J.: State of the art—location on networks: a survey. Part I: the p-center and p-median problems. Manag. Sci. **29**(4), 482–497 (1983)

Complexity Results on Cosecure Domination in Graphs

Kusum[(✉)] and Arti Pandey

Department of Mathematics, Indian Institute of Technology Ropar, Rupnagar,
Punjab 140001, India
{2018maz0011,arti}@iitrpr.ac.in

Abstract. Let $G = (V, E)$ be a simple graph with no isolated vertices.
A dominating set S of G is said to be a cosecure dominating set of G if
for every vertex $v \in S$ there exists a vertex $u \in V \setminus S$ such that $uv \in E$
and $(S \setminus \{v\}) \cup \{u\}$ is a dominating set of G. The MINIMUM COSECURE
DOMINATION PROBLEM is to find a minimum cardinality cosecure dominating set of G. Given a graph G and a positive integer k, the COSECURE
DOMINATION DECISION PROBLEM is to decide whether G has a cosecure
dominating set of cardinality at most k. The COSECURE DOMINATION
DECISION PROBLEM is known to be NP-complete for bipartite, planar,
and chordal graphs. In this paper, we show that the COSECURE DOM-
INATION DECISION PROBLEM remains NP-complete for split graphs, an
important subclass of chordal graphs. On the positive side, we present
a linear-time algorithm to compute the cosecure domination number of
cographs. In addition, we also study the approximation aspects of the
MINIMUM COSECURE DOMINATION PROBLEM. We show that the problem
can be approximated within an approximation ratio of $(\Delta + 1)$ for perfect
graphs with maximum degree Δ. We also prove that the problem cannot
be approximated within an approximation ratio of $(1 - \epsilon)\ln(|V|)$ for any
$\epsilon > 0$, unless P = NP. Moreover, we prove that the MINIMUM COSECURE
DOMINATION PROBLEM is APX-hard for bounded degree graphs.

Keywords: Cosecure Domination · Perfect Graphs · Cographs ·
NP-complete · APX-hard

1 Introduction

Throughout this paper, we consider finite, simple and undirected graphs. A set
$D \subseteq V$ is said to be a *dominating set* of a graph $G = (V, E)$, if every vertex in
V is either in D or is adjacent to some vertex in D. The minimum cardinality
of a dominating set of G is called the *domination number* of G and is denoted
by $\gamma(G)$. Given a graph G, the MINIMUM DOMINATION problem is to find a
dominating set of cardinality $\gamma(G)$. The decision version of this problem is the
DOMINATION DECISION problem which takes a graph G and a positive integer
k as an instance and asks whether there exists a dominating set of cardinality

© The Author(s), under exclusive license to Springer Nature Switzerland AG 2023
A. Bagchi and R. Muthu (Eds.): CALDAM 2023, LNCS 13947, pp. 335–347, 2023.
https://doi.org/10.1007/978-3-031-25211-2_26

at most k. The MINIMUM DOMINATION problem and many of its variations are vastly studied in the literature and a detailed survey of these can be found in the books [12, 13].

One of the important variations of the MINIMUM DOMINATION problem is the secure domination which is defined as follows. A dominating set $S \subseteq V$ of G is called a *secure dominating set* of G, if for every $u \in V \setminus S$, there exists a vertex $v \in S$ adjacent to u such that $(S \setminus \{v\}) \cup \{u\}$ is a dominating set of G. The problem of finding a minimum cardinality secure dominating set of a graph is known as the MINIMUM SECURE DOMINATION PROBLEM. The concept of the secure domination was first introduced by Cockayne et al. [7] in 2005. This problem and its many variants have been extensively studied by several researchers in [2, 5, 7, 15, 16, 21, 22] and elsewhere. A detailed survey on the secure domination and its variant can be found in the book by Haynes et al. [12].

In 2014, another related variation of domination known as cosecure domination was introduced by Arumugam et al. [3]. The concept of cosecure domination was further studied in [14, 19, 25]. For a graph $G = (V, E)$, a dominating set $S \subseteq V$ is called a *cosecure dominating set*, abbreviated as CSDS of G, if for every $u \in S$ there exists a vertex $v \in V \setminus S$ adjacent to u such that $(S \setminus \{u\}) \cup \{v\}$ is a dominating set of G. The minimum cardinality of a cosecure dominating set of G is called the *cosecure domination number* of G and is denoted by $\gamma_{cs}(G)$. Note that if a graph G has isolated vertices then no cosecure dominating set exists for G. Therefore, we will consider graphs without any isolated vertices. Also, observe that the whole vertex set V is never a cosecure dominating set of graph G.

Given a graph G without an isolated vertex, the MINIMUM COSECURE DOMINATION PROBLEM (MCSD problem) is to find a minimum cardinality cosecure dominating set of G. The decision version of this problem is the COSECURE DOMINATION DECISION PROBLEM (CSDD problem) that takes a graph G without isolated vertices and a positive integer k as an instance and asks whether G has a cosecure dominating set of cardinality at most k. Since every cosecure dominating set is a dominating set, we have $\gamma(G) \leq \gamma_{cs}(G)$.

In [3], Arumugam et al. initiated the study of the MINIMUM COSECURE DOMINATION PROBLEM and determined the cosecure domination number for some families of the standard graph classes such as paths, cycles, wheels and complete t-partite graphs. Further, they proved that the CSDD problem is NP-complete even when restricted to bipartite, chordal and planar graphs. In [14], Joseph et al. gave few bounds on the cosecure domination number for certain families of graphs. Later in [19], Manjusha et al. characterized the Mycielski graphs with the cosecure domination number 2 or 3 and gave a sharp upper bound for $\gamma_{cs}(\mu(G))$, where $\mu(G)$ is the mycielski of a graph G. Later in [25], Zou et al. proved that the cosecure domination number of proper interval graphs can be computed in linear-time.

In this paper, we extend the existing literature by investigating some algorithmic and approximation-related results for the MINIMUM COSECURE DOMINATION PROBLEM. To the best of our knowledge, there is no result in the literature regarding the approximation aspects of this problem. The paper is organised as follows. In Sect. 2, we give some pertinent definitions and preliminary results. In Sect. 3, we show that the CSDD problem is NP-complete for split graphs. In Sect. 4, we present a linear-time algorithm for computing the cosecure domination number of cographs. In Sect. 5, we present some approximation related results for the MINIMUM COSECURE DOMINATION PROBLEM. We give an approximation algorithm for the problem with an approximation ratio $(\Delta + 1)$ for perfect graphs with maximum degree Δ. We also show that the MCSD problem cannot be approximated within an approximation ratio of $(1 - \epsilon)\ln(|V|)$ for any $\epsilon > 0$, unless P = NP. Moreover, we prove that the MCSD problem is APX-hard for bounded degree graphs. Finally, Sect. 6 concludes the paper.

2 Preliminaries

2.1 Definitions and Notations

Let $G = (V, E)$ be a graph with the vertex set $V = V(G)$ and the edge set $E = E(G)$. For graph theoretic definitions and notations, we refer to [23]. Let G and H be two graphs such that $V(G) \cap V(H) = \emptyset$. The *disjoint union* of G and H is denoted by $G \cup H$ and is defined as the graph with the vertex set $V(G \cup H) = V(G) \cup V(H)$ and edge set $E(G \cup H) = E(G) \cup E(H)$. The *join* of G and H is the graph $G + H$ with $V(G + H) = V(G) \cup V(H)$ as the vertex set and $E(G + H) = E(G) \cup E(H) \cup \{uv \mid u \subset V(G), v \in V(H)\}$ as the edge set. For $S \subseteq V$, $G[S] = (S, E_S)$ denotes the subgraph induced by S, where the vertex set is S and edge set is $E_S = \{uv \mid u, v \in S \text{ and } uv \in E\}$.

A set $S \subseteq V$ is an *independent set* if for every pair of distinct vertices $u, v \in S$, $uv \notin E$. An independent set of maximum cardinality is called a *maximum independent set* of G. A graph $G = (V, E)$ is said to be a *complete graph* if for any $u, v \in V$, $uv \in E$. Given a graph $G = (V, E)$, a set $C \subseteq V$ is said to be a *clique* if $G[C]$ forms a complete graph. A graph $G = (V, E)$ is said to be a *split graph*, if V can be partitioned into two sets K and I such that K is a clique and I is an independent set, respectively. We represent a split graph G as $G = (K \cup I, E)$. A *cograph* is a graph that can be constructed recursively using the following rules:

1. K_1 is a cograph.
2. Join of two cographs is a cograph.
3. Disjoint union of cographs is a cograph.

A k-coloring of a graph $G = (V, E)$ is a function c from V to S, where S is a set of k colors. Here, $c(u)$ represents the color assigned to a vertex $u \in V$. A k-coloring is said to be a *proper k-coloring* if for every $uv \in E$, $c(u) \neq c(v)$. A graph G is k-colorable if it has a proper k-coloring. The *chromatic number* of G is the least k such that G is k-colorable and is denoted by $\chi(G)$. The order of a largest clique in G is called the *clique number* of G and is denoted by $\omega(G)$. A graph is said to be a *perfect graph* if the chromatic number of every induced subgraph is same as the clique number of that subgraph. That is, $G = (V, E)$ is a perfect graph if and only if for every subset $S \subseteq V$, $\chi(G[S]) = \omega(G[S])$.

For a dominating set S of G and a vertex $u \in S$, if there exists a vertex $v \in V \setminus S$ such that $uv \in E$ and $(S \setminus \{u\}) \cup \{v\}$ is a dominating set of G, then we say that v is a replacement of u for the set S. If there does not exist any vertex which is a replacement of u, then we say that the replacement of u does not exist. Note that S is a CSDS, if every vertex of S has a replacement.

Let $G = C_1 \cup C_2 \cup \cdots \cup C_k$ be a disconnected graph, where $C_1, C_2, \ldots, C_k$ are the connected components of G. Then, $\gamma_{cs}(G) = \sum_{i=1}^{k} \gamma_{cs}(C_i)$. Thus, throughout this manuscript, we will consider only connected graphs.

Proofs of the results marked with $\star$ are omitted due to space constraints.

2.2   Preliminary Results

In this section, we list out some results which are already known in the literature and will be helpful in proving some results in this paper.

Lemma 1 [3]. *If $G = (X, Y, E)$ is a complete bipartite graph with $|X| \leq |Y|$ then*

$$\gamma_{cs}(G) = \begin{cases} |Y| & \text{if } |X| = 1; \\ 2 & \text{if } |X| = 2; \\ 3 & \text{if } |X| = 3; \\ 4 & \text{otherwise.} \end{cases} \tag{1}$$

Lemma 2 [3]. *In a graph $G = (V, E)$, let s be a support vertex and P_s be the set of pendant vertices adjacent to s. If $|P_s| \geq 2$, then every cosecure dominating set S of G contains P_s and does not contain s.*

The following corollary directly follows from the above result.

Corollary 1. *Let $G = (X, Y, E)$ be a star graph having order at least 3, where Y is the set of pendant vertices of G and $x \in X$ is the center of G. Then, every cosecure dominating set S of G contains P and $x \notin S$.*

We know that a cosecure dominating set only exists for graphs without any isolated vertices. The following lemma shows the existence of a cosecure dominating set for any graph with no isolated vertices.

Lemma 3 [3]. *Let $G = (V, E)$ be a graph without any isolated vertices. Then, any maximum independent set of G is also a cosecure dominating set of G.*

3   NP-completeness Result for Split Graphs

In this section, we establish the NP-completeness of CSDD problem for connected split graphs. To prove this result, we make a polynomial-time reduction from the Domination Decision problem, which is already known to be NP-complete for split graphs [4].

First, we present a lemma that tells about some properties of a dominating set of a split graph.

Lemma 4.* *Let $G = (K \cup I, E)$ be a connected split graph and D be a dominating set of G of cardinality k. Then, there exists a dominating set D' of cardinality at most k such that $D' \subseteq K$ and D' satisfies at least one of the following conditions:*

1. *for every vertex $u \in D'$, there exists $v \in I$ such that $uv \in E$.*
2. *$D' \subset K$, that is, D' is properly contained in K.*

With the help of the above lemma, we prove that the decision version of the MCSD problem is NP-complete for connected split graphs.

Theorem 1. *The CSDD problem is NP-complete for connected split graphs.*

Proof. Clearly, the CSDD problem is in NP. Now, to prove the NP-hardness, we provide a reduction from the Domination Decision problem for split graphs to the CSDD problem for split graphs in the following way. Consider a connected split graph $G = (K \cup I, E)$ and a positive integer k as an instance of the Domination Decision problem. Assume that $I = \{v_1, v_2, \ldots, v_r\}$. We construct a graph $H = (V^H, E^H)$ from G as follows. We consider two copies I' and I'' of I, where $I' = \{v'_1, v'_2, \ldots, v'_r\}$ and $I'' = \{v''_1, v''_2, \ldots, v''_r\}$, respectively. Define $V^H = K \cup I \cup I' \cup I''$ and $E^H = E \cup \{uv \mid u \in K \text{ and } v \in I'\} \cup \{v'_i v'_j \mid 1 \le i < j \le r\} \cup \{v_i v'_i, v'_i v''_i \mid 1 \le i \le r\}$. Take $C = K \cup I'$ and $J = I \cup I''$. Note that $V^H = C \cup J$, where C is a clique and J is an independent set. Therefore, H is a connected split graph. Note that H can be constructed from G in polynomial time. Now, we only need to prove the following claim.

Claim 1.* *G has a dominating set of cardinality at most k if and only if H has a cosecure dominating set of cardinality at most $k + |I|$.*

Hence, the theorem is proved.   □

4 Algorithm for Cographs

In this section, we present a linear-time algorithm to find the cosecure domination number of a given cograph. Recall that a *cograph* is a graph that can be constructed recursively using the following rules:

1. K_1 is a cograph.
2. Join of two cographs is a cograph.
3. Disjoint union of cographs is a cograph.

Corresponding to every cograph, there exists a unique rooted tree (cotree) representation upto isomorphism [17]. For a connected cograph G, let the corresponding cotree be denoted by T_G. This cotree T_G satisfies the following properties [18]:

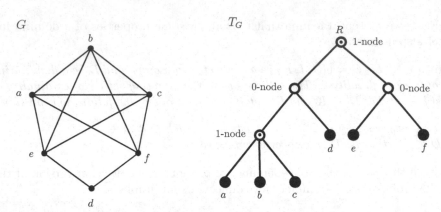

Fig. 1. Illustrating a cograph G and its cotree representation T_G.

P1 Every internal vertex has at least two children.

P2 Each internal vertex of T_G is either labelled as a 1-node or 0-node such that root R is a 1-node and no two adjacent internal vertices get the same label.

P3 Leaves in T_G correspond to the vertices of G. Two vertices x and y are adjacent in G if and only if the lowest common ancestor of x and y is a 1-node in T_G.

First, we give an example illustrating a cograph G and its cotree representation T_G in Fig. 1. As the leaves in T_G correspond to the vertices of G, we remark that the label is same for the leaf in T_G and the corresponding vertex in G. Now, we define some notations related to the cotree T_G of a cograph G. Let R be the root vertex of T_G. For a vertex $x \in V(T_G)$, $ch_{T_G}(x)$ denotes the set of children of x in T_G and $T_G(x)$ denotes the subtree of T_G rooted at x. The set of leaves in $T_G(x)$ is denoted by $L(x)$, where $x \in V(T_G)$. We define $G_{T_G(x)}$ as the subgraph of G induced on $L(x)$. An internal vertex x of T_G with label 0-node (1-node) corresponds to the induced subgraph $G_{T_G(x)}$ of G formed by disjoint union (join) of the induced subgraphs $G_{T_G(x_i)} : 1 \leq i \leq k$ of G, where $ch_{T_G}(x) = \{x_1, x_2, \ldots, x_k\}$. Observe that $G_{T_G(R)}$ is nothing but the cograph G itself. The readers interested in more detailed illustration of the cotree representation corresponding to a cograph may refer to [8,11,17,18]. Now, we prove some lemmas that will help us in proposing the algorithm which determines the cosecure domination number of a given connected cograph.

Consider a connected cograph G and the cotree T_G corresponding to it. Let R be the root vertex of the cotree T_G. Observe that each subtree of T_G represents an induced subgraph of the graph G. The following lemma directly follows from the properties of the cotree of a cograph.

Lemma 5. *If G is a cograph formed by the join of $G_1, G_2, \ldots, G_k$, then for each $i \in [k]$, G_i is either K_1 or a disconnected graph.*

Note that any connected cograph G with at least two vertices can be written as the join of k cographs $G_1, G_2, \ldots, G_k$, where $k \geq 2$. Observe that each G_i corresponds to a subtree of the cotree T_G. In the next lemma, we give a characterization for the graph G to have domination number one.

Lemma 6.* *Let $G = G_1 + G_2 + \cdots + G_k$ be a cograph with $k \geq 2$. Then, $\gamma(G) = 1$ if and only if there exists at least one $i \in [k]$ such that $G_i = K_1$.*

Note that if x is an internal vertex of the cotree T_G which is a 1-node then using Lemma 6, it follows that $\gamma(G_{T_G(x)}) = 1$ if and only if at least one of vertex in $ch_{T_G}(x)$ is a leaf in the cotree T_G. Now in Lemma 7, we give a characterization for cographs to have the cosecure domination number one.

Lemma 7.* *Let $G = G_1 + G_2 + \cdots + G_k$ be a cograph with $k \geq 2$. Then $\gamma_{cs}(G) = 1$ if and only if there exist $p, q \in [k]$ $(p \neq q)$ such that $G_p = G_q = K_1$.*

Consider a cograph G which is the join of $G_1, G_2, \ldots, G_k$, where $k \geq 3$. In the next lemma, we obtain a sufficient condition for the cographs having the cosecure domination number two.

Lemma 8.* *Let $G = G_1 + G_2 + \cdots + G_k, k \geq 3$ be a cograph. If there exist at most one $i \in [k]$ such that $G_i = K_1$, then $\gamma_{cs}(G) = 2$.*

Let G be a cograph formed by the join of two cographs G_1 and G_2. We first prove an upper bound on the cosecure domination number of G, when both G_1 and G_2 contain at least two vertices. Later in Lemma 10, we assume that x_1 is a leaf in cotree T_G and obtain that the cosecure domination number of G is equal to the domination number of G_2, which is the cograph corresponding to the subtree rooted at vertex x_2 in the cotree T_G.

Lemma 9.* *If $G = G_1 + G_2$ is a cograph with $|V(G_1)|, |V(G_2)| \geq 2$, then $\gamma_{cs}(G) \leq 4$.*

Lemma 10.* *If $G = G_1 + G_2$ is a cograph with $G_1 = K_1$, then $\gamma_{cs}(G) = \gamma(G_2)$.*

Let G be a cograph formed by the join of two disconnected graphs G_1 and G_2. In Lemma 11, we assume all the possible cases and determine the cosecure domination number of G in each case. Observe that each connected component of G_i corresponds to either a subtree rooted at a 1-node or a leaf in the cotree T_G.

Lemma 11.* *Let* $G = G_1 + G_2$ *be a cograph where* G_1 *and* G_2 *are disconnected graphs. If* $G_1 = C_1 \cup C_2 \cup \cdots \cup C_r$ *and* $G_2 = C_1' \cup C_2' \cup \cdots \cup C_p'$ *where* $C_i \colon i \in [r]$ *and* $C_j' \colon j \in [p]$ *are the connected components of* G_1 *and* G_2, *respectively. Then, one of the following is true.*

1. *If* $|V(C_i)|, |V(C_j')| = 1$ *for all* $i \in [r]$ *and* $j \in [p]$, *then* G *is a complete bipartite graph.*
2. *If there exists* $i \in [r]$ *and* $j \in [p]$ *such that* $|V(C_i)|, |V(C_j')| \geq 2$ *then* $\gamma_{cs}(G) = 2$.
3. *Let there exist* $i \in [r]$ *such that* $|V(C_i)| \geq 2$ *and* $|V(C_j')| = 1$ *for all* $j \in [p]$. *If* $\gamma(G_1) = 2$ *or* $\gamma(G_2) = 2$ *then* $\gamma_{cs}(G) = 2$. *Otherwise, if* $\gamma(G_1) \geq 3$ *and* $\gamma(G_2) \geq 3$ *then* $\gamma_{cs}(G) = 3$.
4. *Let there exist* $j \in [p]$ *such that* $|V(C_j')| \geq 2$ *and* $|V(C_i)| = 1$ *for all* $i \in [r]$. *If* $\gamma(G_1) = 2$ *or* $\gamma(G_2) = 2$ *then* $\gamma_{cs}(G) = 2$. *Otherwise, if* $\gamma(G_1) \geq 3$ *and* $\gamma(G_2) \geq 3$ *then* $\gamma_{cs}(G) = 3$.

Based on the above lemmas, we design an efficient algorithm **Algorithm 1**, which computes the cosecure domination number of a connected cograph. Observe that a connected cograph $G = (V, E)$ is join of some k cographs, say $G_1, G_2, \ldots, G_k$, where k is at least 2. Using the above fact as a key, we design our algorithm in which depending on the value of k and structure of these k cographs, we consider different cases and compute the value of the cosecure domination number of G (sometimes, using the domination number of G_i's).

A cograph can be recognised in linear-time and its cotree representation can also be computed in linear-time [8,11]. Also, it is also known that the domination number of cographs can be computed in linear-time [20]. Thus, we have the following result.

Theorem 2. *Given a connected cograph* G, *the cosecure domination number of* G *can be computed in linear-time.*

Proof. The correctness of Algorithm 1 directly follows from Lemma 7, Lemma 8, Lemma 10 and Lemma 11. Since, the cotree representation of a cograph can be computed in linear-time and all the steps of the **Algorithm 1** can be executed in linear-time, the cosecure domination number of a cograph can be computed in linear-time. □

Algorithm 1: Cosecure Domination Number of a Cograph

Input: A connected cograph $G = (V, E)$ with the cotree representation T_G of G.
Output: The cosecure domination number of G, $\gamma_{cs}(G)$.
Let R be the root of the cotree T_G and $ch_{T_G}(R) = \{x_1, x_2, \ldots, x_k\}$;
if $(k \geq 3)$ then
   | if *(there are at least two leaves in $ch_{T_G}(R)$)* then
   | | $\gamma_{cs}(G) = 1$;
   | else
   | | $\gamma_{cs}(G) = 2$;

if $(k = 2)$ then
   | Let $ch_{T_G}(R) = \{x_1, x_2\}$;
   | if *(both x_1 and x_2 are leaves)* then
   | | $\gamma_{cs}(G) = 1$;
   | else if *(exactly one of x_1 or x_2 is a leaf)* then
   | | Let x_1 is a leaf and x_2 is an internal vertex;
   | | $\gamma_{cs}(G) = \gamma(G_{T_G(x_2)})$;
   | else if *(both x_1 and x_2 are internal vertices)* then
   | | if *(both $ch_{T_G}(x_1)$ and $ch_{T_G}(x_2)$ has at least one 1-node)* then
   | | | $\gamma_{cs}(G) = 2$;
   | | else if *(exactly one of $ch_{T_G}(x_1)$ or $ch_{T_G}(x_2)$ has at least one 1-node)*
   | | then
   | | | if *($\gamma(G_{T_G(x_1)}) = 2$ or $\gamma(G_{T_G(x_2)}) = 2$)* then
   | | | | $\gamma_{cs}(G) = 2$;
   | | | else
   | | | | $\gamma_{cs}(G) = 3$;
   | | else if *(both $ch_{T_G}(x_1)$ and $ch_{T_G}(x_2)$ are leaves)* then
   | | | Let $p = \min\{|ch_{T_G}(x_1)|, |ch_{T_G}(x_2)|\}$ and
   | | | $q = \max\{|ch_{T_G}(x_1)|, |ch_{T_G}(x_2)|\}$;
   | | | $\gamma_{cs}(G)$ can be obtained using Lemma 1;
return $\gamma_{cs}(G)$;

---

5 Approximation Results

In this section, we find the lower and upper bound on the approximation ratio of the MINIMUM COSECURE DOMINATION PROBLEM. We also show that the problem is APX-hard for graphs with maximum degree 4.

5.1 Upper Bound on Approximation Ratio

In this subsection, we prove that there exists a $(\Delta + 1)$-approximation algorithm for the MCSD problem for the graphs having maximum degree Δ where a maximum independent set can be computed in polynomial time.

Theorem 3.* *Let G be a graph with maximum degree Δ. If a maximum independent set I of G can be computed in polynomial time, then the MCSD problem can be approximated within an approximation ratio of $(\Delta + 1)$.*

Note that the MISP problem is solvable in polynomial time for perfect graphs [10]. Using this and Theorem 3, the following corollary directly follows.

Corollary 2. *The MCSD problem can be approximated within an approximation ratio of $(\Delta + 1)$ for perfect graphs with maximum degree Δ.*

5.2   Lower Bound on Approximation Ratio

In order to obtain a lower bound on the approximation ratio of the MCSD problem, we propose an approximation preserving reduction from the Minimum Domination problem. Before doing that let us recall a result from the literature regarding the lower bound on the approximation ratio of the Minimum Domination problem.

Theorem 4 [6,9]**.** *Given a graph $G = (V, E)$, the Minimum Domination problem cannot be approximated within an approximation ratio of $(1 - \epsilon)ln(n)$ for any $\epsilon > 0$ unless $P = NP$, where $n = |V|$.*

Theorem 5. *Given a graph $G = (V, E)$, the MCSD problem cannot be approximated within an approximation ratio of $(1 - \epsilon)ln(n)$ for any $\epsilon > 0$ unless $P = NP$, where $n = |V|$.*

Proof. We prove this result using contradiction. First, we propose an approximation preserving reduction from the Minimum Domination problem to the MCSD problem as follows. Suppose that a graph $G = (V, E)$ is a given instance of the Minimum Domination problem, where $|V| = n$ and $V = \{v_1, v_2, \ldots, v_n\}$. We construct a new graph $G' = (V', E')$ from G by adding 3 new vertices x, y and z, and making x adjacent to every vertex of $V \cup \{y, z\}$. Formally, $V' = V \cup \{x, y, z\}$ and $E' = E \cup \{xv_i : v_i \in V, 1 \leq i \leq n\} \cup \{xy, xz\}$. Note that $|V'| = |V| + 3$ and $|E'| = |E| + |V| + 2$.

We claim that G has a dominating set of cardinality at most k if and only if G' has a cosecure dominating set of cardinality at most $k + 2$. To see this, first suppose that G has a dominating set D and $|D| \leq k$. Let $S = D \cup \{y, z\}$. As x is a replacement for every vertex of S, S is a cosecure dominating set of G' and $|S| \leq k + 2$. Conversely, assume that G' has a cosecure dominating set S and $|S| \leq k + 2$. Using Lemma 2, it follows that $y, z \in S$ and $x \notin S$. Define a set $D = S \cap V$. Clearly, D is a dominating set of G and $|D| \leq k$. Hence, the claim follows.

Now, suppose that Approx_CSDS is an approximation algorithm that runs in polynomial time and solves the MCSD problem within an approximation ratio of $\alpha = (1 - \epsilon)ln(|V'|)$, for some fixed $\epsilon > 0$. Let t be a fixed integer. Now, we propose the following algorithm Approx_DS to find a dominating set of a given graph G.

Algorithm 2: Approx_DS

Input: A graph $G = (V, E)$.
Output: A dominating set of G.
if *there exists an optimal dominating set D of G of cardinality at most t* **then**
  $\lfloor$ return D;
else
  | Construct a new graph G' using G;
  | Compute a cosecure dominating set S of G' using Approx_CSDS;
  | Define $D = S \cap V$;
  $\lfloor$ return D;

Note that the Approx_DS is a polynomial-time algorithm, as the algorithm Approx_CSDS runs in polynomial-time for G' and every other step of Approx_DS can be computed in polynomial-time. If $|D| \leq t$ then D is an optimal dominating set of G. Now, assume that $|D| > t$.

Suppose that D^* is an optimal dominating set of G and S^* is an optimal cosecure dominating set of G'. Using the above reduction and discussion, it follows that $|S^*| = |D^*| + 2$. Note that $|D^*| > t$. For a graph G, let Approx_DS computes a dominating set D of G and Approx_CSDS computes a cosecure dominating set S of G'. Here, $|D| = |S| - 2 \leq \alpha|S^*| - 2 \leq \alpha|S^*| = \alpha(|D^*|+2) \leq \alpha(1 + \frac{2}{|D^*|})|D^*| < \alpha(1 + \frac{2}{t})|D^*|$. Thus, $|D| \leq \alpha(1 + \frac{2}{t})|D^*|$. Let t be an integer that satisfies $t > \frac{2}{\epsilon}$. Also note that $\ln(n) \approx \ln(n + 3)$, for sufficiently large values of n. Thus, $|D| \leq \alpha(1 + \frac{2}{t})|D^*| \leq (1 - \epsilon)\ln(|V|)(1 + \epsilon)|D^*| \leq (1 - \epsilon')\ln(|V|)|D^*|$ where $\epsilon' = \epsilon^2$. Therefore, Approx_DS approximates the Minimum Domination problem within an approximation ratio of $(1 - \epsilon')\ln(|V|)$ for some $\epsilon' > 0$, which is a contradiction to Theorem 4. Hence, the result follows. $\square$

5.3 APX-hardness

In this subsection, we show that the MINIMUM COSECURE DOMINATION PROBLEM is APX-hard for graphs with maximum degree 4. To prove this result, we give an L-reduction from the Minimum Domination problem for graphs with maximum degree 3, which is already known to be APX-hard [1]. For the definition of L-reduction, we refer to [24].

Now, we define a reduction f from an instance of the Minimum Domination problem to an instance of the MCSD problem in the following way. Given a connected graph $G = (V, E)$ where $V = \{v_1, v_2, \ldots, v_n\}$, define a graph $H = (V^H, E^H)$ as described below. We consider two copies of V, $V' = \{v'_1, v'_2, \ldots, v'_n\}$ and $V'' = \{v''_1, v''_2, \ldots, v''_n\}$. The vertex set of H is $V^H = V \cup V' \cup V''$ and the edge set of H is $E^H = E \cup \{v_i v'_i, v'_i v''_i | 1 \leq i \leq n\}$. Here, $|V^H| = 3n$ and $|E^H| = |E| + 2n$. Note that if the maximum degree of G is 3, then the maximum degree of H is 4 and H can be constructed from G in polynomial time. The following result is obtained by Arumungum et al. [3].

Lemma 12 [3]. *If H is the graph constructed from G using above construction f, then $\gamma_{cs}(H) = \gamma(G) + |V(G)|$.*

Next, to prove that the MCSD problem is APX-hard for graphs with maximum degree 4, we show that the reduction f is an L-reduction.

Claim 2. * f is an L-reduction.*

Now, the following theorem directly follows.

Theorem 6. *The MCSD problem is APX-hard for graphs with maximum degree 4.*

The proof of the following corollary directly follows from Theorem 6 and Theorem 3.

Corollary 3. *The MCSD problem is APX-complete for perfect graph with maximum degree 4.*

6   Conclusion

In this paper, we focused on the algorithmic complexity of the MINIMUM COSECURE DOMINATION PROBLEM on different graph classes. It is known that the decision version of the MCSD problem is NP-complete for bipartite, planar and chordal graphs. We proved that the problem remains NP-complete even when restricted to split graphs. We also proposed a linear-time algorithm to compute the cosecure domination number of cographs. Further, we studied the approximation aspects of the MINIMUM COSECURE DOMINATION PROBLEM and we showed that the problem can be approximated within an approximation ratio of $(\Delta + 1)$ for perfect graphs. In addition, we proved that the MCSD problem cannot be approximated within an approximation ratio of $(1 - \epsilon)\ln(|V|)$ for any $\epsilon > 0$, unless P = NP. Moreover, we proved that the MCSD problem is APX-complete for bounded degree perfect graphs. The complexity status of the MINIMUM COSECURE DOMINATION PROBLEM is still unknown in many important subclasses of bipartite and chordal graphs. It would be an interesting research direction to work on some of these graph classes.

References

1. Alimonti, P., Kann, V.: Some APX-completeness results for cubic graphs. Theoret. Comput. Sci. **237**(1–2), 123–134 (2000)
2. Araki, T., Yamanaka, R.: Secure domination in cographs. Discrete Appl. Math. **262**, 179–184 (2019)
3. Arumugam, S., Ebadi, K., Manrique, M.: Co-secure and secure domination in graphs. Util. Math. **94**, 167–182 (2014)
4. Bertossi, A.A.: Dominating sets for split and bipartite graphs. Inform. Process. Lett. **19**(1), 37–40 (1984)

5. Boumediene Merouane, H., Chellali, M.: On secure domination in graphs. Inform. Process. Lett. **115**(10), 786–790 (2015)
6. Chlebík, M., Chlebíková, J.: Approximation hardness of dominating set problems in bounded degree graphs. Inf. Comput. **206**(11), 1264–1275 (2008)
7. Cockayne, E.J., Grobler, P.J.P., Gründlingh, W.R., Munganga, J., van Vuuren, J.H.: Protection of a graph. Util. Math. **67**, 19–32 (2005)
8. Corneil, D.G., Perl, Y., Stewart, L.K.: A linear recognition algorithm for cographs. SIAM J. Comput. **14**(4), 926–934 (1985)
9. Dinur, I., Steurer, D.: Analytical approach to parallel repetition. In: STOC'14–Proceedings of the 2014 ACM Symposium on Theory of Computing, pp. 624–633. ACM, New York (2014)
10. Grötschel, M., Lovász, L., Schrijver, A.: Polynomial algorithms for perfect graphs. In: Topics on perfect graphs, North-Holland Math. Stud., vol. 88, pp. 325–356. North-Holland, Amsterdam (1984)
11. Habib, M., Paul, C.: A simple linear time algorithm for cograph recognition. Discrete Appl. Math. **145**(2), 183–197 (2005)
12. Haynes, T.W., Hedetniemi, S.T., Henning, M.A. (eds.): Topics in Domination in Graphs. DM, vol. 64. Springer, Cham (2020). https://doi.org/10.1007/978-3-030-51117-3
13. Haynes, T.W., Hedetniemi, S.T., Henning, M.A. (eds.): Structures of Domination in Graphs. DM, vol. 66. Springer, Cham (2021). https://doi.org/10.1007/978-3-030-58892-2
14. Joseph, A., Sangeetha, V.: Bounds on co-secure domination in graphs. Int. J. of Math. Trends Technol. **55**(2), 158–164 (2018)
15. Klostermeyer, W.F., Mynhardt, C.M.: Secure domination and secure total domination in graphs. Discuss. Math. Graph Theory **28**(?), 267–284 (2008)
16. Kumar, J.P., Reddy, P.V.S., Arumugam, S.: Algorithmic complexity of secure connected domination in graphs. AKCE Int. J. Graphs Comb. **17**(3), 1010–1013 (2020)
17. Lerchs, H.: On the clique-kernel structure of graphs. Dept. of Computer Science, University of Toronto 1 (1972)
18. Lin, R., Olariu, S., Pruesse, G.: An optimal path cover algorithm for cographs. Comput. Math. Appl. **30**(8), 75–83 (1995)
19. Manjusha, P., Chithra, M.R.: Co-secure domination in Mycielski graphs. J. Combin. Math. Combin. Comput. **113**, 289–297 (2020)
20. Nicolai, F., Szymczak, T.: Homogeneous sets and domination: a linear time algorithm for distance-hereditary graphs. Networks **37**(3), 117–128 (2001)
21. Poureidi, A.: On computing secure domination of trees. Discrete Math. Algorithms Appl. 13(5), 2150055, 15 (2021)
22. Wang, H., Zhao, Y., Deng, Y.: The complexity of secure domination problem in graphs. Discuss. Math. Graph Theory **38**(2), 385–396 (2018)
23. West, D.B., et al.: Introduction to Graph Theory, vol. 2. Prentice Hall, Upper Saddle River (2001)
24. Williamson, D.P., Shmoys, D.B.: The Design of Approximation Algorithms. Cambridge University Press, Cambridge (2011)
25. Zou, Y.H., Liu, J.J., Chang, S.C., Hsu, C.C.: The co-secure domination in proper interval graphs. Discrete Appl. Math. **311**, 68–71 (2022)

Graph Matching

Latin Hexahedra and Related Combinatorial Structures

Akihiro Yamamura[(✉)]

Department of Mathematical Science and Electrical-Electronic-Computer Engineering,
Akita University, 1-1 Tegata-gakuen, Akita 010-8502, Japan
yamamura@ie.akita-u.ac.jp

Abstract. We introduce a system of Latin rectangles that is a combination of finite number of Latin rectangles that are given by either concrete Latin rectangles or variables representing a Latin rectangle. Using such systems, we prove the existence of a combinatorial object which is considered as a generalization of a Latin square. It is defined on the surface of a regular hexahedron so that any sub-array of any net of the hexahedron is a Latin rectangle. Second, we introduce two combinatorial objects, the existence of which are equivalent to 1-factorizations of the complete tripartite graph $K_{2n,2n,2n}$ and the complete quadripartite graph $K_{n,n,n,n}$, respectively. We also show how to construct 1-factorizations of $K_{2n,2n,2n}$ and $K_{n,n,n,n}$.

Keywords: Latin square · Regular hexahedron · Sudoku puzzle · 1-factorization · Complete tripartite graph · Complete quadripartite graph

1 Introduction

Let A be an $n \times m$ matrix filled with integers in $\{1, 2, 3, \ldots, k\}$, where $k = max(n, m)$. If no integer appears more than once in any row or column, then A is called a *Latin rectangle*. A *Latin square of order n* is an $n \times n$ Latin rectangle [4]. A Latin square can be considered as the multiplication table of a quasigroup, and vice versa. It has been studied from the standpoint of theoretical interests and applications to statistics and experimental designs. In this paper, we study a generalization of a Latin square that is defined on the surface of a regular hexahedron. A sudoku puzzle is an example of a Latin square. Our generalization can be applied to construct another puzzle similar to a sudoku puzzle. In addition, the construction is strongly related to 1-factorization of tripartite graph $K_{2n,2n,2n}$ and the complete quadripartite graph $K_{n,n,n,n}$.

A subgraph of a graph $G = (V, E)$ is called a *factor* of G if it includes all of the vertices of G. If every vertex of a factor has degree h, then it is called an *h-factor*. Therefore, a 1-factor of G is a spanning 1-regular subgraph of G. If E can be partitioned into disjoint 1-factors, then G is called *1-factorizable* [7,9]. For example, a complete bipartite graph $K_{n,n}$ is 1-factorizable because the existence of a 1-factorization is equivalent to that of a Latin square of order n. It is conjectured that every regular graph of order $2n$ is 1-factorizable if degree is $\lambda \times 2n$ where $\lambda \geq \frac{1}{2}$ by Chetwynd-Hilton [2]. Strong [8] shows that every connected Cayley graph on a finite even order Abelian group is 1-factorizable. The complete tripartite graphs $K_{2n,2n,2n}$ is the Cayley graph $\Gamma(S:G)$ with the cyclic group G of order $6n$ and $S = G \setminus H$ with the subgroup H of G generated by the element of order $2n$. The complete qadripartite graph $K_{n,n,n,n}$ is the Cayley graph $\Gamma(S : G)$ with the cyclic group G of order $4n$ and $S = G \setminus H$ with the subgroup H of G generated by the element of order n. Therefore, $K_{2n,2n,2n}$ and $K_{n,n,n,n}$ are 1-factorizable. However, no concrete construction of a 1-factorization of $K_{2n,2n,2n}$ and $K_{n,n,n,n}$ is provided in [8]. In this paper, we provide a concrete construction of a 1-factorization of $K_{2n,2n,2n}$ and $K_{n,n,n,n}$.

2   Latin Hexahedra

A *system of Latin rectangles* filled with integers in $\{1,2,3,\ldots,n\}$ is a finite set of Latin rectangles so that each of them is a combination of concrete Latin rectangles filled with integers in $\{1,2,3,\ldots,n\}$ and variables representing Latin rectangles. We say that a system of Latin rectangles has a solution if we can obtain Latin rectangles by substituting a certain Latin rectangle for each variable appearing in the system.

Example 1. Consider a system of Latin rectangles (1) filled with integers in $\{1,2,3,4\}$, where A and B are 2×2 Latin rectangles and X is a variable representing an 2×2 Latin rectangle.

$$\boxed{A|X} \quad \boxed{B|X} \tag{1}$$

The system of Latin rectangles (1) has a solution if there exists a 2×2 Latin rectangle C filled with integers in $\{1,2,3,4\}$ so that both $\boxed{A|C}$ and $\boxed{B|C}$ are 2×4 Latin rectangles.

Suppose A and B are arrays $\begin{smallmatrix}1&2\\3&4\end{smallmatrix}$ and $\begin{smallmatrix}2&1\\4&3\end{smallmatrix}$, respectively. Let C be an array $\begin{smallmatrix}3&4\\1&2\end{smallmatrix}$. Then C makes both $\boxed{A|C}$ and $\boxed{B|C}$ Latin rectangles and so C is a solution of (1). On the other hand, suppose A and B are arrays $\begin{smallmatrix}1&2\\3&4\end{smallmatrix}$ and $\begin{smallmatrix}4&3\\2&1\end{smallmatrix}$, respectively. Then (1) has no solution.

Example 2. Consider a system of Latin rectangles (2) filled with integers in $\{1,2,3,4\}$, where A is a 2×2 Latin rectangle. We denote the array obtained from A by rotating $\frac{\pi}{2}$ counterclockwise by $\prec$.

$$\boxed{A|X} \quad \boxed{\prec|X} \tag{2}$$

Suppose A is an array $\begin{smallmatrix}1&2\\3&1\end{smallmatrix}$. Then $\prec$ is the array $\begin{smallmatrix}2&1\\1&3\end{smallmatrix}$. Let C be an array $\begin{smallmatrix}3&4\\4&2\end{smallmatrix}$. Then C makes both $\boxed{A|C}$ and $\boxed{\prec|C}$ Latin rectangles and so C is a solution of (2). On the other hand, if A is an array $\begin{smallmatrix}1&2\\3&4\end{smallmatrix}$, then (2) has no solution.

A regular hexahedron of order n is a polyhedron consisting of six faces, each of which forms an $n \times n$ matrix filled with integers in $\{1,2,3,\ldots,4n\}$. A *net* of a hexahedron is an arrangement of a non-overlapping edge-joined polygon which can be folded along edges to become faces of the hexahedron. A *circuit* of a regular hexahedron of order n is a $1 \times 4n$ subarray in one of its nets. A circuit of a regular hexahedron of order 2 is shown in Fig. 1. We note that there exist precisely $3n$ circuits on a regular hexahedron of order n.

A regular hexahedron of order n is called *Latin* if every integer in $\{1,2,3,\ldots,4n\}$ appears exactly once in every circuit. A Latin regular hexahedron of order 2 and its net are shown in Fig. 2.

Fig. 1. Circuit of a regular hexahedron of order 2

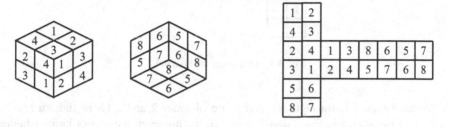

Fig. 2. Latin regular hexahedron of order 2 and its net

A *Sudoku Latin square* is a 9×9 matrix filled with integers in $\{1,2,3,4,5,6,7,8,9\}$ such that each column, each row, and each of the nine 3×3 sub-matrices contain all of the integers from 1 to 9. It appears in the number-placement puzzle. We introduce a similar property into Latin regular hexahedra. Let L be a Latin regular hexahedron of order n. We say L is a *Latin sudoku regular hexahedron* if every integer in $\{1,2,3,\ldots,4n\}$ appears exactly once on each face. For example, a net in Fig. 3 gives a Latin sudoku regular hexahedron of order 4. We also say that L is a *Latin quasi-sudoku regular hexahedron with multiplicity m* if every integer in $\{1,2,3,\ldots,4n\}$ appears exactly m times on each face.

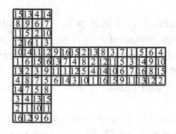

Fig. 3. Net of a Latin sudoku regular hexahedron of order 4

Two circuits of a regular hexahedron L are called *parallel* if they do not intersect. On the other hand, any two distinct non-parallel circuits intersect at exactly two cells on L. Let p be a cell on L. There are precisely two distinct circuits which include p. The other cell in which the two circuits intersect is called the *contraposition* of p and denoted by $\bar{p}$. Therefore, the contraposition $\bar{p}$ is uniquely determined for each cell p. We say two faces A and B of L are in *contraposition* if each cell of A has its contraposition on B. Cells p, q and r and their contrapositions are displayed on a net of regular hexahedron of order 2 in Fig. 4.

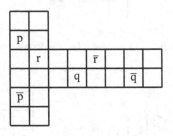

Fig. 4. Contrapositions on a net

We have seen a Latin regular hexahedron of order 2 and a Latin sudoku regular hexahedron of order 4 in Fig. 2 and 3, respectively, however, we do not know whether such objects exist in general. The lemma below gives necessary conditions for such objects to exist.

Lemma 1. *(1) If a Latin regular hexahedron of order k exists, then k is even. (2) If a Latin quasi-sudoku regular hexahedron of order k with multiplicity m exists, then we have $k = 4m$. (3) If a Latin sudoku regular hexahedron of order k exists, then $k = 4$.*

Proof. (1) Let L be a Latin regular hexahedron of order k. Note that L has six faces and each face is filled with integers in $\{1, 2, \ldots, 4k\}$. Suppose L has faces A, B, C, D, E, F and that one of its nets is given below.

Let x be the number of appearance of 1 in A or B, y the number of appearance of 1 in C or E, and z the number of appearance of 1 in D or F, respectively. Note that $\boxed{C\,|\,D\,|\,E\,|\,F}$ is a Latin rectangle by the definition of a Latin regular hexahedron. Since each row contains exactly one 1 and there exist k rows, we have $y + z = k$. Similarly, we obtain $x + y = k$ and $x + z = k$. Then we obtain $x = y = z = k/2$. Since each of x, y and z is an integer, k must be even.

(2) Let L be a Latin quasi-sudoku regular hexahedron of order k with multiplicity m. We have already shown $x = y = z = k/2$. Since x is the number of appearance of 1 in either A or B and the number of appearance of 1 in A and the number of appearance of 1 in B are equal, 1 appears $k/4$ times in A. Therefore, $k = 4m$.

(3) Let L be a Latin sudoku regular hexahedron of order k. Since 1 appears $k/4$ times in each face by (2) and $1 = m = k/4$, we have $k = 4$. □

To prove the existence of a Latin regular hexahedron of order $2n$ and a Latin quasi-sudoku regular hexahedron of order $4m$ with multiplicity m, we relate a Latin regular hexahedron to systems of Latin rectangles.

Lemma 2. *Let n be any positive integer. A system of Latin rectangles with a variable Y, which represents a $2n \times 2n$ array filled with integers in $\{1, 2, \ldots, 4n\}$,*

$$\boxed{Y\,X} \quad \boxed{\succ\,\prec} \tag{3}$$

has a solution, where $\succ$, X and $\prec$ stand for the arrays obtained by rotating Y counterclockwise $\frac{\pi}{2}$, π and $\frac{3\pi}{2}$, respectively.

Proof. Suppose A_1, A_2, A_3 and A_4 are Latin squares of order n such that $|A_i| \cap |A_j| = \emptyset$ if $i \neq j$ and $|A_1| \cup |A_2| \cup |A_3| \cup |A_4| = \{1, 2, \ldots, 4n\}$, where $|A|$ stands for the set of integers appearing on A. Let A be a $2n \times 2n$ array $\begin{array}{|c|c|} \hline A_1 & A_2 \\ \hline A_4 & A_3 \\ \hline \end{array}$ obtained by pasting A_1, A_2, A_3, and A_4. Obviously, the arrays $\boxed{A\,V}$ and $\boxed{\prec\,\succ}$ are Latin rectangles and so A is a solution of (3). □

Theorem 1. *Let n be any positive integer. A system of Latin rectangles with variables (R, T, U, V, W, Y), each of which represents a $2n \times 2n$ array filled with integers in $\{1, 2, \ldots, 8n\}$,*

$$\boxed{U\,V\,W\,Y} \quad \boxed{R\,C\,T\,} \quad \boxed{\,R\,\,L} \tag{4}$$

has a solution. If a sextuple (A, B, C, D, E, F) is a solution of (4), then

$$\begin{array}{|c|c|c|c|} \hline A & & & \\ \hline C & D & E & F \\ \hline B & & & \\ \hline \end{array} \tag{5}$$

provides a net of a Latin regular hexahedron of order $2n$.
    Conversely, if (5) is a net of a Latin regular hexahedron of order $2n$, then the sextuple (A, B, C, D, E, F) is a solution of (4).

Proof. Suppose A and B are $2n \times 2n$ arrays that are solutions of (3) such that $|A| = \{1, 2, \ldots, 4n\}$ and $|B| = \{4n+1, 4n+2, \ldots, 8n\}$. We start with a $4n \times 8n$ array given by (6).

$$\begin{array}{|c|c|c|c|} \hline A & B & V & B \\ \hline B & A & B & A \\ \hline \end{array} \tag{6}$$

Since A and B are solutions of (3) and $|A| \cap |B| = \emptyset$, (6) is a $4n \times 8n$ Latin rectangle with $\{1, 2, \ldots, 8n\}$. Adding a new row to (6) by Hall's marriage theorem, we obtain a Latin rectangle. Suppose the newly added row is formed by four $1 \times 2n$ arrays G, J, K and L as in (7).

$$\begin{array}{|c|c|c|c|} \hline A & B & V & B \\ \hline B & A & B & A \\ \hline G & J & K & L \\ \hline \end{array} \tag{7}$$

We note that $|G| \cap |K| = \emptyset = |J| \cap |L|$. Adding a new row formed by four $1 \times 2n$ arrays $\mathcal{X}$, $\mathcal{T}$, $\mathcal{G}$, and $\mathit{\Gamma}$ above the array (7), we obtain a $(4n+2) \times 8n$ array (8).

$$
\begin{array}{|c|c|c|c|}
\hline
\mathcal{X} & \mathcal{T} & \mathcal{G} & \mathit{\Gamma} \\
\hline
A & \mathcal{B} & V & \mathcal{B} \\
\hline
B & \mathcal{A} & \mathcal{B} & \mathcal{A} \\
\hline
G & J & K & L \\
\hline
\end{array}
\tag{8}
$$

Since $(4n+1) \times 2n$ arrays $\begin{array}{|c|}\hline A\\\hline B\\\hline G\\\hline\end{array}$ and $\begin{array}{|c|}\hline V\\\hline \mathcal{B}\\\hline K\\\hline\end{array}$ are Latin rectangles and $|G| \cap |K| = \emptyset$, the $(4n+$

$2) \times 2n$ array $\begin{array}{|c|}\hline \mathcal{X}\\\hline A\\\hline B\\\hline G\\\hline\end{array}$ is also a Latin rectangle. Likewise, the $(4n+2) \times 2n$ arrays $\begin{array}{|c|}\hline \mathcal{T}\\\hline \mathcal{B}\\\hline \mathcal{A}\\\hline J\\\hline\end{array}$, $\begin{array}{|c|}\hline \mathcal{G}\\\hline V\\\hline \mathcal{B}\\\hline K\\\hline\end{array}$

and $\begin{array}{|c|}\hline \mathit{\Gamma}\\\hline \mathcal{B}\\\hline \mathcal{A}\\\hline L\\\hline\end{array}$ are Latin rectangles. Since the $1 \times 8n$ array $\boxed{G\,J\,K\,L}$ is a Latin rectangle, so

is the array $\boxed{\mathcal{X}\,\mathcal{T}\,\mathcal{G}\,\mathit{\Gamma}}$. It follows that the array (8) is a Latin rectangle. Adding a new row formed by four $1 \times 2n$ arrays M, N, P and Q to (8) by Hall's marriage theorem, we similarly obtain a Latin rectangle (9).

$$
\begin{array}{|c|c|c|c|}
\hline
\mathcal{X} & \mathcal{T} & \mathcal{G} & \mathit{\Gamma} \\
\hline
A & \mathcal{B} & V & \mathcal{B} \\
\hline
B & \mathcal{A} & \mathcal{B} & \mathcal{A} \\
\hline
G & J & K & L \\
\hline
M & N & P & Q \\
\hline
\end{array}
\tag{9}
$$

Adding a new row formed by four $1 \times 2n$ arrays d, $\tilde{\mathcal{O}}$, W, N above the array (9), we obtain a $(4n+4) \times 8n$ array (10).

$$
\begin{array}{|c|c|c|c|}
\hline
d & \tilde{\mathcal{O}} & W & N \\
\hline
\mathcal{X} & \mathcal{T} & \mathcal{G} & \mathit{\Gamma} \\
\hline
A & \mathcal{B} & V & \mathcal{B} \\
\hline
B & \mathcal{A} & \mathcal{B} & \mathcal{A} \\
\hline
G & J & K & L \\
\hline
M & N & P & Q \\
\hline
\end{array}
\tag{10}
$$

As we have seen above, the array (10) is a Latin rectangle. We continue this processes $2n$ times and obtain a Latin square of order $8n$ shown in (11), where C, D, E and F are $2n \times 2n$ arrays, respectively.

$$
\begin{array}{|c|c|c|c|}
\hline
\mathcal{F} & \mathcal{J} & \mathcal{C} & \mathcal{D} \\
\hline
A & \mathcal{B} & V & \mathcal{B} \\
\hline
B & \mathcal{A} & \mathcal{B} & \mathcal{A} \\
\hline
C & D & E & F \\
\hline
\end{array}
\tag{11}
$$

Then, the $8n \times 2n$ arrays 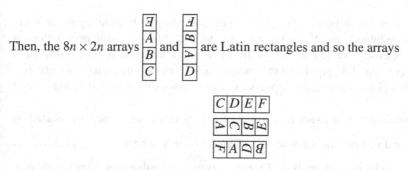 and are Latin rectangles and so the arrays

are $2n \times 8n$ Latin rectangles. Thus, the sextuple (A,B,C,D,E,F) is a solution (4). It is easy to see that (5) is a net of a Latin regular hexahedron of order $2n$. The converse is obvious.   □

We remark that not every Latin regular hexahedron is constructed in the way given by the proof of Theorem 1. It follows from Lemma 1 (1) and Theorem 1 that a Latin regular hexahedron of order k exists if and only if k is even. Next, we shall prove the existence of a Latin quasi-sudoku regular hexahedron of order $4m$ with multiplicity m.

Lemma 3. *Let L be a Latin regular hexahedron of order $2n$. Every integer appears exactly n times on each pair of faces in contraposition and so it appears exactly $3n$ times on L in total.*

Proof. We use the same notation in the proof of Lemma 1. Since $k = 2n$, we have $x = y = z = k/2 = n$. Hence, every integer appears exactly n times on each pair of faces in contrapositions. The total number of appearance of each integer on L is $x + y + z$. Consequently, each integer appears exactly $3n$ times on L in total.   □

Lemma 4. *Let L be a Latin regular hexahedron. Suppose an integer i appears on a cell p and an integer j appears on its contraposition $\overline{p}$. Let L' be a regular hexahedron obtained from L by replacing i by j on p and j by i on $\overline{p}$, respectively. Then L' is also a Latin regular hexahedron.*

Proof. Every circuit not passing p or $\overline{p}$ on L' contains the same integers as the corresponding circuits on L because no cells except for p or $\overline{p}$ are altered. On the other hand, the circuits passing p and $\overline{p}$ on L' contain the same integers as the corresponding circuits on L because p and $\overline{p}$ are located on these two circuits. Therefore every integer appears exactly once on any circuit of L', and therefore, L' is a Latin regular hexahedron.   □

Lemma 5. *Suppose $k_1, k_2, \ldots, k_l$ are positive integers and $k = k_1 + k_2 + \cdots + k_l$. Let A be a $2 \times k$ array on which every integer i in $\{1, 2, \ldots, l\}$ appears exactly $2k_i$ times. By transposing integers on columns of A, we can obtain a $2 \times k$ array so that every integer i in $\{1, 2, \ldots, l\}$ appears exactly k_i times on both the first and the second row.*

Proof. We prove by induction on k. If $k = k_1 = 1$, then 1 appears exactly once on both the first and second row. Suppose the theorem is true for any positive integer less than k. Let A be a $2 \times k$ array on which each integer i appears exactly $2k_i$ times. Suppose there exists a column of A such that a certain integer i appears on both the first and second row of the column. Let A' be a $2 \times (k-1)$ array obtained from A by deleting the column. By the inductive hypothesis, A' can be rearranged to a $2 \times (k-1)$ array on which each integer appears exactly same times in the first row and the second row by transposing integers on a column of A'. Applying the corresponding transpositions to A, we obtain a $2 \times k$ array on which each integer i appears exactly k_i times on both the first and second row.

Next suppose that there exists no column of A such that a same integer is placed on both the first and second row. Choose any column $\boxed{\begin{array}{c} x_1 \\ \hline x_2 \end{array}}$ of A, where $x_1, x_2 \in \{1, 2, \ldots, l\}$. Note that $x_1 \neq x_2$ by our assumption. There exists another column of A on which x_2 is placed because x_2 appears $2k_2$ times on A. It must be either of the forms $\boxed{\begin{array}{c} x_2 \\ \hline x_3 \end{array}}$ or $\boxed{\begin{array}{c} x_3 \\ \hline x_2 \end{array}}$. If a column of the form $\boxed{\begin{array}{c} x_3 \\ \hline x_2 \end{array}}$ is found, then we transpose x_2 and x_3 on the column to obtain $\boxed{\begin{array}{c} x_2 \\ \hline x_3 \end{array}}$. Therefore, we may assume that we found a column of the form $\boxed{\begin{array}{c} x_2 \\ \hline x_3 \end{array}}$. If $x_3 = x_1$, then we have two distinct columns $\boxed{\begin{array}{c} x_1 \\ \hline x_2 \end{array}}$ and $\boxed{\begin{array}{c} x_2 \\ \hline x_1 \end{array}}$ of A. Deleting these two columns from A, we obtain a $2 \times (k-2)$ array A'. By inductive hypothesis, A' can be rearranged by transposing integers on columns so that each integer appears exactly same times in the first and second row. Applying the corresponding rearrangement to A, we obtain a new $2 \times k$ array such that each integer i appears exactly k_i times both in the first and second row.

Now we suppose $x_3 \neq x_1$. Then x_1, x_2, x_3 are distinct from one another. There exists a column in A on which x_3 is placed since x_3 appears $2k_3$ times in A. We can choose a column of the form $\boxed{\begin{array}{c} x_3 \\ \hline x_4 \end{array}}$ by transposing the first row and the second row if necessary, where $x_4 \in \{1, 2, \ldots, l\}$. Note that $x_3 \neq x_4$. If $x_4 \notin \{x_1, x_2\}$, then we analogously choose a column $\boxed{\begin{array}{c} x_4 \\ \hline x_5 \end{array}}$. We continue similar selection of columns. Suppose we have chosen columns $\boxed{\begin{array}{c} x_1 \\ \hline x_2 \end{array}}, \boxed{\begin{array}{c} x_2 \\ \hline x_3 \end{array}}, \boxed{\begin{array}{c} x_3 \\ \hline x_4 \end{array}}, \ldots, \boxed{\begin{array}{c} x_{i-2} \\ \hline x_{i-1} \end{array}}$ and $\boxed{\begin{array}{c} x_{i-1} \\ \hline x_i \end{array}}$, where $x_1, x_2, x_3, \ldots, x_{i-1}$ are distinct from one another and $x_{i-1} \neq x_i$, however, $x_i \in \{x_1, x_2, x_3, \ldots, x_{i-2}\}$. Suppose $x_i = x_j \ (1 \leq j \leq i-2)$. Then we have columns $\boxed{\begin{array}{c} x_j \\ \hline x_{j+1} \end{array}}, \boxed{\begin{array}{c} x_{j+1} \\ \hline x_{j+2} \end{array}}, \boxed{\begin{array}{c} x_{j+2} \\ \hline x_{j+3} \end{array}}, \ldots, \boxed{\begin{array}{c} x_{i-2} \\ \hline x_{i-1} \end{array}}$ and $\boxed{\begin{array}{c} x_{i-1} \\ \hline x_j \end{array}}$. In these columns the integers $x_j, x_{j+1}, \ldots, x_{i-1}$ appears exactly once in both the first and the second row. Deleting these columns from A, we obtain a new $2 \times (k-i+j)$ array A'. By inductive hypothesis, A' can be rearranged by transposing numbers on columns so that each number i appears exactly same times in the first and second row. Applying the corresponding operations to A, we obtain a new $2 \times k$ array so that each number i appears exactly k_i times in the first and second row. □

Theorem 2. *There exists a Latin quasi-sudoku regular hexahedron of order $4n$ with multiplicity n for every positive integer n. In particular, there exists a Latin sudoku regular hexahedron of order 4.*

Proof. Let L be a Latin regular hexahedron of order $4n$, where n is a positive integer. The existence of a Latin regular hexahedron of order $4n$ has been proven in Theorem 1. Every integer appears exactly $2n$ times in each pair of faces in contraposition by Lemma 3. Transposing integers on a cell and its contraposition does not harm the property of a Latin regular hexahedron by Lemma 4. We assume that a pair of faces in contraposition forms a $2 \times 16n^2$ array in which the first row consists of a face and the second row consists of its contraposition. Then we can obtain a new Latin regular hexahedron of order $4n$ so that each integer appears exactly n times in every face by transposing integers on one face and another in contraposition by Lemma 5.   □

3   Latin Three-Axis Design and Latin Four-Axis Design

Applying systems of Latin rectangles, we prove the existence of a 1-factorization of the complete tripartite graph $K_{2n,2n,2n}$ and the complete quadripartite graph $K_{n,n,n,n}$ for any positive integer n.

3.1   1-Factorizations

We consider a system of Latin rectangles (12) with variables (V, W, Y), where V, W and Y represent an $n \times n$ array filled with integers in $\{1, 2, 3, \ldots, 2n\}$.

$$\boxed{\overset{V}{\underset{W}{}}} \quad \boxed{W \mid Y} \quad \boxed{V \mid Y} \tag{12}$$

If a triple (A, C, D) is a solution of (12), then we say that (A, C, D) is a *Latin three-axis design of order* n. The triple (A, C, D) is considered as a complex with three axes op, oq and or shown in Fig. 5. Each face is coordinated by two of its axes. This implies that any subarray of any net of the complex in Fig. 5 is a Latin rectangle.

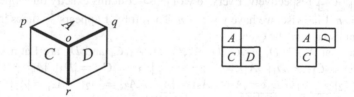

Fig. 5. Latin three-axis design and its net

Equivalently, if every array in (13) is a Latin rectangle, then a triple (A, C, D) is a Latin three-axis design.

$$\boxed{A \mid C} \quad \boxed{C \mid D} \quad \boxed{A \mid D} \tag{13}$$

Next, we consider a system of Latin rectangles (14) with variables (R, T, U, V, W, Y), where R, T, U, V, W and Y represent an $n \times n$ array filled with integers in $\{1, 2, 3, \ldots, 3n\}$.

$$\boxed{R|T|U}\quad\boxed{U|Y|W}\quad\boxed{T|V|Y}\quad\boxed{R|W|V} \tag{14}$$

If a sextuple (A,B,C,D,E,F) is a solution of (14), then we say that (A,B,C,D,E,F) is a *Latin four-axis design of order n*. The sextuple (A,B,C,D,E,F) is considered as a complex with four axes op, oq, or and os shown in Fig. 6.

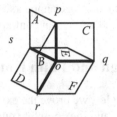

Fig. 6. Latin four-axis design

Each face is coordinated by two of its axes. This implies that if every array in (15) is a Latin rectangle, then a sextuple (A,B,C,D,E,F) is a Latin four-axis design.

$$\boxed{A|B|C}\quad\boxed{C|F|E}\quad\boxed{B|D|F}\quad\boxed{A|E|D} \tag{15}$$

Lemma 6. *(1) There exists no Latin three-axis design of order $2n-1$ for every positive integer n. (2) There exists a Latin three-axis design of order $2n$ for every positive integer n. (3) There exists a Latin four-axis design of order n for every positive integer n.*

Proof. (1) Let (A,C,D) be a solution of (12). Each cell of A,C and D are filled with integers in $\{1,2,3,\ldots,2n\}$. Suppose x,y and z are the numbers of appearance of 1 in $\boxed{A|C}$, $\boxed{C|D}$, $\boxed{A|D}$, respectively. Every row of $\boxed{A|C}$ contains exactly one appearance of 1 and so $x = n$. Likewise, we have $y = z = n$. Therefore, 1 appears $\frac{3n}{2}$ times in A,C and D in total and so n must be even.

(2) Suppose that $A_1,C_1,D_1,A_2,C_2,D_2,A_3,C_3,D_3,A_4,C_4$ and D_4 are Latin squares of order n, $|A_1| = |C_1| = |D_1|$, $|A_2| = |C_2| = |D_2|$, $|A_3| = |C_3| = |D_3|$, $|A_4| = |C_4| = |D_4|$ and $|A_1|\cap|A_2| = |A_1|\cap|A_3| = |A_1|\cap|A_4| = |A_2|\cap|A_3| = |A_2|\cap|A_4| = |A_3|\cap|A_4| = \emptyset$. We set $A = \begin{array}{|c|c|}\hline A_1 & A_2 \\\hline A_4 & A_3 \\\hline\end{array}$, $C = \begin{array}{|c|c|}\hline C_2 & C_4 \\\hline C_3 & C_1 \\\hline\end{array}$ and $D = \begin{array}{|c|c|}\hline D_1 & D_3 \\\hline D_2 & D_4 \\\hline\end{array}$. Then every array in (13) is a Latin rectangle and so (A,C,D) is a Latin three-axis design.

(3) Let A,B,C,D,E and F be Latin squares of order n satisfying $|A| = |F|$, $|B| = |E|$, $|C| = |D|$ and $|A|\cap|B| = |B|\cap|C| = |C|\cap|A| = \emptyset$. It is obvious that every array in (15) is a Latin rectangle and so (A,B,C,D,E,F) is a Latin four-axis design.

$\square$

Lemma 7. *Let n be any positive integer. (1) $K_{2n,2n,2n}$ is 1-factorizable if and only if there exists a Latin three-axis design of order $2n$. (2) $K_{n,n,n,n}$ is 1-factorizable if and only if there exists a Latin four-axis design of order n.*

Proof. (1) Let (A,C,D) be a Latin three-axis design of order $2n$ with axes op, oq, or like in Fig. 5. Suppose op, oq and or are coordinated by disjoint sets X, Y and Z, respectively. We label edges of $K_{2n,2n,2n}$ with independent sets X, Y and Z. Suppose $X = \{x_1, x_2, \ldots, x_{2n}\}$, $Y = \{y_1, y_2, \ldots, y_{2n}\}$ and $Z = \{z_1, z_2, \ldots, z_{2n}\}$. We label the edge $x_i y_j$ by the (x_i, y_j) entry of A, the edge $y_j z_k$ by the (y_j, z_k) entry of D, and the edge $x_i z_k$ by the (x_i, z_k) entry of C. Then the subgraph consisting of edges labeled by an integer $a \in \{1, 2, \ldots, 4n\}$ is a 1-factor of $K_{2n,2n,2n}$, and hence, $K_{2n,2n,2n}$ is partitioned into disjoint 1-factors. Conversely we suppose that $K_{2n,2n,2n}$ is 1-factorizable. Let X, Y and Z be the independent sets of $K_{2n,2n,2n}$, where $X = \{x_1, x_2, \ldots, x_{2n}\}$, $Y = \{y_1, y_2, \ldots, y_{2n}\}$ and $Z = \{z_1, z_2, \ldots, z_{2n}\}$. Every edge is labeled by an integer in $\{1, 2, 3, \ldots, 4n\}$. We now define a Latin three-axis design (A,C,D) as follows. Let A be a $2n \times 2n$ array obtained by coordinating X and Y; if the edge $x_i y_j$ is labeled by an integer p in $\{1, 2, 3, \ldots, 4n\}$, then we place p on the (x_i, y_j)-cell of A. Similarly, let C be a $2n \times 2n$ array obtained by coordinating X and Z and let D be a $2n \times 2n$ array obtained by coordinating Y and Z. It is easy to see that the triple (A,C,D) is a three-axis design. Similarly, we relate a Latin four-axis design (A,B,C,D,E,F) and a 1-factorization of $K_{n,n,n,n}$ by coordinating A by X and W, B by Y and W, C by W and Z, D by X and Y, E by X and Z and F by Y and Z as op is coordinated by W, oq by Z, or by Y and os by X. This implies that a 1-factorization of $K_{n,n,n,n}$ gives a Latin four-axis design and vice versa. □

Theorem 3. $K_{2n,2n,2n}$ and $K_{n,n,n,n}$ are 1-factorizable for every positive integer n.

We show a 1-factorization of $K_{2,2,2}$ and the corresponding Latin three-axis design which is constructed by the method given in the proof of Lemma 6 (2) in Fig. 7. We remark that Lemma 6 and 7 show how to construct 1-factorizations of $K_{2n,2n,2n}$ and $K_{n,n,n,n}$ for any positive integer n.

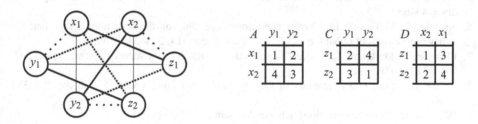

Fig. 7. 1-factorization of $K_{2,2,2}$ and corresponding Latin three-axis design

3.2 Construction of a Latin Regular Hexahedron Using Latin Three-Axis Designs

The existence of a Latin regular hexahedron of order $2n$ is proved in Theorem 1. We shall provide another construction of a Latin regular hexahedron of order $2n$ using Latin three-axis designs. Suppose L_1 and L_2 are Latin three-axis designs such that $|L_1| = \{1, 2, 3, \ldots, 4n\}$ and $|L_2| = \{4n+1, 4n+2, 4n+3, \ldots, 8n\}$, nets of which are given in (16), where A, B, C, D, E and F are $2n \times 2n$ arrays, respectively.

$$L_1 : \begin{array}{|c|} \hline A \\ \hline C\,D \\ \hline \end{array} \qquad\qquad L_2 : \begin{array}{|c|} \hline E \\ \hline B\,F \\ \hline \end{array} \qquad\qquad (16)$$

We can obtain a Latin regular hexahedron pasting L_1 and L_2. Its nets are shown in Fig. 8. The Latin regular hexahedron given in Fig. 2 is constructed in this fashion. Transposing integers on cells in contraposition by Lemma 4, we can construct more Latin regular hexahedra.

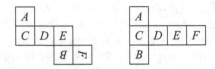

Fig. 8. Nets of Latin regular hexahedron obtained by pasting L_1 and L_2

References

1. Cariolaro, D., Hilton, A.J.W.: An application of Tutte's Theorem to 1-factorization of regular graphs of high degree. Discret. Math. **309**, 4736–4745 (2009)
2. Chetwynd, A.G., Hilton, A.J.W.: Regular graphs of high degree are 1-factorizable. Proc. London Math. Soc. **50**, 193–206 (1985)
3. Chetwynd, A.G., Hilton, A.J.W.: 1-factorizing regular graphs of high degree—an improved bound. Discret. Math. **75**, 103–112 (1989)
4. Dénes, J., Keedwell, A.D.: Latin Squares and their Applications New Developments in the Theory and Applications, 2nd edn. Elsevier, Amsterdam (2015)
5. Mendelsohn, E., Rosa, A.: One-factorizations of the complete graph—a survey. J. Graph Theory **9**, 43–65 (1985)
6. Niessen, T., Volkmann, L.: Class 1 conditions depending on the minimum degree and the number of vertices of maximum degree. J. Graph Theory **14**, 225–246 (1990)
7. Plummer, M.D.: Graph factors and factorization: 1985–2003: a survey. Discret. Math. **307**, 791–821 (2007)
8. Strong, R.A.: On 1-factorizability of Cayley graphs. J. Comb. Theory Ser. B **39**, 298–307 (1985)
9. Wallis, W.D.: One-Factorizations. Kluwer Academic Publishers (1997)

Algorithms and Complexity of Strongly Stable Non-crossing Matchings

B. S. Panda and Sachin[✉]

Department of Mathematics, Indian Institute of Technology Delhi,
New Delhi 110016, India
{bspanda,maz198086}@maths.iitd.ac.in

Abstract. A matching is called *stable* if it has no blocking pair, where a *blocking pair* is a man-woman pair, say (m, w), such that m and w are not matched with each other in the matching but if they get matched with each other, then both of them become better off. A matching is called *non-crossing* if it does not admit any pair of edges that cross each other when all men and women are arranged in two parallel vertical lines with men on one line and women on the other. Two notions of matchings that are stable as well as non-crossing have been identified in the literature, namely (*i*) weakly stable non-crossing matching (WSNM) and (*ii*) strongly stable non-crossing matching (SSNM). An SSNM is a non-crossing matching which is stable in the classical sense, whereas in a WSNM, a blocking pair satisfies an extra condition that it must not cross any matching edge. It is known that the problem of finding a WSNM, which always exists in an SMI instance, is polynomial time solvable. However, the problem of determining the existence of an SSNM in SMTI is known to be NP-complete. We show that this problem is fixed-parameter tractable (FPT) when parameterized by *total length of ties*. We introduce a new notion of stable non-crossing matching, namely semi-strongly stable non-crossing matching (SSSNM). We prove that the problem of determining the existence of an SSSNM in SMI is NP-complete even if size of every man's preference list is at most two. On the positive side, we show that this problem is polynomial time solvable if every man's preference list contains at most one woman.

Keywords: Stable non-crossing matching · Polynomial time algorithm · NP-complete · Fixed-parameter tractable (FPT)

1 Introduction

An instance of the classical Stable Marriage problem (SM) consists of n men and n women with every member (man or woman) having a *preference list* that contains all members of opposite gender in a strict order of preference. A *matching* is a set of man-woman pairs such that no two pairs have a man or woman in common. Our task is to find a matching such that there are no two members of

© The Author(s), under exclusive license to Springer Nature Switzerland AG 2023
A. Bagchi and R. Muthu (Eds.): CALDAM 2023, LNCS 13947, pp. 363–376, 2023.
https://doi.org/10.1007/978-3-031-25211-2_28

the opposite gender who would both prefer each other over their current partners. Such a matching is known as *stable matching*. The variant of SM where preference lists may be incomplete, that is, the preference lists may not contain all members of opposite gender, is denoted by SMI. Note that the members which are not present in some person a's preference list are unacceptable to a, that is, a can not be matched with them. A *tie* is a set of members which are preferred equally by some person. The variant of SM where the preference lists may contain ties as well as may be incomplete, is denoted by SMTI.

A matching is called *non-crossing* if it does not admit any pair of edges that cross each other when all men and women are arranged in two vertical lines. The notion of stable non-crossing matchings has been recently introduced by Ruangwises and Itoh [16]. In their setting, all the members are arranged on two parallel vertical lines with men arranged on one line and women on the other. They established two notions of stability with non-crossingness, namely strongly stable non-crossing matching (SSNM) and weakly stable non-crossing matching (WSNM). An SSNM is a non-crossing matching which is stable in the classical sense, whereas in a WSNM, a blocking pair satisfies an extra condition that it must not cross any matching edge.

The famous Rural Hospitals theorem [6,14,15] guarantees that "once matched, always matched", that is, if a member is matched in one stable matching of an SMI instance, then it remains matched in all stable matchings of that instance. So, while defining blocking pair for the notion of stable non-crossing matching, it is reasonable to always consider those members which are already matched. But a WSNM ignores a blocking pair (if that crosses any edge of the matching), even if it involves already matched members. Therefore, we define a new variant of stable non-crossing matching, namely semi-strongly stable non-crossing matching (SSSNM), that takes matched members into consideration as well and hence is stronger than a WSNM in the sense of stability and an alternative to SSNM. We define it formally in the next section. In Sect. 4, we provide an SMI instance in which SSNM does not exist, but an SSSNM exists.

Ruangwises and Itoh [16] have presented an $O(n^2)$-time algorithm to find a WSNM, which always exists in an SMI instance. Subsequently, Hamada et al. [9] have shown that the problem of determining the existence of an SSNM in an SMI instance is polynomial time solvable. In contrast, they have proved that the problem of determining the existence of an SSNM, given an SMTI instance, is NP-complete (for the weak stability [10]), even for a restricted case when ties are present on only one side and size of each preference list is at most two. Stable non-crossing matchings have not been studied from the parameterized complexity point of view in the literature, but there is some other research dealing with the parameterized complexity of stable matching problems [1,13].

In this paper, we initiate the first study of parameterized complexity of stable non-crossing matchings. Further, we introduce a new variant of stable non-crossing matching, namely semi-strongly stable non-crossing matching (SSSNM), an alternative to SSNM and WSNM. The main contributions of the paper are as follows:

1. We establish that the problem of determining the existence of an SSSNM, given an SMI instance, is NP-complete, even if every man contains at most two women in the preference list.
2. We present a linear time algorithm for finding an SSSNM in an SMI instance for the case when every man contains at most one woman in the preference list. We deduce that an SSSNM always exists in such an instance.
3. We show that the problem of determining the existence of an SSNM in an SMTI instance (for the weak stability) is FPT when parameterized by *total length of ties*.

2   Preliminaries

Let $M = \{m_1, m_2, m_3, \ldots, m_n\}$ and $W = \{w_1, w_2, w_3, \ldots, w_n\}$ be two sets consisting of n men and n women, respectively. Every man (resp. woman) has a *preference list* in which he (resp. she) ranks the women (resp. men) in a decreasing order of preference. The rank of an individual b in a's preference list is denoted by $rank_a(b)$. A *tie* in a person a's preference list is a set of individuals among which a is indifferent, that is, whom a prefers equally. The *length of a tie* T_i, denoted by $l(T_i)$, is the number of persons in T_i. We say that m *prefers* w_1 to w_2 if w_1 and w_2 are not tied, and w_1 proceeds w_2 in m's preference list. A pair (m, w) is said to be *admissible* if w's preference list contains m, and m's preference list contains w. We may use the terms "pair" and "edge" correspondently because a man-woman pair can also be viewed as an edge on the plane. A *matching* M' is a set of disjoint pairs of $M \times W$. A *blocking pair* of a matching M' is an admissible pair $(m_i, w_j) \in (M \times W) \setminus M'$ such that m_i is either unmatched or prefers w_j to his current matched partner, and in return w_j also is either unmatched or prefers m_i to her current matched partner. A matching M' is said to be *stable* if it does not admit any blocking pair. Note that when ties are involved, three notions of stability, named *weak*, *strong*, and *super*, are identified in the literature [10]. In this paper, we refer to the weak notion of stability [10] whenever ties are present. We do not mention this exclusively in the rest of the paper to avoid confusion with the term 'weakly' in WSNM. For additional details on matching with preferences, see [7,11].

Next, we define crossingness. For this, we assume that all men and women are arranged in two vertical lines, with m_i (resp. w_i) lying immediate above to m_{i+1} (resp. w_{i+1}). Two edges cross each other if they cut at an internal point of both segments. Formally, two edges (m_i, w_j) and (m_k, w_l) are said to *cross* each other if $(k - i)(l - j) < 0$. A matching is called *non-crossing* if it does not admit any pair of edges that cross each other. A blocking pair of M' is called a *non-crossing blocking pair*, abbreviated as *n.c.b.p.*, of M' if it does not cross any matching edge. A matching M' is called a *strongly stable non-crossing matching (SSNM)* if (*i*) M' is non-crossing and (*ii*) M' does not admit any blocking pair. A matching M' is called a *weakly stable non-crossing matching (WSNM)* if (*i*) M' is non-crossing and (*ii*) M' does not admit any non-crossing blocking pair. Note that an SSNM is stable in the classical sense, whereas a WSNM need not be. Next, a blocking pair (m, w) of M' is called a *matched crossing blocking pair*, abbreviated as *m.c.b.p.*, of M' if (i) (m, w) crosses some edge (one or more) of M' and (ii) both m and w are matched in M'. We define a new variant of stable non-crossing matching, namely semi-strongly stable non-crossing matching (SSSNM).

Definition 1. *A matching M' is called a semi-strongly stable non-crossing matching (SSSNM) if (i) M' is non-crossing, (ii) M' does not admit any non-crossing blocking pair (n.c.b.p.), and (iii) M' does not admit any matched crossing blocking pair (m.c.b.p.).*

A *parameterized problem* is a pair (Π, k) resulted by associating an integer k known as *parameter*, to each instance of a decision problem Π. Further, a parameterized problem is said to be *fixed-parameter tractable (FPT)* if there is an algorithm that solves it in $O(f(k) \cdot n^c)$ time for some constant c, where f is a computable function depending only on the parameter k, and n is the input size. For further details on parameterized complexity theory, see [3,4].

3   Strongly Stable Non-crossing Matchings

In this section, we study strongly stable non-crossing matchings from the parameterized complexity point of view. In particular, we show that the problem of finding a strongly stable non-crossing matching in an SMTI instance is fixed-parameter tractable if *total length of ties* is considered as the parameter. Our proof technique is based on the proof given in [13]. First, we extend the proposition 1 of [8] to show the uniqueness of an SSNM in an SMI instance, whenever exists.

Proposition 1. *Strongly stable non-crossing matching in an SMI instance, if exists, is unique.*

Proof. Let there be more than one SSNM in an SMI instance I. Since every SSNM is stable in the classical sense. So if a member is matched in one SSNM, then it remains matched in every SSNM because of the Rural Hospitals theorem [6,14,15]. By using the fact that these members can be matched in a non-crossing way in a unique manner only, we get that a strongly stable non-crossing matching in an SMI instance, if exists, is unique.   □

Tie-breaking procedure: Assume that the instance I contains r ties T_1, T_2, $\ldots$, T_r. First, we arrange members of every tie in a fixed order. Let t_{ij} denote the jth element of tie T_i. For every $i = 1$ to r, we define a bijection $\alpha_{i,j} : T_i \to \{1, 2, \ldots, l(T_i)\}$ $\forall$ $j = 1$ to $l(T_i)$, such that $\alpha_{i,j}$ ranks t_{ij} at the first place among all the members of tie T_i, and $\alpha_{i,j}$ ranks the members of $T_i \setminus \{t_{ij}\}$ arbitrarily in a strict order succeeding t_{ij}. We break ties according to each element of $\prod_{i=1}^{r}\{\alpha_{i,1}, \alpha_{i,2}, \ldots, \alpha_{i,l(T_i)}\}$.

Lemma 1. *Suppose a matching M is stable in an SMTI instance I. Then M remains stable in the instance I' created from I by improving the rank of one or more matched persons of M in their matched partner's preference list.*

Proof. This is an immediate consequence of the fact that a blocking pair of M in I' also blocks M in I.   □

Next, let I be an SMTI instance. Let M be a matching of I. We obtain a corresponding SMI instance I_M using M, by breaking ties in I as follows: If for some $k_i \in \{1, 2, \ldots, l(T_i)\}$, the k_i-th element of tie T_i is matched in M with the person whose preference list contains the tie T_i, then we break the ties using the element $(\alpha_{1,k_1}, \alpha_{2,k_2}, \ldots, \alpha_{r,k_r})$. Note that, if none of the element of some tie T_i is matched with the person whose preference list contains T_i, then we arbitrarily select a k_i to get the corresponding bijection α_{i,k_i} for breaking the tie T_i. We call such an SMI instance I_M an *SMI$_M$ instance*, owing to the fact that I_M is created by using the matching M. The following lemma due to [12] shows that the stability of a matching, say M, in an SMTI instance carries over to the corresponding SMI_M instance and vice-versa.

Algorithm 1. SSNM in SMTI

Input: An SMTI instance I with $M = \{m_1, m_2, \ldots, m_n\}$ and $W = \{w_1, w_2, \ldots, w_n\}$ as the sets of men and women, respectively. Without loss of generality, let I contains r ties $T_1, T_2, \ldots, T_r$.

Output: A strongly stable non-crossing matching in I or reports that none exists.

begin

    Set $k = \prod_{i=1}^{r} l(T_i)$, where $l(T_i)$ denotes the length of tie T_i;

    Obtain the k SMI instances from I by applying the *tie-breaking procedure* described before Lemma 1, and order them arbitrarily as $I_1, I_2, \ldots, I_k$;

    Initialize $p = 0$;

    for $j = 1$ to $k+1$ **do**

        if $p=k$ **then**

            └ **return No SSNM;**

        else

            Run Gale-Shapley algorithm [5] on instance I_j to obtain a stable matching, say M_1;

            Let $M' = M \setminus \{m_u \mid m_u$ is unmatched in $M_1\}$ and $W' = W \setminus \{w_v \mid w_v$ is unmatched in $M_1\}$;

            Sort the members of M' and W' in increasing order of indices to obtain permutations α and β, respectively;

            Match l^{th} ($l = 1$ to $|M'|$) member of α with l^{th} member of β. Let the matching so obtained be M_2;

            if M_2 is stable in I_j **then**

                └ **return** M_2;

            else

                └ $p = p + 1$;

Lemma 2. *A matching M is stable in an SMTI instance I iff M is stable in the corresponding SMI$_M$ instance I_M.*

Proof. Let M be a stable matching in I (resp. I_M). Suppose, M is not stable in I_M (resp. I). Let (m_i, w_j) be a blocking pair of M in I_M (resp. I). Then (m_i, w_j) is a blocking pair of M in I (resp. I_M) as well, a contradiction. Therefore, matching M is stable in I_M (resp. I). □

Theorem 1. *Algorithm 1 gives an SSNM in an SMTI instance, say I, or reports that none exists, in $O(k^k \cdot n^2)$ time, where k is the total length of ties.*

Proof. Suppose there exists an SSNM, say M, in the instance I. Then, by Lemma 2, M remains stable and hence strongly stable in the corresponding SMI_M instance, say I_M. Also, non-crossingness carries over to I_M as well. Therefore, M is an SSNM in the instance I_M, created by breaking ties according to some element of $\prod_{i=1}^{r}\{\alpha_{i,1}, \alpha_{i,2}, \ldots, \alpha_{i,l(T_i)}\}$. Also, by proposition 1, SSNM M is unique in I_M. Now, since our algorithm runs through all instances created using every element of $\prod_{i=1}^{r}\{\alpha_{i,1}, \alpha_{i,2}, \ldots, \alpha_{i,l(T_i)}\}$ to find an SSNM until one is established. So, if the algorithm establishes an SSNM in some instance which is checked earlier than I_M, it reports the established SSNM and terminates. Otherwise, it reduces to the instance I_M. The Gale-Shapley algorithm [5] is applied to I_M. Let the matching so obtained be M_1. Due to the Rural Hospitals theorem [6,14,15], the set of matched members in every stable matching of I_M is same as in M_1. Then by using proposition 1 and the fact that the members which are matched in M_1 can be re-matched in a non-crossing manner in only one way, we get that M_2 is same as M, where M_2 is the matching obtained after re-matching the agents of M_1 in the non-crossing manner. Since M is strongly stable, so is M_2 and hence the algorithm outputs M_2, an SSNM.

Conversely, let the algorithm reports a matching, say S. We prove that there exists an SSNM in the instance I. It is sufficient to show that S is an SSNM in I. Since, S is reported by the algorithm, so S is non-crossing and stable in some SMI instance, say I_d, created by breaking ties according to some element of $\prod_{i=1}^{r}\{\alpha_{i,1}, \alpha_{i,2}, \ldots, \alpha_{i,l(T_i)}\}$. So, by Lemma 1, S is stable in I'_d, where I'_d is created from I_d as follows: If in I_d, $(g_i, h_k) \in S$, and h_k is present in some tie T_z of g_i's preference list in the original instance I, then we obtain I'_d by interchanging h_k in g_i's preference list in I_d with that element of T_z which is best ranked in I_d among all the elements of T_z, say with h_j. This process is repeated for every matched agent with respect to S which is present in some tie of his/her matched partner's preference list in the original instance I. Rest of the members of the preference list remain at their respective places as in I_d. The construction is shown below.

$$I \qquad g_i : \ldots \ldots (h_j \ldots h_k \ldots h_l) \ldots \ldots$$
$$I_d \qquad \underline{g_i} : \ldots \ldots h_j \ldots \underline{h_k} \ldots h_l \ldots \ldots$$
$$I'_d \qquad \underline{g_i} : \ldots \ldots \underline{h_k} \ldots h_j \ldots h_l \ldots \ldots$$

Coming back to the proof, since S is stable in I'_d, so by Lemma 2, S is stable in I. Furthermore, S is non-crossing in I_d implies S is non-crossing in I. Therefore, S is a strongly stable non-crossing matching in I.

Next, we analyse the time complexity of the algorithm. For every instance created from I, we run Gale-Shapley algorithm which takes $O(n^2)$ time. Further, the required sorting in our case can be done in time less than $O(n^2)$. Subsequently, the matching M_2 can be obtained in $O(n)$ time and further it can be checked for stability in $O(n^2)$ time. We need to repeat this process for at most $l(T_1).l(T_2).\ldots.l(T_r) \leq k^r \leq k^k$ many instances. So, overall time complexity for the algorithm is $O(k^k \cdot n^2)$. ◻

This theorem leads us to the following corollary.

Corollary 1. *The problem of determining the existence of an SSNM in an SMTI instance is fixed-parameter tractable with 'total length of ties' as the parameter.*

4 Semi-strongly Stable Non-crossing Matchings

In this section, we introduce a new variant of stable non-crossing matching, namely semi-strongly stable non-crossing matching (SSSNM), an alternative to SSNM and WSNM. As defined in Sect. 2, we say a matching M' is *semi-strongly stable non-crossing matching (SSSNM)*, if M' is *non-crossing* and M' neither admits any non-crossing blocking pair nor any matched crossing blocking pair. Consider the following SMI instance involving 4 men, m_1, m_2, m_3, m_4, and 4 women, w_1, w_2, w_3, w_4:

$$m_1 : w_1 \ w_4 \qquad\qquad w_1 : m_1$$
$$m_2 : w_2 \qquad\qquad w_2 : m_4 \ m_2$$
$$m_3 : w_3 \qquad\qquad w_3 : m_3$$
$$m_4 : w_2 \ w_4 \qquad\qquad w_4 : m_4 \ m_1$$

This instance has no SSNM, but contains a SSSNM of size 2, $\{(m_1, w_1), (m_4, w_2)\}$, and a WSNM of size 4, $\{(m_1, w_1), (m_2, w_2), (m_3, w_3), (m_4, w_4)\}$.

We show that the problem of finding a semi-strongly stable non-crossing matching in an SMI instance is NP-complete, even if size of every man's preference list is at most two, whereas linear time solvable if size of every man's preference list is at most one. We denote by *(p,q)-SMI*, a variant of SMI in which size of every man's preference list is at most p, and size of every woman's preference list is at most q. Note that, $p \leq n$ and $q \leq n$. If $p = n$ (resp. $q = n$), then we say that men's (resp. women's) preference list is unrestricted in size.

4.1 (2,n)-SMI

We prove that the problem of finding an SSSNM in a (2,n)-SMI instance is NP-complete by giving a polynomial reduction from the 3-SAT problem which is known to be NP-complete [2].

Theorem 2. *The problem of determining the existence of an SSSNM, given a (2,n)-SMI instance, is NP-complete.*

Proof. First, we show that the problem is in NP. Note that checking whether an edge is n.c.b.p. or m.c.b.p. can be done in $O(n^2)$ time. So, given a matching M of an SMI instance containing n men $m_1, m_2, \ldots, m_n$ and n women $w_1, w_2, \ldots, w_n$, one can easily verify in polynomial time that M is non-crossing, and does not have any n.c.b.p. as well as does not have any m.c.b.p. Hence, the problem of determining the existence of an SSSNM in a (2,n)-SMI instance lies in NP. To show hardness, we give a polynomial reduction from the 3-SAT problem which is known to be NP-complete [2].

Let g be a 3-SAT instance with k clauses C_j $(1 \leq j \leq k)$ and n variables x_i $(1 \leq i \leq n)$. We create an SMI instance I from g as follows.

Construction of Men and Women: Corresponding to each clause C_j, we create two women a_j, b_j and five men c_j^1, c_j^2, c_j^3, d_j^1, d_j^2. Corresponding to each variable x_i, we create four women w_i^-, w_i^+, q_i^-, q_i^+ and four men m_i^-, m_i^+, p_i^-, p_i^+. So, I consists of $2k + 4n$ women and $5k + 4n$ men. Also, we create $3k$ dummy women to equalise the number of men and women in I.

Construction of Preference Lists: Assume that there are n_i negative occurrences of the variable x_i in g. Thus for each l $(1 \leq l \leq n_i)$, let x_i appear lth time negatively in the r_l^ith literal $(1 \leq r_l^i \leq 3)$ of the s_l^ith clause $C_{s_l^i}$. Similarly, assume that there are p_i positive occurrences of the variable x_i in g. Thus for each l $(1 \leq l \leq p_i)$, let x_i appear lth time positively in the u_l^ith literal $(1 \leq u_l^i \leq 3)$ of the v_l^ith clause $C_{v_l^i}$. Next, for each α $(1 \leq \alpha \leq 3)$, assume that the α-th literal of C_j is x_{j_α}. If x_{j_α} appears positively in C_j then let $\sigma_{j_\alpha} = {}$ '+', otherwise if x_{j_α} appears negatively in C_j then let $\sigma_{j_\alpha} = {}$ '−'. Create preference list for each person as follows.

$$(1 \leq i \leq n) \quad m_i^+ : w_i^+ \qquad\qquad w_i^- : p_i^- \; c_{s_1^i}^{r_1^i} \; c_{s_2^i}^{r_2^i} \; \ldots \; c_{s_{n_i}^i}^{r_{n_i}^i} \; m_i^-$$

$$m_i^- : w_i^- \qquad\qquad w_i^+ : p_i^+ \; c_{v_1^i}^{u_1^i} \; c_{v_2^i}^{u_2^i} \; \ldots \; c_{v_{p_i}^i}^{u_{p_i}^i} \; m_i^+$$

$$p_i^+ : q_i^+ \; w_i^+ \qquad\qquad q_i^+ : p_i^+$$

$$p_i^- : q_i^- \; w_i^- \qquad\qquad q_i^- : p_i^-$$

$$(1 \leq j \leq k) \quad c_j^1 : w_{j_1}^{\sigma_{j_1}} \; b_j \qquad\qquad a_j : d_j^2 \; d_j^1$$

$$c_j^2 : w_{j_2}^{\sigma_{j_2}} \; b_j \qquad\qquad b_j : c_j^1 \; c_j^2 \; c_j^3$$

$$c_j^3 : w_{j_3}^{\sigma_{j_3}} \; b_j$$

$$d_j^1 : a_j$$

$$d_j^2 : a_j$$

Note that we take preference list of each dummy woman as empty. Next, alignment of the members is shown in part (a) of Fig. 1. Note that, we can place all the dummy women below the woman b_k. But since their preference lists are empty, so we do not consider them in the alignment and hence in the rest of the proof.

Clearly, the above construction can be completed in polynomial time. Also, we can easily note that the above created instance is an SMI instance. An illustration of the construction of instance I from an example formula g is shown in Fig. 2 in the appendix.

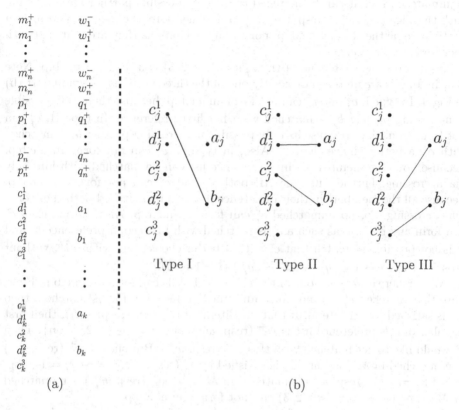

(a)   (b)

Fig. 1. (a) Alignment of members. (b) Matching of c_j^r's in M.

Claim. g is satisfiable iff I admits a semi-strongly stable non-crossing matching.

Proof. ($\Rightarrow$) Let g be satisfiable. Let S be the satisfying assignment of g. We construct an SSSNM M of I using S.

For $1 \le i \le n$, $1 \le j \le k$,

1. Add (m_i^-, w_i^-) to M if $x_i = 1$ with respect to S, otherwise add (m_i^+, w_i^+) to M if $x_i = 0$ with respect to S.
2. Next, add the edges (p_i^-, q_i^-) and (p_i^+, q_i^+) to M.
3. Finally, if C_j is satisfied by the first literal, then add (c_j^1, b_j) to M. Otherwise if C_j is satisfied by the second (resp. third) literal, then add the edges (c_j^2, b_j) and (d_j^1, a_j) (resp. (c_j^3, b_j) and (d_j^2, a_j)) to M. If C_j is satisfied by more than one literal, then select one literal among the satisfying ones arbitrarily.

One can easily observe using the alignment of the members in Fig. 1 that the matching M is non-crossing. It remains to show that M is semi-strongly stable. Note that the men m_i^+ and m_i^- can not form an m.c.b.p. because either they are unmatched or matched with the first preference. Also, m_i^+ (resp. m_i^-) can not form any n.c.b.p. because either that is matched with the first preference or if it is unmatched then it can be matched to his only possible partner via a crossing with the edge (m_i^-, w_i^-) (resp. (m_i^+, w_i^+)). Furthermore, the men p_i^- and p_i^+ can participate neither in any n.c.b.p. nor in any m.c.b.p. as they are matched with their first preference.

Next, we show that none of the c_j^r's ($r = 1, 2, 3$) can form an n.c.b.p. Note that in M, c_j^r's are matched in exactly one of the three types as shown in part (b) of Fig. 1. In type I, c_j^2 and c_j^3 can not form an n.c.b.p. because their only possible non-crossing partner b_j is matched with her first preference. In type II, c_j^3 can not form an n.c.b.p. because its only possible non-crossing partner b_j is matched with c_j^2, a better choice than c_j^3. Also, in type II, c_j^1 can not form an n.c.b.p. because both the members in his preference list can be matched with him only via a crossing. Further, in type III, both c_j^1 and c_j^2 can not form an n.c.b.p. because all the members in their preference lists can be matched with them only via a crossing. So, no unmatched c_j^r can form an n.c.b.p. Also, no matched c_j^r can form any n.c.b.p. as such a c_j^r is matched with his second preference b_j and it is not possible to match that c_j^r with a better choice, i.e., with $w_{jr}^{\sigma_{jr}}$ without crossing the edges (p_i^-, q_i^-) and (p_i^+, q_i^+) ($i = 1$ to n).

We further show that none of the c_j^r's ($r = 1, 2, 3$) can form an m.c.b.p. First, note that matched c_j^r can not form any m.c.b.p. because if c_j^r is matched, then C_j is satisfied by rth literal. If that rth literal of C_j is x_i (resp. $-x_i$), then first member of c_j^r's preference list is w_i^+ (resp. w_i^-) for some $i \in \{1, 2, \ldots, n\}$. Then c_j^r would like to get matched with this w_i^+ (resp. w_i^-). But such a w_i^+ (resp. w_i^-) is unmatched in M (because C_j is satisfied by x_i (resp. $-x_i$) $\implies x_i = 1$ (resp. $x_i = 0$) $\implies w_i^-$ (resp. w_i^+) is matched in M $\implies w_i^+$ (resp. w_i^-) is unmatched in M). Therefore, c_j^r ($r = 1, 2, 3$) can not form any m.c.b.p.

Next, d_j^s ($s = 1, 2$) can not form any n.c.b.p., as from part (b) of Fig. 1 one can note that either (i) both d_j^1 and d_j^2 have to cross an edge to form a blocking pair (Type I) or (ii) d_j^1 is already matched to his first preference, and d_j^2 has to cross an edge to form a blocking pair (Type II), or (iii) d_j^2 is already matched to his first preference, and the only possible partner of d_j^1, that is, a_j is matched with her first preference (Type III). Also, d_j^s ($s = 1, 2$) can not participate in an m.c.b.p. because either they are unmatched or matched with the first preference. Hence no men can form an n.c.b.p. or an m.c.b.p. So, matching M is semi-strongly stable.

($\Leftarrow$) Suppose I admits a semi-strongly stable non-crossing matching, say M. We show that g has a satisfying assignment.

Claim. w's can not be matched with c's or with p's in M. Hence, for $l = $ '$-$', '$+$' and $1 \le i \le n$, p_i^l must match with q_i^l in M.

Proof. First we show that w's can not be matched with c's in M. Let if possible, w_t^l ($t \in \{1, 2, \ldots, n\}$) be the topmost (according to the alignment) among w's which are matched with c's in M. But then (w_t^l, p_t^l) is an n.c.b.p. of M, a contradiction. So, w's can not be matched with c's in M. Next, suppose some p's (one or more) are matched with w's. Let p_b^l ($b \in \{1, 2, \ldots, n\}$) be the bottommost (according to the alignment) among p's which are matched with w's in M. This implies (p_b^l, q_b^l) is an n.c.b.p. of M, a contradiction. So, p's can not be matched with w's in M. Therefore, for $l = $ '$-$', '$+$', p_i^l must match with q_i^l in M for every $i = 1$ to n, else (p_i^l, q_i^l) will be an n.c.b.p. of M.   □

In the above claim, we have shown that w's can not be matched with c's or with p's in M. But since matching M is semi-strongly stable, so exactly one of (m_i^-, w_i^-) and (m_i^+, w_i^+) must be in M, and for each j ($1 \leq j \leq k$), either (i) $(c_j^1, b_j) \in M$ or (ii) $(c_j^2, b_j), (d_j^1, a_j) \in M$, or (iii) $(c_j^3, b_j), (d_j^2, a_j) \in M$.

Now, we construct a satisfying assignment S for g as follows: If $(m_i^-, w_i^-) \in M$ then set $x_i = 1$ in S, otherwise if $(m_i^+, w_i^+) \in M$ then set $x_i = 0$ in S.

Claim. S satisfies g.

Proof. Suppose S does not satisfy g. Let C_j be the unsatisfied clause. Suppose the rth ($r \in \{1, 2, 3\}$) literal of C_j is x_i (resp. $-x_i$). But C_j is unsatisfied, so $x_i = 0$ (resp. $x_i = 1$) with respect to S. This implies $(m_i^+, w_i^+) \subset M$ (resp. $(m_i^-, w_i^-) \in M$) due to construction of S. Now, since for each j ($1 \leq j \leq k$), exactly one of the three (i) $(c_j^1, b_j) \in M$, (ii) $(c_j^2, b_j), (d_j^1, a_j) \in M$, and (iii) $(c_j^3, b_j), (d_j^2, a_j) \in M$ holds. If case (i) holds, then (c_j^1, w_i^+) (resp. (c_j^1, w_i^-)) forms an m.c.b.p. of M if the first literal of C_j is x_i (resp. $-x_i$), a contradiction as the matching M is semi-strongly stable. Similarly, if case (ii) holds, then (c_j^2, w_i^+) (resp. (c_j^2, w_i^-)) forms an m.c.b.p. of M if the second literal of C_j is x_i (resp. $-x_i$), a contradiction. Finally, if case (iii) holds, then (c_j^3, w_i^+) (resp. (c_j^3, w_i^-)) forms an m.c.b.p. of M if the third literal of C_j is x_i (resp. $-x_i$), a contradiction. So, g has no unsatisfied clause with respect to S. Therefore, S satisfies g.   □

□

Hence, the theorem is proved.   □

4.2   (1,n)-SMI

We present a polynomial time algorithm to find a semi-strongly stable non-crossing matching in a (1,n)-SMI instance, which always exists.

Theorem 3. *Algorithm 2 gives a semi-strongly stable non-crossing matching in a (1,n)-SMI instance in $O(n)$ time.*

Proof. Without loss of generality, we assume that there is at least one admissible pair in the instance. Since every man has at most one woman in his preference list, so there are at most n proposals. Hence the algorithm terminates with a matching, say M, and the overall time complexity for the algorithm is $O(n)$.

Algorithm 2. SSSNM in $(1,n)$-SMI

Input: A $(1,n)$-SMI instance I containing n men $m_1, m_2, \ldots, m_n$ and n women w_1, $w_2, \ldots, w_n$.

Output: A semi-strongly stable non-crossing matching in I.

begin

    Initialize $M = \phi$, $max = 0$, and $j = 1$;

    while $j \leq n$ **do**

        if *preference list of w_j is empty* **then**

            $j = j + 1$;

        else Let m_i be the highest ranked man in w_j's preference list

            if $i > max$ **then**

                $M = M \cup \{(m_i, w_j)\}$;

                $max = i$;

                $j = j + 1$;

            else if $i < max$

                Delete m_i from w_j's preference list;

    return M

We show that the matching M, returned by the algorithm is semi-strongly stable non-crossing. Note that, we add an edge to M only if the condition "$i > max$" is satisfied. This implies that the newly added edge does not cross any of the previously added edges of M. This proves that M is non-crossing. Next, M does not have any m.c.b.p. because every matched man is matched with his first preference. It remains to show that M does not have any n.c.b.p. For the sake of contradiction, suppose (m_i, w_j) is an n.c.b.p. of M. Therefore, w_j prefers m_i to her matched partner. Note that w_j prefers m_i even if she is unmatched because an individual always prefers being matched rather than remaining unmatched. Thus m_i must have been deleted from w_j's preference list during the execution of the algorithm. This implies (m_i, w_j) crosses some edge of M, a contradiction. So, M does not have any n.c.b.p. and hence M is SSSNM. $\qquad\square$

This theorem leads us to the following corollary.

Corollary 2. *Semi-strongly stable non-crossing matching always exists in a $(1,n)$-SMI instance.*

5   Conclusion

We have shown that the problem of finding an SSNM in an SMTI instance is fixed-parameter tractable (FPT) when parameterized by *total length of ties*. Further, we have introduced a new variant of stable non-crossing matching, namely semi-strongly stable non-crossing matching (SSSNM). We have shown that the problem of determining the existence of an SSSNM, given a $(2,n)$-SMI instance, is NP-complete. On the positive side, we have presented a linear time algorithm to find an SSSNM in a $(1,n)$-SMI instance, which always exists. It remains open to consider SSSNM on other alignments than vertical lines.

A   Illustration of Construction of Instance in the Proof of Theorem 2

$$g = (x_1 \lor x_2 \lor \neg x_3) \land (\neg x_1 \lor x_3 \lor \neg x_2)$$

$$I$$

m_1^+	:	w_1^+			
m_1^-	:	w_1^-			
m_2^+	:	w_2^+			
m_2^-	:	w_2^-			
m_3^+	:	w_3^+			
m_3^-	:	w_3^-			
p_1^-	:	q_1^-	w_1^-		
p_1^+	:	q_1^+	w_1^+		
p_2^-	:	q_2^-	w_2^-		
p_2^+	:	q_2^+	w_2^+		
p_3^-	:	q_3^-	w_3^-		
p_3^+	:	q_3^+	w_3^+		
c_1^1	:	w_1^+	b_1		
c_1^2	:	w_2^+	b_1		
c_1^3	:	w_3^-	b_1		
d_1^1	:	a_1			
d_1^2	:	a_1			
c_2^1	:	w_1^-	b_2		
c_2^2	:	w_3^+	b_2		
c_2^3	:	w_2^-	b_2		
d_2^1	:	a_2			
d_2^2	:	a_2			

w_1^-	:	p_1^-	c_2^1	m_1^-	
w_1^+	:	p_1^+	c_1^1	m_1^+	
w_2^-	:	p_2^-	c_2^3	m_2^-	
w_2^+	:	p_2^+	c_1^2	m_2^+	
w_3^-	:	p_3^-	c_1^3	m_3^-	
w_3^+	:	p_3^+	c_2^2	m_3^+	
q_1^-	:	p_1^-			
q_1^+	:	p_1^+			
q_2^-	:	p_2^-			
q_2^+	:	p_2^+			
q_3^-	:	p_3^-			
q_3^+	:	p_3^+			
a_1	:	d_1^2	d_1^1		
b_1	:	c_1^1	c_1^2	c_1^3	
a_2	:	d_2^2	d_2^1		
b_2	:	c_2^1	c_2^2	c_2^3	

Fig. 2. An illustration of the construction of instance I from a formula g in the proof of Theorem 2. The matching edges corresponding to a satisfying assignment for g, say $x_1 = 1$, $x_2 = 0$, $x_3 = 0$, are shown in blue colour. (Color figure online)

References

1. Adil, D., Gupta, S., Roy, S., Saurabh, S., Zehavi, M.: Parameterized algorithms for stable matching with ties and incomplete lists. Theoret. Comput. Sci. **723**, 1–10 (2018)
2. Cook, S.A.: The complexity of theorem-proving procedures. In: Proceedings of the Third Annual ACM Symposium on Theory of Computing, pp. 151–158 (1971)

3. Cygan, M., Fomin, F.V., Kowalik, Ł, Lokshtanov, D., Marx, D., Pilipczuk, M., Pilipczuk, M., Saurabh, S.: Parameterized Algorithms. Springer, Cham (2015). https://doi.org/10.1007/978-3-319-21275-3

4. Downey, R.G., Fellows, M.R.: Parameterized Complexity. Monographs in Computer Science, Springer Science & Business Media, New York (2012). https://doi.org/10.1007/978-1-4612-0515-9

5. Gale, D., Shapley, L.S.: College admissions and the stability of marriage. Am. Math. Mon. **69**(1), 9–15 (1962)

6. Gale, D., Sotomayor, M.: Some remarks on the stable matching problem. Discret. Appl. Math. **11**(3), 223–232 (1985)

7. Gusfield, D., Irving, R.W.: The Stable Marriage Problem: Structure and Algorithms. MIT Press, Cambridge (1989)

8. Hamada, K., Miyazaki, S., Okamoto, K.: Strongly stable and maximum weakly stable noncrossing matchings. In: Gasieniec, L., Klasing, R., Radzik, T. (eds.) IWOCA 2020. LNCS, vol. 12126, pp. 304–315. Springer, Cham (2020). https://doi.org/10.1007/978-3-030-48966-3_23

9. Hamada, K., Miyazaki, S., Okamoto, K.: Strongly stable and maximum weakly stable noncrossing matchings. Algorithmica **83**(9), 2678–2696 (2021)

10. Irving, R.W.: Stable marriage and indifference. Discret. Appl. Math. **48**(3), 261–272 (1994)

11. Manlove, D.: Algorithmics of Matching Under Preferences, vol. 2. World Scientific, Singapore (2013)

12. Manlove, D.F., Irving, R.W., Iwama, K., Miyazaki, S., Morita, Y.: Hard variants of stable marriage. Theoret. Comput. Sci. **276**(1–2), 261–279 (2002)

13. Marx, D., Schlotter, I.: Parameterized complexity and local search approaches for the stable marriage problem with ties. Algorithmica **58**(1), 170–187 (2010)

14. Roth, A.E.: The evolution of the labor market for medical interns and residents: a case study in game theory. J. Polit. Econ. **92**(6), 991–1016 (1984)

15. Roth, A.E.: On the allocation of residents to rural hospitals: a general property of two-sided matching markets. Econometrica J. Econ. Soc. **54**, 425–427 (1986)

16. Ruangwises, S., Itoh, T.: Stable noncrossing matchings. In: Colbourn, C.J., Grossi, R., Pisanti, N. (eds.) IWOCA 2019. LNCS, vol. 11638, pp. 405–416. Springer, Cham (2019). https://doi.org/10.1007/978-3-030-25005-8_33

Minimum Maximal Acyclic Matching in Proper Interval Graphs

Juhi Chaudhary[1]($\boxtimes$), Sounaka Mishra[2], and B. S. Panda[3]

[1] Department of Computer Science, Ben-Gurion University of the Negev,
Beersheba, Israel
juhic@post.bgu.ac.il
[2] Department of Mathematics, Indian Institute of Technology Madras, Chennai, India
sounak@iitm.ac.in
[3] Department of Mathematics, Indian Institute of Technology Delhi, New Delhi, India
bspanda@maths.iitd.ac.in

Abstract. Given a graph G, MIN-MAX-ACY-MATCHING is the problem of finding a maximal matching M in G of minimum cardinality such that the set of M-saturated vertices induces an acyclic subgraph in G. The decision version of MIN-MAX-ACY-MATCHING is known to be NP-complete even for planar perfect elimination bipartite graphs. In this paper, we give the first positive algorithmic result for MIN-MAX-ACY-MATCHING by presenting a linear-time algorithm for computing a minimum cardinality maximal acyclic matching in proper interval graphs.

Keywords: Matching · Acyclic matching · Minimum maximal acyclic matching · Linear-time algorithm

1 Introduction

All graphs considered in this paper are simple, connected, and undirected. For a positive integer k, let $[k]$ denote the set $\{1, \ldots, k\}$. For any graph G, let $V(G)$ denote its vertex set and $E(G)$ denote its edge set. For any graph G, the subgraph of G induced by $S \subseteq V(G)$ is denoted by $G[S]$. For a graph G and a set $X \subseteq V(G)$, we use $G - X$ to denote $G[V(G) \setminus X]$. A subset $M \subseteq E(G)$ of edges of a graph G is a *matching* if no two edges of M share a common vertex. Given a matching M of G, a vertex $v \in V(G)$ is M-*saturated* if there exists an edge $e \in M$ incident on v. Given a graph G and a matching M, we use V_M to denote the set of M-saturated vertices of G. For a matching M, if $uv \in M$, then v is the M-*mate* of u and vice versa. A matching M in a graph G is an *acyclic matching* if the subgraph $G[V_M]$ is acyclic (a forest). Given a graph G, ACY-MATCHING asks to find an acyclic matching of maximum cardinality in G [7]. An acyclic matching M of G is a *maximal acyclic matching* if M is not properly contained in any other acyclic matching of G. Given a graph G, MIN-MAX-ACY-MATCHING asks to find a maximal acyclic matching of minimum cardinality in G.

The *minimum maximal acyclic matching number* of G is the minimum cardinality of a maximal acyclic matching among all maximal acyclic matchings in G,

A. Bagchi and R. Muthu (Eds.): CALDAM 2023, LNCS 13947, pp. 377–388, 2023.
https://doi.org/10.1007/978-3-031-25211-2_29

and we denote it by $\mu'_{ac}(G)$. For an example, consider the graph G with vertex set $V(G) = \{a, b, c, d, e, f\}$ and edge set $E(G) = \{ab, bc, cd, de, ae, bd, af, ef\}$. $M_1 = \{ab\}$ and $M_2 = \{af, bc\}$ are two maximal acyclic matchings of G and M_1 is a minimum maximal acyclic matching of G. Therefore, $\mu'_{ac}(G) = 1$.

Related Work. MAX-MIN and MIN-MAX versions of many important optimization problems like DOMINATION [1], FEEDBACK VERTEX SET [6], VERTEX COVER [2], and HITTING SET [5] have recently attracted much interest from many researchers. Considering the MIN-MAX (resp. MAX-MIN) version of maximization (resp. minimization) problems, that is, minimizing (resp. maximizing) the size of a maximal (resp. minimal) solution of the corresponding problem is a natural approach. The initial motivation for studying such problems was an attempt to analyze the worst possible performance of a naive heuristic; however, these problems have gradually been revealed to possess a rich combinatorial structure that makes them interesting in their own right.

Interestingly, unlike the classical MAX-MATCHING problem, the MIN-MAX version of matching is known to be NP-hard, as it is equivalent to the INDEPENDENT EDGE DOMINATION problem [15]. Apart from that, over the years, the MIN-MAX version of many well-known variants of restricted matchings like induced matching and uniquely restricted matching have also been considered in the literature [4,9–11,14].

In 2005, Goddard et al. [7] introduced the concept of *minimum maximal acyclic matching* and proved that MIN-MAX-ACY-MATCHING is NP-hard for general graphs. Recently, we proved that the decision version of MIN-MAX-ACY-MATCHING is NP-complete for planar perfect elimination bipartite graphs [3]. We also proved that MIN-MAX-ACY-MATCHING cannot be approximated within a ratio of $n^{1-\epsilon}$, for any $\epsilon > 0$ unless P = NP, even for bipartite graphs, and MIN-MAX-ACY-MATCHING is APX-hard for 4-regular graphs [3].

Our Contribution. In Sect. 2, we propose a linear-time algorithm to compute a minimum cardinality maximal acyclic matching in proper interval graphs. The brief idea behind our greedy algorithm is given below. Proper interval graphs admit a vertex ordering that has been useful in the past in designing many linear-time algorithms. Apart from this, note that acyclic matchings also admit an ordering in proper interval graphs. In other words, if $M = \{e_1, \ldots, e_k\}$ is an acyclic matching in a proper interval graph G, then it is possible to define an ordering on M such that the endpoints of edges in M are also ordered with respect to the "given" vertex ordering of G. To design our algorithm, we exploit this property of acyclic matchings and characterize the "first edge" that belongs to some minimum maximal acyclic matching of the given connected proper interval graph.

Here we note without proof that the decision version of MIN-MAX-ACY-MATCHING is NP-complete for dually chordal graphs which is a superclass of proper interval graphs. Proofs of the results marked with (∗) are omitted due to lack of space.

2   Algorithm for Proper Interval Graphs

Given a graph G, where $V(G) = \{v_1, \ldots, v_n\}$, a vertex $v_i \in V(G)$ is *simplicial* in G, if $G[N[v_i]]$ is a clique in G. An ordering $\alpha = (v_1, \ldots, v_n)$ of $V(G)$ is a *perfect elimination ordering* (PEO) of G if v_i is simplicial in $G_i = G[\{v_i, v_{i+1}, \ldots, v_n\}]$ for each $i \in [n]$. A PEO $\alpha = (v_1, \ldots, v_n)$ of G is a *bi-compatible elimination ordering* (BCO) if $\alpha^{-1} = (v_n, \ldots, v_1)$, i.e., the reverse of α, is also a PEO of G. It has been characterized in [8] that a graph is a proper interval graph if and only if it has a BCO. Furthermore, given a proper interval graph G, a BCO of G can be computed in linear time [13].

Consider the following observation, which shows that acyclic matchings in proper interval graphs admit an ordering with respect to a given BCO.

Observation 1 [12]. *Let G be a connected proper interval graph with a BCO $\sigma(G) = (v_1, \ldots, v_n)$ and let M be an acyclic matching of G. If $v_a v_b, v_c v_d \in M$ such that $v_a < v_b$ and $v_c < v_d$ in $\sigma(G)$, then either $v_b < v_c$ or $v_d < v_a$ in $\sigma(G)$.*

Notations Used. Given a proper interval graph G, let $\sigma(G) = (v_1, \ldots, v_n)$ be a BCO of G. For each $i \in [n]$ and integer $k \geq 2$, let $L[v_i](= L^1[v_i])$ denote the maximum indexed neighbor of v_i in $\sigma(G)$, and let $L^k[v_i]$ denote the maximum indexed neighbor of $L^{k-1}[v_i]$ in $\sigma(G)$. Furthermore, let v^- (resp. v^+) denote the consecutive vertex just before (resp. after) a vertex v in $\sigma(G)$. Note that $v_n^+ = v_n$ and $v_1^- = v_1$ in $\sigma(G)$. Also, let $L[v_n] = v_n$. For each $e \in E(G)$, let $l(e)$ and $r(e)$ denote the two endpoints of e. Without loss of generality, for every $e \in E(G)$, assume that $l(e) < r(e)$ in $\sigma(G)$. By Observation 1, if $M_k = \{e_1, \ldots, e_k\}$ is an acyclic matching of G, then, without loss of generality, we assume that $r(e_i) < l(e_{i+1})$ in $\sigma(G)$ for each $i \in [k-1]$. Here, we say that for each $j \in [k]$, e_j is the jth edge in M_k with respect to $\sigma(G)$. In other words, given a proper interval graph G with a BCO $\sigma(G)$, the edge $e \in M_k$ is said to be the *first edge* of M_k with respect to $\sigma(G)$ if $r(e) < l(e')$ for every $e'(\neq e) \in M_k$. Let us name such a matching M_k a *standard acyclic matching*.

Overview. Given a connected proper interval graph G, we use a BCO of G to define an ordering on its vertices. We then check whether $L[v_1] = L^2[v_1]$ or not. If $L[v_1] = L^2[v_1]$, then by Lemma 5, we say that G is a clique, and picking any one edge in matching is sufficient. So we assume without loss of generality that throughout this section, G is not a clique, i.e., $L[v_1] \neq L^2[v_1]$ (see Lemma 5). Next, we consider two cases based on whether $L[v_1] = L[v_2]$ or not. If $L[v_1] = L[v_2]$, then in Theorem 19, we show that $L[v_1]L^2[v_1]$ is the first edge of some minimum maximal acyclic matching of G. Otherwise, if $L[v_1] \neq L[v_2]$, then in Theorem 20, we show that either $L[v_1]L^2[v_1]$ or $L[v_2]^-L[v_2]$ is the first edge of some minimum maximal acyclic matching of G. Moreover, Theorem 23 characterizes the conditions under which $L[v_1]L^2[v_1]$ or $L[v_2]^-L[v_2]$ is chosen (as the first edge in our desired minimum maximal acyclic matching). Once the first edge of some minimum maximal acyclic matching is chosen, Lemmas 21 and 22 help us in getting rid of those vertices that cannot be saturated by our desired matching later in the algorithm. In this way, we obtain a subgraph of the

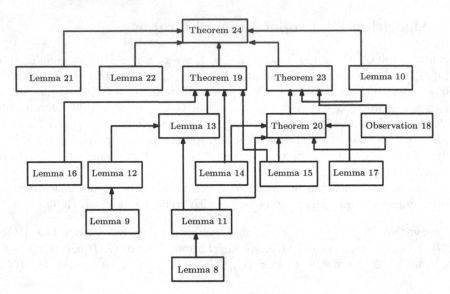

Fig. 1. Roadmap of proof of Theorem 24. The rest of the observations and lemmas are used throughout this section and has not been shown in the roadmap.

given proper interval graph, which, by the hereditary property (of proper interval graphs), is again a proper interval graph. We then again compute the first edge of some minimum maximal acyclic matching in this subgraph. Finally, Theorem 24 shows that combining the first edge and the matching edges obtained recursively by applying the same procedure to the subgraph obtained after carefully removing the vertices from G gives an optimal matching of G. A roadmap of proof of Theorem 24 is given in Fig. 1.

Now, let us describe everything discussed above in a formal manner.

The next two observations (Observations 2 and 3) follow from the definitions of a BCO and an acyclic matching.

Observation 2 [12]. *Let G be a proper interval graph with a BCO $\sigma(G) = (v_1, \ldots, v_n)$. Then, the following hold.*

(a) If $v_i v_j \in E(G)$, then $v_k v_j \in E(G)$ for all $k, i \leq k \leq j - 1$.
(b) If $v_i < v_j$ in $\sigma(G)$, then $L[v_i] \leq L[v_j]$ in $\sigma(G)$.

Observation 3 [12]. *Let G be a proper interval graph with a BCO $\sigma(G) = (v_1, \ldots, v_n)$. For each $i \in [n]$, let $S_i = \{v_i, v_{i+1}, \ldots, L[v_i]\}$. If M is an acyclic matching of G, then $|V_M \cap S_i| \leq 2$ for each $i \in [n]$.*

The following lemma characterizes the condition under which two edges in a proper interval graph form an acyclic matching.

Lemma 4. *Let G be a proper interval graph with a BCO $\sigma(G) = (v_1, \ldots, v_n)$ and let $v_a v_b, v_c v_d \in E(G)$ such that $v_a < v_b < v_c < v_d$ in $\sigma(G)$. Then, $M = \{v_a v_b, v_c v_d\}$ is an acyclic matching of G if and only if $v_a v_c, v_b v_d \notin E(G)$.*

Proof. Let $M = \{v_a v_b, v_c v_d\}$ be an acyclic matching of G. Targeting a contradiction, suppose that $v_a v_c \in E(G)$ (resp. $v_b v_d \in E(G)$). By Observation 2(a), $v_b v_c \in E(G)$. This implies that v_a, v_b, v_c, v_a (resp. v_b, v_c, v_d, v_b) forms a cycle in $G[V_M]$, a contradiction to the fact that M is an acyclic matching of G.

Conversely, since $v_a v_b, v_c v_d \in E(G)$ and $v_a v_c, v_b v_d \notin E(G)$, $G[V_M]$ is either a $P_4{}^1$ or a $2K_2$ depending on whether $v_b v_c \in E(G)$ or not, respectively. This is true because by Observation 2(a), $v_a v_d \notin E(G)$ as $v_b v_d \notin E(G)$. Thus M is an acyclic matching of G.   □

The next lemma characterizes proper interval graphs, which are also a clique.

Lemma 5. *Let G be a connected proper interval graph with a BCO $\sigma(G) = (v_1, \ldots, v_n)$. Then, G is a complete graph if and only if $L[v_1] = L^2[v_1]$.*

Proof. If G is a complete graph, then $v_1 v_n \in E(G)$. This implies that $L[v_1] = v_n$. Also, $L^2[v_1] = L[L[v_1]] = L[v_n] = v_n$.

Conversely, let $L[v_1] = v_k$. If $k = n$, then, by Observation 2(a), we are done. So assume that $k < n$. Since $L[v_1] = v_k$ and $L[v_1] = L^2[v_1]$, $v_k v_{k+1} \notin E(G)$. By Observation 2(a), $v_i v_j \notin E(G)$ for any $i \leq k$ and $j \geq k + 1$. It leads to a contradiction to our assumption that G is a connected graph. Thus $k = n$, which implies that G is a complete graph.   □

Consider the following lemma, which is crucial to proceed further.

Lemma 6. *Let G be a proper interval graph with a BCO $\sigma(G) = (v_1, \ldots, v_n)$. Let e_i, e_j, e_k be distinct edges in G such that $l(e_i) \leq l(e_j) \leq l(e_k)$ and $r(e_i) \leq r(e_j) \leq r(e_k)$ in $\sigma(G)$. If $\{e_i, e_k\}$ is not an acyclic matching of G, then neither $\{e_i, e_j\}$ nor $\{e_j, e_k\}$ is an acyclic matching of G.*

Proof. We will prove the contrapositive statement of Lemma 6. Let $\{e_i, e_j\}$ be an acyclic matching of G. By Lemma 4, $l(e_i)l(e_j)$, $r(e_i)r(e_j) \notin E(G)$. By Observation 2, note that $l(e_i)l(e_k)$, $r(e_i)r(e_k) \notin E(G)$. Since $G[\{l(e_i), l(e_k), r(e_i), r(e_k)\}]$ is an acyclic graph, $\{e_i, e_k\}$ is an acyclic matching of G. Note that the discussion for $\{e_j, e_k\}$ is similar.   □

The next lemma is a generalization of Lemma 6.

Lemma 7. *Let G be a proper interval graph with a BCO $\sigma(G) = (v_1, \ldots, v_n)$. Let $M_k = \{e_1, \ldots, e_k\}$ be a standard acyclic matching of G. Then, the following hold.*

(a) *If $r(e_k) < l(e) < r(e)$ for some $e \in E(G)$, then $\{e_k, e\}$ is an acyclic matching of G if and only if $M_k \cup \{e\}$ is an acyclic matching of G.*

(b) *If $l(\bar{e}) < r(\bar{e}) < l(e_1)$ for some $\bar{e} \in E(G)$, then $\{\bar{e}, e_1\}$ is an acyclic matching of G if and only if $M_k \cup \{\bar{e}\}$ is an acyclic matching of G.*

---

[1] Let K_n and P_n denote a *complete graph* and a *path graph* on n vertices, respectively.

Proof. *Proof of* (*a*) Since M_k is an acyclic matching, $\{e_i, e_j\}$ is an acyclic matching for every distinct $i, j \in [k]$. If $\{e_k, e\}$ is an acyclic matching of G, then by
Lemma 6, $\{e_i, e\}$ is an acyclic matching for each $i \in [k-1]$. This implies that
$M_k \cup \{e\}$ is an acyclic matching of G. Conversely, if $M_k \cup \{e\}$ is an acyclic
matching, then since $\{e_k, e\} \subseteq M_k \cup \{e\}$, $\{e_k, e\}$ is an acyclic matching of G.

Proof of (*b*) If $\{\bar{e}, e_1\}$ is an acyclic matching of G, then by Lemma 6, $\{\bar{e}, e_i\}$ is
an acyclic matching for each $i \in [k]$. This implies that $M_k \cup \{\bar{e}\}$ is an acyclic
matching of G. Next, if $M_k \cup \{\bar{e}\}$ is an acyclic matching of G, then since $\{\bar{e}, e_1\} \subseteq$
$M_k \cup \{\bar{e}\}$, $\{\bar{e}, e_1\}$ is an acyclic matching of G. □

The following two lemmas (Lemmas 8 and 9) are used to describe the situation where a particular edge of a given acyclic matching is replaced with some
specific edge in a proper interval graph.

Lemma 8 (*). *Let G be a connected proper interval graph with a BCO $\sigma(G) =$
$(v_1, \ldots, v_n)$. Let $M_k = \{e_1, \ldots, e_k\}$ be a standard acyclic matching of G. For
some $j \in [k-1]$ and $v \in V(G)$ such that $l(e_j)v \in E(G)$ and $r(e_j) < v$ in $\sigma(G)$,
let $M = (M_k \backslash \{e_j\}) \cup \{l(e_j)v\}$. Then, M is acyclic if and only if $vr(e_{j+1}) \notin E(G)$.
Moreover, if M is not an acyclic matching, then $G[V_M]$ has only one cycle of
the form $v, l(e_{j+1}), r(e_{j+1}), v$.*

Lemma 9 (*). *Let G be a connected proper interval graph with a BCO $\sigma(G) =$
$(v_1, \ldots, v_n)$. Let $M_k = \{e_1, \ldots, e_k\}$ be a standard acyclic matching of G. For
some $u \in V(G)$ such that $l(e_1) < u < r(e_1)$ in $\sigma(G)$, let $M = (M_k \backslash \{e_1\}) \cup$
$\{ur(e_1)\}$. Then, M is acyclic if and only if $ul(e_2) \notin E(G)$. Moreover, if M is not
an acyclic matching, then $G[V_M]$ has only one cycle of the form $u, r(e_1), l(e_2), u$.*

Lemma 10 (*). *Let G be a connected proper interval graph with a BCO $\sigma(G) =$
$(v_1, \ldots, v_n)$. If $G_i = G[\{v_i, v_{i+1}, \ldots, v_n\}]$ and $G_{i'} = G[\{v_{i'}, v_{i'+1}, \ldots, v_n\}]$, where
$i < i'$, then $\mu'_{ac}(G_{i'}) \leq \mu'_{ac}(G_i)$.*

The following three lemmas (Lemmas 11–13) describe the conditions under
which we can replace the first edge of a given minimum maximal acyclic matching
of a proper interval graph with our desired edge.

Lemma 11 (*). *Let G be a connected proper interval graph with a BCO $\sigma(G) =$
$(v_1, \ldots, v_n)$ and let $ab, ab' \in E(G)$ such that $a < b < b'$ in $\sigma(G)$. Let ab be the
first edge in some minimum maximal acyclic matching M of G and ab' be the
first edge in some maximal acyclic matching of G. Then, there exists a minimum
maximal acyclic matching M_{ac} of G such that ab' is the first edge in M_{ac} with
respect to $\sigma(G)$.*

Lemma 12 (*). *Let G be a connected proper interval graph with a BCO
$\sigma(G) = (v_1, \ldots, v_n)$ and let $ab, a'b \in E(G)$ such that $a < a' < b$ in $\sigma(G)$.
Let ab be the first edge in some minimum maximal acyclic matching M of G
and $a'b$ be the first edge in some maximal acyclic matching of G. Then, there
exists a minimum maximal acyclic matching M_{ac} of G such that $a'b$ is the first
edge in M_{ac} with respect to $\sigma(G)$.*

Lemma 13 (∗). *Let G be a connected proper interval graph with a BCO $\sigma(G) = (v_1, \ldots, v_n)$ and let $ab, a'b' \in E(G)$ such that $a < a'$ and $b < b'$ in $\sigma(G)$. Let ab be the first edge in some minimum maximal acyclic matching M of G and $a'b'$ be the first edge in some maximal acyclic matching of G. Then, there exists a minimum maximal acyclic matching M_{ac} of G such that $a'b'$ is the first edge in M_{ac} with respect to $\sigma(G)$.*

Note that in order to apply Lemmas 11–13, we need to show that our desired edge (i.e., the edge that we want to show is the first edge of some minimum maximal acyclic matching) is also the first edge of some maximal (not necessarily minimum) acyclic matching of G. Therefore, we need the following lemma.

Lemma 14 *Let G be a connected proper interval graph with a BCO $\sigma(G) = (v_1, \ldots, v_n)$. Then, there exists a maximal acyclic matching M of G such that the following hold.*

(a) $L[v_1]L^2[v_1]$ is the first edge in M with respect to $\sigma(G)$.
(b) $L[v_2]^-L[v_2]$ is the first edge in M with respect to $\sigma(G)$.

Proof. To prove (a), we only need to show that there does not exist any edge, say $e \in E(G)$, such that i) $\{e, L[v_1]L^2[v_1]\}$ is an acyclic matching of G, and ii) $l(e) < r(e) < L[v_1] < L^2[v_1]$ in $\sigma(G)$. Let $M_1 = \{v_1v_2, L[v_1]L^2[v_1]\}$. By Lemma 6, it is enough to show that M_1 is not an acyclic matching of G. Note that if M_1 is an acyclic matching of G, then it contradicts Observation 3.   □

To prove (b), we only need to show that there does not exist any edge, say $e \in E(G)$, such that i) $\{e, L[v_2]^-L[v_2]\}$ is an acyclic matching of G, and ii) $l(e) < r(e) < L[v_2]^- < L[v_2]$ in $\sigma(G)$. If we show that $M_1 = \{v_1v_2, L[v_2]^-L[v_2]\}$ is not an acyclic matching of G, then by Lemma 6, we are done. Note that if M_1 is an acyclic matching of G, then it contradicts Observation 3.   □

Now, consider the following lemma.

Lemma 15. *Let G be a connected proper interval graph with a BCO $\sigma(G) = (v_1, \ldots, v_n)$. Furthermore, let M be a maximal acyclic matching of G. If e is the first edge in M with respect to $\sigma(G)$, then $l(e), r(e) \in N(L[v_1])$.*

Proof. Without loss of generality, let $l(e) = a$, $r(e) = b$ and $a < b$ in $\sigma(G)$. If $a, b \in N(L[v_1])$, then we are done. So, assume that either $a \notin N(L[v_1])$ or $b \notin N(L[v_1])$. First, suppose that $a \notin N(L[v_1])$. This implies that $L^2[v_1] < a$ in $\sigma(G)$. Since $a < b$ in $\sigma(G)$, $L^2[v_1] < b$ in $\sigma(G)$. This implies that $b \notin N(L[v_1])$. Now, define $M' = \{v_1L[v_1]\} \cup M$. Since $v_1a, L[v_1]b \notin E(G)$, by Lemma 4, $\{v_1L[v_1], ab\}$ is an acyclic matching of G. By Lemma 7, M' is an acyclic matching of G. This leads to a contradiction to the fact that M is maximal in G. Hence, $a \in N(L[v_1])$.

Next, suppose that $a \in N(L[v_1])$ and $b \notin N(L[v_1])$. If $a \in N(v_1)$, then $v_1 \leq a \leq L[v_1] < L^2[v_1] < b$ in $\sigma(G)$. By Observation 2(b), there is no edge between a and b, which is absurd. Hence, $a \notin N(v_1)$, and thus $L[v_1] < a \leq L^2[v_1] < b$ in $\sigma(G)$. Next, define $M' = M \cup \{v_1L[v_1]\}$. Since $v_1a, L[v_1]b \notin E(G)$, by Lemma 4, $\{v_1L[v_1], ab\}$ is an acyclic matching of G. By Lemma 7, M' is an acyclic matching of G. It leads to a contradiction to the fact that M is maximal in G. Hence, $b \in N(L[v_1])$. □

Next, we discuss some results depending on whether $L[v_1] = L[v_2]$ or not.

Lemma 16. *Let G be a connected proper interval graph with a BCO $\sigma(G) = (v_1, \ldots, v_n)$ such that $L[v_1] = L[v_2]$. Furthermore, let M be a maximal acyclic matching of G. If e is the first edge in M with respect to $\sigma(G)$, then $l(e) \in N(v_1)$.*

Proof. Without loss of generality, let $l(e) = a$, $r(e) = b$ and $a < b$ in $\sigma(G)$. If $a \in N(v_1)$, then we are done. So assume that $a \notin N(v_1)$. Define $M' = M \cup \{v_1 v_2\}$. Since $v_1 < v_2 < L[v_1] = L[v_2] < a < b$ in $\sigma(G)$, $v_1 a, v_2 b \notin E(G)$. Thus, by Lemma 4, $\{v_1 v_2, ab\}$ is an acyclic matching. By Lemma 7, M' is an acyclic matching of G, a contradiction to the fact that M is maximal in G. Hence, $a \in N(v_1)$. □

Lemma 17. *Let G be a proper interval graph with a BCO $\sigma(G) = (v_1, \ldots, v_n)$ such that $L[v_1] \neq L[v_2]$, and let M be a maximal acyclic matching of G. If e is the first edge in M with respect to $\sigma(G)$, then either $l(e) \in N(v_1)$ or $r(e) \in N(v_2)$.*

Proof. Without loss of generality, let $l(e) = a$, $r(e) = b$ and $a < b$ in $\sigma(G)$. By Observation 2(b) and the assumption that $L[v_1] \neq L[v_2]$, it is clear that $L[v_1] < L[v_2]$ in $\sigma(G)$. If either $a \in N(v_1)$ or $b \in N(v_2)$, then we are done. So assume that $a \notin N(v_1)$ and $b \notin N(v_2)$. This implies that $L[v_1] < a$ and $L[v_2] < b$ in $\sigma(G)$. Define $M' = M \cup \{v_1 v_2\}$. Since $v_1 a, v_2 b \notin E(G)$, by Lemma 4, $\{v_1 v_2, ab\}$ is an acyclic matching. By Lemma 7, M' is an acyclic matching, a contradiction to the fact that M is maximal in G. Hence, either $a \in N(v_1)$ or $b \in N(v_2)$. □

Observation 18. *Let G be a connected proper interval graph with a BCO $\sigma(G) = (v_1, v_2, \ldots, v_n)$ such that $L[v_1] \neq L[v_2]$. Then, $v_1 < v_2 \leq L[v_1] \leq L[v_2]^- < L[v_2] \leq L^2[v_1]$ in $\sigma(G)$.*

Proof. By Observation 2(b), $L[v_1] \leq L[v_2]$ in $\sigma(G)$. Since $L[v_1] \neq L[v_2]$, $L[v_1] < L[v_2]$ in $\sigma(G)$. Since $v_2 \leq L[v_1]$, $L[v_2] \leq L^2[v_1]$ (by Observation 2(b)). Further, note that $L[v_2]^- \neq L^2[v_1]$. Else, if $L[v_2]^- = L^2[v_1]$, then $L^2[v_1] < L[v_2]$ in $\sigma(G)$, a contradiction. □

Now, we are ready to prove one of the main theorems in this section.

Theorem 19. *Let G be a connected proper interval graph with a BCO $\sigma(G) = (v_1, \ldots, v_n)$ such that $L[v_1] = L[v_2]$. Then, there exists a minimum maximal acyclic matching M_{ac} of G such that $L[v_1]L^2[v_1]$ is the first edge in M_{ac} with respect to $\sigma(G)$.*

Proof. Let M be a minimum maximal acyclic matching of G, and let ab be the first edge in M with respect to $\sigma(G)$. Without loss of generality, let $a < b$ in $\sigma(G)$. If $ab = L[v_1]L^2[v_1]$, then we are done. So let us assume that $ab \neq L[v_1]L^2[v_1]$. By Lemma 16, $a \in N(v_1)$, so $a \leq L[v_1]$ in $\sigma(G)$. By Lemma 15, $b \in N(L[v_1])$, so $b \leq L^2[v_1]$ in $\sigma(G)$. Also, by Lemma 14, there exists a maximal acyclic matching of G containing $L[v_1]L^2[v_1]$ as the first edge with respect to $\sigma(G)$. Therefore, by Lemmas 11 to 13, there exists a minimum maximal acyclic matching M_{ac} of G such that $L[v_1]L^2[v_1]$ is the first edge in M_{ac} with respect to $\sigma(G)$.   □

Now, consider the following theorem.

Theorem 20. *Let G be a connected proper interval graph with a BCO $\sigma(G) = (v_1, \ldots, v_n)$ such that $L[v_1] \neq L[v_2]$. Then, there exists a minimum maximal acyclic matching M_{ac} of G such that either $L[v_1]L^2[v_1]$ or $L[v_2]^-L[v_2]$ is the first edge in M_{ac} with respect to $\sigma(G)$.*

Proof. Let M be a minimum maximal acyclic matching of G, and let ab be the first edge in M with respect to $\sigma(G)$. Without loss of generality, let $a < b$ in $\sigma(G)$. By Lemma 17, it is clear that either $a \in N(v_1)$ or $b \in N(v_2)$. Now, consider the following cases:

Case 1: $a \in N(v_1)$ and $b \notin N(v_2)$. Since $a \in N(v_1)$ and $b \notin N(v_2)$, $a \leq L[v_1]$ and $L[v_2] < b$ in $\sigma(G)$. By Lemma 15, $b \leq L^2[v_1]$ in $\sigma(G)$. Since $L[v_1] \neq L[v_2]$, by Observation 2(b), $L[v_1] < L[v_2]$ in $\sigma(G)$. Therefore, $a \leq L[v_1] \leq L[v_2]^- < L[v_2] < b \leq L^2[v_1]$ in $\sigma(G)$. By Lemmas 11–14, there exists a minimum maximal acyclic matching M_{ac} of G such that $L[v_1]L^2[v_1]$ is the first edge in M_{ac} with respect to $\sigma(G)$.

Case 2: $a \notin N(v_1)$ and $b \in N(v_2)$. Since $a \notin N(v_1)$ and $b \in N(v_2)$, $L[v_1] < a$ and $b \leq L[v_2]$ in $\sigma(G)$. Note that in this case $L[v_1] \neq L[v_2]^-$, else it will imply that $L[v_2]^- < a < b \leq L[v_2]$ in $\sigma(G)$. This is not possible as $L[v_2]^-$ and $L[v_2]$ are consecutive in $\sigma(G)$. Therefore, $L[v_1] < a \leq L[v_2]^- < L[v_2] \leq L^2[v_1]$ in $\sigma(G)$. By Lemmas 11–14, there exists a minimum maximal acyclic matching M_{ac} of G such that $L[v_2]^-L[v_2]$ is the first edge in M_{ac} with respect to $\sigma(G)$.

Case 3: $a \in N(v_1)$ and $b \in N(v_2)$. Since $a \in N(v_1)$ and $b \in N(v_2)$, $a \leq L[v_1]$ and $b \leq L[v_2]$ in $\sigma(G)$. Further, by Observation 18, $a \leq L[v_1] \leq L[v_2]^- < L[v_2] \leq L^2[v_1]$ in $\sigma(G)$. Now, by Lemmas 11–14, there exist minimum maximal acyclic matchings M_1 and M_2 of G such that $L[v_1]L^2[v_1]$ is the first edge in M_1 and $L[v_2]^-L[v_2]$ is the first edge in M_2 with respect to $\sigma(G)$.   □

Next, Lemma 21 (resp. Lemma 22) helps us in characterizing those vertices that can be "safely" removed from the input graph under the assumption that $L[v_1]L^2[v_1]$ (resp. $L[v_2]^-L[v_2]$) is the first edge with respect to a BCO in some minimum maximal acyclic matching M_{ac} of G.

Lemma 21. *Let G be a connected proper interval graph with a BCO $\sigma(G) = (v_1, \ldots, v_n)$. Let M_{ac} be a minimum maximal acyclic matching of G such that $L[v_1]L^2[v_1]$ is the first edge in M_{ac} with respect to $\sigma(G)$. If $S_1 = \{v_1, v_2, \ldots, L^2[v_1]\}$ and $S_2 = \{v_j \mid v_j > L^2[v_1]$ and $L[v_j] = L^3[v_1]\}$, then none of the vertices from the set $(S_1 \cup S_2) \setminus \{L[v_1], L^2[v_1]\}$ is saturated by M_{ac}.*

Proof. By Observation 3, it is clear that to maintain the acyclic property of M_{ac}, none of the vertices from the set $S_1 = \{v_1, v_2, \ldots, L^2[v_1]\} \setminus \{L[v_1], L^2[v_1]\}$

is saturated by M_{ac}. Next, let us assume that $v_j \in S_2$ is saturated by M_{ac}. Let $v_{j'}$ be the M_{ac}-mate of v_j. Since $L[v_j] = L^3[v_1]$, $v_{j'} \leq L^3[v_1]$. It leads to a contradiction to Observation 3. Thus, none of the vertices from the set $(S_1 \cup S_2) \setminus \{L[v_1], L^2[v_1]\}$ is saturated by M_{ac}.   □

Algorithm 1. MMAM-PIG(G)

Input: A proper interval graph G with BCO $\sigma(G) = (v_1, \ldots, v_n)$;
Output: A Min-Max-Acy-Matching M;
$M \leftarrow \emptyset$, $i \leftarrow 1$, $k \geq 2$;
$G^1 \leftarrow G$;
$v^+ = $ vertex next to vertex v in $\sigma(G^i)$, $i \geq 1$;
$\alpha(G^i) = $ first vertex in the BCO $\sigma(G^i)$, $i \geq 1$;
$\beta(G^i) = $ second vertex in the BCO $\sigma(G^i)$, $i \geq 1$;
$L[v] = $ maximum indexed neighbor of vertex v in $\sigma(G^i)$, $i \geq 1$;
$L^k[v] = $ maximum indexed neighbor of vertex $L^{k-1}[v]$ in $\sigma(G^i)$, $i \geq 1$;

if $(|V(G^i)| \leq 1)$ then
    return M;
else
    if $(L[\alpha(G^i)] = L^2[\alpha(G^i)])$ then
        $M \leftarrow M \cup \{\alpha(G^i)L[\alpha(G^i)]\}$;
        $i \leftarrow i + 1$;
        $G^i \leftarrow G^{i-1} - \{\alpha(G^{i-1}), \ldots, L^2[\alpha(G^{i-1})]\}$;
    else
        if $(L[\alpha(G^i)] \neq L[\beta(G^i)]$ and $L^3[\alpha(G^i)] \neq L^2[L[\beta(G^i)]^-]))$ then
            $M \leftarrow M \cup \{L[\beta(G^i)]^- L[\beta(G^i)]\}$;
            $i \leftarrow i + 1$;
            $G^i \leftarrow G^{i-1} - \{\alpha(G^{i-1}), \ldots, L[L[\beta(G^i)]^-]\}$;
            while $(L^2[\beta(G^{i-1})] = L[\alpha(G^i)])$ do
                $x \leftarrow \alpha(G^i)$;
                $\alpha(G^i) \leftarrow \alpha^+(G^i)$;
                $G^i \leftarrow G^i - \{x\}$;
        else
            $M \leftarrow M \cup \{L[\alpha(G^i)]L^2[\alpha(G^i)]\}$;
            $i \leftarrow i + 1$;
            $G^i \leftarrow G^{i-1} - \{\alpha(G^{i-1}), \ldots, L^2[\alpha(G^{i-1})]\}$;
            while $(L^3[\alpha(G^{i-1})] = L[\alpha(G^i)])$ do
                $x \leftarrow \alpha(G^i)$;
                $\alpha(G^i) \leftarrow \alpha^+(G^i)$;
                $G^i \leftarrow G^i - \{x\}$;

Lemma 22. *Let G be a connected proper interval graph with a BCO $\sigma(G) = (v_1, \ldots, v_n)$. Let M_{ac} be a minimum maximal acyclic matching of G such that*

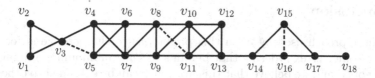

Fig. 2. A proper interval graph G. The dashed edges represent a minimum maximum acyclic matching obtained by algorithm MMAM-PIG(G).

$L[v_2]^- L[v_2]$ *is the first edge in* M_{ac} *with respect to* $\sigma(G)$. *If* $S_1 = \{v_1, v_2, \ldots, L[L[v_2]^-]\}$ *and* $S_2 = \{v_j \mid v_j > L[L[v_2]^-]$ *and* $L[v_j] = L^2[v_2]\}$, *then none of the vertices from the set* $(S_1 \cup S_2) \setminus \{L[v_2]^-, L[v_2]\}$ *is saturated by* M_{ac}.

Proof. By Observation 3, it is clear that to maintain the acyclic property of M_{ac}, none of the vertices from the set $S_1 = \{v_1, v_2, \ldots, L[L[v_2]^-]\}$ is saturated by M_{ac}. Next, let us assume that $v_j \in S_2$ is saturated by M_{ac}. Let $v_{j'}$ be the M_{ac}-mate of v_j. Since $L[v_j] = L^2[v_2]$, $v_{j'} \leq L^2[v_2]$. It leads to a contradiction to Observation 3. Thus, none of the vertices from the set $(S_1 \cup S_2) \setminus \{L[v_2]^-, L[v_2]\}$ is saturated by M_{ac}. □

Now, we are ready to prove the following theorem.

Theorem 23 (∗). *Let* G *be a connected proper interval graph with a BCO* $\sigma(G) = (v_1, \ldots, v_n)$ *such that* $L[v_1] \neq L[v_2]$. *Then, there exists a minimum maximal acyclic matching* M *of* G *such that the following hold.*

(a) If $L^3[v_1] = L^2[L[v_2]^-]$, *then* $L[v_1]L^2[v_1] \in M$.
(b) If $L^3[v_1] \neq L^2[L[v_2]^-]$, *then* $L[v_2]^- L[v_2] \in M$.

Based on the lemmas and theorems discussed above, we present a linear-time algorithm (MMAM-PIG(G)) to compute a maximal acyclic matching of minimum cardinality in a proper interval graph G in Algorithm 1.

Theorem 24 (∗). *Given a proper interval graph* G *with a BCO* $\sigma(G)$, *Algorithm 1 correctly computes a minimum maximal acyclic matching of* G.

We illustrate the execution of algorithm MMAM-PIG(G) on the proper interval graph G shown in Fig. 2 in Table 1.

Table 1. Illustration of algorithm MMAM-PIG(G) on graph G shown in Fig 2.

i	$\lvert G^i \rvert \leq 1$	$\sigma(G^i)$	$L[\alpha(G^i)]$ $=L^2[\alpha(G^i)]$	$L[\alpha(G^i)] =$ $L[\beta(G^i)]$	$L^3[\alpha(G^i)] =$ $L^2[L[\beta(G^i)]^-]$	Update of M	Remove vertices
1	No	$(v_1, \ldots, v_{18})$	No	Yes	-	Add $v_3 v_5$	$\{v_1, \ldots, v_5\}$
2	No	$(v_6, \ldots, v_{18})$	No	No	Yes	Add $v_8 v_{11}$	$\{v_6, \ldots, v_{12}\}$
3	No	$(v_{13}, \ldots, v_{18})$	No	No	No	Add $v_{15} v_{16}$	$\{v_{13}, \ldots, v_{18}\}$
4	Yes	-	-	-	-	-	-

388 J. Chaudhary et al.

3 Conclusion

The main approach used by us in this paper was to characterize an edge with respect to some vertex ordering that belongs to some minimum maximal acyclic matching of G, and we believe that this approach can be extended to other graph classes as well. Also, since many recent papers talk about the parameterized complexity of restricted matchings, it is a promising direction for future research.

References

1. Bazgan, C., et al.: The many facets of upper domination. Theor. Comput. Sci. **717**, 2–25 (2018)
2. Boria, N., Croce, F.D., Paschos, V.T.: On the max min vertex cover problem. Discret. Appl. Math. **196**, 62–71 (2015)
3. Chaudhary, J., Mishra, S., Panda, B.S.: On the complexity of minimum maximal acyclic matching. In: Proceedings of the 28th International Computing and Combinatorics Conference (COCOON) (to appear)
4. Chaudhary, J., Panda, B.S.: On the complexity of minimum maximal uniquely restricted matching. Theor. Comput. Sci. **882**, 15–28 (2021)
5. Damaschke, P.: Parameterized algorithms for double hypergraph dualization with rank limitation and maximum minimal vertex cover. Discret. Optim. **8**(1), 18–24 (2011)
6. Dublois, L., Hanaka, T., Ghadikolaei, M.K., Lampis, M., Melissinos, N.: (In)approximability of maximum minimal FVS. In: Proceedings of the 31st International Symposium on Algorithms and Computation (ISAAC), Leibniz International Proceedings in Informatics, vol. 181, pp. 3:1–14 (2020)
7. Goddard, W., Hedetniemi, S.M., Hedetniemi, S.T., Laskar, R.: Generalized subgraph-restricted matchings in graphs. Discret. Math. **293**(1–3), 129–138 (2005)
8. Jamison, R.E., Laskar, R.: Elimination orderings of chordal graphs. In: Proceedings of the Seminar on Combinatorics and Applications, pp. 192–200 (1982)
9. Lepin, V.V.: A linear algorithm for computing of a minimum weight maximal induced matching in an edge-weighted tree. Electron. Notes Discret. Math. **24**, 111–116 (2006)
10. Orlovich, Y.L., Finke, G., Gordon, V., Zverovich, I.: Approximability results for the maximum and minimum maximal induced matching problems. Discret. Optim. **5**(3), 584–593 (2008)
11. Orlovich, Y. L., Zverovich, I. E.: Maximal induced matchings of minimum/ maximum size. Technical report, DIMACS TR 2004–26 (2004)
12. Panda, B.S., Chaudhary, J.: Acyclic matching in some subclasses of graphs. In: Gasieniec, L., Klasing, R., Radzik, T. (eds.) IWOCA 2020. LNCS, vol. 12126, pp. 409–421. Springer, Cham (2020). https://doi.org/10.1007/978-3-030-48966-3_31
13. Panda, B.S., Das, S.K.: A linear time recognition algorithm for proper interval graphs. Inf. Process. Lett. **87**(3), 153–161 (2003)
14. Panda, B.S., Pandey, A.: On the complexity of minimum cardinality maximal uniquely restricted matching in graphs. In: Arumugam, S., Bagga, J., Beineke, L.W., Panda, B.S. (eds.) ICTCSDM 2016. LNCS, vol. 10398, pp. 218–227. Springer, Cham (2017). https://doi.org/10.1007/978-3-319-64419-6_29
15. Yannakakis, M., Gavril, F.: Edge dominating sets in graphs. SIAM J. Appl. Math. **38**(3), 364–372 (1980)

Graph Partition and Graph Covering

Transitivity on Subclasses of Chordal Graphs

Subhabrata Paul$^{(\boxtimes)}$ and Kamal Santra

Department of Mathematics, IIT Patna, Patna, India
{subhabrata,kamal_1821ma04}@iitp.ac.in

Abstract. Let $G = (V, E)$ be a graph, where V and E are the vertex and edge sets, respectively. For two disjoint subsets A and B of V, we say A *dominates* B if every vertex of B is adjacent to at least one vertex of A in G. A vertex partition $\pi = \{V_1, V_2, \ldots, V_k\}$ of G is called a *transitive k-partition* if V_i dominates V_j for all i, j, where $1 \leq i < j \leq k$. The maximum integer k for which the above partition exists is called *transitivity* of G and it is denoted by $Tr(G)$. The MAXIMUM TRANSITIVITY PROBLEM is to find a transitive partition of a given graph with the maximum number of partitions. It was known that the decision version of MAXIMUM TRANSITIVITY PROBLEM is NP-complete for chordal graphs [Iterated colorings of graphs, *Discrete Mathematics*, 278, 2004]. In this paper, we first prove that this problem can be solved in linear time for *split graphs* and for the *complement of bipartite chain graphs*, two subclasses of chordal graphs. We also discuss Nordhaus-Gaddum type relations for transitivity and provide counterexamples for an open problem posed by J. T. Hedetniemi and S. T. Hedetniemi [The transitivity of a graph, *J. Combin. Math. Combin. Comput*, 104, 2018]. Finally, we characterize transitively critical graphs having fixed transitivity.

Keywords: Transitivity · Split graphs · Complement of bipartite chain graphs · Nordhaus-Gaddum relations · Transitively critical graphs

1 Introduction

Graph partitioning is one of the classical problems in graph theory. In a partitioning problem, the goal is to partition the vertex set (or edge set) into some parts with desired properties, such as independence, having minimum edges across partite sets, etc. In this article, we are interested in partitioning the vertex set into some parts such that the partite sets follow some domination relation among themselves. For a graph $G = (V, E)$, the *neighbourhood* of a vertex $v \in V$ is the set of all adjacent vertices of v and is denoted as $N_G(v)$. The *degree* of a vertex v in G, denoted as $\deg_G(v)$, is the number of edges incident to v. A vertex v is said to *dominate* itself and all its neighbouring vertices. A *dominating set* of $G = (V, E)$ is a subset of vertices D such that every vertex $x \in V \setminus D$ has a neighbour $y \in D$, that is, x is dominated by some vertex y of D. For two disjoint subsets A and B of V, we say A *dominates* B if every vertex of B is adjacent to at least one vertex of A.

A. Bagchi and R. Muthu (Eds.): CALDAM 2023, LNCS 13947, pp. 391–402, 2023.
https://doi.org/10.1007/978-3-031-25211-2_30

Graph partitioning problems, based on a domination relation among the partite sets, have been extensively studied in literature. Cockayne and Hedetniemi, in 1977, introduced the notion of *domatic partition* of a graph $G = (V, E)$, where the vertex set is partitioned into k parts, $\pi = \{V_1, V_2, \ldots, V_k\}$, such that each V_i is a dominating set of G [3]. The maximum order of such a domatic partition is called *domatic number* of G and it is denoted by $d(G)$. Another similar type of partitioning problem is *the Grundy partition*. Christen and Selkow introduced a Grundy partition of a graph $G = (V, E)$ in 1979 [2]. In the Grundy partitioning problem, the vertex set is partitioned into k parts, $\pi = \{V_1, V_2, \ldots, V_k\}$, such that each V_i is an independent set and for all $1 \leq i < j \leq k$, V_i dominates V_j. The maximum order of such a partition is called *the Grundy number* of G and it is denoted by $\Gamma(G)$. In 2004, Hedetniemi et al. introduced another such partitioning problem, namely *upper iterated domination partition* [4]. In an upper iterated domination partition, the vertex set is partitioned into k parts, $\pi = \{V_1, V_2, \ldots, V_k\}$, such that for each $1 \leq i \leq k$, V_i is a minimal dominating set of $G \setminus (\cup_{j=1}^{i-1} V_j)$. The *upper iterated domination number*, denoted by $\Gamma^*(G)$, is equal to the maximum order of such a vertex partition. Recently, in 2018, Haynes et al. generalized the idea of domatic partition and introduced the concept of *upper domatic partition* of a graph G, where the vertex set is partitioned into k parts, $\pi = \{V_1, V_2, \ldots, V_k\}$, such that for each i, j, with $1 \leq i < j \leq k$, either V_i dominates V_j or V_j dominates V_i or both [7]. The maximum order of such an upper domatic partition is called *upper domatic number* of G and it is denoted by $D(G)$. All these problems, domatic number [1], Grundy number [9,15], upper iterated number [4], upper domatic number [7] have been extensively studied both from an algorithmic and structural point of view.

In this article, we study a similar graph partitioning problem, namely *transitive partition*. In 2018, Hedetniemi et al. [8] have introduced this notion as a generalization of Grundy partition. A *transitive k-partition* is defined as a partition of the vertex set into k parts, $\pi = \{V_1, V_2, \ldots, V_k\}$, such that for all $1 \leq i < j \leq k$, V_i dominates V_j. The maximum order of such a transitive partition is called *transitivity* of G and is denoted by $Tr(G)$. The MAXIMUM TRANSITIVITY PROBLEM (MTP) is to find a transitive partition of a given graph with the maximum number of parts. Note that a Grundy partition is a transitive partition with the additional restriction that each partite set must be independent. In a domatic partition $\pi = \{V_1, V_2, \ldots, V_k\}$ of G, since each partite set is a dominating set of G, we have domination property in both directions, that is, V_i dominates V_j and V_j dominates V_i for all $1 \leq i < j \leq k$. However, in a transitive partition $\pi = \{V_1, V_2, \ldots, V_k\}$ of G, we have domination property in one direction, that is, V_i dominates V_j for $1 \leq i < j \leq k$. In an upper domatic partition $\pi = \{V_1, V_2, \ldots, V_k\}$ of G, for all $1 \leq i < j \leq k$, either V_i dominates V_j or V_j dominates V_i or both. The definition of each vertex partitioning problem ensures the following inequalities for any graph G. For any graph G, $1 \leq \Gamma(G) \leq \Gamma^*(G) \leq Tr(G) \leq D(G) \leq n$.

In the introductory paper, J. T. Hedetniemi and S. T. Hedetniemi [8] showed, that the upper bound on the transitivity of a graph G is $\Delta(G) + 1$, where $\Delta(G)$

is the maximum degree of G. They also gave two characterizations for graphs with $Tr(G) = 2$ and for graphs with $Tr(G) \geq 3$. They further showed that transitivity and Grundy number are the same for trees. Therefore, the linear-time algorithm for finding the Grundy number of a tree, presented in [9], implies that we can find the transitivity of a tree in linear time as well. Also, for a subclass of bipartite graphs, namely bipartite chain graphs, MTP can be solved in linear time [13]. Moreover, for any graph, transitivity is equal to upper iterated domination number, that is, $\Gamma^*(G) = Tr(G)$ [8], and the decision version of the upper iterated domination problem is known to be NP-complete for chordal graphs [10]. Therefore, MTDP is NP-complete for chordal graphs as well. MTDP is also known to be NP-complete for perfect elimination bipartite graphs [13]. It is also known that every connected graph G with $Tr(G) = k \geq 3$ has a transitive partition $\pi = \{V_1, V_2, \ldots, V_k\}$ such that $|V_k| = |V_{k-1}| = 1$ and $|V_{k-i}| \leq 2^{i-1}$ for $2 \leq i \leq k-2$ [6]. This implies that MTP is fixed-parameter tractable [6]. Also, graphs with transitivity at least t, for some integer t, have been characterized in [13].

In this article, we study the computational complexity of the transitivity problem in subclasses of chordal graphs. The organization and main contributions of this article are summarized as follows. Section 2 contains basic definitions and notations that are followed throughout the article. Sections 3 and 4 describe two linear-time algorithms for split and for the complement of bipartite chain graphs, respectively. Section 5 deals with Nordhaus-Gaddum type relations for transitivity. In Sect. 6, we present a characterization of transitively vertex-edge critical graphs having fixed transitivity. Finally, Sect. 7 concludes the article.

2 Notation and Definition

Let $G = (V, E)$ be a graph with V and E as its vertex and edge sets, respectively. A graph $H = (V', E')$ is said to be a *subgraph* of a graph $G = (V, E)$, if and only if $V' \subseteq V$ and $E' \subseteq E$. For a subset $S \subseteq V$, the *induced subgraph* on S of G is defined as the subgraph of G whose vertex set is S and edge set consists of all of the edges in E that have both endpoints in S and it is denoted by $G[S]$. The *complement* of a graph $G = (V, E)$ is the graph $\overline{G} = (\overline{V}, \overline{E})$, such that $\overline{V} = V$ and $\overline{E} = \{uv | uv \notin E\}$.

A subset of $S \subseteq V$, is said to be an *independent set* of G, if every pair of vertices in S are non-adjacent. A subset of $K \subseteq V$, is said to be a *clique* of G, if every pair of vertices in K are adjacent. The cardinality of a clique of maximum size is called *clique number* of G and it is denoted by $\omega(G)$. A graph $G = (V, E)$ is said to be a *split graph* if V can be partitioned into an independent set S and a clique K.

A graph is called *bipartite* if its vertex set can be partitioned into two independent sets. A bipartite graph $G = (X \cup Y, E)$ is called a *bipartite chain graph* if there exists an ordering of vertices of X and Y, say $\sigma_X = (x_1, x_2, \ldots, x_{n_1})$ and $\sigma_Y = (y_1, y_2, \ldots, y_{n_2})$, such that $N(x_{n_1}) \subseteq N(x_{n_1-1}) \subseteq \ldots \subseteq N(x_2) \subseteq N(x_1)$ and $N(y_{n_2}) \subseteq N(y_{n_2-1}) \subseteq \ldots \subseteq N(y_2) \subseteq N(y_1)$. Such ordering of X and Y is called a *chain ordering* and it can be computed in linear time [11]. A graph G is said to be a $2K_2$-free, if it does not contain a pair of independent edges as an induced subgraph. It is well-known that the class of bipartite chain graphs and $2K_2$-free bipartite graphs are the same. An edge between two non-consecutive vertices of a cycle is called a *chord*. If every cycle in G of length at least four has a chord, then G is called a *chordal graph*.

3   Transitivity in Split Graphs

In this section, we design a linear-time algorithm for finding the transitivity of a given split graph. To design the algorithm, we first prove that the transitivity of a split graph G can be either $\omega(G)$ or $\omega(G) + 1$, where $\omega(G)$ is the size of a maximum clique in G. Further, we characterize the split graphs with the transitivity equal to $\omega(G) + 1$.

Lemma 1. *Let $G = (S \cup K, E)$ be a split graph, where S and K are an independent set and a clique of G, respectively. Also, assume that K is the maximum clique of G, that is, $\omega(G) = |K|$. Then $\omega(G) \leq Tr(G) \leq \omega(G) + 1$. Further, $Tr(G) = \omega(G) + 1$ if and only if every vertex of K has a neighbour in S.*

Proof. Note that $Tr(G) \geq \omega(G)$. As we can make a transitive partition $\pi = \{V_1, V_2, \ldots, V_{\omega(G)}\}$ of size $\omega(G)$ by considering each V_i contains exactly one vertex from maximum clique and all the other vertices in V_1. To prove that $Tr(G) \leq \omega(G) + 1$, suppose $Tr(G) \geq \omega(G) + 2$. Let $\pi = \{V_1, V_2, \ldots, V_{\omega(G)+2}\}$ be a transitive partition of G. Since $|K| = \omega(G)$, there exist at least two sets in π, say V_i and V_j with $i < j$, such that V_i and V_j contains only vertices from S. Note that, in this case V_i cannot dominate V_j as S is an independent set of G. Therefore, we have a contradiction. Hence, $\omega(G) \leq Tr(G) \leq \omega(G) + 1$.

Let every vertex of K have a neighbour in S. Now consider a vertex partition of G, say $\pi = \{V_1, V_2, \ldots, V_{\omega(G)+1}\}$, such that $V_1 = S$ and for each $i > 1$, V_i contains exactly one vertex from K. Since every vertex of K has a neighbour in S, V_1 dominates every other partition in π. Moreover, as K is a clique, each V_i, with $i > 1$ dominates V_j for all $2 \leq i < j \leq \omega + 1$. Hence, π is a transitive partition of G. Now, since $Tr(G) \leq \omega(G) + 1$, we have $Tr(G) = \omega(G) + 1$.

Conversely, let $Tr(G) = \omega(G) + 1$ and $\pi = \{V_1, V_2, \ldots, V_{\omega(G)+1}\}$ be a transitive partition of G. Note that if there exist two sets in π, that contain only vertices from S, then using similar arguments as before, we have a contradiction. Therefore, there exists exactly one set in π, say V_l, that contains only vertices from S as $\omega(G) = |K|$. Hence, each set of π, except V_l, contains exactly one vertex from K. Suppose there exists a vertex, say x, in K that has no neighbour in S and also let $x \in V_p$ for some set V_p in π. Note that there is no edge between the vertices of V_p and V_l. Therefore, neither V_p dominates V_l nor V_l dominates V_p. This contradicts the fact that π is a transitive partition. Hence, every vertex of K has a neighbour in S.

Based on the above lemma, we have the following algorithm for finding transitivity of a given split graph.

Algorithm 1. TRANSITIVITY_SPLIT(G)

1: **Input:** A split graph $G = (V, E)$
2: **Output:** The transitivity of G, that is, $Tr(G)$
3: Find a vertex partition of V into S and K, where S and K are an independent set and a clique of G, respectively and $\omega(G) = |K|$.
4: **for all** $v \in K$ **do**
5:     **if** v has no neighbour in S **then**
6:         $t = |K|$.
7:         **break**
8: $t = |K| + 1$.
9: **return** (t)

Note that the required vertex partition in line 3 of Algorithm 1 can be computed in linear time [5]. Also, the for loop in line $4 - 7$ runs in O(n+m) time. Hence, we have the following theorem:

Theorem 2. *The* MAXIMUM TRANSITIVITY PROBLEM *can be solved in linear time for split graphs.*

4   Transitivity in the Complement of Bipartite Chain Graphs

In this section, we find the transitivity of the complement of a bipartite chain graph, say G, by showing that the transitivity of G is equal to the Grundy number of G. To do that, first, we show two essential properties, one of the transitive partitions and the second of the Grundy partitions for the complement of a bipartite chain graph. The proofs of these lemmas are omitted due to space constraints.

Lemma 3. *Let* $G = (X \cup Y, E)$ *be the complement of a bipartite chain graph and* $Tr(G) = k$. *Then there exists a transitive partition* $\pi = \{V_1, V_2, \ldots, V_k\}$ *of* G *such that either* $|V_1| = 1$ *or* $V_1 = \{x, y\}$, *where* $x \in X$ *and* $y \in Y$.

Lemma 4. *Let* $G = (X \cup Y, E)$ *be the complement of a bipartite chain graph. Also, let* $\pi = \{V_1, V_2, \ldots, V_k\}$ *be a Grundy partition of* G *with* $\Gamma(G) = k$ *and* $X'_G = \{x \in X | x \in V_i \text{ and } |V_i| = 1\}$ *and* $Y'_G = \{y \in Y | y \in V_i \text{ and } |V_i| = 1\}$. *Then exactly one of the following two cases is true:*

(i) Both $|X'_G|$ *and* $|Y'_G|$ *cannot be empty simultaneously.*
(ii) The graph G *is the disjoint union of* $K_{|X|}$ *and* $K_{|Y|}$ *and* $|X| = |Y|$.

Now, we are ready to show that the transitivity and the Grundy number are equal for the complement of a bipartite chain graph.

Theorem 5. *Let $G = (X \cup Y, E)$ be the complement of a bipartite chain graph. Then $\Gamma(G) = Tr(G)$.*

Proof. We use induction on n, where n is the number of vertices of G. If $n = 1$, then $\Gamma(G) = Tr(G) = 1$ trivially. For $n = 2$, G is either K_2 or $\overline{K}_2$ and therefore, $\Gamma(G) = Tr(G)$. Let us assume that the induction hypothesis is true, that is, $\Gamma(G) = Tr(G)$ for the complement of every bipartite chain graph having less than n vertices. Let us consider a transitive partition $\pi = \{V_1, V_2, \ldots, V_k\}$ of G with $Tr(G) = k$. By Lemma 3, we can assume that $|V_1| = 1$ or $V_1 = \{x, y\}$ for some $x \in X$ and $y \in Y$. Let $H = G \setminus V_1$. Note that H is also the complement of a bipartite chain graph, since deleting a vertex from X (or Y) does not change the chain ordering of the remaining vertices. By induction hypothesis, we have $\Gamma(H) = Tr(H)$. Moreover, note that $Tr(H) = k - 1$. Hence, we have $\Gamma(H) = Tr(H) = k - 1$. Let $\pi' = \{V_1', V_2', \ldots, V_{k-1}'\}$ be a Grundy partition of H. Now, if $V_1 = \{x\}$ (or $\{y\}$), then x (correspondingly y) is adjacent to every vertex of G because π is a transitive partition of G. Therefore, $\pi'' = \{V_1, V_1', V_2', \ldots, V_{k-1}'\}$ forms a Grundy partition of G which implies $\Gamma(G) \geq k = Tr(G)$. Also, for any graph we know that $\Gamma(G) \leq Tr(G)$, hence $\Gamma(G) = Tr(G)$. So, let us assume that $V_1 = \{x, y\}$ for some $x \in X$ and $y \in Y$. Now, if $xy \notin E$, that is, V_1 is an independent set, then from a Grundy partition of H of order $(k - 1)$ we can construct a Grundy partition of G of order k by appending V_1. Then by similar argument, we have $\Gamma(G) = Tr(G)$. So, we assume that $xy \in E$. Since, $H = G \setminus V_1$ is the complement of a bipartite chain graph, by induction hypothesis $\Gamma(H) = Tr(H)$. Now, by Lemma 4, we can assume that H has a Grundy partition, say $\pi' = \{V_1', V_2', \ldots, V_{k-1}'\}$, such that either $|X_H'| \neq \phi$ or $|Y_H'| \neq \phi$, where X_H' and Y_H' is defined in a similar way as in Lemma 4 or H is the disjoint union of $K_{|X_H|}$ and $K_{|Y_H|}$ and $|X_H| = |Y_H|$, where $X_H = X \cap H$ and $Y_H = Y \cap H$. If H is the disjoint union of $K_{|X_H|}$ and $K_{|Y_H|}$ and $|X_H| = |Y_H|$, then $\pi'' = \{V_1', V_2', \ldots, V_{k-1}', \{x\}, \{y\}\}$ forms a transitive partition of G of order $(k + 1)$. This is a contradiction to the fact that $Tr(G) = k$. So, we assume that H has a Grundy partition, say $\pi' = \{V_1', V_2', \ldots, V_{k-1}'\}$, such that either $|X_H'| \neq \phi$ or $|Y_H'| \neq \phi$. The remaining proof is done by dividing into the following four cases:

Case 1. *Every vertex of X_H' and Y_H' are adjacent to y and x, respectively*

In this case, consider the vertex partition $\pi'' = \{V_1', V_2', \ldots, V_{k-1}', \{x\}, \{y\}\}$. Clearly, π'' is a transitive partition of G of order $(k + 1)$. This is a contradiction to the fact that $Tr(G) = k$.

Case 2. *Every vertex of X_H' is adjacent to y but there exists a vertex $y_t \in Y_H'$ such that $xy_t \notin E$*

Let $y_t \in V_p'$. In this case, let us consider the vertex partition $\pi'' = \{U_1, U_2, \ldots, U_k\}$, where $U_1 = \{x, y_t\}$, $U_i = V_{i-1}'$, for $2 \leq i \leq p$, $U_{p+1} = \{y\}$ and

$U_j = V'_{j-1}$, for all $p+2 \leq j \leq k$. Clearly, the partition π'' forms a Grundy partition of G which implies $\Gamma(G) \geq k = Tr(G)$. Since, for any graph $\Gamma(G) \leq Tr(G)$, therefore, $\Gamma(G) = Tr(G)$.

Case 3. *There exists a vertex $x_s \in X'_H$ such that $yx_s \notin E$ but every vertex of Y'_H is adjacent to x*

This case is similar to Case 2.

Case 4. *There exists a vertex $x_s \in X'_H$ such that $yx_s \notin E$ and there exists a vertex $y_t \in Y'_H$ such that $xy_t \notin E$*

In this case, $\{x, y_t, y, x_s\}$ induces a $2K_2$ in $\overline{G}$. This is a contradiction to the fact that $\overline{G}$ is a bipartite chain graph.

Hence, for the complement of a bipartite chain graph G, $\Gamma(G) = Tr(G)$.

It was proved in [15] that for the complement of a bipartite graph, $\Gamma(G) = n-p$, where n is the number of vertices of G and p is the cardinality of a minimum edge dominating set of $\overline{G}$. We also know that the minimum edge dominating set of a bipartite chain graph can be computed in linear time [14]. Therefore, we have the following corollary:

Corollary 6. *The transitivity of the complement of bipartite chain graphs can be computed in linear time.*

Remark 1. *Identifying graphs with equal transitivity and Grundy number was posed as an open question in [8]. Theorem 5 partially answers this question by showing that the complement of bipartite chain graphs form such a graph class.*

5 Nordhaus-Gaddum Type Bounds for Transitivity

Let $G = (V, E)$ be a simple graph. A proper k-coloring of G is a function c from V to $\{1, 2, \ldots, k\}$ such that $c(u) \neq c(v)$ if and only if $uv \in E$. The minimum value of k for which a proper coloring exists is called *chromatic number* of G and it is denoted by $\chi(G)$. In 1956, Nordhaus and Gaddum [12] studied the chromatic number of a graph G and its complement $\overline{G}$. They established lower and upper bound for the product and the sum of $\chi(G)$ and $\chi(\overline{G})$ in terms of the number of vertices of G. Since then, any bound on the sum or the product of a parameter of a graph G and its complement $\overline{G}$ is called a Nordhaus-Gaddum type inequality. In this section, we study Nordhaus and Gaddum type relations for transitivity.

From [13], it is known that for a bipartite chain graph G, $Tr(G) = t + 1$, where t is the maximum integer such that G contains either $K_{t,t}$ or $K_{t,t} - \{e\}$ as an induced subgraph. Let $\sigma_X = (x_1, x_2, \ldots, x_{n_1})$ and $\sigma_Y = (y_1, y_2, \ldots, y_{n_2})$ be the chain ordering of G. Because of this chain ordering if $x_p y_p \in E$ for some

p, then $\{x_1, x_2, \ldots, x_p\}$ and $\{y_1, y_2, \ldots, y_p\}$ induces a complete bipartite graph. Therefore, it follows that if j is the maximum index such that $x_j y_j \in E$, then

$$Tr(G) = \begin{cases} j+2 & x_{j+1}y_j, x_j y_{j+1} \in E \\ j+1 & \text{otherwise} \end{cases}$$

We know that for the complement of a bipartite graph G, $\Gamma(G) = n - p$, where n is the number of vertices of G and p is the cardinality of a minimum edge dominating set of $\overline{G}$ [15]. Therefore, from Theorem 5 we have, for a bipartite chain graph G, $Tr(\overline{G}) = n - p$, where n is the number of vertices of G and p is the cardinality of a minimum edge dominating set of G. Also from [14], we know that for a bipartite chain graph, p is equal to the maximum index j such that $x_j y_j \in E$. Therefore, we have the following Nordhaus-Gaddum relation for transitivity in bipartite chain graphs:

Theorem 7. *Let $G = (X \cup Y, E)$ be a bipartite chain graph with the ordering $\sigma_X = (x_1, x_2, \ldots, x_{n_1})$ and $\sigma_Y = (y_1, y_2, \ldots, y_{n_2})$ as its chain ordering, that is, $N(x_{n_1}) \subseteq N(x_{n_1-1}) \subseteq \ldots \subseteq N(x_2) \subseteq N(x_1)$ and $N(y_{n_2}) \subseteq N(y_{n_2-1}) \subseteq \ldots \subseteq N(y_2) \subseteq N(y_1)$. Also assume that j is the maximum index such that $x_j y_j \in E$. Then,*

$$Tr(G) + Tr(\overline{G}) = \begin{cases} n+2 & x_{j+1}y_j, x_j y_{j+1} \in E \\ n+1 & \text{otherwise} \end{cases}$$

Based on Lemma 1 and the fact that the complement of a split graph is also a split graph, we have the following Nordhaus-Gaddum relation for transitivity in split graphs, whose proof is omitted due to space constraints.

Theorem 8. *Let $G = (S \cup K, E)$ be a split graph, where S and K are the independent set and clique of G, respectively. Also, assume that K is the maximum clique of G, that is, $\omega(G) = |K|$. Then,*

$$Tr(G) + Tr(\overline{G}) = \begin{cases} n+2 & \text{if, in } G, \text{ every vertex of } K \text{ has a neighbour in } S \\ n+1 & \text{otherwise} \end{cases}$$

Remark 2. *In [8], Hedetniemi and Hedetniemi posed the following open question about the sum of $Tr(G)$ and $Tr(\overline{G})$: for any graph G, is $Tr(G) + Tr(\overline{G}) = n + 1$ if and only if $G = K_n$ or $G = \overline{K}_n$? Theorem 7 and 8 show the existence of some bipartite chain graph and split graph, respectively, for which $Tr(G) + Tr(\overline{G}) = n + 1$. Moreover, also for $K_{n,n}$, $Tr(G) + Tr(\overline{G}) = 2n + 1$ which shows another counter example for the above mentioned open question.*

6   Transitively Critical Graphs

The concept of transitively critical graph was introduced by Haynes et al. in [6]. A graph $G = (V, E)$ is said to be *transitively vertex critical (transitively edge critical*) if deleting any vertex from V (respectively, edge from E) results in a graph whose transitivity is less than $Tr(G)$. A transitively vertex critical (transitively edge critical) graph G with $Tr(G) = k$ is called by Tr_k^v-critical (respectively, Tr_k^e-critical). Characterizations of vertex critical graph have been studied in [6]for some small values of k. In this section, we introduce a generalization of transitively critical graphs, namely *transitively vertex-edge critical* graphs and give characterization of such graphs for every fixed value of k. Using this characterization, we then characterize transitively edge critical graphs for every fixed value of k.

A transitively vertex-edge critical graph is basically a graph which is both transitively vertex and edge critical. The formal definition is as follows:

Definition 9. *A graph $G = (V, E)$ is said to be a transitively vertex edge-critical graph if deleting any element from $V \cup E$ results in a graph whose transitivity is less than $Tr(G)$. A transitively vertex edge-critical graph G with $Tr(G) = k$ is called $Tr_k^{(v,e)}$-critical.*

Note that unlike transitively edge critical graphs, every transitively vertex-edge critical graph is connected. The graph K_1 is the only connected graph with $Tr(G) = 1$ and it is both transitively edge and vertex critical. Therefore, the only $Tr_1^{(v,e)}$-critical is K_1.

The following proposition characterizes the $Tr_2^{(v,e)}$-critical graphs.

Proposition 10. *The only $Tr_2^{(v,e)}$-critical graph is K_2.*

Proof. Clearly, transitivity of K_2 is 2. Also, if we remove any edge or any vertex from K_2, we are left with $\overline{K}_2$ or K_1 and transitivity of those graphs is 1. Hence, K_2 is a $Tr_2^{(v,e)}$-critical.

Let G be a $Tr_2^{(v,e)}$-critical graph with n vertices. Since $Tr(G) = 2$, G is a disjoint union of stars which is shown by Hedetniemi et al. [8]. As G is a vertex critical graph, which implies G must be a connected graph [6]. Therefore, G is a star. Now, if G contains more than one edge, then removal of that edge from G does not decrease the transitivity, which contradicts the fact that G is a transitively edge-critical. Therefore, G can only be K_2. Hence, the only $Tr_2^{(v,e)}$-critical graph is K_2.

Next, we generalize the characterization for $Tr_k^{(v,e)}$-critical graphs for $k \geq 3$. To this end, we recall the concept of t-atom, which was introduced by Zaker in [15]. For the sake of completeness, we give the definition of t-atom here.

Definition 11 ([15]). *A t-atom is defined in a recursive way as follows:*

1. *The only 1-atom is K_1.*
2. *Let $H = (V, E)$ be any $(t-1)$-atom with n vertices. Consider an independent set I_r on r vertices for any $r \in \{1, 2, \ldots n\}$. For that fixed r, consider a r vertex subset W of V and add a perfect matching between the vertices of I_r and W. Then join an edge between each vertex of $V \setminus W$ and an (and to only one) arbitrary vertex of I_r. The resultant graph G is a t-atom.*

The set of t-atoms is denoted by $\mathcal{A}_t$. The following lemma describes the transitively vertex edge-critical graph with transitivity k.

Lemma 12. *If G is $Tr_k^{(v,e)}$-critical, then $G \in \mathcal{A}_k$.*

Proof. For an integer t, $Tr(G) \geq t$ if and only if G contains a t-atom as a subgraph, which is shown by Paul and Santra [13]. Since the transitivity of G is k, therefore, G contains a k-atom as a subgraph. Let $H \in \mathcal{A}_k$ and G contains H as a subgraph. Since the $Tr(H) \geq k$ and $Tr(G) \geq Tr(H)$, then $Tr(H) = k$. If G has an edge other than edges of H, then removal of that edge from G does not decrease the transitivity, which contradicts the fact that G is a transitively edge-critical. Also, G cannot contain more vertex than H, as G is a transitively vertex critical graph too. Therefore, $G = H$.

The only 3-atoms are K_3 and P_4. Also the graphs K_3 and P_4 are both Tr_3^v-critical and Tr_3^e-critical. Therefore, the converse of Lemma 12 is true for $k = 3$. Hence, we have the following corollary.

Corollary 13. *The only $Tr_3^{(v,e)}$-critical graphs are K_3 or P_4.*

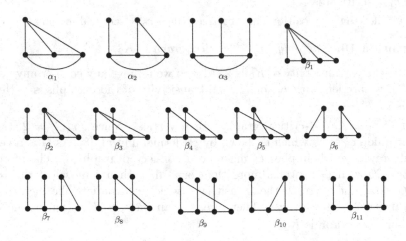

Fig. 1. The class $\mathcal{A}_4$.

For $k = 4$, the converse of Lemma 12 is not true. The class of graphs $\mathcal{A}_4$ is illustrated in Fig. 1. Note that every graph in $\mathcal{A}_4$, has transitivity equal to 4 but only β_2 is not transitively edge-critical. Therefore, we have the following corollary.

Corollary 14. *A graph G is $Tr_4^{(v,e)}$-critical if and only if $G \in \mathcal{A}_4' = (\mathcal{A}_4 \setminus \{\beta_2\})$.*

Generalizing this result, we have the following main theorem.

Theorem 15. *Let $\mathcal{A}_k$ be the set of all k-atoms and $\mathcal{B}_k$ be the set of k-atoms which are neither $Tr_k^{(v,e)}$-critical nor have transitivity equal to k. A graph G is $Tr_k^{(v,e)}$-critical if and only if $G \in \mathcal{A}'_k = (\mathcal{A}_k \setminus \mathcal{B}_k)$.*

Proof. Let G be a $Tr_k^{(v,e)}$-critical. Since the transitivity of G is k, then G contains a k-atom as a subgraph, as for an integer t, $Tr(G) \geq t$ if and only if G contains a t-atom as a subgraph, which is shown by Paul and Santra [13]. Let H be a k-atom and G contains H as a subgraph. Since the $Tr(H) \geq k$ and $Tr(G) \geq Tr(H)$, then $Tr(H) = k$. If G has an edge other than edges of H, then removal of that edge from G does not decrease the transitivity, which contradicts the fact that G is a transitively edge critical. Also, G cannot contain more vertex than H, as G is also a transitively vertex critical graph. Hence, $G \in \mathcal{A}'_k = \mathcal{A}_k \setminus \mathcal{B}_k$.

Next, we characterize the Tr_k^e-critical graphs for a fixed value of k. For this characterization, we first show the following relation between $Tr_k^{(v,e)}$-critical and Tr_k^e-critical graphs. The proof is omitted due to space constraints.

Theorem 16. *A graph G with n vertices and $Tr(G) = k$ is Tr_k^e-critical if and only if $G = H \cup \overline{K}_{n-n_H}$, where H is a $Tr_k^{(v,e)}$-critical graph having n_H vertices.*

The characterization of Tr_k^e-critical graphs follows immediately from the above theorem.

Corollary 17. *Let $\mathcal{A}_k$ be the set of k-atoms and $\mathcal{B}_k$ be the set of k-atoms which are neither $Tr_k^{(v,e)}$-critical nor have transitivity equal to k. A graph G with n vertices, is Tr_k^e-critical if and only if $G = H \cup \overline{K}_{n-n_H}$, where $H \in \mathcal{A}'_k = \mathcal{A}_k \setminus \mathcal{B}_k$.*

7   Conclusion

In this paper, we have proved that the transitivity of a given split and the complement of bipartite chain graphs can be computed in linear time. Then, we have discussed Nordhaus-Gaddum type relations for transitivity in split graphs and bipartite chain graphs and have given counter-examples to an open question posed in [8]. We have also studied transitively vertex-edge critical graphs. It would be interesting to investigate the complexity status of this problem in other subclasses of chordal graphs. Designing an approximation algorithm for this problem would be another challenging open problem.

Acknowledgements. Subhabrata Paul is supported by the SERB MATRICS Research Grant (No. MTR/2019/000528). The work of Kamal Santra is supported by the Department of Science and Technology (DST) (INSPIRE Fellowship, Ref No: DST/INSPIRE/ 03/2016/000291), Govt. of India.

References

1. Chang, G.J.: The domatic number problem. Discret. Math. **125**(1-3), 115–122 (1994)
2. Christen, C.A., Selkow, S.M.: Some perfect coloring properties of graphs. J. Comb. Theory Ser. B **27**(1), 49–59 (1979)
3. Cockayne, E.J., Hedetniemi, S.T.: Towards a theory of domination in graphs. Networks **7**(3), 247–261 (1977)
4. Erdös, P., Hedetniemi, S.T., Laskar, R.C., Prins, G.C.E.: On the equality of the partial Grundy and upper ochromatic numbers of graphs. Discret. Math. **272**(1), 53–64 (2003)
5. Hammer, P.L., Simeone, B.: The splittance of a graph. Combinatorica **1**(3), 275–284 (1981)
6. Haynes, T.W., Hedetniemi, J.T., Hedetniemi, S.T., McRae, A., Phillips, N.: The transitivity of special graph classes. J. Comb. Math. Comb. Comput. **110**, 181–204 (2019)
7. Haynes, T.W., Hedetniemi, J.T., Hedetniemi, S.T., McRae, A., Phillips, N.: The upper domatic number of a graph. AKCE Int. J. Graphs Comb. **17**(1), 139–148 (2020)
8. Hedetniemi, J.T., Hedetniemi, S.T.: The transitivity of a graph. J. Comb. Math. Comb. Comput. **104**, 75–91 (2018)
9. Hedetniemi, S.M., Hedetniemi, S.T., Beyer, T.: A linear algorithm for the Grundy (coloring) number of a tree. Congr. Numer. **36**, 351–363 (1982)
10. Hedetniemi, S.M., Hedetniemi, S.T., McRae, A.A., Parks, D., Telle, J.A.: Iterated colorings of graphs. Discret. Math. **278**(1–3), 81–108 (2004)
11. Heggernes, P., Kratsch, D.: Linear-time certifying recognition algorithms and forbidden induced subgraphs. Nordic J. Comput. **14**(1–2), 87–108 (2007)
12. Nordhaus, E.A., Gaddum, J.W.: On complementary graphs. Am. Math. Mon. **63**(3), 175–177 (1956)
13. Paul, S., Santra, K.: Transitivity on subclasses of bipartite graphs. J. Comb. Optim. **45**, 27 (2022)
14. Verma, S., Panda, B.S.: Grundy coloring in some subclasses of bipartite graphs and their complements. Inf. Process. Lett. **163**, 105999 (2020)
15. Zaker, M.: Results on the Grundy chromatic number of graphs. Discret. Math. **306**(23), 3166–3173 (2006)

Maximum Subgraph Problem for 3-Regular Knödel graphs and its Wirelength

R. Sundara Rajan[1], Remi Mariam Reji[1(✉)], and T. M. Rajalaxmi[2]

[1] Department of Mathematics, Hindustan Institute of Technology and Science,
Chennai 603 103, India
remimariamreji@gmail.com
[2] Department of Mathematics, Sri Sivasubramaniya Nadar College of Engineering,
Chennai 603 110, India

Abstract. The maximum subgraph problem (MSP) of a graph is the estimation of the greatest number of edges in the induced subgraph of all subsets of the vertex set of the same cardinality. The Knödel graph $W_{\Delta,n}$ of n vertices and the highest degree Δ lies in $[1, \lfloor \log_2(n) \rfloor]$, where n is even, is a minimum linear gossip graph and with minimum broadcasting. In this paper, we will obtain the maximum subgraph of the 3-regular Knödel graph $W_{3,n}, n \geq 16$, n is even, and thereby obtain its wirelength.

Keywords: Maximum subgraph problem · Edge congestion · Wirelength · Linear arrangement · Knödel graph

1 Introduction

Graph theory has wide applications in computer science. Mainly, in multiprocessor architectures graphs are used as the interconnection topology. Graph databases are used in database designing. Graphs are also used for designing communication networks. In image processing, graph search algorithms are used to find edge boundaries, and graphs are used to calculate the alignment of the picture. The symmetry and regularity of graphs simplify the algorithms for different network-related problems.

For a graph $G(V, E)$, two versions of the edge isoperimetric problem are given in the literature [1] and the problems are NP-complete [4]. They are applied in solving the exact wirelength of graph embeddings for parallel networks, data structures, and biological models. It is also applied in computing the bisection width of a network and graph partitioning problems. In this paper, we focus on the *Maximum Subgraph Problem* (MSP) and it is defined as follows: For a given graph, find a subset of vertices so that the induced subgraph of this subset has the maximum number of edges when taken over all subgraphs having the same number of vertices. That is, for a given ℓ, find $S \subseteq V$ such that if $I_G(\ell) = \max\limits_{S \subseteq V, |S| = \ell} |I_G(S)|$ where $I_G(S) = \{(x, y) \in E : x, y \in S\}$, then $|S| = \ell$

A. Bagchi and R. Muthu (Eds.): CALDAM 2023, LNCS 13947, pp. 403–414, 2023.
https://doi.org/10.1007/978-3-031-25211-2_31

and $I_G(\ell) = |I_G(S)|$ [10]. A subset of vertices $S \subseteq V$ is called *optimal* if $|S| = \ell$ and $|I_G(S)| = I_G(\ell)$ for a given ℓ with $\ell = 1, 2, \ldots, n$ [10].

The paper is organised as follows. The next section contains some preliminaries. Section 3 gives an introduction about the Knödel graphs, its properties, and labeling of its vertices. In Sect. 4, we solve the maximum subgraph problem of 3-regular Knödel graphs $W_{3,n}, n \geq 16$, n is even. Also, we obtain its minimum linear arrangement and the wirelength of embedding $W_{3,2^n}, n \geq 4$, n is even into 1-rooted complete binary tree T_n^1. Finally, the paper is concludes with an open problem in Sect. 6 .

2   Preliminaries

The process of mapping a guest graph A into a host graph B (usually an interconnection network) is called a *graph embedding*. Consider two finite graphs A and B. An embedding of A into B is a pair (g, P_g) stated as below [8]:

1. g is a $1-1$ map from $V(A)$ to $V(B)$.
2. P_g is a $1-1$ map from $E(A)$ to $\{P_g(e) : P_g(e)$ is a path in B joining $g(u)$ and $g(v)$ where $e = (uv) \in E(A)\}$.

For brevity, we use g to denote the pair (g, P_g). For an edge $e \in E(B)$, the count of edges $(uv) \in E(A)$ such that $e \in P_g(uv)$ between $g(u)$ and $g(v)$ in B is known as the edge congestion of e, denoted by $EC_g(e)$ [8]. Then

$$EC_g(A, B) = \max_{e \in E(B)} EC_g(e), \text{and}$$

$$EC(A, B) = \min_{g:A \to B} EC_g(A, B).$$

Also, the edge congestion of any subset T of $E(B)$ is given by

$$EC_g(T) = \sum_{e \in T} EC_g(e).$$

For an embedding g of A into B, the wirelength [8] is defined as

$$WL_g(A, B) = \sum_{e \in E(B)} EC_g(e).$$

Therefore, the wirelength of A embedded into B is,

$$WL(A, B) = \min_{g:A \to B} WL_g(A, B).$$

Lemma 1 (Modified Congestion Lemma [8]). *Consider an embedding g of A into B. Assume that X is an edge cut of B such that when removing the edges of X, B breaks into two components B_1 and B_2 and let $A_1 = G[g^{-1}(V(B_1))]$ and $A_2 = G[g^{-1}(V(B_2))]$. Further, X fulfills the below conditions:*

(i) *For all $(uv) \in E(A_i)$, $i = 1, 2$, $P_g(uv)$ has no edges in X.*

(ii) For all $(uv) \in E(A)$, where $u \in V(A_1)$ and $v \in V(A_2)$, $P_g(uv)$ contains strictly an edge in X.

(iii) $V(A_1)$ and $V(A_2)$ are optimal sets.

Then $EC_g(X)$ is minimum throughout all embeddings g from A into B and

$$EC_g(X) = \sum_{a \in V(A_1)} deg_A(a) - 2|E(A_1)| = \sum_{a \in V(A_2)} deg_A(a) - 2|E(A_2)|.$$

Remark 1 [8]. When A is regular, in Modified Congestion Lemma it is sufficient to find out that either $V(A_1)$ or $V(A_2)$ will be an optimal set.

Lemma 2 (Partition Lemma [8]). *Consider an embedding g of A into B. Let $E(B)$ have the partition $X_1, X_2, \ldots, X_q$ where each X_i is an edge cut of B. Let X_i fulfill all three conditions of Modified Congestion Lemma. Then*

$$WL_g(A, B) = \sum_{i=1}^{q} EC_g(X_i).$$

Remark 2 [8]. For an embedding g of A into B which fulfills the Partition Lemma, we have

$$WL(A, B) = WL_g(A, B).$$

3   The Knödel graphs

In 1975, W. Knödel [7] introduced the Knödel graph as the topology underlying a time optimal algorithm for gossipping among n vertices, where n is even, while Fraigniaud and Peters defined the family of Knödel graphs [3]. The Knödel graphs, denoted by $W_{\Delta,n}$, are regular graphs with order n, where n is even, and the highest degree Δ lies in $[1, \lfloor log_2(n) \rfloor]$. The Knödel graph $W_{\Delta,2^\Delta}$ have become competitors for the hypercubes and recursive circulant graphs of the same order and degree, mainly in the field of gossiping and broadcasting.

Definition 1 [3]. *The Knödel graph, denoted by $W_{\Delta,n}$, contains $n \geq 2$ vertices, where n is even, and the highest degree Δ lies in $[1, \lfloor log_2(n) \rfloor]$. The vertices are denoted by (a, b) where $a = 1, 2$ and $0 \leq b \leq \frac{n}{2} - 1$. The vertices $(1, b)$ and $(2, b + 2^k - 1 \ mod(\frac{n}{2}))$ will be connected by an edge for every b, $0 \leq b \leq \frac{n}{2} - 1$, $k = 0, \ldots, \Delta - 1$.*

In $W_{\Delta,n}$, the edge which joins the vertices $(1, b)$ and $(2, b + 2^k - 1 \ mod(\frac{n}{2}))$ is said to have dimension k, for $0 \leq k \leq \Delta - 1$. For $\Delta = 1$, $W_{1,n}$ is made up of $\frac{n}{2}$ disjoint copies of K_2. When $\Delta \geq 2$, a Hamiltonian cycle is generated by the alternate edges in dimensions 0 and 1. Hence $W_{\Delta,n}$ is connected iff $\Delta \geq 2$. The edge connectivity is $\lambda(W_{\Delta,n}) = \Delta$, vertex connectivity is $\frac{2\Delta}{3} < \kappa(W_{\Delta,n}) \leq \Delta$ and it is vertex-transitive for any even n and $1 \leq \Delta \leq \lfloor log_2(n) \rfloor$. Also $W_{\Delta+1,2n}$ can be formed from two copies of $W_{\Delta,n}$ [3].

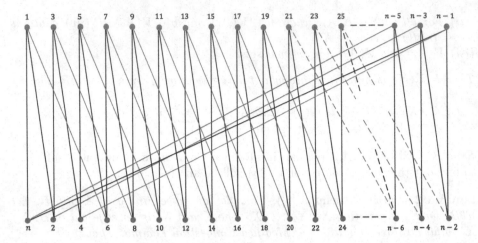

Fig. 1. Labeling of Knödel graph $W_{3,n}$

In this paper, we compute the maximum subgraph of 3-regular Knödel graphs $W_{3,n}, n \geq 16$, n is even. The vertex set of the Knödel graph is partitioned into V_1 and V_2, since it is bipartite, where V_1 consist of vertices of the form $(1, x)$, $0 \leq x \leq \frac{n}{2} - 1$ and V_2 consist of vertices of the form $(2, y)$, $0 \leq y \leq \frac{n}{2} - 1$. They are labeled as follows: The vertices of V_1 will be labeled as $f((1, x)) = 1 + 2x$, $0 \leq x \leq \frac{n}{2} - 1$ and the vertices of V_2 will be labeled as $f((2, 0)) = n$ and $f((2, y)) = 2y$, $1 \leq y \leq \frac{n}{2} - 1$. In Fig. 1 the labeling of $W_{3,32}$ is shown. The vertex set V_1 consists of vertices with odd label (denoted in red) and the set V_2 consists of vertices with even label (denoted in blue). We use the above labeling for our entire study.

The Knödel graph $W_{\Delta,n}$, $n \geq 2^{\Delta+1}$ and n is even, of any dimension has a 6-cycle. This is the smallest cycle contained in the graph; hence the girth of the graph is 6. Also, every 6-cycle begins from a vertex with odd label when taken in an increasing order (a set $A = \{a_1, a_2, \ldots, a_n\}$ is said to be an *increasing order* if $a_p < a_q$ for all $p < q$, $p, q \in \{1, 2, \ldots, n\}$). Note that Fig. 1 and 2(a) are isomorphic by taking $f(i) = i, \forall\, i \in \{1, 2, \ldots, n\}$. Further, in Fig. 2(a), the consecutive vertices in increasing order are in the outer cycle with a clockwise direction.

4   Results

In this section, we will initially solve the maximum subgraph problem for the 3-regular Knödel graph $W_{3,n}, n \geq 16$, n is even. Then we will compute its minimum linear arrangement. Later we will obtain the wirelength of embedding $W_{3,2^n}, n \geq 4$ into 1-rooted complete binary tree T_n^1.

4.1   Maximum Subgraph Problem

Consider a graph $G(V, E)$ with a non-empty subset of the vertex set $S \subseteq V$. Then, the subgraph of G having S as the vertex set and edge set consisting of edges of G having both ends in S is called *subgraph induced* by S, denoted by $G[S]$ [11].

Lemma 3. *Let K_1 be a subset of the vertex set of the 3-regular Knödel graph $W_{3,n}, n \geq 16$, n is even, consists of k_1 consecutive vertices taken in increasing order and starting from a vertex with an odd label. Let K_2 be another subset of the vertex set of $W_{3,n}, n \geq 16$, n is even, consists of k_2 consecutive vertices taken in increasing order and starting from a vertex with even label, so that $k_1 = k_2$. Then $|E(G[K_2])| \leq |E(G[K_1])|$.*

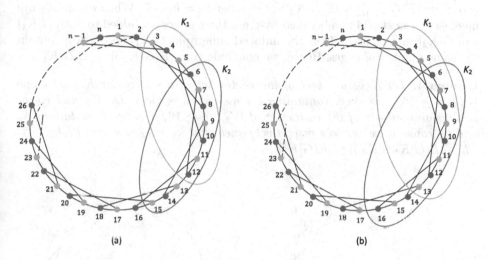

Fig. 2. (a) When $k_1(= k_2)$ is even (b) When $k_1(= k_2)$ is odd

Proof. We prove this lemma with the following two cases.

Case 1 ($k_1(= k_2)$ is even): Let the set K_1 consist of k_1 consecutive vertices of the 3-regular Knödel graph $W_{3,n}, n \geq 16$, n is even, taken in increasing order and starting from a vertex with an odd label. Since k_1 is even, the end vertex will have even label, see Fig. 2(a). Let the set K_2 contain k_2 consecutive vertices of $W_{3,n}, n \geq 16$, n is even, taken in increasing order and starting from a vertex with an even label. Since k_2 is even, the end vertex will have an odd label. For $k_1 = k_2 = 2, 4$, the subgraph induced is a path since the girth is 6. Here $|E(G[K_1])| = |E(G[K_2])|$. When $k_1 = k_2 = 6$, the subgraph induced by K_1 is a 6-cycle, but K_2 is not. Therefore, $|E(G[K_2])| < |E(G[K_1])|$.

Consider $k_1 = k_2 > 6$. When k_1 and k_2 are increased to next even number step-by-step, three edges are added to both $G[K_1]$ and $G[K_2]$ in each step. Hence the number of edges in the induced subgraphs $G[K_1]$ and $G[K_2]$ depends only on the number of edges induced by the first six vertices. Therefore, $|E(G[K_2])| < |E(G[K_1])|$. Since the last vertex of K_2 is odd, if we take in the reverse order we get a subset of vertices, say $\{a_{k_2}, a_{k_2-1}, \ldots, a_1\}$ starting with an odd label. However, the induced subgraph on the first six vertices doesn't form a cycle C_6, since the girth is obtained only when the vertices are taken in increasing order. Therefore, $|E(G[K_2])| \leq |E(G[K_1])|$.

Case 2 ($k_1 (= k_2)$ is odd): Let K_1 be the set containing k_1 consecutive vertices of the 3-regular Knödel graph $W_{3,n}, n \geq 16$, n is even, taken in an increasing order and starting from a vertex with an odd label. Since k_1 is odd, the end vertex will have odd label, see Fig. 2(b). Let the set K_2 consist of k_2 consecutive vertices of $W_{3,n}, n \geq 16$, n is even, taken in increasing order and starting from a vertex with an even label. Since k_2 is also odd, the end vertex will have an even label. For $k_1 = k_2 = 3, 5$, the subgraph induced is a path since the girth of $W_{3,n}$ is 6. Here $|E(G[K_1])| = |E(G[K_2])|$. Consider $k_1 = k_2 > 5$. When k_1 and k_2 are increased to next odd number step-by-step, three edges are added to both $G[K_1]$ and $G[K_2]$ in each step. Thus the induced subgraphs $G[K_1]$ and $G[K_2]$ contain the same number of edges. Hence, we conclude that $|E(G[K_2])| = |E(G[K_1])|$.

Lemma 4. *Let K be a subset of the vertex set of the 3-regular Knödel graph $W_{3,n}, n \geq 16$, n is even, containing k consecutive vertices. Let K_1 and K_2 be two disjoint subsets of the vertex set of $W_{3,n}, n \geq 16$, n is even, containing k_1 and k_2 consecutive vertices respectively such that $k_1 + k_2 = k$ and $k_1, k_2 \geq 1$. Then $|E(G[K_1 \cup K_2])| \leq |E(G[K])|$.*

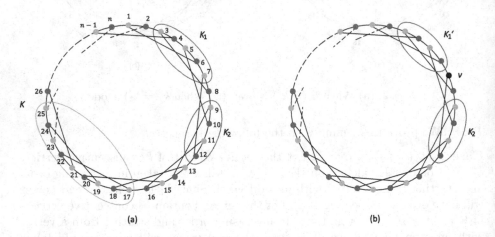

(a)   (b)

Fig. 3. (a) The subset K, K_1 and K_2 in $W_{3,n}$ (b) The subset $K_1 \backslash \{v\}$ and K_2 in $W_{3,n}$

Proof. We prove this result by the method of induction on k. Let $k = 2$, K is a subset of 2 consecutive vertices, and it is easy to see that $|E(G[K])| = 1$. If $k_1 = 1$ and $k_2 = 1$, then $k_1 + k_2 = 2$. By the labeling of Knödel graphs, there is no edge between two vertices with odd labels as well as two vertices with even labels in a Knödel graph since the graph is bipartite. In this case, $G[K_1 \cup K_2]$ is a disconnected subgraph of the 3-regular Knödel graph $W_{3,n}, n \geq 16$, n is even, and we have $|E(G[K_1 \cup K_2])| = 0$. If K_1 has a vertex with an odd label and K_2 has a vertex with an even label or vice versa, then there is a chance of an edge between K_1 and K_2. Thus we get $|E(G[K_1 \cup K_2])| = 1$. Therefore, $|E(G[K_1 \cup K_2])| \leq |E(G[K])|$.

Assume that the result is true for the case of $k - 1$ consecutive vertices. Now consider the case when K is a subset of k consecutive vertices. Then K_1 and K_2 are two disjoint subsets of k_1 and k_2 consecutive vertices respectively such that $k_1 + k_2 = k$, see Fig. 3(a). Without loss of generality, let v be the last vertex of K_1 having $\deg_{G[K_1 \cup K_2]}(v) \leq 3$. Deletion of the vertex v from $K_1 \cup K_2$ will obtain $K_1' \cup K_2$ which contains $k - 1$ vertices, as shown in Fig. 3(b). By induction hypothesis we have $|E(G[K_1' \cup K_2])| \leq |E(G[K'])|$, where K' is a subset of $W_{3,n}, n \geq 16$, n is even, on $k - 1$ consecutive vertices. Therefore, $|E(G[K_1 \cup K_2])| = |E(G[K_1' \cup \{v\} \cup K_2])| = |E(G[K_1' \cup K_2])| + \deg_{G[K_1 \cup K_2]}(v) \leq |E(G[K'])| + \deg_{G[K_1 \cup K_2]}(v) \leq |E(G[K])|$, since any vertex of K except the end vertices have degree 3.

Proceeding with a similar argument, we obtain the following result.

Lemma 5. *Let K be a subset of the vertex set of the 3-regular Knödel graph $W_{3,n}, n \geq 16$, n is even, containing k consecutive vertices. Let $K_1, K_2, \ldots, K_\ell$ be ℓ disjoint subsets of the vertex set of the Knödel graph containing $k_1, k_2, \ldots, k_\ell$ consecutive vertices respectively such that $k_1 + k_2 + \ldots + k_\ell = k$ and $k_1, k_2, \ldots, k_\ell \geq 1$. Then $|E(G[K_1 \cup K_2 \cup \ldots \cup K_\ell])| \leq |E(G[K])|$.*

Theorem 1. *A set of k, $1 \leq k \leq n$, consecutive vertices of the 3-regular Knödel graph $W_{3,n}, n \geq 16$, n is even, taken in increasing order and starting from a vertex with an odd label, induces a maximum number of edges.*

Proof. Let K be the set of k consecutive vertices of the 3-regular Knödel graph $W_{3,n}, n \geq 16$, n is even, taken in increasing order and starting from a vertex with an odd label. Let K' be the set of k non-consecutive vertices of $W_{3,n}, n \geq 16$, n is even, such that $K' = \bigcup_{i=1}^{p} K_i$, $p \geq 2$, where K_i's are disjoint sets of consecutive vertices, taken in increasing order and each starting from a vertex with an odd label which gives $\sum_{i=1}^{p} |K_i| = k$.

Now we will prove this result by induction on p. Let $p = 2$, then by Lemma 4, the result is true. Let $L' = \bigcup_{i=1}^{\ell} K_i$, where $|L'| = \sum_{i=1}^{\ell} |K_i|$ and L is the set of $\sum_{i=1}^{\ell} |K_i|$ consecutive vertices. Then by assumption for $p = \ell$ and by Lemma 5, we have $|E(G[L'])| \leq |E(G[L])|$. Let $L'' = L' \cup K_{\ell+1}$. By induction hypothesis $|E(G[L'])|$ is not maximum. Again by Lemma 4, $G[L'']$ will not induce a maximum number of edges on $|L''| = |K_1| + |K_2| + \cdots + |K_{\ell+1}|$ vertices. Hence the proof.

Remark 3. For $k \leq 5$, the maximum subgraph of k vertices of the 3-regular Knödel graph $W_{3,n}, n \geq 16$, n is even, contains $k - 1$ edges, irrespective of taking the vertex subset starting with an odd or even label.

Corollary 1. *The number of edges in a maximum subgraph on k vertices of the 3-regular Knödel graph $W_{3,n}, n \geq 16$, n is even, where $2 \leq k < n$, is given by*

$$I_G(k) = \begin{cases} k-1 & \text{if } k = 2, 3, 4, 5 \\ \frac{1}{2}(3k-6) & \text{if } 6 \le k \le n-4 \text{ and } k \text{ is even} \\ \frac{1}{2}(3k-7) & \text{if } 7 \le k \le n-5 \text{ and } k \text{ is odd} \\ \frac{3n}{2} - 2i - 1 & \text{if } k = n-i, i = 1, 2, 3. \end{cases}$$

Proof. Label the vertices of the 3-regular Knödel graph $W_{3,n}$, $n \ge 16$, n is even, according to the labeling given in Sect. 3. Let K be the subset of the vertex set of the Knödel graph $W_{3,n}$, $n \ge 16$, n is even, consisting of k consecutive vertices, $2 \le k < n$, taken in increasing order and starting from a vertex with an odd label. Then we have the following cases:

Case 1 ($k = 2, 3, 4, 5$): Clearly, the induced subgraph is a path for $k = 2, 3, 4$, and 5 since the girth of the $W_{3,n}$, $n \ge 16$, n is even, is 6. Therefore, $I_G(k) = k-1$ when $k = 2, 3, 4$ and 5.

Case 2 ($7 \le k \le n-5$ and k is odd): For any odd k, $7 \le k \le n-5$, the subgraph induced by k vertices consist of one pendant vertex, five vertices of degree 2 and $k-6$ vertices of degree 3. Then the number of edges induced by k vertices is given by

$$I_G(k) = \frac{1}{2} \sum_{v \in K} deg_{G[K]}(v) = \frac{1}{2}(1 \cdot 1 + 2 \cdot 5 + 3 \cdot (k-6)) = \frac{1}{2}(3k-7).$$

Case 3 ($6 \le k \le n-4$ and k is even): When $k = 6$, the induced subgraph is a cycle, which is the girth of $W_{3,n}$, $n \ge 16$, n is even. Therefore, $I_G(6) = 6$. When k is even and $6 \le k \le n-4$, the induced subgraph contains six vertices of degree 2 and $k-6$ vertices of degree 3. Therefore,

$$I_G(k) = \frac{1}{2} \sum_{v \in K} deg_{G[K]}(v) = \frac{1}{2}(3k-6).$$

Case 4 ($k = n-i$, $i = 1, 2, 3$): For $i = 1$, let v be the last vertex of $W_{3,n}$, $n \ge 16$, n is even. Remove v from the graph so that the resultant subgraph has $n-1$ consecutive vertices, starting from a vertex with an odd label. Let this subset of vertices be K. Then,

$$I_G(n-1) = |E| - deg_{W_{3,n}}(v) = \frac{3n}{2} - 3.$$

Consider $k = n-2$. Here we remove the last vertex v' from the subgraph $G[K]$ such that the resultant subgraph consists of $n-2$ consecutive vertices starting from a vertex with an odd label. Let this subset of vertices be K'. Then,

$$I_G(n-2) = |E(G[K])| - deg_{G[K]}(v') = \frac{3n}{2} - 5.$$

Similarly for $k = n-3$, we have

$$I_G(n-3) = \frac{3n}{2} - 7.$$

Hence, we can conclude that for $k = 2^n - i$, $i = 1, 2, 3$

$$I_G(k) = |E| - (2i+1) = \frac{3n}{2} - 2i - 1.$$

4.2 Minimum Linear Arrangement

Consider an undirected graph $G(V, E)$ having n vertices. A bijective mapping $g : V \to \{1, 2, \ldots, n\}$ [6] is called a linear of G. The minimum linear arrangement (MinLA) is defined as finding a linear arrangement g which minimises the sum of values of $|g(u) - g(v)|$ for every edges (uv) in G [5]. The particular case of embedding graphs into k-dimensional grids is a linear arrangement. The wirelength of embedding any graph G into a path is called the minimum linear arrangement of G, denoted by $MinLA(A)$ [6]. Mathematically, $WL(A, B) = MinLA(A)$ when B is a path with $|V(A)| = |V(B)|$.

Theorem 2. *Let A be the 3-regular Knödel graph $W_{3,n}$, $n \geq 16$, n is even, and B be the path on n vertices P_n. Then*

$$MinLA(A) = WL(A, B) = \frac{13n}{2} - 22.$$

Proof. Since the host graph is a path on n vertices, the vertices can be labeled from left to right as $1, 2, \ldots, n$. Let $g(u) = u$ for every $u \in V(A)$ and for $(xy) \in E(A)$, let $P_g(xy)$ be the shortest path from $g(x)$ to $g(y)$ in P_n.

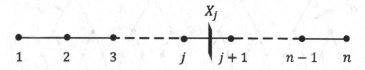

Fig. 4. The edge cut of P_n

Let X_j be an edge cut of P_n given as $X_j = \{j, j + 1\}$, $1 \leq j \leq n - 1$, see Fig. 4. For $1 \leq j \leq n - 1$, $E(P_n) \setminus X_j$ has components B_j and $\overline{B_j}$. Let $V(B_j) = \{1, 2, \ldots, , j\}$. Let $A_j = G[V(g^{-1}(B_j))]$ and $\overline{A_j} = G[V(g^{-1}(\overline{B_j}))]$. By Lemma 1, A_j is optimal with j vertices. Therefore, X_j fulfills the three conditions of Modified Congestion Lemma. Then, for $1 \leq j \leq n - 1$, $EC_g(X_j)$ is minimum and is given by

$$EC_g(X_j) = 3j - 2I_A(j).$$

Thus, the wirelength is minimum by Partition Lemma and we have

$$WL(A, B) = \sum_{j=1}^{n-1} EC_g(X_j) = \sum_{j=1}^{n-1} (3 \cdot j - 2I_A(j))$$

$$= 3 \cdot \sum_{j=1}^{n-1} j - 2 \left(\sum_{j=2}^{5} (j-1) + \sum_{\substack{j=6 \\ \& \ j \text{ is even}}}^{n-4} \frac{1}{2}(3j-6) \right.$$

$$\left. + \sum_{\substack{j=7 \\ \& \ j \text{ is odd}}}^{n-5} \frac{1}{2}(3j-7) + \frac{9n}{2} - 15 \right)$$

$$= \frac{13n}{2} - 22.$$

4.3   Wirelength of Embedding $W_{3,2^n}$, $n \geq 4$ into 1-rooted Complete Binary Tree T_n^1

The complete binary tree (CBT) T_n [9] having n levels is a binary tree which has n levels where n is any non-negative integer. Each level j, $1 \leq j \leq n$, contains 2^{j-1} vertices in such a way that all the leaves lie at the same level and there will be exactly two children for each internal vertex. Therefore, T_n has exactly $2^n - 1$ vertices. By joining a pendant vertex to the root of a complete binary tree T_n we get a 1-rooted complete binary tree, denoted by T_n^1. Now, the root of T_n^1 is the new vertex, say t, and it is at level 0. Clearly, $|V(T_n^1)| = 2^n, n \geq 1$.

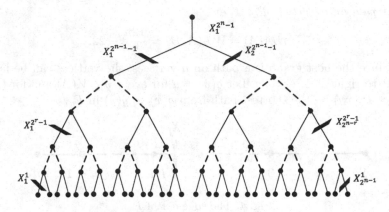

Fig. 5. Cut edges of 1-rooted complete binary tree T_n^1

Theorem 3. *Let A be the 3-regular Knödel graph $W_{3,2^n}$, $n \geq 4$, n is even, and B be the 1-rooted complete binary tree T_n^1. Then*

$$WL(A, B) = 9 \cdot 2^{n-1} - 11.$$

Proof. Using inorder traversal [2], we label the vertices of T_n^1 from 1 to 2^n. Let $g(a) = a$ for all $a \in V(W_{3,2^n})$ and let $P_g(uv)$ be a shortest path joining $g(u)$ and $g(v)$ in T_n^1, for every $(uv) \in E(W_{3,2^n})$.

Let $X_m^{2^r-1}$, $r = 1, 2, \ldots, n$, $m = 1, 2, \ldots, 2^{n-r}$, be the cut edge of T_n^1, see Fig. 5. Then the end vertices of $X_m^{2^r-1}$ lies in $(n-r)^{\text{th}}$ level and $(n-r+1)^{\text{th}}$ level. Thus when $X_m^{2^r-1}$ is removed, T_n^1 separates into two components $B_m^{2^r-1}$ and $\overline{B_m^{2^r-1}}$ with $V(B_m^{2^r-1}) = \{2^r(m-1)+1, 2^r(m-1)+2, \ldots, 2^r(m-1)+2^r-1\}$. Let $G[g^{-1}(V(B_m^{2^r-1}))] = A_m^{2^r-1}$ and $G[g^{-1}(V(\overline{B_m^{2^r-1}}))] = \overline{A_m^{2^r-1}}$. By Lemma 1, $A_m^{2^r-1}$ is optimal with $2^r - 1$ vertices. Therefore, the conditions of Modified Congestion Lemma are satisfied by $X_m^{2^r-1}$. Then, for $r = 1, 2, \ldots, n$, $m = 1, 2, \ldots, 2^{n-r}$, $EC_g(X_m^{2^r-1})$ is minimum and is given by

$$EC_g(X_m^{2^r-1}) = 3(2^r - 1) - 2I_A(2^r - 1).$$

Thus, the wirelength is minimum by Partition Lemma and

$$WL(A,B) = \sum_{r=1}^{n} \sum_{m=1}^{2^{n-r}} EC_g(X_m^{2^r-1}) = \sum_{r=1}^{n} \sum_{m=1}^{2^{n-r}} (3(2^r - 1) - 2I_A(2^r - 1))$$

$$= 3\left(1 + (n-1)2^n\right) - 2\sum_{r=1}^{n} 2^{n-r} I_A(2^r - 1)$$

$$= 9 \cdot 2^{n-1} - 11.$$

5 Implementation

Harper established the minimum linear arrangement problem in 1964 [5] as a method for developing error-correcting codes with the fewest possible average absolute errors. The problem is used extensively in software diagram layout, entity relationship models, UML sequence diagram layout, and data flow diagram layout. It is also used to resolve wiring problems and the single-machine job scheduling problem. An application of minimum linear arrangement in computational biology is an oversimplified model of some neural activity in the cortex.

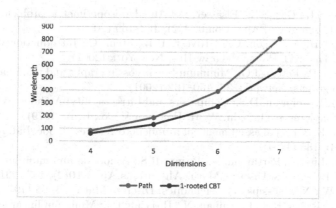

Fig. 6. Comparison of wirelengths of embedding path P_{2^n} and 1-rooted complete binary tree T_n^1 into 3-regular Knödel graph $W_{3,2^n}$, $n \geq 4$, n is even

A good presentation of a graph relies on the total length of all edges. The field of a graph drawing is where the minimum linear arrangement finds its most widespread application. To provide a graph representation with good comprehension and readability, we have to reduce the number of crossing edges. The bipartite graphs are represented using bipartite drawing in which the vertex partitions form two parallel lines and the straight line between them will form the edges. The bipartite crossing number is the least number of edge crossings over all bipartite drawings. This is similar to the minimum linear arrangement problem in that it can be simplified by minimizing the total edge length. Hence, the solution to the bipartite crossing number problem can be approximated from the solution of the minimum linear arrangement problem.

In Fig. 6, the wirelengths for embedding the Knödel graph $W_{3,2^n}, n \geq 4$ into the path and 1-rooted complete binary tree is shown and we have observed that when the number of pendent vertices in a graph increases, the wirelength becomes small.

6   Concluding Remarks

In this article, we have solved the maximum subgraph problem for 3-regular Knödel graph $W_{3,n}$, $n \geq 16$, n is even and we obtained the minimum linear arrangement. Further we calculated the wirelength of embedding $W_{3,n}$, $n \geq 16$, n is even into 1-rooted complete binary tree T_n^1. Finding the wirelength of embedding Knödel graphs into other host graphs are under investigation. Solving the maximum subgraph problem for the Knödel graph $W_{\Delta,n}, n \geq 2^{\Delta+1}$ is challenging for $4 \leq \Delta \leq \lfloor \log_2 n \rfloor$.

References

1. Bezrukov, S.L., Das, S.K., Elsässer, R.: An edge-isoperimetric problem for powers of the petersen graph. Ann. Comb. **4**(2), 153–169 (2000)
2. Cormen, T.H., Leiserson, C.E., Rivest, R.L., Stein, C.: Introduction Algorithms, 2nd edn. The MIT Press, McGraw-Hill, New York (2001)
3. Fraigniaud, P., Peters, J.G.: Minimum linear gossip graphs and maximal linear (Δ, k)-gossip graphs. Networks **38**, 150–162 (2001)
4. Garey, M.R., Johnson, D.S.: Computers and Intractability; A Guide to the Theory of NP-Completeness, 1st edn. W. H. Freeman & Co., USA (1979)
5. Harper, L.H.: Optimal assignments of numbers to vertices. J. Soc. Ind. Appl. Math. **12**(1), 131–135 (1964)
6. Jiang, X., Liu, Q., Parthiban, N., Rajan, R.S.: A note on minimum linear arrangement for BC graphs. Discrete Math. Algorithms Appl. **10**(2), 1–7 (2018)
7. Knödel, W.: New gossips and telephones. Discrete Math. **13**, 95 (1975)
8. Miller, M., Rajan, R.S., Parthiban, N., Rajasingh, I.: Minimum linear arrangement of incomplete hypercubes. Comput. J. **58**(2), 331–337 (2015)
9. Rajasingh, I., Manuel, P., Rajan, B., Arockiaraj, M.: Wirelength of hypercubes into certain trees. Discrete Appl. Math. **160**(18), 2778–2786 (2012)
10. Rajasingh, I., Sundara Rajan, R., Manuel, P.: A linear time algorithm for embedding christmas trees into certain trees. Parallel Process. Lett. **25**(4), 1–17 (2015)
11. Xu, J.: Topological Structure and Analysis of Interconnection Networks. Kluwer Academic Publishers, Dordrecht (2001)

Graph Covering Using Bounded Size Subgraphs

Barun Gorain, Shaswati Patra, and Rishi Ranjan Singh(✉)

Indian Institute of Technology Bhilai, Sejbahar, Raipur, India
{barun,shaswatip,rishi}@iitbhilai.ac.in

Abstract. A variant of graph covering problem demands to find a set
of sub-graphs when the union of sub-graphs contain all the edges of G.
Another variant of graph covering problem requires finding a collection of
subgraphs such that the union of the vertices of subgraphs forms a vertex
cover. We study the later version of the graph covering problem. The
objective of these problems is to minimize the size/cost of the collection
of subgraphs. Covering graphs with the help of a set of edges, set of
vertices, tree or tour has been studied extensively in the past few decades.
In this paper, we study a variant of the graph covering problem using
two special subgraphs. The first problem is called *bounded component
forest cover* problem. The objective is to find a collection of minimum
number of edge-disjoint bounded weight trees such that the vertices of the
forest, i.e., collection of edge-disjoint trees, cover the graph. The second
problem is called *bounded size walk cover* problem. It asks to minimize
the number of bounded size walks which can cover the graph. Walks
allow repetition of vertices/edges. Both problems are a generalization
of classical vertex cover problem, therefore, are NP-hard. We give 4ρ
and 6ρ factor approximation algorithm for bounded component forest
cover and bounded size walk cover problems respectively, where ρ is an
approximation factor to find a solution to the tree cover problem.

Keywords: Graph Covering · Vertex Cover Problem · Tree Cover
Problem · Approximation Algorithm

1 Introduction

A set of vertices are said to cover an edge if at least one end vertex of that edge
is present in that set of vertices. Graph covering problems aim to find a subset
of graph vertices such that all edges are covered by that subset while minimizing
some objective function. The classical vertex cover problem is a graph covering
problem which requires finding a minimum size/cost subset of graph vertices that
covers all the edges of a given graph. The problem of covering graphs with specific
subgraphs is studied by several researchers in the past four decades [1–3,5]. The
objective in such problems is to determine an optimal size/cost collection of
subgraphs of a graph such that the union of vertices of the subgraphs covers

© The Author(s), under exclusive license to Springer Nature Switzerland AG 2023
A. Bagchi and R. Muthu (Eds.): CALDAM 2023, LNCS 13947, pp. 415–426, 2023.
https://doi.org/10.1007/978-3-031-25211-2_32

all edges of the original graph. Based on the topology of subgraphs, several variations of the problem are defined. As most of these variations are NP-hard, the main goal is to design efficient approximation schemes.

In this paper, we aim to study two variants of graph covering with bounded size subgraphs. The problems aim to cover the graph with a minimum number of subgraphs each of whose weight is bounded by a given real number. Formally, let $G = (V, E, w)$ be a weighted graph where $w : E \rightarrow R^+$. A forest cover of G is a collection of disjoint trees $\{T_1, T_2, \cdots, T_j\}$ such that the union of the vertices in all the trees in the collection will be a vertex cover. Note that disjoint trees do not have any common vertices/edges. The cardinality of a forest cover is j, the number of trees in the forest cover. We define a problem named *bounded component forest cover* (BCFC) problem as follows.

Definition 1. *For a given weighted graph $G = (V, E, w)$, and a non-negative real number λ, find a forest cover of minimum cardinality such that the weight of each tree in the forest cover is at most λ.*

Note that when $\lambda = 0$, the problem is reduced to the minimum vertex cover problem. Hence, we have the following result.

Theorem 1. *The BCFC problem over (G, λ) is NP-hard.*

The second problem is motivated by a real-life application problem monitoring a large art gallery. A guard can see the entire corridor from one of its end junction points. The objective is to place a minimum number of mobile guards in such a way that every corridor can be under the scrutiny of at least one guard in t time period, for a given time $t > 0$. If every guard moves with a constant average velocity v, then the movement routes of the guards decompose the graph into subgraphs, each of which has a length at most vt, and the set of vertices covered by the guards must form a vertex cover. In this case, each subgraph is a walk of length at most vt, and the walks in the solution may be intersecting, i.e., may have common edges/vertices.

Formally, let $G = (V, E)$ be a weighted graph with the weight function $w : E \rightarrow R^+$. A walk cover of G is a collection of walks $\{P_1, P_2, \cdots, P_j\}$ which are allowed to intersect, i.e., may have common edges/vertices such that the union of the vertices on all the walks in the collection, forms a vertex cover. The cardinality of a walk cover is j, the number of walks in the walk cover. Analogous to BCFC, we define a problem named *bounded size walk cover* (BSWC) as follows.

Definition 2. *For a given weighted graph $G = (V, E, w)$, and a non-negative real number λ, find a walk cover of minimum cardinality such that the weight of each walk in the walk cover is at most λ.*

If $\lambda = 0$, BSWC problem is reduced to the minimum vertex cover problem. Therefore, the following result holds.

Theorem 2. *The BSWC problem over (G, λ) is NP-hard.*

2   Related Work

In this section, we briefly mention works related to BCFC, BSWC and graph covering problems. The tree(tour) cover problem was first defined by Arkin et al. [1] in 1993. The tree(tour) cover problem of an edge weighted graph deals with finding a minimum weight tree(tour) in the graph such that the vertices of the tree(tour) are the vertices of some vertex cover of the graph. These two problems are NP-hard as an instance of vertex cover problem, and traveling salesperson problem [6] can be reduced to an instance of tree and tour cover problem, respectively. Arkin et al. have designed a 3.55 and 5.5-factor approximation algorithm for tree cover and tour cover problems, respectively. In [2,3], researchers have studied the tree cover problem and proposed improved approximation algorithms. Koneman et al. [2] gave a linear programming formulation for the tree cover problem and derived a 3-factor rounding algorithm. Fujito [3] gave a 2-factor approximation algorithm to find a minimum tree cover. Viet Hung Nguyen [4] established a 3.5 approximation factor for the tour cover using a compact linear program which is weaker as compared to 3-factor proposed by Konemann et al. [2]. Researchers have studied a similar problem called edge dominating set problem [5,7,8] that finds a subset of edges E_1 in a graph $G = (V, E)$ such that for each edge not in E_1 has at least one common end vertex with some edges of E_1. It is a minimization problem.. This problem is a special case of BSWC problem when $\lambda = 1$ and the graph is unweighted. Researchers [5,7,8] have proposed various approximation algorithms to solve the edge dominating set problem and the best-known algorithm has an approximation factor 2 [5]. Fujito and Nagamochi [5] and Parekh [9] have proposed 2-factor approximation algorithms to find minimum vertex cover, minimum edge dominating set, and some related problems. Monien and Speckeumeyer [11] have established an approximation factor ≤ 1.8 for finding a minimum vertex cover in all graphs with ≤ 146000 nodes. To find a minimum vertex cover in graphs authors [12,13] have proposed different approximation algorithms whose approximation factors are lesser than 2. In [14], authors have proved that it is NP-hard to establish an approximation factor lesser than 1.36067 for a vertex cover problem in a graph.

The problem of graph covering using walks is related to a well-studied problem of graph exploration by mobile agents. If the mobile agents have to monitor the edges of the network by visiting at least one of its end vertices, they have to visit walks containing all vertices of some vertex cover of the graph. The optimal number of agents required for edge exploration is a related problem to BSWC; therefore, we briefly mention a few results on edge exploration. In a graph exploration problem single or multiple mobile agents have to visit the nodes or edges of a graph. Many research works are concerned with the exploration of the graph by a single mobile agent, as discussed in [15–18,20]. In [21], authors have assumed that in the deterministic exploration of the graph by multiple agents, the movement of the agents are coordinated centrally. In [22], authors have designed different approximation algorithms for the collective exploration of an arbitrary graph by a group of mobile agents. In [19], authors have studied the problem graph exploration where starting from a node, a mobile agent has to visit all the vertices of an anonymous graph where the nodes do not have ids, but the edges incident on a node are labeled with port numbers. Dhar et al. [23,24] studied the edge exploration of an anonymous graph by a mobile agent.

3   Results

In this section, we present constant factor approximation schemes for both considered problems.

3.1   Constant Factor Approximation Algorithm for BCFC

Recall a tree cover problem in an edge weighted graph deals with finding a minimum weight tree such that the vertices of the tree form a vertex cover of the graph. First, we show that a constant factor approximation algorithm for the tree cover problem can be used to design a constant factor approximation algorithm to BCFC. The general idea is to find a tree cover of a given graph and then split the tree into bounded size components such that all vertices of the tree cover are preserved in the process of splitting. The resulting forest is a solution of BCFC problem on the given graph. The tree cover problem is NP-Hard; therefore, the proposed scheme obtains an approximated tree cover solution using some ρ-factor approximation algorithm. The following lemma from [25] helps us to find a solution of BCFC from a given solution of tree cover problem.

Lemma 1 ([25]). *Let $\beta > 0$ be a positive real number and let T be any tree with vertex set V_T and the edge set E_T. If for each $e \in E_T$, $w(e) \leq \beta$, then T can be split into sub-trees $\zeta_1, \zeta_2, \cdots, \zeta_k$ where $k \leq \max\{\lceil \frac{w(T)}{\beta} \rceil, 1\}$ such that $w(\zeta_i) \leq 2\beta$ for each $1 \leq i \leq k$.*

The procedure of how to split the tree into sub-trees is explained in [26].

Let $G = (V, E, w)$ be a given weighted graph. We define a weight function $w_{\frac{1}{2}}$ as follows:

$$w_{\frac{1}{2}}(e) = \begin{cases} \frac{2w(e)}{\lambda} & \text{if } w(e) \leq \frac{\lambda}{2}, \\ 1 & \text{otherwise.} \end{cases}$$

Let $G' = (V, E, w_{\frac{1}{2}})$ where $G = (V, E, w)$. Let $w_{\frac{1}{2}}(X)$ denote the sum of the weights of the edges in a subgraph X of graph G'. Similarly, $w(X)$ is defined for a subgraph X of graph G. The following lemma establishes a relationship between the optimal tree cover of G' and the number of trees in the optimal solution of BCFC.

Lemma 2. *Let OPT_{BCFC} be the number of sub-trees in the optimal solution of BCFC problem over (G, λ) and let OPT_{TC} be the optimal tree cover of G'. Then $w_{\frac{1}{2}}(OPT_{TC}) \leq 4OPT_{BCFC} - 2$.*

Proof. Let $\xi_1, \xi_2, \cdots, \xi_{OPT_{BCFC}}$ be the trees in an optimal solution of BCFC for (G, λ). Then by the definition of $w_{\frac{1}{2}}$, $w_{\frac{1}{2}}(\xi_i) \le 2$, for each i, $1 \le i \le OPT_{BCFC}$. Construct a graph $H = (V_H, E_H)$ with OPT_{BCFC} many vertices as follows.

For every tree ξ_i, take a vertex u_i in V_H. Add an edge $(u_i, u_j) \in E_H$, if there exists a vertex $v_i \in V_{\xi_i}$ and there exist a vertex $v_j \in V_{\xi_j}$ such that v_i and v_j are connected by a path with at most two edges in G. Assign $w(u_i, u_j) = \min_{\{P_{xy} | x \in V_{\xi_i}, y \in V_{\xi_j}\}} \{w(P_{xy})\}$, where P_{xy} is a path between x and y with at most two edges in G. Since G is connected, and the vertices of OPT_{BCFC} forms a vertex cover, therefore the graph H is also connected. Let τ be the minimum spanning tree of H with respect to w and E_τ be the set of edges in the τ. Note that for every edge in $e \in E_\tau$, $w_{\frac{1}{2}}(e) \le 2$, as there can be at most two edges in G corresponding to one edge in τ and the weight of an edge in G with respect to $w_{\frac{1}{2}}$ is at most 1. Let $Z = (\bigcup_{i=1}^{OPT_{BCFC}} \xi_i) \bigcup \tau$. Clearly, Z is a tree cover of G and $w_{\frac{1}{2}}(Z) \le \sum_{i=1}^{OPT_{BCFC}} w_{\frac{1}{2}}(\xi_i) + \sum_{e \in E_\tau} w_{\frac{1}{2}}(e)$. Recall, $w_{\frac{1}{2}}(\xi_i) \le 2$ and $|E_\tau| = |V_\tau| - 1 = |V_H| - 1 = OPT_{BCFC} - 1$, therefore, we have $w_{\frac{1}{2}}(Z) \le 2OPT_{BCFC} + 2(OPT_{BCFC} - 1) = 4OPT_{BCFC} - 2$.

Since, OPT_{TC} is an optimal tree cover of G', we have $w_{\frac{1}{2}}(OPT_{TC}) \le w_{\frac{1}{2}}(Z) \le 4OPT_{BCFC} - 2$ ∎

Next, we describe our approach to find a solution for BCFC problem over (G, λ). Let A be an approximation algorithm with ρ-factor approximation guarantee for the tree cover problem. Let APX_{TC} be the tree cover returned by A for the input graph $G' = (v, E, w_{\frac{1}{2}})$. First, we obtain an approximated tree cover APX_{TC} of G'. Then each edge e in APX_{TC} for which $w(e) > \frac{\lambda}{2}$ is deleted from APX_{TC}. After deletion of such edges, let APX_{TC} splits into h sub-trees $\chi_1, \chi_2, \cdots, \chi_h$. For $i = 1$ to h, a set of sub-trees S_i is computed from χ_i using the tree splitting strategy proposed in [26] such that weight of each sub-tree in S_i has weight at most 2. Finally, forest cover $APX_{BCFC} = \bigcup_{i=1}^h S_i$ is returned as the solution to the BCFC problem.

Theorem 3. *Let $|APX_{BCFC}|$ be the number of trees in the forest cover APX_{BCFC}, then $|APX_{BCFC}| \le 4 \cdot \rho \cdot OPT_{BCFC}$, when we have a ρ-factor approximation algorithm for the tree cover problem.*

Proof. Let T be a sub-tree in a set of sub-trees $S_i \subseteq APX_{BCFC}$. Then, $w_{\frac{1}{2}}(T) \le 2$. Furthermore, for each edge $e \in T$, $w(e) \le \frac{\lambda}{2}$ and $w(T) \le \lambda$. According to Lemma 1, the number of trees in the set of sub-trees S_i, $|S_i| \le \max\{\lceil w_{\frac{1}{2}}(\chi_i) \rceil, 1\} = \max\{\lceil \frac{2w(\chi_i)}{\lambda} \rceil, 1\} \le w_{\frac{1}{2}}(\chi_i) + 1$. Recall, $APX_{BCFC} = \bigcup_{i=1}^h S_i$, therefore, $|APX_{BCFC}| = \sum_{i=1}^h |S_i| \le \sum_{i=1}^h (w_{\frac{1}{2}}(\chi_i) + 1) = \sum_{i=1}^h w_{\frac{1}{2}}(\chi_i) + h$. Note that, $w_{\frac{1}{2}}(APX_{TC}) = \sum_{i=1}^h w_{\frac{1}{2}}(\chi_i) + h - 1$, therefore, we have $APX_{BCFC} \le w_{\frac{1}{2}}(APX_{TC}) + 1 \le \rho \cdot w_{\frac{1}{2}}(OPT_{TC}) + 1$. Using Lemma 2, we have $APX_{BCFC} \le 4\rho \cdot OPT_{BCFC}$. ∎

Theorem 4 ([3]). *There exist a 2 factor approximation algorithm for tree cover problem.*

In view of Theorem 3 and Theorem 4 we have the final result in this subsection.

Theorem 5. *There exists an 8-factor approximation algorithm for BCFC.*

3.2 Constant Factor Approximation Algorithm for BSWC

Recall that a solution to BCFC can be easily converted to a solution of BSWC by doubling the edges in each component of the forest cover, breaking the tour into two bounded size walks, which may intersect. Let $|APX_{BCFC}|$ be the number of trees in the forest cover APX_{BCFC}, which is a solution of BCFC given by the approximation algorithm given in the above section. Then, after doubling the trees in APX_{BCFC} and cutting the formed tour due to doubling into two walks, we would have $2|APX_{BCFC}|$ walks. Therefore, the number of walks would be less than or equal to $8 \cdot \rho \cdot OPT_{BCFC}$ when we have a ρ factor approximation algorithm for the tree cover problem.

In this section, we show that an approximation scheme for the tree cover problem can be used to design a constant factor approximation scheme for BSWC. The general idea is to find a tree cover of a given graph and then delete high-cost edges. This process may result in a forest. The edges in the resulting forest are doubled to form tours over vertices in all respective components of the forest. Splitting these tours into bounded-size walks results in a collection of walks which is a feasible solution for BSWC. We prove that this approximation approach guarantees to give $6 \cdot \rho \cdot OPT_{BSWC}$ solution for BSWC problem.

Let $G = (V, E, w)$ be a weighted undirected graph, where every edge $e \in E$ has a positive real weight. We define a weight function w' on the graph G such that for each edge $e \in G$, $w'(e) = \frac{w(e)}{\lambda}$ if $w(e) \leq \lambda$ else $w'(e) = 1$. Let $G' = (V, E, w')$ where $G = (V, E, w)$. Let λ be a non-negative real number. Note that two walks may intersect and may have common vertices/edges. A set of walks $\{P_1, P_2, \cdots, P_j\}$, such that each walk is of weight at most λ, is called bounded size walk cover if union of vertices in all the walks forms a vertex cover of G. For any real $\lambda \geq 0$, the objective of BSWC problems is to find the minimum cardinality walk cover of G such that the weight of each walk in the walk cover is at most λ.

The above problem is NP-hard. To solve this problem, we design an approximation algorithm that finds a tree cover of the graph and splits the tree cover into sub-trees of smaller size by deleting high-weight edges, but all the vertices of the tree cover must be present in the sub-trees. Deletion of high-weight edges is a classical mechanism to break a tree in problems that have bounded size constraints [26]. The proposed algorithm finds walk cover from a tree cover following the idea of constructing sub-trees from a minimum spanning tree given in the Algorithm 1 in the paper [27]. We have modified Algorithm 1 from [27] according to our requirement to result walks which may intersect. For the sake

---

Algorithm 1: Bounded Size Walk Cover Algorithm

---

1 Find an approximate tree cover APX_{TC} in G' using ρ-approximation tree cover algorithm .

2 From the tree cover APX_{TC} delete each edge e with cost $w(e) \geq \lambda$. Let k be the number of edges deleted from APX_{TC}. It splits APX_{TC} into $k + 1$ sub-trees denoted as $T_0, T_1, \cdots, T_k$.

3 **for** $i = 0$ *to* k **do**

4   Find a tour ET_i on T_i by doubling the edges.

5   Delete an arbitrary edge from ET_i, to get a path C_i.

6 **end**

7 Define $APX_{BSWC} = \emptyset$.

8 **for** $i = 0$ *to* k **do**

9   **while** $w(C_i) > \lambda$ **do**

10   Let $C_i = u_i^1 u_i^2 \cdots u_i^{|V(C_i)|}\}$.

11   Let u_i^j be the first vertex on C_i such that $w(u_i^j \cdots u_i^{j+1}) > \lambda$.

12          $APX_{BSWC} = APX_{BSWC} \bigcup (u_i^1 \cdots u_i^j)$, $C_i = C_i \setminus (u_i^1 \cdots u_i^j u_i^{j+1})$.

13   Delete all edges of the path $(u_i^1 \cdots u_i^j u_i^{j+1})$ from C_i. To delete the path $(u_i^1 \cdots u_i^j u_i^{j+1})$, we delete the vertices $\{u_i^1, \cdots, u_i^j\}$ and the edges $\{(u_i^1, u_i^2), \cdots, (u_i^{j-1}, u_i^j)\}$ from C_i.

14   **end**

15      $APX_{BSWC} = APX_{BSWC} \bigcup C_i$

16 **end**

17 Return APX_{BSWC}.

---

of completeness, the modified algorithm is summarized as Algorithm 1 in this paper and its working procedure is explained as follows. The algorithm finds an approximated tree cover APX_{TC} in the graph G using some ρ-factor approxima-tion algorithm. It deletes all heavy edges with weight more than λ from APX_{TC}. Let k number of edges are deleted which splits APX_{TC} into $k + 1$ different com-ponents $T_0, \cdots, T_k$. In each T_i, $i = 0, \cdots, k$, the algorithm finds a tour ET_i by doubling edges of T_i and at the end, it deletes one arbitrary edge from ET_i to get a walk C_i. Note that an edge may appear more than one time in such tour. For $i = 1, \cdots, k$, each walk C_i, is split into sub-walks of weight less than or equal to λ and those sub-walks are added to the solution APX_{BSWC}. Let the walk C_i is represented as a sequence of vertices $(u_i^1 u_i^2 \cdots u_i^{|V(C_i)|})$. In the walk C_i let u_i^j be the first vertex such that $w(u_i^1 \cdots u_i^{j+1}) > \lambda$, then it adds the walk $(u_i^1 \cdots u_i^j)$ to the set APX_{BSWC} and deletes the walk $(u_i^1 \cdots u_i^j u_i^{j+1})$ from C_i. To delete any walk $(u_i^1 \cdots u_i^j u_i^{j+1})$ from C_i , the algorithm deletes the vertices $\{u_i^1 \cdots u_i^j u_i^j\}$ and the edges $\{(u_i^1, u_i^2) \cdots (u_i^j, u_i^{j+1})\}$. It continues this process until $w(C_i) \leq \lambda$. Finally, we add the truncated walk C_i to the set APX_{BSWC}.

The execution steps of the Algorithm 1 are depicted with the help of an example, as shown in Fig. 1. Let $G = (V, E, w)$ be a positive edge-weighted graph as shown in Fig. 1a and let the bound on the weight of each walk be $\lambda = 20$. The algorithm first computes an approximate tree cover APX_{TC} in the graph G using

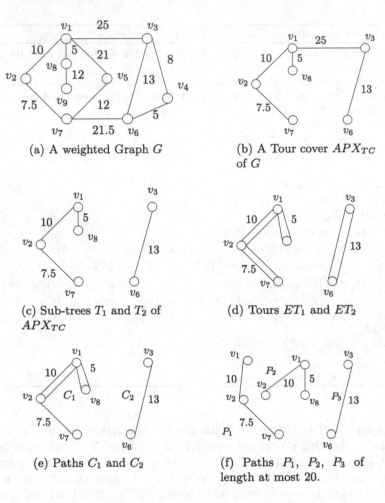

(a) A weighted Graph G

(b) A Tour cover APX_{TC} of G

(c) Sub-trees T_1 and T_2 of APX_{TC}

(d) Tours ET_1 and ET_2

(e) Paths C_1 and C_2

(f) Paths P_1, P_2, P_3 of length at most 20.

Fig. 1. Example: Construction of bounded size walk cover.

some existing tree cover algorithm. Let Fig. 1b be the approximated tree cover APX_{TC}. Then, the algorithm deletes the edge (v_1, v_3) whose weight is greater than $\lambda = 20$ from APX_{TC}. Deletion of (v_1, v_3) splits APX_{TC} into two different sub-trees T_1 and T_2 as shown in Fig. 1c. Next, the algorithm doubles the edges of sub-trees T_1 and T_2 and finds tours ET_1 and ET_2, respectively, as depicted in Fig. 1d. From each tour ET_1 and ET_2, the algorithm deletes an arbitrary edge and gets open walks C_1 and C_2 as shown in Fig. 1e. In this example (v_2, v_7) and (v_3, v_6) are deleted from ET_1 and ET_2 respectively. Note that an edge may occur twice in these walks due to doubling. In the walk C_1, the algorithm starts from node v_7 and visits up to the node v_1. Since the walk $P_1 = (v_7, v_2, v_1)$ is the largest visited walk with weight at most 20, it adds the sub-walk $P_1 = (v_7, v_2, v_1)$ into the solution and deletes the walk P_1 along with the edge (v_1, v_8) from C_1. Then it finds the sub-walk $P_2 = (v_8, v_1, v_2)$ from the remaining walk of P_1. Similarly, it finds sub-walk P_3 from the walk C_2. All the walks have a weight at most 20 as shown in Fig. 1f. Note that P_1 and P_2 are intersecting and have a common edge (v_1, v_2).

The following Lemmas and Theorem give correctness and derive the approximation factor of Algorithm 1. Recall APX_{BSWC} is the output of Algorithm 1, which is a set of walks.

Lemma 3. APX_{BSWC} *is a bounded size walk cover of graph G.*

Proof. The walks in APX_{BSWC} are constructed by deleting some edges of a tree cover of the graph G. So the walks in APX_{BSWC} include all the vertices of the tree cover of G. Therefore, the vertices of all the walks in APX_{BSWC} still form a vertex cover of graph G. According to Algorithm 1, the weight of each walk in APX_{BSWC} are bounded to be less or equal to λ. Hence, APX_{BSWC} is a feasible solution to BSWC problem in the graph G. $\square$

To establish the approximation factor of Algorithm 1, we define certain variables. Let OPT_{BSWC} be the minimum number of bounded size walks, which forms the optimal solution of $BSWC$ problem over a graph G. Let OPT_{TC} be the optimal tree cover of the graph G'. We establish a relation between OPT_{BSWC} and $w'(OPT_{TC})$.

Lemma 4. $w'(OPT_{TC}) \leq 3.OPT_{BSWC} - 2$

Proof. Let $\{Q_1, Q_2, \cdots, Q_{OPT_{BSWC}}\}$ be the set of walks which forms the optimal solution of $BSWC$ as shown in Fig. 2. Weight of each Q_i is less than or equal to λ, i.e. $w(Q_i) \leq \lambda$ and $w'(Q_i) \leq 1$. We construct a graph $H = (V_H, E_H)$, similarly to how we constructed a graph H in the proof of Lemma 2. The graph H contains all walks $Q_1, \cdots, Q_{OPT_{BSWC}}$ as a subgraph and contains a few extra edges/vertices from G to connect these walks into a single connected component. In the graph G, if two walks Q_i and Q_j have common vertices or edges, then join them into a single component by taking the union of those walks so that each edge/vertex appears exactly once. We start with graph G, and then, we contract each component from the previous step, which is a subgraph of G, into a single vertex by contracting all the edges and respective vertices to obtain a graph $G_c = (V_c, E_c)$. Afterward, we find a minimum spanning tree $MST\ T$ on the contracted graph G_c. Graph H is constructed from $MST\ T$ by reversing the contraction of the components. The graph H is a sub-graph of G containing all the vertices of walks in the set $\{Q_1, Q_2, \cdots, Q_{OPT_{BSWC}}\}$. The sub-graph H is a tree cover of G as depicted in Fig. 3.

Fig. 2. $Q_1, \cdots, Q_{OPT_{BSWC}}$ be the walks in the optimal solution of $BSWC$

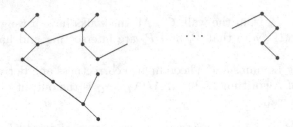

Fig. 3. Sub-graph H of G formed by edges of Q_i and MST T on G_c

In the graph G_c, consider two nodes u_x and u_y that represent two components that may be formed by union of few walks from the set $\{Q_1, Q_2, \cdots, Q_{OPT_{BSWC}}\}$. Let u_x and u_y are connected in G_c through a path p_{xy}, which is also present in MST T. Then, the number of edges on path p_{xy} is at most two. Otherwise, if three edges are present in p_{xy}, then any end vertex of middle edge can not be covered by any vertex of Q_i, for $1 \le i \le OPT_{BSWC}$, and hence the covering of all edges is not guaranteed. Therefore, weight of each such path p_{xy} in G', $w'(p_{xy}) \le 2$. Note that the vertices H form a vertex cover of G. As per the construction, the graph H contains all walks from the set $\{Q_1, Q_2, \cdots, Q_{OPT_{BSWC}}\}$ and at most $OPT_{BSWC} - 1$ many paths (of p_{xy} type) to connect all walks. The weight of each walk Q_i in G' is $w'(Q_i) \le 1$. Hence, the weight of sub-graph H in G' is given as $w'(H) \le OPT_{BSWC} + 2.(OPT_{BSWC} - 1) \le 3OPT_{BSWC} - 2$. Since H is also a tree cover of the graph G, weight of the optimal tree cover in G', $w'(OPT_{TC}) \le w'(H) \le 3OPT_{BSWC} - 2$.   □

Theorem 6. *Let $Y = |APX_{BSWC}|$ be the number of walks of a bounded weight in the set APX_{BSWC} resulted by Algorithm 1. Then, $Y \le 6 \cdot \rho \cdot OPT_{BSWC}$, when we have a ρ-factor approximation algorithm for the tree cover problem.*

Proof. The proposed algorithm obtains an approximate tree cover APX_{TC} of G' using some ρ-factor approximation algorithm, i.e. weight of APX_{TC} in G' is $w'(APX_{TC}) \le \rho.w'(OPT_{TC})$, where OPT_{TC} is the optimal tree cover of G'. The algorithm deletes the edges from APX_{TC} whose weight is greater than λ in G. After deletion of heavy edges, let APX_{TC} be split into $\{T_1, T_2, ..., T_m\}$ sub-trees. The algorithm doubles all edges in each sub-tree for a set of sub-tours $\{ET_1, ET_2, ..., ET_m\}$. After the deletion of an arbitrary edge from each sub-tour, the algorithm finds open walks $\{C_1, C_2, ..., C_m\}$. Each walk C_i is then splitted into bounded size sub-walks which are kept in the solution as walks that may intersect. Note that $Y \le \sum_{i=1}^{m}\lceil w'(C_i)\rceil \le \sum_{i=1}^{m} \frac{w(C_i)}{\lambda} + m$. We have $w'(APX_{TC}) = \sum_{i=1}^{m} w'(T_i) + m - 1$ which can be rewritten as $2w'(APX_{TC}) \ge \sum_{i=1}^{m} \frac{w(C_i)}{\lambda} + 2m - 2$. Hence $Y \le 2w'(APX_{TC}) - m + 2 \le 2\rho \cdot w'(OPT_{TC}) - m + 2$. Using Lemma 4, we have $Y \le 6 \cdot \rho \cdot OPT_{BSWC}$, as $\rho > 1$.   □

In view of Theorem 6 and Theorem 4, we have the final result in this subsection.

Theorem 7. *There exists a 12-factor approximation algorithm for BSWC.*

4   Conclusion and Future Work

In this paper, we have studied two graph covering problems: bounded component forest cover (BCFC) problem and bounded size walk cover (BSWC) problem. The problems are NP-hard due to a trivial reduction to the classical vertex cover problem when the bound on weight is 0. We designed $4.\rho$ factor approximation algorithm for the bounded component forest cover problem, where ρ is the approximation factor for finding a solution of tree cover problem. We further give a $6.\rho$ factor approximation algorithm for bounded size walk cover problem. Using 2-factor approximation algorithm given by Fujito [3] for tree cover problem, we have 8-factor and 12-factor approximation algorithm for BCFC and BSWC respectively.

Reducing these approximation factors is the first obvious direction to work on. One possible such improvement may be due to starting of with a subgraph other than a solution to the tree cover problem, which may bring down the final approximations factors. Studying bounded size path cover and bounded size intersecting sub-tour cover are other alternatives that we plan to look in future. Another future direction is to study the graph covering problem using other types of bounded size subgraphs.

Acknowledgements. The authors would like to thank the referees for the helpful comments. Barun Gorain and Rishi Ranjan Singh acknowledges the support of the Research Initiation Grant awarded by IIT Bhilai, India. Barun Gorain acknowledges the support of the Science and Engineering Research Board (SERB), Department of Science and Technology, Govt. of India (Grant Number: CRG/2020/-005964 and Grant Number: MTR/2021/000118).

References

1. Arkin, E.M., Halldorsson, M.M., Hassin, R.: Approximating the tree and tour covers of a graph. Inf. Process. Lett. **47**(6), 275–282 (1993)
2. Könemann, J., Konjevod, G., Parekh, O., Sinha, A.: Improved approximations for tour and tree covers. Algorithmica **38**(3), 441–449 (2004)
3. Fujito, T.: How to trim a MST: a 2-approximation algorithm for minimum cost-tree cover. ACM Trans. Algorithms (TALG) **8**(2), 1–11 (2012)
4. Nguyen, V.H.: Approximating the minimum tour cover with a compact linear program. In: van Do, T., Thi, H.A.L., Nguyen, N.T. (eds.) Advanced Computational Methods for Knowledge Engineering. AISC, vol. 282, pp. 99–104. Springer, Cham (2014). https://doi.org/10.1007/978-3-319-06569-4_7
5. Fujito, T., Nagamochi, H.: A 2-approximation algorithm for the minimum weight edge dominating set problem. Discrete Appl. Math. **118**(3), 199–207 (2002)
6. Christofides, N.: Worst-case analysis of a new heuristic for the traveling salesman problem. Technical report, GSIA, Carnegie Mellon University (1976)
7. Carr, R., Fujito, T., Konjevod, G., Parekh, O.: A 2 1/10-approximation algorithm for a generalization of the weighted edge-dominating set problem. In: Paterson, M.S. (ed.) ESA 2000. LNCS, vol. 1879, pp. 132–142. Springer, Heidelberg (2000). https://doi.org/10.1007/3-540-45253-2_13
8. Fujito, T.: On approximability of the independent/connected edge dominating set problems. Inf. Process. Lett. **79**(6), 261–266 (2001)

9. Parekh, O.: Edge dominating and hypomatchable sets. In: Proceedings of the Thirteenth Annual ACM-SIAM Symposium on Discrete Algorithms, pp. 287–291 (2002)
10. Edmonds, J., Johnson, E.L.: Matching: a well-solved class of integer linear programs. In: Jünger, M., Reinelt, G., Rinaldi, G. (eds.) Combinatorial Optimization — Eureka, You Shrink! LNCS, vol. 2570, pp. 27–30. Springer, Heidelberg (2003). https://doi.org/10.1007/3-540-36478-1_3
11. Monien, B., Speckenmeyer, E.: Ramsey numbers and an approximation algorithm for the vertex cover problem. Acta Informatica 22(1), 115–123 (1985)
12. Halperin, E.: Improved approximation algorithms for the vertex cover problem in graphs and hypergraphs. SIAM J. Comput. 31(5), 1608–1623 (2002)
13. Karakostas, G.: A better approximation ratio for the vertex cover problem. In: Caires, L., Italiano, G.F., Monteiro, L., Palamidessi, C., Yung, M. (eds.) ICALP 2005. LNCS, vol. 3580, pp. 1043–1050. Springer, Heidelberg (2005). https://doi.org/10.1007/11523468_84
14. Dinur, I., Safra, S.: The importance of being biased. In: Proceedings of the Thirty-Fourth Annual ACM Symposium on Theory of Computing, pp. 33–42 (2002)
15. Bender, M. A., Fernández, A., Ron, D., Sahai, A., Vadhan, S.: The power of a pebble: exploring and mapping directed graphs. In: Proceedings of the Thirtieth Annual ACM Symposium on Theory of Computing, pp. 269–278 (1998)
16. Fleischer, R., Trippen, G.: Exploring an unknown graph efficiently. In: Brodal, G.S., Leonardi, S. (eds.) ESA 2005. LNCS, vol. 3669, pp. 11–22. Springer, Heidelberg (2005). https://doi.org/10.1007/11561071_4
17. Panaite, P., Pelc, A.: Exploring unknown undirected graphs. J. Algorithms 33(2), 281–295 (1999)
18. Duncan, C.A., Kobourov, S.G., Kumar, V.A.: Optimal constrained graph exploration. ACM Trans. Algorithms (TALG) 2(3), 380–402 (2006)
19. Gorain, B., Pelc, A.: Deterministic Graph Exploration with Advice, ACM Trans. Algorithms 15, 8:1–8:17 (2018)
20. Diks, K., Fraigniaud, P., Kranakis, E., Pelc, A.: Tree exploration with little memory. J. Algorithms 51(1), 38–63 (2004)
21. Bender, M.A., Slonim, D.K.: The power of team exploration: two robots can learn unlabeled directed graphs. In: Proceedings 35th Annual Symposium on Foundations of Computer Science, pp. 75–85. IEEE (1994)
22. Frederickson, G.N., Hecht, M.S., Kim, C.E.: Approximation algorithms for some routing problems. In: 17th Annual Symposium on Foundations of Computer Science (SFCS 1976), pp. 216–227. IEEE (1976)
23. Dhar, A.K., Gorain, B., Mondal, K., Patra, S., Singh, R.R.: Edge exploration of a graph by mobile agent. In: Li, Y., Cardei, M., Huang, Y. (eds.) COCOA 2019. LNCS, vol. 11949, pp. 142–154. Springer, Cham (2019). https://doi.org/10.1007/978-3-030-36412-0_12
24. Dhar, A.K., Gorain, B., Mondal, K., Patra, S., Singh, R.R.: Edge exploration of anonymous graph by mobile agent with external help. Computing (2022). https://doi.org/10.1007/s00607-022-01136-8
25. Khani, M.R., Salavatipour, M.R.: Improved approximation algorithms for the min-max tree cover and bounded tree cover problems. Algorithmica 69(2), 443–460 (2014)
26. Even, G., Garg, N., Könemann, J., Ravi, R., Sinha, A.: Min-max tree covers of graphs. Oper. Res. Lett. 32(4), 309–315 (2004)
27. Gorain, B., Mandal, P.S., Mukhopadhyaya, K.: Generalized bounded tree cover of a graph. J. Graph Algorithms Appl. 21(3), 265–280 (2017)

Axiomatic Characterization of the Toll Walk Function of Some Graph Classes

Lekshmi Kamal K. Sheela[1] , Manoj Changat[1]([✉]) , and Iztok Peterin[2]

[1] Department of Futures Studies, University of Kerala, Trivandrum 695 581, India
lekshmisanthoshgr@gmail.com, mchangat@keralauniversity.ac.in
[2] University of Maribor, FEECS, Koroška 46, 2000 Maribor, Slovenia
iztok.peterin@um.si

Abstract. A toll walk $W = w_1 \ldots w_k$ in a graph G is a walk in which w_1 is adjacent only to w_2 and w_k is adjacent only to w_{k-1} among all vertices of W. The toll walk interval $T(u, v)$ between $u, v \in V(G)$ consists of all vertices that belong to a toll walk that starts in u and ends in v. We consider several standard axioms with respect to $T(u, v)$ and characterize some graph classes defined by forbidden induced subgraphs. In particular, we present a new characterization of interval graphs and some subclasses of asteroidal triples-free graphs.

Keywords: Transit function · Toll walk · *AT*-free graphs · Interval graphs · Axiomatic characterization

1 Introduction

A toll walk W from a vertex u to a different vertex v of a graph G is a special walk that contains exactly one neighbor of u, the second vertex of W, and exactly one neighbor of v, the for-last vertex of W. One can model a toll walk with the entrance fee or toll that is payed only once, that is at the first vertex when entering the rest of a system, that is represented by a graph. Similar one exit out of the system exactly once, at the neighbor of the final vertex.

Toll walks were first introduced by Alcon [1] as a tool to characterize dominating pairs in interval graphs. Later (despite the publication year), Alcon et al. [2] observed all vertices that belong to toll walks between u and v as the toll interval $T(u, v)$. This gives rise to the toll walk function $T : V(G) \times V(G) \to 2^{V(G)}$ of a graph G and to toll convexity. The main result from [2] is that G is a convex geometry (i.e. satisfies the Minkowski-Krein-Milman property stating that any convex subset is the convex hull of its extreme vertices) with respect to toll convexity if and only if G is an interval graph. Beside that also toll convexity on the standard products was studied and some classic convexity connected invariants were considered. Later, Gologranc and Repolusk [11,12] studied the toll number of standard graph products. Recently Dourado [10] considered the hull number for toll convexity.

© The Author(s), under exclusive license to Springer Nature Switzerland AG 2023
A. Bagchi and R. Muthu (Eds.): CALDAM 2023, LNCS 13947, pp. 427–446, 2023.
https://doi.org/10.1007/978-3-031-25211-2_33

In this work, we study the toll walk function T of a graph from axiomatic point of view. More accurately, we consider several well-known betweenness axioms and axioms connected with the induced path function together with some new axioms connected with the toll walk function. For the toll walk function we show the connection between the mentioned axioms and the forbidden induced subgraphs, which gives rise to a new axiomatic characterization of interval graphs and of subclass of asteroidal triples-free graphs.

2   Preliminaries

Let G be a graph with vertex set $V(G)$ and edge set $E(G)$. We consider only finite simple connected graphs, that is graphs without multiple edges and loops. The *open neighborhood* $N(v)$ of $v \in V(G)$ is the set $\{u \in V(G) : uv \in E(G)\}$ and the *closed neighborhood* $N[v]$ is $N(v) \cup \{v\}$. The *complement graph* of G is denoted as usual by $\overline{G}$. A *walk* W_k in a graph G is a sequence of k vertices $w_1, \ldots, w_k$ where $w_i w_{i+1} \in E(G)$ for every $i \in \{1, \ldots, k-1\}$. We simply write $W_k = w_1 \cdots w_k$. Notice that some vertices of W_k can repeat in W_k. If all vertices of a walk differ, then we say that W_k is a *path* P_k of G. A path $P_k = v_1 \cdots v_k$ will be also denoted as v_1, v_k-path and we say that P_k starts in v_1 and ends in v_k and $u \xrightarrow{P} x$ denotes the subpath of a path P with end vertices u and x. The distance $d(u, v)$ between $u, v \in V(G)$ is the minimum number of edges on a u, v-path or infinite if such a path does not exists. Any u, v-path of length $d(u, v)$ is called a u, v-*geodesic*.

A set of three vertices in a graph G such that each pair is joined by a path that avoids the neighborhood of the third vertex is known as an *asteroidal triple* in G. Graph G is called as *AT-free graph* if G has no asteroidal triple.

A *join* of graphs G_1 and G_2 is a graph $G_1 \vee G_2$ that is obtained by disjoint copies of G_1 and G_2 together with all the possible edges between the mentioned copies. Several well known graph families are joins, like a complete bipartite graph $K_{m,n} \cong \overline{K}_m \vee \overline{K}_n$ and wheels $W_n \cong K_1 \vee C_n$. In particular, for $n \geq 1$, we will be interested in joins $F_2^{n+1} \cong K_1 \vee P_{n+1}$ that are also called *fans* and $F_3^n \cong K_2 \vee P_n$ and $F_4^n \cong \overline{K}_2 \vee P_n$.

2.1   Transit Functions and Axioms

A *transit function* on a set V is a function $R : V \times V \longrightarrow 2^V$ such that for every $u, v \in V$ the following three conditions hold:

(t1) $u \in R(u, v)$;
(t2) $R(u, v) = R(v, u)$;
(t3) $R(u, u) = \{u\}$.

The *underlying graph* G_R of a transit function R is a graph with vertex set V, where distinct vertices u and v are adjacent if and only if $R(u, v) = \{u, v\}$.

The well studied transit functions in graphs are the interval function I_G, induced path function J_G and the all path function A_G. The *interval function* I_G of a connected graph G is defined with respect to the standard distance d in G as $I : V \times V \longrightarrow 2^V$ where

$$I_G(u,v) = \{w \in V(G) : w \text{ lies on some } u, v\text{-geodesic in } G\}$$

which can also be expressed by the distances as

$$I_G(u,v) = \{w \in V(G) : d(u,w) + d(w,v) = d(u,v)\}.$$

An induced path is a chordless path, where a chord of a path is an edge joining two non-consecutive vertices of a path. The *induced path transit function* $J(u,v)$ of G is a natural generalization of the interval function and is defined as $J(u,v) = \{w \in V(G) : w \text{ lies on an induced } u, v\text{-path}\}$. Well known is also the *all-paths transit function* $A(u,v) = \{w \in V(G) : w \text{ lies on some } u, v\text{-path}\}$, which consists of the vertices lying on all u, v-paths. For any two vertices u and v of a connected graph G, it is clear that $I(u,v) \subseteq J(u,v) \subseteq A(u,v)$.

Probably first approach to axiomatic description of a transit function I_G for a tree G goes back to Sholander [17]. His work was later improved by Chvátal et al. [8]. A full characterization of I_G for a connected graph G was presented by Mulder and Nebeský [15]. The argument $x \in R(u,v)$ can be interpreted as x is between u and v. All the above mentioned characterizations are framed using a set of first order axioms on a transit function. It is proved in [7] that the all paths function A of a connected graph G also possess a first order axiomatic characterization similar to that of the interval function I. In this paper, we consider only first order axioms on a transit function R.

Axiom (b1). If there exists elements $u, v, x \in V$ such that $x \in R(u,v), x \neq v$, then $v \notin R(x,u)$.

Axiom (b2). If there exists elements $u, v, x \in V$ such that $x \in R(u,v)$, then $R(u,x) \subseteq R(u,v)$.

These two Axioms are enough to assure the connectedness of G_R of a transit function R as shown in [6].

Lemma 1. *[6] If the transit function R on a non-empty set V satisfies Axioms (b1) and (b2), then G_R is connected.*

It is proved by Nebeský [16] that a first order axiomatic characterization of the induced path function J of an arbitrary connected graph G in a similar fashion to that of the interval function I is not possible. For a transit function satisfying the betweenness Axioms (b1) and (b2) axiomatic characterizations are presented in [6] with the additional axioms defined as follows.

Axiom (J0). If there exist different elements $u, x, y, v \in V$ such that $x \in R(u,y)$ and $y \in R(x,v)$, then $x \in R(u,v)$.

Axiom (J2). If there exist elements $u, v, x \in V$ such that $R(u, x) = \{u, x\}$, $R(x, v) = \{x, v\}, u \neq v$ and $R(u, v) \neq \{u, v\}$, then $x \in R(u, v)$.

Axiom (J3). If there exist elements $u, v, x, y \in V$ such that $x \in R(u, y), y \in R(x, v), x \neq y, R(u, v) \neq \{u, v\}$, then $x \in R(u, v)$.

There is a nice explanation of Axiom (J2) for G_R. The conditions demand that x is a common neighbor of nonadjacent different vertices u and v in G_R. So, Axiom (J2) yields that a common neighbor of two nonadjacent vertices u and v belong to $R(u, v)$. Notice also that Axiom (J0) is a relaxation of Axiom (J3).

2.2 Toll Walks

A *toll walk* between two different vertices w_1 and w_k of a finite connected graph G are vertices $w_1, \ldots, w_k$ that satisfy the following conditions:

- $w_i w_{i+1} \in E(G)$ for every $i \in \{1, \ldots, k-1\}$,
- $w_1 w_i \in E(G)$ if and only if $i = 2$,
- $w_k w_i \in E(G)$ if and only if $i = k - 1$.

That is, a toll walk W from u to v is a walk in which u is adjacent only to the second vertex of W, and v is adjacent only to the for-last vertex of W. Notice that if $uv \in E(G)$, then the only toll walk between u and v is uv. In addition we define a toll walk that starts and ends in the same vertex w as w itself.

The function $T : V \times V \to 2^V$ defined as

$$T_G(u, v) = \{x \in V(G) : x \text{ lies on a toll walk between } u \text{ and } v\}$$

is called the *toll walk function* on G. Clearly T is a transit function since T fulfills all three transit axioms. It is clear from the definition of the toll walk function T on G that G and G_T are isomorphic.

The following lemma by Liliana Alcon et al. [2] gives a characterization of vertices that belong to a toll walk in a graph, which we use frequently.

Lemma 2. *[2] A vertex v is in some toll walk between two different non-adjacent vertices x and y if and only if $N[x] - \{v\}$ does not separate v from y and $N[y] - \{v\}$ does not separate v from x.*

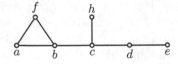

Fig. 1. $T(a, e) = \{a, b, c, d, e, h\}$ and $A(a, e) = \{a, b, c, d, e, f\}$ are incomparable.

For any two vertices u and v of a connected graph G, it is clear that $I(u,v) \subseteq J(u,v) \subseteq T(u,v)$. But the all path transit function A and toll walk transit function T of a graph are incomparable. In Fig. 1 $afbcde$ is a path, however it is not a toll walk, since a is adjacent to both f and b. So, a path need not to be a toll walk. While $abchcde$ is a toll walk that is not a path. So, a toll walk need not to be a path.

Let the three vertices u, v, w be an asteroidal triple. Then there exist a u, v-path denoted by $P_{u,v}$ that does not contain a neighbor of w, a u, w-path which avoids a neighbor of v and a v, w-path that does not contain a neighbor of u. That is $N[w]$ does not separate u from v, since $P_{u,v}$ does not contain a neighbor of w. Similarly $N[v]$ does not separate u from w and $N[u]$ does not separate v from w. That is, we have $u \in T(v,w), v \in T(u,w)$ and $w \in T(u,v)$ by Lemma 2.

Notice that the toll walk function satisfies Axiom (J2) but not Axioms (b1) and (b2) on a connected graph G. We define new Axioms (TW1), (TW2) and (TW3) for a transit function $R : V \times V \to 2^V$ and the toll walk function satisfies these axioms for any connected graph G as proved in Sect. 3

Axiom (TW1). If there exist different elements u, v, x, y, z such that $x, y \in R(u,v), x \neq y$ and $R(x,z) = \{x,z\}$ and $R(z,y) = \{z,y\}$ and $R(x,v) \neq \{x,v\}$ and $R(u,y) \neq \{u,y\}$, then $z \in R(u,v)$.

Axiom (TW2). If there exist different elements u, v, x, z such that $x \in R(u,v)$ and $R(u,x) \neq \{u,x\}$ and $R(x,v) \neq \{x,v\}$ and $R(x,z) = \{x,z\}$, then $z \in R(u,v)$.

Axiom (TW3). If there exist different elements u, v, x such that $x \in R(u,v)$, then there exist $v_1 \in R(x,v), v_1 \neq x$ with $R(x,v_1) = \{x,v_1\}$ and $R(u,v_1) \neq \{u,v_1\}$.

2.3   Characterizations by Forbidden Induced Subgraphs

Let G and $G_1, G_2, \ldots, G_k$ be connected graphs. We say that G is $(G_1 G_2 \cdots G_k)$-free graph if G has no induced subgraphs isomorphic to $G_1, G_2, \ldots, G_k$. The list of forbidden subgraphs can be infinite and we can write also a whole family of graphs instead of one graph. There are several graph classes that are defined by forbidden (induced) subgraphs. Probably the most famous example are $K_{1,3}$-free graphs, better known as claw-free graphs. Otherwise, a characterization with respect to forbidden (induced) subgraphs of a graph class is always desirable. Such a characterizations yield that a graph class is hereditary. Hereditary properties were studied extensively, see [3,4] for a flavor. One of the best known such characterizations are *bipartite graphs* that are C_{2k+1}-free graphs for $k \geq 1$. Another graph class that can be define as some induced cycle free graphs are *chordal graphs* which are C_k-free graphs for every $k \geq 4$. It was shown in [5] that Axiom (J0) for J_G is characteristic for chordal graphs. An induced cycle C_k, $k \geq 5$, is also called a *hole*. If we forbid all holes in G, then we say that G is *hole-free* or H-free for short.

Interval graphs are the intersection graphs of the intervals on the real line. This means that the vertices of an interval graph represent a set of intervals on the real line and two intervals are adjacent whenever they intersect. Interval graphs have a representation with infinite forbidden induced subgraphs, however this can be expressed in more dense style if we forbid asteroidal triples. The following classic result was proven by Lekkerkerker and Boland in [14].

Theorem 1. *[14] Graph G is an interval graph if and only if G is chordal AT-free graph.*

One can consider many different forbidden subgraphs, but often forbidden subgraphs have a small number of vertices. Some examples of such graphs that will be important later in this work are at Fig. 2. At this figure are from left to right graphs *house* denoted in this paper by H, cycle C_5, *domino* D and graph P. Notice a small notation discrepancy in notation as H-free can mean a house-free and hole-free graph as well. We solve this by never using just hole-free graphs, but always when we will have hole-free graph, it will also be a house-free and we will denote this by HH-free graph. If there is only one H, then it means house. In particular, (HC_5)-free graph is a graph without an induced house and induced C_5 and (HHD)-free is a graph without a house, a hole and a domino graph as an induced subgraph. Also (PAT)-free graph is a graph without a graph P and without asteroidal triples.

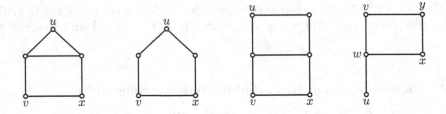

Fig. 2. Graphs house H, C_5, domino D and P (from left to right).

There are several connections in literature between special transit functions that fulfill certain axioms and forbidden induced subgraphs characterization. See for instance [5] where it was shown that I fulfilles Axiom (J3) on graph G if and only if G is $HHD(K_1 \vee P_4)$-free graph among many other results of this type. The following result is from [6].

Lemma 3. *[6] If R is a transit function on a non-empty finite set V satisfying Axioms (b1), (J2) and (J3), then G_R is HHP-free.*

Recall the families of graphs $F_2^{n+1} \cong K_1 \vee P_{n+1}$ and $F_3^n \cong K_2 \vee P_n$ and $F_4^n \cong \overline{K}_2 \vee P_n$ for $n \geq 1$. We will extend each of them into a new families XF_2^{n+1}, XF_3^n and XF_4^n, respectively.

We obtain XF_2^{n+1} from $F_2^{n+1} \cong K_1 \vee P_{n+1}$ by adding three new vertices u, v, x where u is adjacent only to the first vertex of path P_{n+1}, v is adjacent only to the last vertex of path P_{n+1} and x is adjacent only to the vertex of K_1. The family XF_2^{n+1} is represented in the middle of the last line on Fig. 3.

Both XF_3^n and XF_4^n can be obtained from $F_3^n \cong K_2 \vee P_n$ and $F_4^n \cong \overline{K}_2 \vee P_n$, respectively, by the same procedure. For this let $P_n = p_1 \ldots p_n$ and $V(K_2) = \{y_1, y_2\} = V(\overline{K}_2)$. We again add three new vertices u, v, x where u is adjacent only to p_1 and y_1, v is adjacent only to p_n and y_2 and x is adjacent only to y_1 and y_2. The families XF_3^n and XF_4^n are represented in the for-last and last place, respectively, of the last line on Fig. 3.

The following characterization of AT-free graphs with forbidden induced subgraphs from [13], see also [18], will be important later. All the forbidden induced subgraphs are depicted on Fig. 3. We use the same notation as presented in [18].

Theorem 2. *[13] A graph G is $(C_k T_2 X_2 X_3 X_{30} \ldots X_{41} XF_2^{n+1} XF_3^n XF_4^n)$-free for $k \geq 6$ and $n \geq 1$ if and only if G is AT-free graph.*

3 Axioms on the Toll Walk Function

In this section, we discus the axioms satisfied by the toll function in an arbitrary connected graph as well as those satisfied by the special classes of graphs that are the subclasses of AT-free graphs. We begin with axioms satisfied by the toll walk function for an arbitrary connected graph. The proof is presented in Appendix A.

Proposition 1. *The toll walk function satisfies the Axioms (TW1), (TW2) and (TW3).*

The toll walk function need not satisfy Axiom (b1) for arbitrary connected graphs. The following theorem characterizes the graph class in which the toll walk function satisfies Axiom (b1).

Theorem 3. *The toll walk function T of a graph G satisfies Axiom (b1) if and only if G is (HC_5DAT)-free graph.*

Proof. First assume that a graph G is not (HC_5DAT)-free graph. Then G contains an asteroidal triple or a domino or a C_5 or a house as an induced subgraph. If G contains an asteroidal triple u, v, w, then there exist at least one path joining each pair that avoids the neighbors of the third vertex. That is u, v, w are independent and $N[u]$ does not separate w from v, $N[v]$ does not separate w from u and $N[w]$ does not separate u from v. By Lemma 2 we have $w \in T(u, v)$, $u \in T(w, v)$ and $v \in T(w, u)$. This means that T does not satisfy Axiom (b1). If G contains a domino, a C_5 or a house as an induced subgraph, then in each case, we can find vertices u, v, x as shown at Fig. 2, such that v and x are adjacent and u is not adjacent to v and x. Clearly $x \in T(u, v)$ and $v \in T(x, u)$, hence T does not satisfy Axiom (b1).

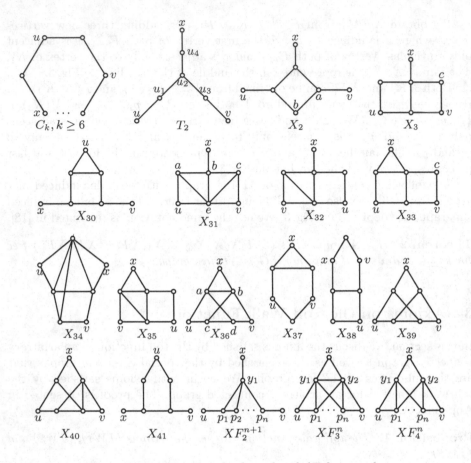

Fig. 3. Forbidden induced subgraphs of AT-free graphs.

Conversely suppose that Axiom (b1) does not hold for T for vertices u, v, x where $x \in T(u, v)$ and $x \neq v$ and $v \in T(x, u)$. Clearly $x \neq u$, because otherwise we have $v \in T(x, x) = \{x\}$, a contradiction. Similar $u \neq v$ as $v \neq x \in T(u, v)$. Since $x \in T(u, v)$ and $v \in T(x, u)$, $N[u] - \{x\}$ does not separate x from v, $N[v] - \{x\}$ does not separate x from u and $N[x] - \{v\}$ does not separate u from v by Lemma 2. If u, x, v are independent vertices, then, they form an asteroidal triple.

To avoid this, we may assume that one of ux, uv, vx must be an edge. If $uv \in E(G)$, then we get a contradiction with $x \in T(u, v)$ and $u \neq x \neq v$. Similar, $ux \in E(G)$ yields a contradiction with $v \in T(x, u) = \{u, x\}$. So, $vx \in E(G)$. Since $x \in T(u, v)$ there exists a u, v-walk W_1 without any neighbors of v beside x. Similar, as $v \in T(x, u)$ there exists a u, x-walk W_2 without any neighbors of x beside v. In W_2 we can find a v, u-induced path $P = vv_1 \ldots v_p u$ without any neighbors of x different than v and in W_1 an x, u-induced path $Q = xx_1 \ldots x_q u$ without any neighbors of v different than x. Notice that $v_1 \neq u$, because otherwise $uv \in E(G)$, a contradiction. Also $x_1 \neq u$ as $ux \notin E(G)$. Hence, $p \geq 1$ and $q \geq 1$. Therefore $v \xrightarrow{P} u \xrightarrow{Q} xv$ yields a cycle C_n where $n \geq 5$. If C_n is induced, then G contains a cycle of length at least five. If $n = 5$, then G is not a C_5-free graph and if $n > 5$, then G contains an asteroidal triple. So, we may assume that C_n is not induced and there exists a chord between x_i, $i \in \{1, \dots, q\}$, and v_j, $j \in \{1, \dots, p\}$. To avoid induced C_n, $n \geq 5$, there exists a chord $x_1 v_1$. If $p = 1 = q$, then u, x, v_1, v, x_1 induce a house H and G is not an H-free graph. Next we may assume without loss of generality that $p = 1$ and $q > 1$. If $v_1 x_2 \in E(G)$, then v, x, v_1, x_1, x_2 induce a house H again and G is not an H-free graph. Now either $x_2 u \in E(G)$ (that is $q = 2$) or $v_1 x_3 \in E(G)$ (when $q > 2$) and $v_1 x_2 \notin E(G)$ to avoid an induced C_n, $n \geq 5$. In the first case v, x, v_1, x_1, x_2, u and in the second case v, x, v_1, x_1, x_2, x_3 induce a domino D and G is not a D-free graph. Finally, let both $p, q \geq 2$. Again we have some chords to avoid induced C_n, $n \geq 5$. If $x_1 v_2 \in E(G)$ or $x_2 v_1 \in E(G)$, then we obtain an induced house H on v, x, v_1, x_1, z where $z = v_2$ or $z = x_2$, respectively. Otherwise one of $x_1 v_3, v_1 x_3, x_2 v_2$ must be an edge (here it is possible that $v_3 = u$ or $x_3 = u$). In all three cases we obtain an induced domino D on vertices v, x, v_1, x_1 and v_2, v_3 or x_2, x_3 or x_2, v_2, respectively. Therefore G is not an H- or a D-free graph and the proof is concluded.   □

We continue with a slightly modified Axiom (b1) for a transit function R for independent elements as follows to obtain the sequential characterization of AT-free graphs.

Axiom (b1'). If there exist elements $u, v, x \in V$ such that $x \in R(u, v), v \neq x, R(v, x) \neq \{v, x\}$, then $v \notin R(u, x)$.

It is clear that in a house and C_5 if we take u, v, x in such a way that $x \in R(u, v), v \neq x, R(v, x) \neq \{v, x\}$, then $R(u, x) = \{u, x\}$. That is in house and C_5, T satisfies Axiom (b1'). Also, in a domino, if the three vertices u, v, x are independent, then T fulfills Axiom (b1') for them. That is, if we replace Axiom (b1) by Axiom (b1') in Theorem 3, then the underlying graph G_T is AT-free.

Corollary 1. *The toll walk function T on a graph G satisfies Axiom (b1') if and only if G is AT-free.*

Notice that Axiom (b1) does not contain condition $R(v, x) \neq \{v, x\}$ and therefore it implies Axiom (b1'). In other words Axiom (b1') represents a more relaxed condition than Axiom (b1).

The following two propositions will be needed later. Their proofs are quite standard and can be found in Appendix A.

Proposition 2. *Let T be a tolled walk function on a connected graph G. If T satisfies Axiom (b1) on G then T satisfies Axiom (b2).*

Proposition 3. *Let T be a tolled walk function on a graph G which does not contain a P-graph as an induced subgraph. If T satisfies Axiom (b1) on G, then T satisfies Axiom (J3).*

4   A Characterization of the Toll Walk Function of Subclasses of AT-Free Graphs

In this section, we define a set of first order axioms for an arbitrary transit function R to be the toll walk function of the concerned graph class.

The following easy lemma will often be handy. Its proof is in Appendix A.

Lemma 4. *Let R be a transit function on a non-empty finite set V satisfying Axioms (J2), (J3) and (TW2). If P_n, $n \geq 2$, is an induced u, v-path in G_R, then $V(P_n) \subseteq R(u, v)$. Moreover, if z is adjacent to an inner vertex of P_n that is nonadjacent to u or to v in G_R, then $z \in R(u, v)$.*

We continue with a result that is basically one implication of a characterization presented in Theorem 6. The proof contains similar ideas as the proof of Theorem 3 and can be found in Appendix A.

Theorem 4. *If R is a transit function on a non-empty finite set V satisfying Axioms (b1), (J2), (J3), (TW1) and (TW2), then G_R is (HC_5PAT)-free graph.*

Next goal is to derive which axioms are needed for an arbitrary transit function R to be forced to be the toll walk transit function.

Theorem 5. *If R is a transit function on a non-empty finite set V that satisfies Axioms (b1), (b2), (J2), (J3), (TW1), (TW2) and (TW3), then $R = T$ on G_R.*

Proof. Let u and v be two distinct vertices of G_R and first assume that $x \in R(u, v)$. We have to show that $x \in T(u, v)$ on G_R. Clearly $x \in T(u, v)$ whenever $x \in \{u, v\}$. So assume that $x \notin \{u, v\}$. If $R(u, x) = \{u, x\}$ and $R(x, v) = \{x, v\}$, then uxv is a toll walk of G_R and $x \in T(u, v)$ follows. Suppose next that $R(x, v) \neq \{x, v\}$. We will construct an x, v-path Q in G_R without a neighbor of u (except possibly x). For this let $x = v_0$. By Axiom (TW3) there exist a neighbor of v_0, say v_1, and $v_1 \in R(v_0, v)$ with $R(u, v_1) \neq \{u, v_1\}$. Since $v_1 \in R(v_0, v)$, by Axioms (b1) and (b2) we have $R(v_1, v) \subseteq R(v_0, v)$ and $v_0 \notin R(v_1, v)$. Hence, $R(v_1, v) \subset R(v_0, v)$ follows. Moreover, $v_1 \in R(v_0, v) \subseteq R(u, v)$ by Axiom (b2). If $v_1 \neq v$, then we can continue with the same procedure to get v_2, where $R(v_1, v_2) = \{v_1, v_2\}$, $R(u, v_2) \neq \{u, v_2\}$, $R(v_2, v) \subset R(v_1, v)$ and $v_2 \in R(u, v)$. By repeating this step we obtain a sequence of vertices $v_0, v_1, \ldots, v_q$, $q \geq 2$, such that

1. $R(v_i, v_{i+1}) = \{v_i, v_{i+1}\}, i \in \{0, 1, \ldots, q - 1\}$,
2. $R(u, v_i) \neq \{u, v_i\}, i \in \{1, \ldots, q\}$,
3. $R(v_{i+1}, v) \subset R(v_i, v), i \in \{0, 1, \ldots, q - 1\}$.

Clearly this sequence needs to stop, because V is finite and by the last condition. Hence, we may assume that $v_q = v$. Now, if $R(u, x) = \{u, x\}$, then we have a toll u, v-walk $uxv_1 \ldots v_{q-1}v$ and $x \in T(u, v)$. Otherwise, $R(u, x) \neq \{u, x\}$ and we can symmetrically build a sequence $u_0, u_1, \ldots, u_r$, where $u_0 = x$, $u_r = u$ and $u_0u_1 \ldots u_r$ is an x, u-path in G_R that avoids $N[v]$. Clearly, $uu_{r-1}u_{r-2} \ldots u_1xv_1 \ldots v_{q-1}v$ is a toll u, v-walk and $x \in T(u, v)$.

Now suppose that $x \in T(u, v)$ and $x \notin \{u, v\}$. We have to show that $x \in R(u, v)$. By Lemma 2 $N[u] - x$ does not separate x and v and $N[v] - x$ does not separate u and x. Let W be a toll u, v-walk containing x. Clearly W contains an induced u, v-path, say Q. If x belongs to Q, then $x \in R(u, v)$ by Lemma 4. So, we may assume that x does not belongs to Q. The underlying graph G_R is AT-free by Theorem 4. Thus, Q contains a neighbor of x, say x'. If $R(u, x') = \{u, x'\}$ and $R(x', v) = \{x', v\}$, then we have a contradiction with W being a toll u, v-walk containing x. Without loss of generality, we may assume that $R(x', v) \neq \{x', v\}$. If also $R(u, x') \neq \{u, x'\}$, then $x \in R(u, v)$ by the second claim of Lemma 4. So, let now $R(u, x') = \{u, x'\}$. Since x and v are not separated by $N[u] - \{x\}$ by Lemma 2, there exists an induced x, v-path S without a neighbor of u. Let $S = s_0 s_1 \cdots s_k$, $s_0 = x$ and $s_k = v$ and let s_j be the first vertex of S that belongs also to Q. Notice that s_j can equal to v but it is different than x' and that $j > 0$. We may choose S such that it minimally differ from Q. This means that $s_0, \ldots, s_{j-2}$ are adjacent only to x' on Q. If $x' s_{j-1} \notin E(G_R)$, then $x' \xrightarrow{Q} s_j s_{j-1} s_{j-2} x'$ form an induced cycle. This yields a hole or a P graph together with u, a contradiction because G_R is $(HC_5 PAT)$-free by Theorem 4. Hence, $x' s_{j-1} \in E(G_R)$. We have $x', s_j \in R(u, v)$ by Lemma 4 and clearly $x' \neq s_j$, $R(s_{j-1}, s_j) = \{s_{j-1}, s_j\}, R(x', s_{j-1}) = \{x', s_{j-1}\}$, which gives $s_{j-1} \in R(u, v)$ by Axiom (TW1). We continue with the same step $j - 1$ times, only that we lower the indexes of s_{j-1} and s_j at each step by $i \in \{1, \ldots, j-1\}$ in the natural order and get that consequently $s_{j-2}, s_{j-3}, \ldots, s_0 \in R(u, v)$. This completes the proof because $s_0 = x \in R(u, v)$. $\qquad\square$

The toll walk function satisfy Axioms (J2), (TW1), (TW2) and (TW3) for a graph G. So, using Theorems 3, 4 and 5 and Propositions 2 and 3, we have the following characterization of toll walk function of $(HC_5 PAT)$-free graph because a P-free graph implies also a D-free graph.

Theorem 6. *Let R be a transit function on a finite set V and G_R be the underlying graph of R. Then R satisfy Axioms (b1), (b2), (J2), (J3), (TW1), (TW2) and (TW3) if and only if G_R is $(HC_5 PAT)$-free graph and $R = T$ on G_R.*

In Appendix B we present several examples that establishes the independence between Axioms (b1), (b2), (J2), (J3), (TW1), (TW2) and (TW3).

We continue with a similar approach with more relax conditions. For this we use Axiom (b1') instead of Axiom (b1). The results and their proofs (that can be found in Appendix A) are somewhat similar as in the case of Axiom (b1).

Theorem 7. *If R is a transit function on a non-empty finite set V satisfying Axioms (b1'), (J2), (J3), (TW1) and (TW2), then G_R is PAT-free graph.*

Also Lemma 1 can be generalized if we replace Axiom (b1) with Axiom (b1').

Proposition 4. *If a transit function R on a non-empty finite set V satisfies Axioms (b1') and (b2), then G_R is connected.*

The following result is an attempt to get a similar statement as in Theorem 5, where the conditions are relaxed by replacing Axiom (b1) with Axiom (b1'). However we can show only that $T \subseteq R$, which means $T(u, v) \subseteq R(u, v)$ for any elements $u, v \in V$. This is not surprising due to Theorem 5

Theorem 8. *If R is a transit function on a non-empty finite set V satisfying Axioms (b1'), (b2), (J2), (J3), (TW1), (TW2) and (TW3), then $T \subseteq R$ on G_R.*

We end this section with a remark that we do not have the equality in above theorem.

5   A New Characterization of Interval Graphs with a Help of the Toll Walk Function

In this section, we present a new characterization of interval graphs that involves the toll walk function T. The proofs are in Appendix A.

Theorem 9. *The toll walk function T on a graph G satisfies Axioms (b1) and (J0) if and only if G is an interval graph.*

Theorem 10. *If R is a transit function on a non-empty finite set V satisfying Axioms (b1), (J2), (J0), (TW1) and (TW2), then G_R is an interval graph.*

By Theorem 5 the following result holds since Axiom (J0) implies Axiom (J3).

Theorem 11. *If R is a transit function on a non-empty finite set V satisfying Axioms (b1), (b2), (J2), (J0), (TW1), (TW2) and (TW3), then $R = T$ on G_R.*

We have for a graph G, the toll walk function satisfy Axioms (J2), (TW1), (TW2) and (TW3). By Thoerems 9, 10 and 11 and Proposition 2 we have the following characterization of the toll walk function of an interval graph.

Theorem 12. *A transit function R on a finite set V satisfy Axioms (b1), (b2), (J2), (J0), (TW1), (TW2) and (TW3) if and only if G_R is an interval graph and $R = T$ on G_R.*

The independence of the Axioms (b1), (b2), (J2), (J0), (TW1), (TW2) and (TW3) will easily follow from the Examples in Appendix B.

Acknowledgments. Lekshmi Kamal K.S. acknowledges the financial support from CSIR, Government India for providing CSIR Senior Research Fellowship (CSIR-SRF) (No 09/102(0260)/2019-EMR-I).

Appendix A

In this appendix we have detailed proofs of the results that use standard approaches or are similar to the proofs presented in the main part of the paper.

Proof of Proposition 1. (TW1): Suppose $x, y \in T(u, v)$, $x \neq y$ and $T(x, z) = \{x, z\}$ and $T(z, y) = \{z, y\}$ and $T(x, v) \neq \{x, v\}$ and $T(u, y) \neq \{u, y\}$. Since $x \in T(u, v)$ and $T(x, v) \neq \{x, v\}$ there exist a u, x-path P that avoids the neighborhood of v. Since $y \in T(u, v)$ and $T(u, y) \neq \{u, y\}$ there exists a y, v-path Q that avoids the neighborhood of u. Also $T(x, z) = \{x, z\}$ and $T(z, y) = \{z, y\}$. If z is not adjacent to both u and v, then the walk $u \xrightarrow{P} xzy \xrightarrow{Q} v$ is a u, v-toll walk containing z. If z is adjacent to atleast one of u or v (say u), then the walk $uzy \xrightarrow{Q} v$ is a u, v-toll walk containing z. If z is adjacent to both u and v, then the walk uzv is a u, v-toll walk containing z. That is $z \in T(u, v)$.

(TW2): Suppose $x \in T(u, v)$, $T(u, x) \neq \{u, x\}$, $T(x, v) \neq \{x, v\}$ and $T(x, z) = \{x, z\}$. Since $x \in T(u, v)$ and $T(x, v) \neq \{x, v\}$ there exist a u, x-path P that avoids the neighborhood of v and and an x, v-path Q that avoids the neighborhood of u. If $T(u, z) = \{u, z\}$ and $T(v, z) = \{v, z\}$, then $z \in T(u, v)$. Suppose either $T(u, z) \neq \{u, z\}$ or $T(v, z) \neq \{v, z\}$. With out loss of generality we may assume $T(u, z) \neq \{u, z\}$. The walk formed by P and the edge zv is a toll walk containing z when $T(v, z) = \{v, z\}$. Otherwise also $T(v, z) \neq \{v, z\}$ and the walk formed by concatenation of P and Q is a toll walk containing z. That is in all cases $z \in T(u, v)$.

(TW3): Suppose $x \in T(u, v)$. There exists an x, v-path P that avoids the neighborhood of u with possible exception of x. For the neighbor v_1 of x on P it follows that $v_1 \in T(x, v), v_1 \neq x$ with $T(x, v_1) = \{x, v_1\}$ and $T(u, v_1) \neq \{u, v_1\}$. □

Proof of Proposition 2. Suppose T satisfies Axiom (b1). If T does not satisfy Axiom (b2), then there exist u, v, x, y such that $x \in T(u, v), y \in T(u, x)$ and $y \notin T(u, v)$. Since $x \in T(u, v)$, there exist a x, v-path, say P, without a neighbor of u and since $y \in T(u, x)$, there exist a u, y-path, say Q, without a neighbor of x and y, x-path, say R, without a neighbor of u. Since $y \notin T(u, v)$, either neighbor of u separates y from v or neighbor of v separates y from u. But $y \xrightarrow{R} x \xrightarrow{P} v$ is a y, v-path that does not contain a neighbor of u. So the only possibility is neighbor of v separates y from u. Therefore Q contains a neighbor of v, say v'. Now $u \xrightarrow{Q} v'v \xrightarrow{P} x$ is a u, x-toll walk containing v. That is $v \in T(u, x)$, a contradiction with Axiom (b1). □

Proof of Proposition 3. Suppose T satisfies Axiom (b1). If T does not satisfy Axiom (J3), then there exist u, v, x, y such that is $x \in T(u, y), y \in T(x, v)$ but $x \notin T(u, v)$. Since $x \in T(u, y)$, there exists a u, x-path, say P, without a neighbor of y and there exists an x, y-path, say Q, without a neighbor of u. Also since $y \in T(x, v)$, there exists an x, y-path, say R, without a neighbor of v and there exists a y, v-path, say S, without a neighbor of x. But $x \notin T(u, v)$ implies either

$N[v]$ separate x from u or $N[u]$ separate x from v. Without loss of generality, we may assume that $N[v]$ separates x from u. Let v' be the neighbor of v closest to x on P. Since G is P-free at least one of $v'x, xy$ and yv is not an edge. If yv is not an edge, then $y \xrightarrow{R} x \xrightarrow{P} v'v$ is a y,v-toll walk containing x. That is $x \in T(y,v)$. If xy is not an edge, then $y \xrightarrow{S} vv' \xrightarrow{P} x$ is a y,x-toll walk containing v. That is $v \in T(y,x)$. Finally, if xv' is not an edge, then $xyvv' \xrightarrow{P} u$ is a x,u-toll walk containing y. That is $y \in T(x,u)$, and we have a contradiction to Axiom (b1).$\square$

Proof of Lemma 4. If $n = 2$, then $P_2 = uv$ and $R(u,v) = \{u,v\}$ by the definition of G_R. If $n = 3$, then $P_3 = uxv$ and $x \in R(u,v)$ by Axiom (J2). Let now $n = 4$ and $P_4 = uxyv$. By Axiom (J2) we have $x \in R(u,y)$ and $y \in R(x,v)$. Now, Axiom (J3) (together with Axiom (t2)) implies that $x, y \in R(u,v)$. For a longer path $P_n = uxx_3 \ldots x_{n-2}yv, n > 4$, we continue by induction. By induction hypothesis we have $\{u, x, x_3, \ldots, x_{n-2}, y\} \subseteq R(u,y)$ and $\{x, x_3, \ldots, x_{n-2}, y, v\} \subseteq R(x,v)$. In particular, $x, x_3, \ldots, x_{n-2} \in R(u,y)$ and $y \in R(x,v)$. By Axiom (J3) we get $x, x_3, \ldots, x_{n-2} \in R(u,v)$ and symmetrically we also have $y \in R(u,v)$ by the same axiom.

For the second part let z be a neighbor of x_i, $i \in \{3, \ldots, n-2\}$ that is not adjacent to u, v. Clearly, in such a case $n \geq 5$. By the first part of the proof, we have $x_i \in R(u,v)$ and we have $z \in R(u,v)$ by Axiom (TW2). $\square$

Proof of Theorem 4. Let R be a transit function satisfying Axioms (b1), (J2), (J3), (TW1) and (TW2). Graph G_R is HHP-free by Lemma 3 and with this also C_5-free. Now we have to prove that G_R is also AT-free. By Theorem 2 it is enough to prove that G_R does not contain as an induced subgraph any of the graphs $C_k, T_2, X_2, X_3, X_{30}, \ldots, X_{41}, XF_2^{n+1}, XF_3^n, XF_4^n, k \geq 6, n \geq 1$, depicted on Fig. 3. Since the graphs $C_k, X_2, X_3, X_{30}, X_{32}, \ldots, X_{41}, XF_4^n$ contains a house, a hole or a P-graph as an induced subgraph, it is sufficient to prove that G_R is $T_2, X_{31}, XF_2^{n+1}, XF_3^n$-free for $n \geq 1$.

Suppose first that T_2 is an induced subgraph of G_R with vertices as shown on Fig. 3. By Lemma 4 we have $u_1, u_2, u_3, u_4 \in R(u,v) \cap R(u,x)$. In particular, $u_4 \in R(u,v)$, $R(u, u_4) \neq \{u, u_4\}$, $R(v, u_4) \neq \{v, u_4\}$ and $R(x, u_4) = \{x, u_4\}$. By Axiom (TW2) this yields that $x \in R(u,v)$. By symmetric arguments we get that also $v \in R(u,x)$, a contradiction with Axiom (b1).

We continue with the same approach also when G_R contains X_{31}, XF_2^{n+1} or XF_3^n for $n \geq 1$ as an induced subgraph. By this we rely on the notation from Fig. 3. There is an induced path $uabcv$ or $up_1y_1p_nv$ if G_R contains X_{31} or XF_2^{n+1} for $n > 1$, respectively, as an induced subgraph. By the second claim of Lemma 4 we have $x \in R(u,v)$ in both mentioned cases. If G_R contains XF_2^2 or XF_3^n as an induced subgraph, then paths up_1p_2v and uy_1y_2v, respectively, are induced. Hence, $p_1, p_2 \in R(u,v)$ and $y_1, y_2 \in R(u,v)$, respectively, by Lemma 4. We can use Axiom (TW1) and obtain $y_1 \in R(u,v)$ and $x \in R(u,v)$, respectively. Finally, we get also $x \in R(u,v)$ in the case of XF_2^2 by Axiom (TW2). So, in all the remaining cases we have $x \in R(u,v)$.

We will obtain a contradiction with Axiom (b1) by showing that $v \in R(u,x)$ as well in all the remaining cases. Let first X_{31} be an induced subgraph of G_R. A path $uebx$ is induced and we have $e, b \in R(u,x)$ by Lemma 4. By Axiom (TW1) we get first $c \in R(u,x)$ and later also $v \in R(u,x)$. Next let XF_2^{n+1}, $n \geq 1$, be an induced subgraph of G_R. A path up_1y_1x is induced in G_R and $p_1, y_1 \in R(u,x)$ by Lemma 4. By consecutive use of Axiom (TW1) we get $p_i \in R(u,x)$ for $i \in \{2,\ldots,n\}$. At the end we have $v \in R(u,x)$ by Axiom (TW2). Finally, let XF_3^n, $n \geq 1$, be an induced subgraph of G_R. A path up_1y_2x is induced in G_R and $p_1, y_2 \in R(u,x)$ by Lemma 4. By consecutive use of Axiom (TW1) we get $p_i \in R(u,x)$ for $i \in \{2,\ldots,n\}$. At the end we have $v \in R(u,x)$ by Axiom (TW2). Hence, in all the cases we have $x \in R(u,v)$ and $v \in R(u,x)$ which yields a contradiction with Axiom (b1). □

Proof of Theorem 7. Let R be a transit function on V satisfying Axioms (b1'), (J2), (J3), (TW1) and (TW2). We have to show that G_R is PAT-free. We prove that G_R does not contains any of the graphs $C_k, T_2, X_2, X_3, X_{30}, \ldots, X_{41}$, $XF_2^{n+1}, XF_3^n, XF_4^n$ from Fig. 3 and P-graph from Fig. 2 as an induced subgraph for $k \geq 6$ and $n \geq 1$. We use the notation of vertices u, v, x as shown on Figs. 2 and 3 for all the mentioned graphs. Clearly, u, v and x are independent on all the graphs from Fig. 3 as well as for P. We will show that $x \in R(u,v)$ and $v \in R(u,x)$ in all the mentioned graphs, which yields a contradiction with Axiom (b1'). We can use the same proof as for Theorem 4 to show that G_R is $(T_2X_{31}XF_2^{n+1}XF_3^n)$-free. It is easy to see that $x \in R(u,v)$ for graphs $C_k, X_2, X_3, X_{32}, X_{33}, X_{35}, X_{38}, X_{39}, X_{40}$ and XF_4^n for $k \geq 6$ and $n \geq 1$ by Lemma 4. Similar is $x \in R(u,v)$ for graphs X_{34}, X_{36} and X_{37} by Lemma 4 and Axiom (TW1). Finally, $x \in R(u,v)$ for graphs X_{30} and X_{41} by Lemma 4 and Axiom (TW2). If G_R contains P as an induced subgraph, then by Lemma 4 and by Axiom (J3), we have that $x \in R(u,v)$.

Next we show that $v \in R(u,x)$ holds for all the graphs from Fig. 3. We have $v \in R(u,x)$ for graphs C_k, X_{37} and X_{38} for $k \geq 6$ by Lemma 4. Also $v \in R(u,x)$ for graph X_{30} by Lemma 4 and Axiom (TW2). We get $v \in R(u,x)$ for graphs $X_{32}, X_{34}, X_{35}, X_{36}, X_{39}, X_{40}, X_{41}$ and XF_4^n by Lemma 4 and Axiom (TW1). Here we use Axiom $TW1$ once for X_{35}, n-times for XF_4^n and twice for the rest. Let us observe X_{36} in every detail (the same steps are also for X_{39} and X_{40}). Paths $ucbx$ and uax are induced and we have $a, b, c \in R(u,x)$ by Lemma 4. Now d is adjacent to $a, c \in R(u,x)$ and $d \in R(u,x)$ by Axiom (TW1). Finally $v \in R(u,x)$ by Axiom (TW2) again since v is adjacent to $b, d \in R(u,x)$. For X_2 we have $a \in R(x,b)$ and $b \in R(a,u)$ by Lemma 4. Next $a \in R(u,x)$ by Axiom (J3) and finally $v \in R(u,x)$ follows by Axiom (TW2). We end with X_3 and X_{33}. Here we have $c \in R(x,v)$ and $v \in R(c,u)$ by Lemma 4. Now $v \in R(u,x)$ by Axiom (J3). For P we have $v \in R(u,y)$ and $y \in R(v,x)$ by Lemma 4. Now, $v \in R(u,x)$ follows by Axiom (J3). In all the cases we have $R(u,x) \neq \{u,x\}$, $R(x,v) \neq \{x,v\}$, a contradiction since R satisfies Axiom (b1'). □

Proof of Proposition 4. Suppose $u, v \in V$. We have to show that there exist a path between u and v. For this we use induction on $|R(u,v)|$. If $|R(u,v)| = 1$, then $u = v$ and there is nothing to show. If $|R(u,v)| = 2$, then $R(u,v) = \{u,v\}$ and uv is an edge of G_R. Let now $|R(u,v)| = m > 2$. Suppose $x \in R(u,v)$ with $R(u,x) = \{u,x\}$ and $R(x,v) = \{x,v\}$, then both ux and vx are edges of G_R and u and v are joined by a path. If $R(u,x) \neq \{u,x\}$ and $R(x,v) \neq \{x,v\}$, we have $|R(u,x)| < m$ and $|R(x,v)| < m$ by Axioms (b1') and (b2). By induction hypothesis there is a path in G_R between u and x and also a path between x and v. A u,v-walk is obtained by concatenating these two paths and hence u and v are connected by a path. Let $R(u,x) = \{u,x\}$ and $R(x,v) \neq \{x,v\}$, then there exist a $z_1 \in R(x,v)$. If $R(x,z_1) \neq \{x,z_1\}$ and $R(z_1,v) \neq \{z_1,v\}$, then $|R(x,z_1)| < m$ and $|R(z_1,v)| < m$ by Axioms (b1') and (b2). Again by induction, there is a path between x and z_1 and between z_1 and v and by concatenating both paths together with $ux \in E(G_R)$, we obtain a u,v-walk. Hence a u,v-path exists and we are done. Suppose next that $R(x,z_1) = \{x,z_1\}$ and $R(z_1,v) \neq \{z_1,v\}$. We can find a $z_2 \in R(z_1,v)$ and either we can find a u,v-path when $R(z_1,z_2) \neq \{z_1,z_2\}$ and $R(z_2,v) \neq \{z_2,v\}$ or we can find a $z_3 \in R(z_2,v)$ when $R(z_1,z_2) = \{z_1,z_2\}$ and $R(z_2,v) \neq \{z_2,v\}$. Continuing this procedure we obtain a sequence of vertices $u,x,z_1,z_2,\ldots$ where $ux, xz_1, z_iz_{i+1} \in E(G_R)$ and $R(z_i,v) \neq \{z_i,v\}$ for every $i \geq 1$. Moreover $R(u,v) \supset R(x,v) \supset R(z_1,v) \supset \cdots$ holds by Axioms (b1') and (b2). This sequence must be finite because V is finite and this yields the desired u,v-path. □

Proof of Theorem 8. Suppose that $x \in T(u,v)$ and $x \notin \{u,v\}$. We have to show that $x \in R(u,v)$. By Lemma 2 $N[u] - x$ does not separate x and v and $N[v] - x$ does not separate u and x. Let W be a toll u,v-walk containing x. Clearly W contains an induced u,v-path, say Q. By Lemma 4 we have $V(Q) \subseteq R(u,v)$. If x belongs to Q, then $x \in R(u,v)$. So, we may assume that x does not belongs to Q. Moreover, we may assume that x does not belong to any induced u,v-path. The underlying graph G_R is AT-free by Theorem 7. Thus, Q contains a neighbor of x, say x'. If $R(u,x') = \{u,x'\}$ and $R(x',v) = \{x',v\}$, then we have a contradiction with W being a toll u,v-walk containing x. Without loss of generality, we may assume that $R(x',v) \neq \{x',v\}$. If also $R(u,x') \neq \{u,x'\}$, then $x \in R(u,v)$ by the second claim of Lemma 4. So, let now $R(u,x') = \{u,x'\}$. Since x and v are not separated by $N[u] - \{x\}$ by Lemma 2, there exists an induced x,v-path S without a neighbor of u. Let $S = s_0s_1\cdots s_k$, $s_0 = x$ and $s_k = v$ and let s_j be the first vertex of S that belongs also to Q. Notice that s_j can equal to v but it is different than x' and that $j > 0$. We may choose S such that it minimally differ from Q. This means that $s_0,\ldots,s_{j-2}$ are adjacent only to x' on Q. If $x's_{j-1} \notin E(G_R)$, then $s_{j-2} \in R(u,s_{j-1})$ and $s_{j-1} \in R(s_{j-2},v)$ by Lemma 4. By Axiom (J3) we get $s_{j-2} \in R(u,v)$. Finally, we use Axiom (TW1) for $j - 2$ times to get $s_{j-3}, s_{j-4},\ldots,s_0 \in R(u,v)$, one after each use of Axiom (TW1). We are done since $x = s_0$. We are left with the case $x's_{j-1} \in E(G_R)$. We have $x', s_j \in R(u,v)$ by Lemma 4 and clearly $x' \neq s_j$, $R(s_{j-1},s_j) = \{s_{j-1},s_j\}, R(x',s_{j-1}) = \{x',s_{j-1}\}$, which gives $s_{j-1} \in R(u,v)$ by Axiom (TW1). We continue with the same step $j - 1$ times, only that we lower the indexes of s_{j-1} and s_j at each step by $i \in \{1,\ldots,j-1\}$ in the natural order and get that consequently $s_{j-2}, s_{j-3}, \ldots, s_0 \in R(u,v)$. This completes the proof because $s_0 = x \in R(u,v)$. □

Proof of Theorem 9. If the toll walk function T of a graph G satisfies Axiom (b1), then G is (HC_5DAT)-free graph by Theorem 3. Next we show that if T satisfies $(J0)$, then G is chordal. Suppose that G contains an induced cycle C_n, $n \geq 4$ with three consecutive vertices x, u, v and y is any other vertex of C_n. Clearly $x \in T(u,y)$ and $y \in T(x,v)$ but $x \notin T(u,v)$ since uv is a edge in G. That is, if T satisfy Axiom $(J0)$, then G is a chordal graph. By Theorem 1 G is an interval graph.

To prove the converse, suppose G is an interval graph and assume the toll walk function T of G does not satisfy Axiom (b1) or $(J0)$. If T does not satisfies Axiom (b1), then G has an asteroidal triple or has a domino or C_5 or a house as an induced subgraph by Theorem 3. This is a contradiction with Theorem 1 since G is an interval graph. Now suppose Axiom $(J0)$ is not satisfied. There exist distinct vertices u, x, y, v such that $x \in T(u,y)$ and $y \in T(x,v)$ and $x \notin T(u,v)$. Now, $x \notin T(u,v)$ implies that $N(u) - x$ separate x from v or $N(v) - x$ separate u from x. With out loss of generality, we may assume that $N(u) - x$ separates x from v. So, every x,v-path contains at least one neighbor of u. But x belongs to a u,v-walk, say W, formed by an induced u,x-path P, an induced x,y-path Q and an induced y,v-path S. Since $x \notin T(u,v)$, there exist a neighbor of u, say u_1, that belongs to Q or to S. Because $x \in T(u,y)$, u_1 does not belongs to Q and u_1 belongs to S. We may choose u_1 to be the first vertex which is adjacent to u after the vertex y. The path $uu_1 \overset{S}{\longrightarrow} y$ is an induced u,y-path which avoids the neighbors of x. Suppose $T(u,x) \neq \{u,x\}$ and $T(x,y) \neq \{x,y\}$, then the vertices u, x, y form an asteroidal triples, a contradiction with G being an interval graph. So, $T(u,x) = \{u,x\}$ or $T(x,y) = \{x,y\}$ holds.

Suppose first that $T(u,x) = \{u,x\}$. The path $ux \overset{Q}{\longrightarrow} y \overset{S}{\longrightarrow} u_1$ and the edge uu_1 form a cycle of length at least four. Since G is interval graph, there an edge between a vertex of Q and a vertex from S. Let a the first vertex on Q after x which is adjacent to some vertex, say b, in the u_1,y-subpath of S. We have a cycle $ux \overset{Q}{\longrightarrow} ab \overset{S}{\longrightarrow} u_1u$ of length at least four, a contradiction. Hence, we may assume that $T(u,x) \neq \{u,x\}$ which yields $T(x,y) = \{x,y\}$. To avoid the induced cycle $u \overset{P}{\longrightarrow} xy \overset{S}{\longrightarrow} u_1u$, there must be an edge from a vertex of P to a vertex of S. Let a the last vertex on P before x that is adjacent to some vertex, say b, on the u_1,y-subpath of S. Cycle $a \overset{P}{\longrightarrow} xy \overset{S}{\longrightarrow} ba$ is induced of length at least four, a final contradiction. □

Proof of Theorem 10. From the definitions of Axioms $(J0)$ and $(J3)$, it follows that if R satisfies Axiom $(J0)$, then R satisfies Axiom $(J3)$. From Theorem 4, if R satisfies the Axioms (b1), $(J2)$, $(J3)$, $(TW1)$ and $(TW2)$, then G_R is HPC_5AT-free. We need to prove that G_R is chordal by Theorem 1. Even more, it is enough to prove that G_R is C_4-free, because it is C_5-free and C_k-free for $k \geq 6$ as an AT-free graph. Suppose conversely that G_R contains $C_4 = uxyv$. We have

$x \in R(u, y)$ and $y \in R(x, v)$ by Lemma 4 and $x \in R(u, v)$ follows by Axiom (J0). This is a contradiction to $R(u, v) = \{u, v\}$. Hence, G_R is an interval graph. □

Appendix B

In this appendix we show the independence of the axioms used in this paper. In all the examples we have $R(a, a) = \{a\}$ for all $a \in V$.

Example 1. There exists a transit function that satisfies Axioms (b2), (J2), (J3), (TW1), (TW2) and (TW3), but not (b1).
Let $V = \{u, v, w, x, y\}$ and define a transit function R on V as follows: $R(u, v) = R(u, x) = R(y, v) = R(y, w) = R(x, w) = V$, $R(u, y) = \{u, y\}$, $R(u, w) = \{u, w\}$, $R(x, v) = \{x, v\}$, $R(x, y) = \{x, y\}$ and $R(w, v) = \{w, v\}$. It is straightforward but tedious to see that R satisfies Axioms (b2), (J2), (J3), (TW1), (TW2) and (TW3). In additions $x \in R(u, v), x \neq v$ and $v \in R(u, x)$ and R does not satisfy Axiom (b1).

Example 2. There exists a transit function that satisfies Axioms (b1), (J2), (J3), (TW1), (TW2) and (TW3), but not (b2).
Let $V = \{u, v, x, y\}$ and define a transit function R on V as follows: $R(u, v) = \{u, x, v\}$, $R(u, x) = \{u, y, x\}$, $R(u, y) = \{u, v, y\}$, $R(x, y) = \{x, y\}$, $R(x, v) = \{x, v\}$ and $R(y, v) = \{y, v\}$. It is straightforward but tedious to see that R satisfies Axioms (b1), (J2), (J3), (TW1), (TW2) and (TW3). Axiom (b2) does not hold, because $x \in R(u, v)$, $y \in R(u, x)$ and $x \notin R(u, v)$.

Example 3. There exists a transit function that satisfies Axioms (b1), (b2), (J3), (TW1), (TW2) and (TW3), but not (J2).
Let $V = \{u, v, x, y\}$ and define a transit function R on V as follows: $R(u, v) = \{u, x, v\}$, $R(u, y) = \{u, y\}$, $R(u, x) = \{u, x\}$, $R(x, y) = \{x, y\}$, $R(x, v) = \{x, v\}$ and $R(y, v) = \{y, v\}$. It is straightforward but tedious to see that R satisfies Axioms (b1), (b2), (J3), (TW1), (TW2) and (TW3). On the other hand $R(u, y) = \{u, y\}$, $R(y, v) = \{y, v\}$, $R(u, v) \neq \{u, v\}$ and $y \notin R(u, v)$ and R does not satisfy Axiom (J2).

Example 4. There exists a transit function that satisfies Axioms (b1), (b2), (J2), (TW1), (TW2) and (TW3), but not (J3).
Let $V = \{u, v, w, x, y\}$ and define a transit function R on V as follows: $R(u, v) = \{u, w, v\}$, $R(u, y) = \{u, x, w, y\}$, $R(u, w) = \{u, w\}$, $R(u, x) = \{u, x\}$, $R(x, y) = \{x, y\}$, $R(x, w) = \{x, w\}$, $R(x, v) = \{x, y, w, v\}$, $R(w, y) = \{w, y\}$, $R(w, v) = \{w, v\}$ and $R(y, v) = \{y, v\}$. It is straightforward but tedious to see that R satisfies Axioms (b1), (b2), (J2), (TW1), (TW2) and (TW3). Beside that $x \in R(u, y)$, $y \in R(x, v)$, $R(u, v) \neq \{u, v\}$ but $x \notin R(u, v)$. So R does not satisfy Axiom (J3).

Example 5. There exists a transit function that satisfies Axioms (b1), (b2), (J2), (J3), (TW2) and (TW3), but not (TW1).

Let $V = \{u, v, w, x, y\}$ and define a transit function R on V as follows: $R(u, v) = \{u, x, y, v\}$, $R(u, y) = \{u, x, y\}$, $R(u, w) = \{u, x, w\}$, $R(u, x) = \{u, x\}$, $R(x, v) = \{x, y, v\}$, $R(x, w) = \{x, w\}$, $R(x, y) = \{x, y\}$, $R(w, y) = \{w, y\}$, $R(w, v) = \{w, y, v\}$ and $R(y, v) = \{y, v\}$. It is straightforward but tedious to see that R satisfies Axioms (b1), (b2), (J2), (J3), (TW2) and (TW3). Since $x, y \in R(u, v)$, $x \neq y$, $R(x, w) = \{x, w\}$, $R(w, y) = \{w, y\}$, $R(x, v) \neq \{x, v\}$ and $R(u, y) \neq \{u, y\}$, but $w \notin R(u, v)$, R does not satisfies Axiom (TW1).

Example 6. There exists a transit function that satisfies Axioms (b1), (b2), (J2), (J3), (TW1) and (TW3), but not (TW2).

Let $V = \{u, v, w, x, y, z\}$ and define a transit function R on V as follows: $R(u, v) = \{u, w, x, y, v\}$, $R(u, y) = \{u, w, x, y\}$, $R(u, x) = \{u, w, x\}$, $R(u, z) = \{u, w, x, z\}$, $R(u, w) = \{u, w\}$, $R(w, v) = \{w, x, y, v\}$, $R(w, y) = \{w, x, y\}$, $R(w, z) = \{w, x, z\}$, $R(w, x) = \{w, x\}$, $R(x, v) = \{x, y, v\}$, $R(x, z) = \{x, z\}$, $R(x, y) = \{x, y\}$, $R(z, y) = \{z, x, y\}$, $R(z, v) = \{z, x, y, v\}$ and $R(y, v) = \{y, v\}$. It is straightforward but tedious to see that R satisfies Axioms (b1), (b2), (J2), (J3), (TW1) and (TW3). But R does not satisfies axiom (TW2), since $x \in R(u, v)$, $R(u, x) \neq \{u, x\}$, $R(x, v) \neq \{x, v\}$, $R(x, z) = \{x, z\}$ but $z \notin R(u, v)$.

Example 7. There exists a transit function that satisfies Axioms (b1), (b2), (J2), (J3), (TW1) and (TW2), but not (TW3).

Let $V = \{u, v, x, y\}$ and define a transit function R on V as follows: $R(u, v) = \{u, x, y, v\}$, $R(u, y) = \{u, y\}$, $R(u, x) = \{u, x\}$, $R(x, y) = \{x, y\}$, $R(x, v) = \{x, y, v\}$ and $R(y, v) = \{y, v\}$. It is straightforward but tedious to see that R satisfies Axioms (b1), (b2), (J2), (J3), (TW1) and (TW2). But R does not hold for Axiom (TW3), since $x \in R(u, v)$ and there does not exist a v_1 such that $v_1 \in R(x, v)$, $v_1 \neq x$ with $R(x, v_1) = \{x, v_1\}$ and $R(u, v_1) \neq \{u, v_1\}$.

Appendix C

At the end of this work we would like to mention two open problems. The first one is a general one and it seems to be to ambitious at the moment.

Problem 1. Is there an axiomatic characterization of the toll walk transit function of an arbitrary connected graph G?

It seems that it is much more realistic to attack above problem by smaller steps. A natural first possible step is to deal first with chordal graphs.

Problem 2. Is there a characterization of the toll walk transit function of chordal graph?

References

1. Alcon, L.: A note on path domination. Discuss. Math. Graph Theory **36**, 1021–1034 (2016)
2. Alcon, L., et al.: Toll convexity. Eur. J. Comb. **46**, 161–175 (2015)
3. Bollobás, B., Thomason, A.: Hereditary and monotone properties of graphs. In: Graham, R.L., Nešetřil, J. (eds) The Mathematics of Paul Erdős II. Algorithms and Combinatorics, vol. 14, pp. 77–78. Springer, Heidelberg (1997). https://doi.org/10.1007/978-3-642-60406-5_7
4. Borowiecki, M., Broere, I., Frick, M., Mihók, P., Semanišin, G.: A survey of hereditary properties of graphs. Discuss. Math. Graph Theory **17**, 5–50 (1997)
5. Changat, M., et al.: A forbidden subgraph characterization of some graph classes using betweenness axioms. Discrete Math. **313**, 951–958 (2013)
6. Changat, M., Mathew, J., Mulder, H.M.: The induced path function, monotonicity and betweenness. Discrete Appl. Math. **158**(5), 426–433 (2010)
7. Changat, M., Klavzar, S., Mulder, H.M.: The all-paths transit function of a graph. Czechoslov. Math. J. **51**, 439–448 (2001). https://doi.org/10.1023/A:1013715518448
8. Chvátal, V., Rautenbach, D., Schäfer, P.M.: Finite Sholander trees, trees, and their betweenness. Discrete Math. **311**, 2143–2147 (2011)
9. Corneil, D.G., Olariu, S., Stewart, L.: Asteroidal triple-free graphs. In: van Leeuwen, J. (ed.) WG 1993. LNCS, vol. 790, pp. 211–224. Springer, Berlin, Heidelberg (1994). https://doi.org/10.1007/3-540-57899-4_54
10. Dourado, M.C.: Computing the hull number in toll convexity. Ann. Oper. Res. **315**, 121–140 (2022). https://doi.org/10.1007/s10479-022-04694-4
11. Gologranc, T., Repolusk, P.: Toll number of the Cartesian and the lexicographic product of graphs. Discrete Math. **340**, 2488–2498 (2017)
12. Gologranc, T., Repolusk, P.: Toll number of the strong product of graphs. Discrete Math. **342**, 807–814 (2019)
13. Köhler, E.: Graphs without asteroidal triples. Ph.D. thesis, Technische Universität Berlin, Cuvillier Verlag, Göttingen (1999)
14. Lekkerkerker, C.G., Boland, J.C.: Representation of a finite graph by a set of intervals on the real line. Fundamenta Math. **51**, 45–64 (1962)
15. Mulder, H.M., Nebeský, L.: Axiomatic characterization of the interval function of a graph. Europ. J. Combin. **30**, 1172–1185 (2009)
16. Nebeský, L.: The induced paths in a connected graph and a ternary relation determined by them. Math. Bohem. **127**(3), 397–408 (2002). https://doi.org/10.21136/MB.2002.134072
17. Sholander, M.: Trees, lattices, order and betweenness. Proc. Amer. Math. Soc. **3**, 369–381 (1952)
18. Information system on graph classes and their inclusions, graphclass: AT-free. https://www.graphclasses.org/classes/gc_61.html. Accessed 06 Feb 2022

Structural Parameterization of Alliance Problems

Sangam Balchandar Reddy and Anjeneya Swami Kare[(✉)]

School of Computer and Information Sciences, University of Hyderabad,
Hyderabad, India
{21mcpc14,askcs}@uohyd.ac.in

Abstract. The alliance problems have been studied extensively during the last couple of decades. In this paper, we approach the problems from the standpoint of parameterized complexity. We study three problems, namely the defensive, offensive and powerful alliances. Given a graph $G = (V, E)$, a non-empty set $S \subseteq V$ is a defensive alliance, if for every vertex $v \in S : |N_G(v) \cap S| \geq |N_G(v) \setminus S| - 1$ and a non-empty set $S \subseteq V$ is an offensive alliance, if for every vertex $v \in N[S] \setminus S : |N_G(v) \cap S| \geq |N_G(v) \setminus S| + 1$. A powerful alliance is both defensive and offensive simultaneously. We majorly focus on a set of parameters for which the complexity of the problems is unknown. Our main results are as follows: (1) All the three alliance problems are *fixed parameter tractable* parameterized by distance to clique of the input graph. (2) All the three alliance problems are *fixed parameter tractable* parameterized by the combined parameter twin cover and the number of cliques outside the twin cover. (3) All the three alliance problems are *fixed parameter tractable* parameterized by the combined parameter twin cover and the size of the largest clique outside the twin cover.

Keywords: Defensive alliance · Offensive alliance · Powerful alliance · Twin cover · Distance to clique

1 Introduction

An alliance is a connection between multiple individuals, states or parties. The union of individuals is considered to be stronger than the individual. The concept of alliances in graphs was first introduced by Kristiansen, Hedetniemi and Hedetniemi [18]. They have studied three problems, namely the defensive, offensive and powerful alliances. The initial algorithmic results of the problem were given by Jamieson [15]. Alliances in graphs have been well studied [3] and generalizations such as r-alliances are also studied [10,20]. Let $G = (V, E)$ be a finite, simple and undirected graph with V as the vertex set and E as the edge set. We define the closed neighbourhood of a vertex v by $N_G[v]$ and the open neighbourhood by $N_G(v)$. We denote the boundary of a set $S \subseteq V$ by ∂S which represents the set of vertices in the neighbourhood of S, excluding S ($\partial S =$

© The Author(s), under exclusive license to Springer Nature Switzerland AG 2023
A. Bagchi and R. Muthu (Eds.): CALDAM 2023, LNCS 13947, pp. 447–459, 2023.
https://doi.org/10.1007/978-3-031-25211-2_34

$N[S] \setminus S$). A defensive alliance is a non-empty set $S \subseteq V$, such that for every vertex $v \in S : |N_G(v) \cap S| \geq |N_G(v) \setminus S| - 1$. An offensive alliance is a non-empty set $S \subseteq V$, such that for every vertex $v \in \partial S : |N_G(v) \cap S| \geq |N_G(v) \setminus S| + 1$. A powerful alliance is both defensive and offensive simultaneously. An alliance is global if it is also a dominating set.

The problem definitions for the alliance problems are stated as follows:

DEFENSIVE r-ALLIANCE:
A non-empty set $S \subseteq V(G)$ is a defensive r-alliance if for each $v \in S$, $|N(v) \cap S| \geq |N(v) \setminus S| + r$.

Input: A simple, undirected graph $G = (V, E)$, and a positive integer k.
Question: Is there a defensive r-alliance $S \subseteq V$ such that $|S| \leq k$?

DEFENSIVE ALLIANCE:
A non-empty set $S \subseteq V(G)$ is a defensive alliance if it satisfies the condition for defensive (-1)-alliance.

OFFENSIVE r-ALLIANCE:
A non-empty set $S \subseteq V(G)$ is an offensive r-alliance if for each $v \in \partial S, |N(v) \cap S| \geq |N(v) \setminus S| + r$.

Input: A simple, undirected graph $G = (V, E)$, and a positive integer k.
Question: Is there an offensive r-alliance $S \subseteq V$ such that $|S| \leq k$?

OFFENSIVE ALLIANCE:
A non-empty set $S \subseteq V(G)$ is an offensive alliance if it satisfes the condition for an offensive (1)-alliance.

POWERFUL r-ALLIANCE:
A powerful r-alliance is simultaneously a defensive r-alliance and an offensive $(r + 2)$-alliance.

Input: A simple, undirected graph $G = (V, E)$, and a positive integer k.
Question: Is there a powerful r-alliance $S \subseteq V$ such that $|S| \leq k$?

Given a graph $G = (V, E)$, a problem is considered to be *fixed parameter tractable* w.r.t a parameter k, if there exists an algorithm with running time $\mathcal{O}((|V| + |E|)^{\mathcal{O}(1)} \cdot f(k))$, where f is a computable function. We use $\mathcal{O}^*(f(k))$ to denote the time complexity of the form $\mathcal{O}((|V| + |E|)^{\mathcal{O}(1)} \cdot f(k))$. For more information on *parameterized complexity* we refer the reader to [5].

1.1 Previous Work

The decision version of the problems is known to be NP-complete. Jamieson et al. [14] showed that the defensive alliance is NP-complete even when restricted to split, chordal and bipartite graphs. The defensive r-alliance [21] and global defensive r-alliance problems [9] are NP-complete for any r. Fernau et al. [10] showed that the offensive r-alliance and global offensive r-alliance problems are

NP-complete for any fixed r. There are polynomial time algorithms for finding minimum alliances in trees [2,14,15]. There is a polynomial time algorithm for finding a minimum defensive alliance in series parallel graphs [6]. Gaikwad and Maity [12] proved that the defensive alliance is NP-complete on circle graphs.

In terms of parameterized complexity, Fernau and Raible [8] proved that all the three problems, including their global versions are fixed parameter tractable when parameterized by the solution size k. Using integer linear programming framework, Kiyomi and Otachi [17] showed that all the alliance problems are fixed parameter tractable when parameterized by vertex cover number of the input graph. Gaikwad and Maity [12] showed that the defensive alliance is W[1]-hard parameterized by a wide range of parameters such as the feedback vertex set, treewidth, cliquewidth, treedepth and pathwidth. Bliem and Woltran [1] proved that the defensive alliance is W[1]-hard parameterized by treewidth of the input graph. Recently, both the defensive and offensive alliances were also shown to be fixed parameter tractable parameterized by neighbourhood diversity of the input graph [11].

1.2 Our Results

We study the problems for structural parameters distance to clique and twin cover. Our results are as follows:

- All the three alliance problems are *fixed parameter tractable* parameterized by distance to clique of the input graph.
- All the three alliance problems are *fixed parameter tractable* parameterized by the combined parameter twin cover and the number of cliques outside the twin cover.
- All the three alliance problems are *fixed parameter tractable* parameterized by the combined parameter twin cover and the size of the largest clique outside the twin cover.

2 FPT Algorithms

In this section, we present FPT algorithms for alliance problems. The parameters we consider are distance to clique and twin cover. We show that all the three alliance problems are in FPT for the parameter distance to clique. When it comes to twin cover, we show that all the three problems are in FPT for the combined parameters: (1) twin cover and the number of cliques outside the twin cover. (2) twin cover and the size of the largest clique outside the twin cover.

2.1 Alliances Parameterized by Distance to Clique

Definition 1. For a graph $G = (V, E)$, the parameter *distance to clique* is the cardinality of the smallest set $D \subseteq V$ such that $V \setminus D$ is a clique.

Using a simple branching algorithm, we can compute set D of size at most k in $\mathcal{O}^*(2^k)$ time, if such a set exists.

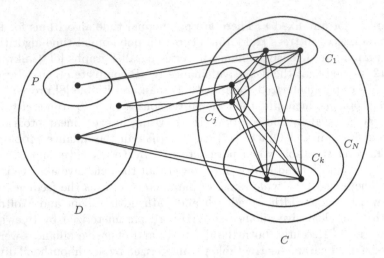

Fig. 1. Partitioning of the vertex set V into sets D and C, where $|D|$ is the distance to clique and C is a clique.

Theorem 1. Given a graph $G = (V, E)$ and $D \subseteq V$ such that $V \setminus D$ is a clique, the DEFENSIVE r-ALLIANCE problem can be solved in $\mathcal{O}^*(f(|D|))$ time.

Consider a graph $G = (V, E)$ and $D \subseteq V$ such that $|D|$ is the distance to clique of G and $C = V \setminus D$. We partition the vertices of C into t twin classes which are represented by $C_1, C_2, ..., C_t$ ($t \leq 2^{|D|}$), such that all the vertices in a twin class C_i have same adjacency in D. Let $S \subseteq V$ be a defensive r-alliance of G. We guess the vertex sets $P = S \cap D$ and compute $S_C = S \cap C$. We also guess a subset of twin classes $C_N \subseteq C$ from which no vertices are picked in the solution. See Fig. 1 for an illustration. After the guess of P and C_N, we compute S_C using integer linear programming. For each of $u \in D$, we define $demand(u) = \frac{1}{2}(deg(u) + r) - |N(u) \cap P|$. For each $u \in D$, we denote by $M(u)$ the set of indices i such that $C_i \subseteq N(u)$. In other words, $M(u)$ represents the indices of the twin classes from C that u is adjacent to. x_i represents the number of vertices in $C_i \cap S$. In our ILP formulation, there are t variables that are $x_1, x_2, ..., x_t$.

Lemma 1. The set S is a defensive r-alliance if and only if

1. For each $u \in P$, $\sum_{i \in M(u)} x_i \geq demand(u)$.
2. Each $v \in C \setminus C_N$ has to satisfy $|N(v) \cap P| + \sum_{i \in \{1,2,...,t\}} x_i \geq |N[v] \setminus P| - \sum_{i \in \{1,2,...,t\}} x_i + r$.

Proof.

1. For each $u \in P$, $deg(u) = |N(u) \cap S| + |N(u) \setminus S|$ and $|N(u) \cap S| \geq |N(u) \setminus S| + r$ holds if and only if $2 * |N(u) \cap S| \geq deg(u) + r$, which is equivalent to $|N(u) \cap S_C| \geq demand(u)$ implies $\sum_{i \in M(u)} x_i \geq demand(u)$.

2. For each $v \in C \setminus C_N$, $|N(v) \cap S| = |N(v) \cap P| + |S_C|$, which is indeed $|N(v) \cap P| + \sum_{i \in \{1,2,...,t\}} x_i$; $|N(v) \setminus S| + r = |(N(v) \cap D) \setminus P| + |C| - \sum_{i \in \{1,2,...,t\}} x_i + r$, which equals $|N[v] \setminus P| - \sum_{i \in \{1,2,...,t\}} x_i + r$.

□

The ILP formulation for the defensive r-alliance is given as

Minimize $\sum_{i \in \{1,2,...,t\}} x_i$

Subject to

- $\sum_{i \in M(u)} x_i \geq demand(u)$, for each $u \in P$.
- $|N(v) \cap P| + \sum_{i \in \{1,2,...,t\}} x_i \geq |N[v] \setminus P| - \sum_{i \in \{1,2,...,t\}} x_i + r$, for every $v \in C \setminus C_N$.
- $x_i \leq |C_i|$, for each $i \in \{1, 2, ..., t\}$.

The ILP will output the optimal values of x_i for all $i \in \{1, 2, ..., t\}$. If $x_i > 0$, we need to pick x_i vertices from C_i. As all the vertices in C_i have same neighbourhood, we can pick any x_i vertices. Hence, we obtain the vertex set S_C.

ILP formulation
Integer Linear Programming is a framework used to formulate a given problem using a finite number of variables. The problem definition is given as follows:

Problem. p-Opt-ILP
Instance: A matrix $A \in \mathbb{Z}^{m*p}$, and vectors $b \in \mathbb{Z}^m$ and $c \in \mathbb{Z}^p$.
Objective: Find a vector $x \in \mathbb{Z}^p$ that minimizes $c^\top x$ and satisfies that $Ax \geq b$.
Parameter: p, the number of variables.

Lenstra [19] showed that deciding the feasibilty of a p-ILP is fixed parameter tractable with running time doubly exponential in p, where p is the number of variables. Later, Kannan [16] gave a p^p algorithm for p-ILP. Fellows et al. [7] proved that p-Opt-ILP, the optimization version of the problem is also fixed parameter tractable.

Theorem 2. [7] p-Opt-ILP can be solved using $\mathcal{O}(p^{2.5p+o(p)} \cdot L \cdot log(MN))$ arithmetic operations and space polynomial in L, where L is the number of bits in the input, N is the maximum absolute value any variable can take, and M is an upper bound on the absolute value of the minimum taken by the objective function.

In our ILP formulation, we have at most $2^{|D|}$ variables. The values of all the variables and the objective function are bounded by n. The constraints can be represented using $\mathcal{O}(4^{|D|} \cdot logn)$ bits. With the help of Theorem 2, we will be able to solve the problem with the guess (P, C_N) in FPT time. There are $2^{|D|}$ candidates for P and $2^{2^{|D|}}$ candidates for C_N. To obtain S_C, we solve $8^{|D|}$

ILP formulas where each formula can be computed in $\mathcal{O}^*(f(|D|))$ time. This concludes the proof of Theorem 1.

Theorem 3 $(\star^1)$**.** Given a graph $G = (V, E)$ and $D \subseteq V$ such that $V \setminus D$ is a clique, the OFFENSIVE r-ALLIANCE problem can be solved in $\mathcal{O}^*(f(|D|))$ time.

Theorem 4 $(\star)$**.** Given a graph $G = (V, E)$ and $D \subseteq V$ such that $V \setminus D$ is a clique, the POWERFUL r-ALLIANCE problem can be solved in $\mathcal{O}^*(f(|D|))$ time.

2.2 Alliances Parameterized by Twin Cover and the Number of Cliques Outside the Twin Cover

Definition 2. For a graph $G = (V, E)$, the parameter *twin cover* is the cardinality of the smallest set $T \subseteq V$ such that $V \setminus T$ is a disjoint union of cliques wherein all the vertices in each clique have the same adjacency in the twin cover.

Theorem 5. [4,13] If a minimum twin cover in G has size at most k, then it is possible to compute a twin cover of size at most k in time $\mathcal{O}(|E||V| + k|V| + 1.2738^k)$.

From Theorem 5, it is possible to find a twin-cover of size k in FPT time. In this subsection, we consider the combined parameter twin cover and the number of cliques outside the twin cover.

Theorem 6. Given a graph $G = (V, E)$, $T \subseteq V$ is a twin cover of G and y is the number of cliques outside the twin cover, the DEFENSIVE r-ALLIANCE problem can be solved in $\mathcal{O}^*(f(|T|, y))$ time.

Consider a graph $G = (V, E)$. Let $T \subseteq V$ be a twin cover of G and $C = V \setminus T$. We partition the vertices of C into y cliques which are represented by $C_1, C_2, ..., C_y$, such that all the vertices in a clique C_i have same adjacency in T. Let $S \subseteq V$ be a defensive r-alliance of G. We guess the vertex sets $P = S \cap T$ and compute $S_C = S \cap C$. We also guess a subset of cliques $C_N \subseteq C$ from which no vertices are picked in the solution. See Fig. 2 for an illustration. After the guess of P and C_N, we compute S_C using integer linear programming. For each of $u \in T$, we define $demand(u) = \frac{1}{2}(deg(u) + r) - |N(u) \cap P|$. For each $u \in T$, we denote by $M(u)$ the set of indices i such that $C_i \subseteq N(u)$. In other words, $M(u)$ represents the indices of the cliques from C that u is adjacent to. x_i represents the number of vertices in $C_i \cap S$. In our ILP formulation, there are y variables that are $x_1, x_2, ..., x_y$.

Lemma 2. The set S is a defensive r-alliance if and only if

1. For each $u \in P$, $\sum_{i \in M(u)} x_i \geq demand(u)$.
2. Each $v \in (C \setminus C_N) \cap C_i$ has to satisfy $|N(v) \cap P| + x_i \geq |N[v] \setminus P| - x_i + r$.

---

[1] Due to the space limit, the proofs of statements marked with a $\star$ have been omitted.

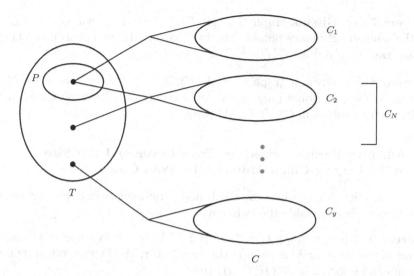

Fig. 2. Partitioning of the vertex set V into sets T and C, where T is a twin cover and C is a disjoint union of cliques.

Proof.

1. For each $u \in P$, $\deg(u) = |N(u) \cap S| + |N(u) \setminus S|$ and $|N(u) \cap S| \geq |N(u) \setminus S| + r$ holds if and only if $2 * |N(u) \cap S| \geq \deg(u) + r$, which is equivalent to $|N(u) \cap S_C| \geq demand(u)$ implies $\sum_{i \in M(u)} x_i \geq demand(u)$.
2. For each $v \in (C \setminus C_N) \cap C_i$, $|N(v) \cap S| = |N(v) \cap P| + x_i$; $|N(v) \setminus S| + r = |N[v] \setminus P| - x_i + r$. □

The ILP formulation for the defensive r-alliance is given as

Minimize $\sum_{i \in \{1,2,\ldots,y\}} x_i$

Subject to

- $\sum_{i \in M(u)} x_i \geq demand(u)$, for each $u \in P$.
- $|N(v) \cap P| + x_i \geq |N[v] \setminus P| - x_i + r$, for every $v \in (C \setminus C_N) \cap C_i$.
- $x_i \leq |C_i|$, for each $i \in \{1, 2, \ldots, y\}$.

In our ILP formulation, we have y variables, where y is the number of cliques outside the twin cover. The values of all the variables and the objective function are bounded by n. The constraints can be represented using $\mathcal{O}(y \cdot |T| \cdot log n)$ bits. With the help of Theorem 2, we will be able to solve the problem with the guess (P, C_N) in FPT time. There are $2^{|T|}$ candidates for P and 2^y candidates for C_N. To obtain S_C, we solve $2^{|T|} * 2^y$ ILP formulas where each formula can be computed in $\mathcal{O}^*(f(|T|, y))$ time. This concludes the proof of Theorem 6.

Theorem 7 $(\star)$**.** Given a graph $G = (V, E)$, $T \subseteq V$ is a twin cover of G and y is the number of cliques outside the twin cover, the OFFENSIVE r-ALLIANCE problem can be solved in $\mathcal{O}^*(f(|T|, y))$ time.

Theorem 8 $(\star)$**.** Given a graph $G = (V, E)$, $T \subseteq V$ is a twin cover of G and y is the number of cliques outside the twin cover, the POWERFUL r-ALLIANCE problem can be solved in $\mathcal{O}^*(f(|T|, y))$ time.

2.3 Alliances Parameterized by Twin Cover and the Size of the Largest Clique Outside the Twin Cover

In this subsection, we consider the combined parameter twin cover and the size of the largest clique outside the twin cover.

Theorem 9. Given a graph $G = (V, E)$, $T \subseteq V$ is a twin cover of G and z is the size of the largest clique outside the twin cover, the DEFENSIVE r-ALLIANCE problem can be solved in $\mathcal{O}^*(f(|T|, z))$ time.

Consider a graph $G = (V, E)$. Let $T \subseteq V$ be a twin cover of G and $C = V \setminus T$. We partition the vertices of C into t clique sets which are represented by $C_1, C_2, ..., C_t$ ($t \leq 2^{|T|}$), such that all the vertices in a clique set C_i have same adjacency in T. Let $S \subseteq V$ be a defensive r-alliance of G. We guess the vertex sets $P = S \cap T$ and compute $S_C = S \cap C$. For each of $u \in T$, we define $demand(u) = \frac{1}{2}(deg(u) + r) - |N(u) \cap P|$. For each $u \in T$, we denote by $M(u)$ the set of indices i such that $C_i \subseteq N(u)$. In other words, $M(u)$ represents the indices of the clique sets from C that u is adjacent to. We have at most z different size cliques in each clique set whose sizes range from 1 to z. We represent the cliques of size l in the clique set C_i as C_i^l. We define the deficiency for a clique set C_i as, $d_i = |N(C_i) \cap (T \setminus P)| - |N(C_i) \cap P|$.

We place cliques of all sizes from each clique set into one of the following three types: *full*, *partial* and *null*. *full* cliques have all of its vertices picked in the solution, *partial* cliques have some of its vertices picked whereas the *null* cliques have no vertices picked. $C_i^{l,F}$ represents the union of all *full* cliques in C_i^l. $C_i^{l,P}, C_i^{l,N}$ represents the union of all *partial* cliques and union of all *null* cliques in C_i^l respectively. We denote each *partial* clique in $C_i^{l,P}$ by C_i^{l,P_j}, where j denotes the index of a partial clique. See Fig. 3 for an illustration of the clique types in the clique set C_i. x_i^{l,P_j} represents the number of vertices in $C_i^{l,P_j} \cap S$. In the ILP formulation, we need to assign an individual variable to every *partial* clique. Here, the idea is to limit the number of partial cliques in each clique set C_i, which results in formulating the ILP using a significantly lesser number of variables.

Lemma 3. Consider a set C_i^l from a clique set C_i with a non-positive deficiency d_i, there exists an optimal solution with at most two *partial* cliques from C_i^l.

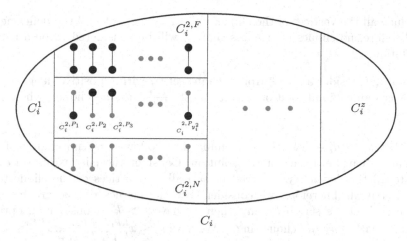

Fig. 3. Representation of cliques of type *full*, *partial* and *null* of length two in the clique set C_i. The black highlighted vertices belong in the alliance set.

Proof.

Case 1: $d_i = 0$. Let S be an optimal solution containing p *partial* cliques and f *full* cliques from C_i^l. Let C_i^{l,P_1}, C_i^{l,P_2} and C_i^{l,P_3} be three *partial* cliques from C_i^l and we have, $\frac{l}{2} \leq x_i^{l,P_1}, x_i^{l,P_2}, x_i^{l,P_3} < l$ and $\frac{3l}{2} \leq x_i^{l,P_1} + x_i^{l,P_2} + x_i^{l,P_3} \leq 3l - 3$. We obtain another solution(S') with only two *partial* cliques as follows. We try to convert one among C_i^{l,P_1}, C_i^{l,P_2} and C_i^{l,P_3} to *full* clique by placing all of its vertices in the solution and we claim that this also gives an optimal solution. Let C_i^{l,P_1} be the *full* clique in S', we get $\frac{l}{2} \leq x_i^{l,P_2}, x_i^{l,P_3} < l$ and $\frac{l}{2} \leq x_i^{l,P_2} + x_i^{l,P_3} \leq 2l - 3$. It is easy to verify that we have values for x_i^{l,P_2} and x_i^{l,P_3} satisfying the above constraints. If $x_i^{l,P_2} + x_i^{l,P_3} \in [\frac{l}{2}, l)$, then we need only one *partial* clique among x_i^{l,P_2}, x_i^{l,P_3} as we can set another value to zero. If $x_i^{l,P_2} + x_i^{l,P_3} \in [l, 2l - 3]$, then both x_i^{l,P_2}, x_i^{l,P_3} lie in $[\frac{l}{2}, l)$. We claim that S is an optimal solution if and only if S' is an optimal solution. It is clear that all the *partial* cliques in S' satisfy the defensive alliance property. Now, consider any vertex $u \in P$ which is adjacent to the clique set C_i, there is no change in the total number of adjacent vertices to u from C_i that goes into the solution in both S and S'. So, u satisfies the defensive alliance property in S if and only if it does in S'. We perform the conversion of all triplets from *partial* cliques until only two *partial* cliques remain.

Case 2: $d_i < 0$. It can be easily inferred from case 1 that even for any negative value of d_i, there exists an optimal solution with at most two *partial* cliques. □

Lemma 4. Consider a clique set C_i with a positive deficiency d_i, then no cliques of size less than d_i from C_i can be a part of the solution.

Proof. If C_i has a positive deficiency of d_i, then for any clique to be a part of the solution from C_i, we need to include d_i more neighbours from C_i. But, the only neighbours of any clique from C_i are the vertices of the clique itself and even if we include all the vertices of the clique, we won't be able to fill the deficiency of C_i. Therefore, no cliques of size less than d_i will be considered to be a part of the solution. $\qquad\square$

Lemma 5. Consider a set C_i^l from a clique set C_i with a positive deficiency d_i, there exists an optimal solution with at most $\lceil \frac{l-1}{l-k_i} \rceil$ *partial* cliques, where $k_i = d_i + \frac{l-d_i}{2}$.

Proof. Let $k_i = d_i + \frac{l-d_i}{2}$. From Lemma 4, it follows that no cliques of size less than d_i can be a part of the solution. Consider the cliques of size d_i, in order to fill the deficiency, we need to add all the vertices of the clique to S, which is trivial. Therefore, we consider only the cliques of size greater than d_i from each clique set. If $l > d_i$, then we have $l > k_i$. Consider an optimal solution S with p *partial* cliques in C_i^l. We have $k_i \leq x_i^{l,P_1}, x_i^{l,P_2}, ..., x_i^{l,P_p} < l$ and $k_i \cdot p \leq x_i^{l,P_1} + x_i^{l,P_2} + ... + x_i^{l,P_p} \leq p(l-1)$. Let $N = x_i^{l,P_1} + x_i^{l,P_2} + ... + x_i^{l,P_p}$. Now, we transform the instance S into another instance S' with at most p' *partial* cliques, where $p' \leq \lceil \frac{l-1}{l-k_i} \rceil$. Let $C_i^{l,P_1}, C_i^{l,P_2}, ..., C_i^{l,P_{p'}}$ be p' *partial* cliques from C_i^l and we have,

$$N = (\lceil \tfrac{N}{l} \rceil - p')l + p' \cdot m, \text{ where } m \in [k_i, l-1].$$

$$p' \cdot l - p' \cdot m = \lceil \tfrac{N}{l} \rceil \cdot l - N$$

$$p' \cdot (l - (l-1)) \leq (\lceil \tfrac{N}{l} \rceil - \tfrac{N}{l})l$$

$$p' \leq \tfrac{l-1}{l} \cdot l$$

$$p' \leq l - 1 \tag{1}$$

We have to absorb the maximum value of $\lceil \frac{N}{l} \rceil \cdot l - N$, that is $l - 1$, using p' partial cliques. This gives us the following tighter bound.

$$p' \leq \lceil \frac{l-1}{l-k_i} \rceil \tag{2}$$

From (1) and (2), We have that $p' \leq \lceil \frac{l-1}{l-k_i} \rceil$. We claim that S is an optimal solution if and only if S' is an optimal solution. It is clear that all the *partial* cliques in S' satisfy the defensive alliance property. Now, consider any vertex $u \in P$ which is adjacent to the clique set C_i, there is no change in the total number of adjacent vertices to u from C_i that goes into the solution in both S and S'. So, u satisfies the defensive alliance property in S if and only if it does in S'. Hence, we conclude that there is an optimal solution with at most $\lceil \frac{l-1}{l-k_i} \rceil$ *partial* cliques. $\qquad\square$

Let y_i^l be the number of *partial* cliques in C_i^l. From Lemma 3 and Lemma 5, it is clear that there is an optimal solution with at most $\lceil \frac{l-1}{l-k_i} \rceil$ *partial* cliques from C_i^l and we have $y_i^l \leq \lceil \frac{l-1}{l-k_i} \rceil$. We guess the cliques from C_i^l that goes into $C_i^{l,P}$ in $y_i^l + 1$ ways and from the remaining $|C_i - C_i^{l,P}|$ cliques, $C_i^{l,F}$ can be guessed in $m - y_i^l + 1$ ways, where m is the number of cliques in C_i^l. As we have guessed $P, C_i^{l,F}$ and $C_i^{l,P}$, we compute S_C using integer linear programming. $x_i^{l,P}$ represents the sum of $x_i^{l,P_1}, x_i^{l,P_2}, ..., x_i^{l,P_{y_i^l}}$. In our ILP formulation, there are at most $\sum_{l=1}^{z} \sum_{i=1}^{t} y_i^l$ variables that are $x_1^{1,P_1}, ..., x_1^{z,P_{y_1^z}}, x_2^{1,P_1}, ..., x_2^{z,P_{y_2^z}}, ..., x_l^{z,P_{y_l^z}}$.

Lemma 6. The set S is a defensive r-alliance if and only if

1. For each $u \in P$, $\sum_{i \in M(u)} \sum_{l=1}^{l=z} |C_i^{l,F}| + \sum_{i \in M(u)} \sum_{l=1}^{l=z} \sum_{j=1}^{j=y_i^l} x_i^{l,P_j} \geq demand(u)$.
2. Each $v \in C_i^{l,F}$ has to satisfy $|N(v) \cap P| + l \geq |N[v] \setminus P| - l + r$.
3. Each $v \in C_i^{l,P_j}$ has to satisfy $|N(v) \cap P| + x_i^{l,P_j} \geq |N[v] \setminus P| - x_i^{l,P_j} + r$.

Proof.

1. For each $u \in P$, $deg(u) = |N(u) \cap S| + |N(u) \setminus S|$ and $|N(u) \cap S_C| = \sum_{i \in M(u)} \sum_{l=1}^{l=z} |C_i^{l,F}| + \sum_{i \in M(u)} \sum_{l=1}^{l=z} \sum_{j=1}^{j=y_i^l} x_i^{l,P_j}$. $|N(u) \cap S| \geq |N(u) \setminus S| + r$ holds if and only if $2 * |N(u) \cap S| \geq deg(u) + r$, which implies $\sum_{i \in M(u)} \sum_{l=1}^{l=z} |C_i^{l,F}| + \sum_{i \in M(u)} \sum_{l=1}^{l=z} \sum_{j=1}^{j=y_i^l} x_i^{l,P_j} \geq demand(u)$.
2. For each $v \in C_i^{l,F}$, $|N(v) \cap S| = |N(v) \cap P| + l$; $|N(v) \setminus S| + r = |N[v] \setminus P| - l + r$.
3. For each $v \in C_i^{l,P_j}$, $|N(v) \cap S| = |N(v) \cap P| + x_i^{l,P_j}$; $|N(v) \setminus S| + r = |N[v] \setminus P| - x_i^{l,P_j} + r$. □

The ILP formulation for the defensive r-alliance is given as

Minimize $\sum_{i \in \{1,2,...,t\}} \sum_{l=1}^{l=z} \sum_{j=1}^{j=y_i^l} x_i^{l,P_j}$

Subject to

- $\sum_{i \in M(u)} \sum_{l=1}^{l=z} |C_i^{l,F}| + \sum_{i \in M(u)} \sum_{l=1}^{l=z} \sum_{j=1}^{j=y_i^l} x_i^{l,P_j} \geq demand(u)$, for each $u \in P$.
- $|N(v) \cap P| + l \geq |N[v] \setminus P| - l + r$, for each $v \in C_i^{l,F}$.
- $|N(v) \cap P| + x_i^{l,P_j} \geq |N[v] \setminus P| - x_i^{l,P_j} + r$, for every $v \in C_i^{l,P_j}$.
- $x_i^{l,P_j} < l$, for each $i \in \{1,2,...,t\}$, $l \in \{1,2,...,z\}$ and $j \in \{1,2,...,y_i^l\}$.

In our ILP formulation, we have at most $\sum_{l=1}^{z} \sum_{i=1}^{t} y_i^l$ variables, where z is the size of the largest clique outside the twin cover and $y_i^l \leq \lceil \frac{l-1}{l-k_i} \rceil$. The values of all the variables and the objective function are bounded by n. The constraints can be represented using $\mathcal{O}(\sum_{l=1}^{z} \sum_{i=1}^{t} y_i^l \cdot |T| \cdot logn)$ bits. With the help of Theorem 2, we will be able to solve the problem with the guess $(P, C_i^{l,F}$ and $C_i^{l,P})$ in FPT time. There are $2^{|T|}$ candidates for P and there are $\sum_{l=1}^{z} \sum_{i=1}^{t} y_i^l \cdot \mathcal{O}(n)$ candidates for $(C_i^{l,F}$ and $C_i^{l,P})$. To obtain S_C, we solve $2^{|T|} \cdot \sum_{l=1}^{z} \sum_{i=1}^{t} y_i^l \cdot \mathcal{O}(n)$ ILP formulas, where each formula can be computed in $\mathcal{O}^*(f(|TC|, z))$ time. This concludes the proof of Theorem 9.

Theorem 10 ($\star$). Given a graph $G = (V, E)$, $T \subseteq V$ is a twin cover of G and z is the size of the largest clique outside the twin cover, the OFFENSIVE r-ALLIANCE problem can be solved in $\mathcal{O}^*(f(|T|, z))$ time.

Theorem 11 ($\star$). Given a graph $G = (V, E)$, $T \subseteq V$ is a twin cover of G and z is the size of the largest clique outside the twin cover, the POWERFUL r-ALLIANCE problem can be solved in $\mathcal{O}^*(f(|T|, z))$ time.

3   Conclusion

In this work, we have proved that all three alliance problems are fixed parameter tractable when parameterized by distance to clique. We have that all three alliance problems are fixed parameter tractable parameterized by twin cover and the number of cliques outside the twin cover. We also proved that all three alliance problems are fixed parameter tractable parameterized by twin cover and the size of the largest clique outside the twin cover.

As a future direction, it will be interesting to check whether the idea presented in Sect. 2.3. can be extended to prove the tractability of the problem parameterized by twin cover. The hardness results for the parameters distance to cluster and twin cover can be considered. The problem still remains unsolved for the parameter modular width. It is also interesting to initiate the study on connected and independent versions of the offensive alliance problem.

References

1. Bliem, B., Woltran, S.: Defensive alliances in graphs of bounded treewidth. Discret. Appl. Math. **251**, 334–339 (2018)
2. Chang, C.W., Chia, M.L., Hsu, C.J., Kuo, D., Lai, L.L., Wang, F.H.: Global defensive alliances of trees and cartesian product of paths and cycles. Discret. Appl. Math. **160**(4), 479–487 (2012)
3. Chellali, M., Haynes, T.W.: Global alliances and independence in trees. Discuss. Math. Graph Theory **27**(1), 19–27 (2007)
4. Chen, J., Kanj, I.A., Xia, G.: Improved upper bounds for vertex cover. Theoret. Comput. Sci. **411**(40), 3736–3756 (2010)
5. Cygan, M., et al.: Parameterized Algorithms. Springer, Cham (2015). https://doi.org/10.1007/978-3-319-21275-3

6. Enciso, R.: Alliances in graphs: parameterized algorithms and on partitioning series-parallel graphs. Ph.D. thesis, USA (2009)
7. Fellows, M.R., Lokshtanov, D., Misra, N., Rosamond, F.A., Saurabh, S.: Graph layout problems parameterized by vertex cover. In: Hong, S.-H., Nagamochi, H., Fukunaga, T. (eds.) ISAAC 2008. LNCS, vol. 5369, pp. 294–305. Springer, Heidelberg (2008). https://doi.org/10.1007/978-3-540-92182-0_28
8. Fernau, H., Binkele-Raible, D.: Alliances in graphs: a complexity-theoretic study. In: Proceeding Volume II of the 33rd International Conference on Current Trends in Theory and Practice of Computer Science, pp. 61–70 (2007)
9. Fernau, H., Rodriguez-Velazquez, Alberto, J., Sigarreta, J.: Global r-alliances and total domination. In: 7th Cologne-Twente Workshop on Graphs and Combinatorial Optimization, CTW 2008, pp. 98–101 (2008)
10. Fernau, H., Rodríguez, J.A., Sigarreta, J.M.: Offensive r-alliances in graphs. Discret. Appl. Math. **157**(1), 177–182 (2009)
11. Gaikwad, A., Maity, S., Tripathi, S.K.: Parameterized complexity of defensive and offensive alliances in graphs. In: Proceedings of the 17th International conference on Distributed Computing and Internet Technology, pp. 175–187 (2021)
12. Gaikwad, A., Maity, S.: Defensive alliances in graphs. Theoret. Comput. Sci. **928**, 136–150 (2022)
13. Ganian, R.: Twin-cover: Beyond vertex cover in parameterized algorithmics. In: Proceedings of the 6th International Conference on Parameterized and Exact Computation, vol. 7112, pp. 259–271 (2011)
14. Jamieson, L., Hedetniemi, S., Mcrae, A.: The algorithmic complexity of alliances in graphs. JCMCC. J. Comb. Math. Comb. Comput. **68**, 137–150 (2009)
15. Jamieson, L.H.: Algorithms and Complexity for Alliances and Weighted Alliances of Various Types. Ph.D. thesis, USA (2007)
16. Kannan, R.: Minkowski's convex body theorem and integer programming. Math. Oper. Res. **12**(3), 415–440 (1987)
17. Kiyomi, M., Otachi, Y.: Alliances in graphs of bounded clique-width. Discret. Appl. Math. **223**, 91–97 (2017)
18. Kristiansen, P., Hedetniemi, M., Hedetniemi, S.: Alliances in graphs. J. Comb. Math. Comb. Comput. **48**, 157–177 (2004)
19. Lenstra, H.W.: Integer programming with a fixed number of variables. Math. Oper. Res. **8**(4), 538–548 (1983)
20. Shafique, K., Dutton, R.: Maximum alliance-free and minimum alliance-cover sets. Congr. Numer. **162**, 139–146 (2003)
21. Sigarreta, J., Bermudo, S., Fernau, H.: On the complement graph and defensive k-alliances. Discret. Appl. Math. **157**(8), 1687–1695 (2009)

Author Index

Printed in the United States
by Baker & Taylor Publisher Services

Printed in the United States
by Baker & Taylor Publisher Services